# Shorebirds
An illustrated behavioural ecology

**Jan van de Kam**
The Netherlands

**Bruno Ens**
Alterra, Wageningen, The Netherlands

**Theunis Piersma**
Royal Netherlands Institute for Sea Research (NIOZ), Texel,
The Netherlands; University of Groningen, The Netherlands

**Leo Zwarts**
RIZA, Lelystad, The Netherlands

# Shorebirds
## An illustrated behavioural ecology

Translated by Petra de Goeij (The Netherlands)
and Suzanne J. Moore (New Zealand)

KNNV Publishers

**Shorebirds. An illustrated behavioural ecology**

Authors: Bruno Ens, Jan van de Kam, Theunis Piersma and Leo Zwarts
Photography: Jan van de Kam
Translated from Dutch to English by: Petra de Goeij and Suzanne J. Moore
Edited by: Suzanne J. Moore and Phil F. Battley
Graphic design and production: Teo van Gerwen - Design, Leende, The Netherlands
Lithography: Real Concepts, Doetinchem, The Netherlands
Printer: DZS, Ljubljana, Slovenia

This publication was supported financially by:
The African-Eurasian Waterbird Agreement
Alterra
Flemish Institute of Nature Conservation Belgium
RIZA
Wetlands International
WWF Germany
WWF Netherlands

Original title: *Ecologische atlas van de Nederlandse wadvogels*
© Dutch edition: Schuyt & Co Uitgevers en Importeurs BV, Haarlem, The Netherlands, 1999
ISBN 90-6097-509-X
© English updated edition: KNNV Publishers, Utrecht, The Netherlands, 2004
ISBN 90-5011-192-0

www.knnvuitgeverij.nl

KNNV Publishers is a foundation of the Royal Dutch Society for Natural History.

No part of this book may be reproduced in any form by print, photocopy, microfilm or any other means without the written permission of the publisher.

**The African-Eurasian Waterbird Agreement**
The Agreement on the Conservation of African-Eurasian Migratory Waterbirds (AEWA) was concluded at the Negotiation Meeting in The Hague, The Netherlands in 1995. After the required number of ratifications was accrued, the AEWA entered into force on November 1st 1999. Since then the number of contracting parties has increased steadily and is currently 46.

The AEWA emcompasses Europe, Central Asia, the Middle East and Africa. The agreement covers in total 235 species of birds ecologically dependent on wetlands for at least part of their annual cycle, including many species of pelicans, storks, flamingoes, swans, geese, ducks, waders, gulls and terns. The Wadden Sea is an extremely important breeding, wintering and feeding area for many of them.

WWF Netherlands
Postbus 7
3700 AA Zeist
The Netherlands

WWF Germany
Rebstöcker Straße 55
60326 Frankfurt
Germany

# Contents

Foreword, *by Peter Prokosch* 7

Preface 8

1. Tidal areas 11
    1.1 The tidal landscape 11
    1.2 Tidal movements 14
    1.3 Sediment characteristics 17
    1.4 Food for birds 19

2. Portrait Gallery 27
    2.1 Species and subspecies 27
    2.2 Waterbirds by species 34
    2.3 Characteristic attributes of waterbirds 84

3. Migration 95
    3.1 Wadden Sea: migration crossroads 95
    3.2 Flight travels 98
    3.3 Havens in a hostile world 118
    3.4 Balancing the books 125
    3.5 Migration strategies 134
    3.6 The origin of waterbird migration 141

4. Food 147
    4.1 Decisions 147
    4.2 Food requirements 154
    4.3 Time budgets 157
    4.4 Food availability 169
    4.5 Variation in food accessibility 177
    4.6 Exploiting the available food 188
    4.7 The distribution of waders across the tidal flats 205
    4.8 Predation pressure of waterbirds on benthic animals 227

5. Reproduction 231
    5.1 No reproduction, no life 231
    5.2 Reproduction: a broad overview 241
    5.3 The best place to breed 261
    5.4 The best time to breed 273
    5.5 Choosing a partner 279
    5.6 Variations on a reproductive theme 287
    5.7 Career planning 299

6. Looking to the future 307
    6.1 Counters and twitchers 307
    6.2 Where and how are numbers limited? 308
    6.3 Effects of summer conditions on populations 314
    6.4 Effects of winter conditions on populations 317
    6.5 What is carrying capacity? 321
    6.6 Natural selection and population dynamics 321
    6.7 Competition between species 324
    6.8 Are we leaving the birds enough room? 324
    6.9 From species protection to habitat protection 341
    6.10 The convention circus 341
    6.11 Scientific research and monitoring 343

References 345

Index 363

# Foreword

I have had the privilege of sharing many wonderful experiences with the birds described in this book. I have seen the breeding knot in the Arctic tundra on the Taimyr Peninsula (p. 115). I have studied them on migration (p. 142), including the one on my balance (p. 107) in North Friesland which had sufficient fat stores to make the 4000 km journey to Siberia in a single flight. I have held the bar-tailed godwit in Northern Norway (p. 278 and back cover), and grey plovers and dotterels (p. 291), recording information of scientific interest before photographing and releasing them. I have spent significant periods of time in their Wadden Sea and Arctic tundra habitats, often in the company of the authors, good old friends and colleagues.

It was not just because of my involvement in coastal bird research and conservation in the globally important Wadden Sea region that I was impressed with the Dutch version of this book. It was also because I was convinced that it would appeal to and inform a far wider audience than ornithologists in the three Wadden Sea countries of Denmark, Germany, and The Netherlands. I am happy therefore that it has been possible to produce an English version; this will make the book accessible to a far wider international readership. I am sure that the book will induce enthusiasm and fascination worldwide for the most amazing group of migratory birds on earth, the shorebirds.

The photographs are incomparable. The amazing 'biological eye' of Jan van de Kam, nurtured by his patience and love for the Wadden Sea birds, has enabled him to capture so much action in brilliant detail. That he gets so many complicated things exactly right makes it a privilege for scientists to work with Jan.

The science in the book is first class. Bruno Ens, Theunis Piersma and Leo Zwarts are part of a long-standing Dutch tradition of studies on coastal waterbirds and behavioural ecology; for a long time this has involved benthic work as well. Only rarely have scientists made the effort to produce their findings and knowledge in such an educational way.

Finally I would like to consider the conservation effects of such an endeavour. The results of basic research have been important in raising awareness of what has been happening in the Wadden Sea and has led to the level of international protection that the Wadden Sea enjoys today. This success story may motivate conservationists in other parts of the world. It may also remind those in the region itself to remain open-minded about new and important threats to the Wadden Sea's ecological wellbeing, for example the dreadful mechanical dredging for cockles and mussels allowed in The Netherlands for the last 20 years. This book should do much to help raise the international Wadden Sea area, already a good example of an "international park", to the appropriate status of World Heritage Site.

It is an honour for WWF, the international conservation organisation, to support the publication of this great book. We will use the publication as an additional tool for our own engagement in the Wadden Sea, the Northeast Atlantic, the Arctic, the west coast of Africa, and throughout the world.

Peter Prokosch
CEO WWF Germany

# Preface

This is an illustrated behavioural ecology of birds that forage on intertidal shores. Generally, these birds prefer to search for food whilst walking near the retreating tide, although some are capable of swimming or diving for food. In Dutch, such birds are known as 'wadvogels', a term that can be literally translated as 'mudflat birds'. This book was first published in Dutch as the 'Ecologische Atlas van de Nederlandse Wadvogels'. In this English edition, we use the term 'waterbirds' to refer collectively to the large group of waders, gulls, herons, ducks, and geese that depend on the tidal zone for food. Much of the focus in this book is on the long-legged waterbirds known as waders in Europe and Australasia, and as shorebirds in much of the rest of the world. This focus is reflected in the title of the book.

Although much of our research has been carried out in Northwestern Europe and connected coastal wetlands such as those in West Africa, the accounts should be relevant to waterbird ecology worldwide. The most important tidal area in Northwestern Europe is the Wadden Sea, which extends from The Netherlands east and north through Germany to Denmark. However, waterbirds are also abundant in the deltas in the southwest of The Netherlands. This book describes the annual cycles of Dutch waterbirds. As many of these birds, and especially the shorebirds, breed in the high north and winter under the tropical sun, we also cover tundras and tropical mudflats.

The first chapter provides background information about the locations and sizes of major tidal flat areas. We look at how tides, climate, vegetation, sediment characteristics, and food availability affect the birds' foraging, and how different species use the tidal flats through the year. We also discover that it is not really possible to precisely locate the areas used throughout the year, as we don't know exactly where 'our' birds are when they are not in the Wadden Sea. Sometimes, these vague boundaries can be a problem for systematic researchers. However, the birds can live happily without clear boundaries, and many nature lovers are comfortable with such uncertainty. It is also not possible to say unequivocally which species belong in the book, as many have diverse connections to the different tidal areas along the migration routes. The 40 species described in Chapter 2, will at least give the reader a good idea of the range of species that can be seen in the tidal zones of Northwestern Europe.

In the following three chapters, we successively discuss the most important themes in the lives of waterbirds: migration, food, and reproduction. Some waterbird species are seen infrequently, in small numbers, or in very localised areas. Others appear to dominate the entire tidal area: oystercatchers, dunlins, knots, bar-tailed godwits, redshanks, and curlews are almost always present in large numbers. These more numerous species have been the subjects of much of the pioneering research that has been done in the last few years. Chapters 3, 4, and 5 all deal with the question 'How do birds survive in this highly dynamic environment?'

Chapter 6 describes waterbird populations and the factors influencing them. We also discuss the likelihood that the large numbers of birds currently breeding in and migrating through the Dutch, German and Danish Wadden Sea, will continue to find a safe haven in the years to come. The future of this exceptional natural area lies completely in our hands.

Waterbirds almost always operate in groups and the researchers studying them do much the same. We can only produce good research through intensive cooperation with kindred spirits locally and in the other countries along the migration routes of our waterbirds. The text of this book came into existence through the productive cooperation of the four authors and with help from many others, who shared data, analysed, commented on, and edited parts of the text.

This English language edition was translated by Petra de Goeij and Suzanne Moore and edited by Suzanne Moore and Phil Battley. Special thanks to Simon Delany of Wetlands International and Ben Koks for their help in updating the bird numbers in the present edition. Thanks are also due to Dick Visser for turning the authors' complicated scribblings into the beautiful maps and figures that are an essential part of this book. We thank Anne-Marie Blomert for organising the references and Yvonne Verkuil for producing the index for this edition. For their assistance during the preparation of this book we thank Phil Atkinson, Joop Bakker, Albert Beintema, Jan Beukema, Anne-Marie Blomert, Bert Brinkman, Stephen Browne, Jenny Cremer, Nick Davidson, Simon Delaney, Piet Duiven, Klaas van Dijk, Karel Essink, Petra de Goeij, John Goss-Custard, Gudmundur A. Gudmundsson, Lieuwe Haanstra, Tom van der Have, Dik Heg, Jacob Höglund, Hermann Hötker, Joop Jukema, Jan van der Kamp, Ben Koks, Anita Koolhaas, Andreas Kannen, Marcel Kersten, Janosz Kis, Kate Lessells, Åke Lindström, Peter

Meininger, Ron Mes, Gabriel Nève, Ken Norris, Albert Oost, Katja Phillippart, Jouke Prop, the Rogers family, Elze de Ruiter, Roland Sandberg, Hans Schekkerman, Michael Soloviev, Bernard Spaans, Bill Sutherland, Tamas Szekèly, Pavel Tomkovich, Ingrid Tulp, Yvonne Verkuil, Wetlands International, Popko Wiersma and Yuri Zharikov.

Some more personal acknowledgments are also warranted. Jan van de Kam has spent many years photographing waterbirds in the Wadden Sea as well as in wintering areas in Africa, and on the breeding grounds in Lapland, Siberia and Arctic Canada. It is only through the invaluable help of numerous enthusiasts in and outside of The Netherlands, that these expeditions could produce the photographs displayed in this book. Bruno Ens especially thanks Bart Ebbinge, who took the first steps to enable Alterra, formerly known as IBN (Dutch Institute for Forestry and Nature Conservation), to participate in this project. The inspired enthusiasm of Bert Jansen in combination with his persuasiveness was decisive. Many thanks to Anneke, Arnoud and Hedwig, who patiently accepted the many extra hours he spent working on this book at night and on weekends. Theunis Piersma thanks his research group for their helpful attitude during the writing of this book. His research on shorebirds and intertidal ecology at the Royal NIOZ on Texel and at the University of Groningen was supported by the PIONIER grant from the NWO (Netherlands Organisation for Scientific Research). Leo Zwarts thanks Bart Fokkens, head of the Department 'Inrichting en Herstel' of RIZA, who stimulated Leo to move from writing articles for scientific journals to take the time to make this knowledge more easily accessible.

Finally, we would like to express our gratitude to Rudi Drent and Wim Wolff, who, over the last 25 years, have immensely stimulated research on waterbirds and tidal areas.

*Jan van de Kam*
*Bruno Ens*
*Theunis Piersma*
*Leo Zwarts*

**Jan van de Kam** (1938) spends his time photographing, filming, and writing about the natural world. Jan has written and made a wide range of books and films on the ecology of intertidal areas worldwide, and was the initiator of this book. The outstanding photos displayed here result from the thousands of hours Jan has spent sitting in hides on the tidal flats of Europe and Africa, and his journeys to waterbird breeding grounds in the Canadian and Siberian tundra. Like the other three authors, Jan lives in The Netherlands.

**Bruno Ens** (1956) is a product of the University of Groningen, The Netherlands, where he followed a masters degree with a PhD on the social behaviour of the oystercatcher (capitalising on his interest in competition between animals). Bruno is presently employed by Alterra (Research Institute for a Green World), on Texel, the westernmost island in the Wadden Sea. There he models the effects of human impacts such as climate change, commercial shellfish harvesting, disturbance and subsidence on waterbird populations. His guiding principle is that changes in abundance can only be understood by considering the choices being made by individual birds.

**Theunis Piersma** (1958) works on habitat choice and migration strategies of shorebirds, particularly by exploiting the ecological contrasts to be found among species and sites around the world. Taking an evolutionary viewpoint in his research, Theunis leads research teams at the NIOZ (Royal Netherlands Institute for Sea Research) and at the University of Groningen. His studies have focused on shorebird energetics at the individual level, predator-prey relationships on tidal flats, and the link between resource abundance and population dynamics of shorebirds. This work has been carried out in areas ranging from the far northern tundras to the tidal flats of southern South America, and often has clear implications for management.

**Leo Zwarts** (1946) has one of the longer histories of research into the relationship between shorebirds and their food, which he studied for 20 years. This research initially focused on the tidal flats of the Wadden Sea, but later Leo became increasingly involved in studies on African tidal areas. In recent years he has been working at RIZA (Netherlands Institute for Inland Water Management and Waste Water Treatment), where he studies the birds that inhabit large freshwater systems.

# 1. Tidal areas

## 1.1 The tidal landscape

### 1.1.1 Introduction

As any Dutch geologist will tell you, 'wad' or its plural 'wadden' is the Dutch word for an area of flat shore bordering a shallow coastal sea with a chain of islands not far offshore. The sand and silt flats that lie between the islands and the coast are exposed at low tide and flooded at high tide. In English, we don't have a specific name for these areas; the closest we can come is to use the general term intertidal or tidal flats. Tidal flats are made up of fine sediment particles that have settled out of the water. The fine particles are moved by the tides in a continuous process of deposition, erosion and removal. This endless moving of sand and silt, of channels, and even of whole islands makes tidal flats a highly dynamic landscape.[807]

A tidal flat can be defined most simply as an area of sea floor that is exposed by the receding tide. The sea floor is only exposed in places where the water depth at high tide is less than the tidal range (the difference in water level between high and low tides). The tidal range in the Wadden Sea is small, just a few metres, so its sea floor is only exposed where the sea is very shallow.

Tidal flats occur along the edges of shallow coastal seas. They often form in estuaries, areas where river mouths widen and become tidal, such as at the mouth of the River Elbe in Germany. All British tidal flats are estuarine. The waterbirds discussed in this book are equally at home in the mouth of the Thames, in a shallow bay like the Wash, or in a true 'wadden' area like the Wadden Sea. The birds are all feeding on the same habitat, tidal flats.

Although tidal flats are areas of periodically exposed sea floor, not all exposed sea floor is a tidal flat. Like tidal flats, sandy beaches and rocky coasts are often exposed at low tide and submerged at high tide. One difference between them is that there may be surf waves breaking on beaches, but not on tidal flats. The stronger wave action and water currents only allow coarse sediment particles to settle out on beaches. But the differences between beaches and tidal flats aren't always this obvious. Some very exposed tidal areas are made up of the same coarse sand as many beaches. Conversely, some beaches are quite sheltered and siltier, which can make the beach look like a tidal flat. The difference between a tidal flat and a rocky coast is more obvious, but there are still many transitional forms. In many places, tidal flats are made up of sand or silt mixed with pebbles or eroded volcanic rock.

Tidal landscapes are strongly influenced by the sea and are mostly, but not always, saline. High and low tides can also occur in river mouths where the water is brackish, or further upstream, exposing freshwater flats at low tide. Nowadays, few large brackish tidal flats and hardly any freshwater tidal flats remain in Northwestern Europe. The once extensive freshwater and brackish tidal areas at the mouths of the Maas and the Rhine were lost when sea arms in the Southwestern Netherlands were closed off. Smaller freshwater tidal areas may still be found along the Oude Maas in The Netherlands, downstream of the Western Schelde in Belgium, along the Elbe and the Weser in Germany, and in the mouths of some British rivers, such as the Firth of Forth.

Freshwater tidal areas are often covered in rushes, reeds and willows. Saline tidal areas are usually largely barren, apart from the upper tidal zones. In temperate areas, these upper zones form salt marshes and are covered in low-growing salt-tolerant plants. In the tropics, mangrove forests often grow in these zones. Huge areas of salt marsh have been lost from densely populated parts of the world, particularly over recent years. Mangroves are also disappearing rapidly, especially in Southeast Asia.

Worldwide, the lower tidal zone is usually devoid of vegetation, apart from some patches of seagrass. Here and there, small patches of mussel or oyster beds may develop. These shellfish beds provide a hard substrate, which animals and plants normally found on a rocky coast, such as barnacles and seaweed, can establish themselves on. In some intertidal areas, such as the Banc d'Arguin in Mauritania, tidal flats are almost completely covered in seagrass. On the east coast of the United States, large areas of the intertidal zone are covered with *Spartina*. When rivers deposit high nutrients loads on the mudflat, the entire area may be covered in a dense mat of sea lettuce, *Enteromorpha* and other large algae.

Figure 1.1. The Dutch, German and Danish Wadden Sea with tidal flats shown in yellow.

### 1.1.2 The locations of intertidal areas

Intertidal areas are found in all climate zones, including the Arctic Circle and around the equator (Figure 1.2).[214] The biggest tidal areas in the world lie in the temperate zone: in Northwestern Europe (the North Sea and the Irish Sea), along the east coast of Canada (the Bay of Fundy and Hudson Bay), in Eastern Asia (the Yellow Sea) and Western Asia (the Persian Gulf), and in South America. There are huge intertidal areas in the tropics, but most of these are covered in mangrove forest. There are also extensive tropical intertidal areas where the mangroves only grow along the edges of the mudflats in West Africa (Mauritania and Guinea-Bissau) Southeast Asia (China, Vietnam and Irian Jaya), Northern Australia and along the north coast of South America (Surinam). The biggest tidal area of all is on the coast of the Yellow Sea, in Eastern Asia, and is estimated to contain one million hectares of mud! Half of this lies along the west coast of North and South Korea, the other half is in China on the other side of the Yellow Sea, mainly around the mouth of the Yellow River.[946]

We can calculate the sizes of intertidal areas from coastal charts. Although out on the tidal flats the location of the low water line changes daily, charts show a precise low water line, chart datum. Chart datum is the lowest low water line reached during spring tides (not the average low water line). So using charts to calculate tidal areas tends to overestimate the area involved, but if we wish to compare different areas, we can use charts to calculate the maximum tidal surface.

The Wadden Sea is one of the biggest intertidal areas in the world. It covers over 450 km from Den Helder in The Netherlands to Esbjerg in Denmark. It is 7 - 10 km or even 20 km wide in some places. During the spring low tides, a total of 490 000 ha of intertidal mud is exposed; this is more than half the total surface between Den Helder and Esbjerg (Figure 1.1). The second biggest intertidal area in Europe is just one sixteenth the size of the Wadden Sea. This area, Morecambe Bay on the English west coast, contains 33 700 ha of tidal flat. The other large intertidal areas in Northwestern Europe are also in Great Britain: the Wash on the English east coast (29 800 ha of tidal flat) and Solway Firth in Scotland (24 600 ha of tidal flat). Smaller tidal areas are found along the English coast in the southwest (Severn, 16 900 ha), the west (Dee, 13 000 ha) and the east (Humber, 13 500 ha). There are many small intertidal areas at the mouth of the Thames, which cover 25 500 ha in total. If you add the intertidal areas of all 155 British estuaries and bays, they have a surface area of 300 000 ha.[162] The Wadden Sea is more than half as big again as the total area of all British intertidal sites.

Once the river deltas of the Southwestern Netherlands were blocked off by dams and dykes, half of all the Dutch silt tidal flats were lost. Only the 8300 ha Western Schelde and 11 400 ha of the Eastern Schelde were saved. Other European tidal flats are found in the river mouths and bays of Iceland and Ireland (80 000 ha) and along the Atlantic coast of France (70 000 ha). France also has tidal flats in bays on the coasts of Normandy, Brittany and the Vendée, and in the mouths of large rivers like the Somme and Gironde. The most famous French tidal areas are the Bay of Mont St Michel (25 000 ha) and the Gulf of Arcachon (15 000 ha). Southwestern Europe has one large intertidal area in the mouth of the Tagus, in Portugal (12 000 ha[74]). Further south, along the Moroccan coast of Africa are many small intertidal areas and two large bays: Merja Zerga (2200 ha) in the north and the Bay of Dakhla (2500 ha) in the far south. On the edge of the Sahara, in Mauritania lie the Bay d'Arguin (3300 ha mud) and the Banc d'Arguin (42 700 ha).[948, 997] Although there are hardly any mud or tidal flats at the mouth of the Senegal River, there are tidal flats downstream of three other rivers flowing out of Gambia and Senegal: Saloum (6600 ha[777]), Gambia and Casamance. Still further south, along the coasts of Guinea-Bissau,

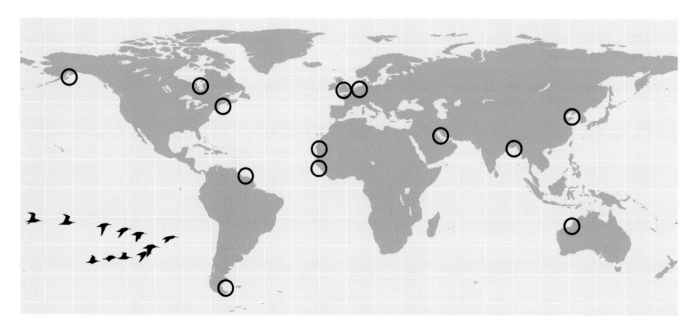

Figure 1.2 The most important intertidal areas worldwide.

Guinea and Sierra Leone, is Africa's biggest intertidal area, a total of 284 300 ha of intertidal flats. More than half of this area lies in Guinea-Bissau (157 000 ha[967]), a quarter in Guinea-Conakry (68 300 ha[16]) and the rest in Sierra Leone (59 000 ha[883]). There are no other large intertidal areas in Africa, but there are smaller tidal areas in Namibia, South Africa, Mozambique, Tanzania, and on the island of Madagascar.

Millions of waders and other waterbirds which breed in the high north and winter in the tropics or in the temperate zone feed in tidal areas on migration and during winter. Birds that breed in Eastern Canada, and winter in Surinam, fly via the west coast of the Atlantic Ocean, refuelling at Hudson Bay and in the Bay of Fundy. Similarly, intertidal areas in Northwestern Europe are essential for the millions of birds that breed in the vast area from Northeastern Canada to Northwestern Asia, and that winter in Western Europe and West Africa. These birds require an estimated total of 1.3 million hectares of intertidal area, three quarters of which is in Western Europe and one quarter in West Africa.

During the last ice age, 20 000 years ago, the sea level in the North Sea was 60 m lower than it is now. Great Britain was connected to the European mainland, and the North Sea didn't extend further south than Scotland and Southern Norway. It's difficult to say where tidal areas existed then, but they certainly were not where they are now. It's very likely that back then, intertidal areas were less extensive than the 900 000 ha now found along the European coast. When sea level rises over areas that were once tidal flat, it seems obvious that the tidal flats would be replaced by a shallow sea. But this only happens if the amount of sand and silt being deposited remains constant as sea level rises, which doesn't seem to be the case. Rising sea levels lead to increased sediment inflow that, if the sea level doesn't rise too quickly, can allow the tidal flats to rebuild. Rising sea levels have greatest impact on the lower coastal zone. Changes higher in the coastal zone, in the salt marshes, are likely to occur much more slowly. Without sea walls and dykes, it's likely that intertidal areas would just move inland and rebuild themselves.

When sea level falls, the inflow of new sediment decreases. This may lengthen the lifespan of the intertidal flats, which in the long-term will become dry land. In the meantime, new intertidal areas may develop along the new coastline. The total surface area of tidal flats has been very variable for thousands of years, probably with larger intertidal areas when sea level is high and less when sea level falls. Unfortunately, we can only speculate on what effects this would have had on birds dependent on the intertidal flats for most of the year (Chapter 3).

We do know something of how sedimentation currently occurs in the Wadden Sea. Strangely enough, most of the sediment comes not from rivers, but directly from the North Sea, which has become an enormous sand reservoir after years of glaciers scouring the sea floor. Waves and tidal currents have moved this North Sea sand and formed the sandy Wadden Sea islands. The Wadden Sea islands are now the most important source of sand for the tidal areas behind them: to raise these tidal flats sand must come from the islands' coasts.

The Rhine only carries very fine silt, which remains in suspension and is transported north through the Wadden Sea. Eroding silt banks in the Southern North Sea also contribute to the Wadden Sea's silt load. It's clear that some of this silt is deposited in the Wadden Sea, but how much exactly is unknown.

### 1.1.3 Alternatives to tidal areas

Many of the same bird species that live on tidal areas also use non-tidal sand and silt flats. At some such sites, the sea is so shallow that wind can temporarily blow the water away, exposing the sea floor without any assistance from the tide. This happens along the south coast of the Baltic Sea and the north coast of the Black Sea. These areas have many islands and peninsulas in a very shallow coastal zone, so that in most wind directions part of the sea floor will be exposed somewhere. Although by definition these are not intertidal areas, they are still very important to migratory birds and interesting for researchers to compare with real intertidal areas as the local availability of these areas is much less predictable than the regular tidal ebb and flow of tidal flats.[181, 896]

Coastal lagoons, such as those in tropical Africa along the coast of the Gulf of Guinea, are also not true intertidal areas.[615] Many of these lagoons are permanently closed off from the sea, although some are connected to the sea for part of the year. In the rainy season, the lagoons fill up with freshwater. The pressure of this water can then break the dunes between the lagoon and the sea, creating a temporary tidal flat with a daily tidal rhythm and little flats that fall dry. The lagoons may also develop areas similar to tidal flats if the dunes don't breach, and evaporation causes the lagoons to become shallow enough to provide attractive foraging sites for long-legged, often fish-eating, waders. If the water level drops further other waders may also forage there. However, these areas aren't always available for the birds. Whether these lagoon flats appear sooner or later, or not at all, depends on a combination of rainfall and whether the dunes break through. This makes coastal lagoons an unpredictable environment with feeding grounds that vary in size weekly and annually.

Saltpans can also attract large numbers of waders, depending on the water depth. People construct these ponds to evaporate seawater so that they can harvest the salt, indirectly creating artificial coastal lagoons. When the salt concentration gets too high many of the aquatic and benthic animals die. The few species that can survive, such as mosquito larvae and brine shrimps, may reach extremely high densities and become a very important food source for flamingoes and other waders.

### 1.1.4 Climate

The intertidal areas of Northwestern Europe are very different from those in Africa. Most of these differences are due to climate. In Northwestern Europe the surface temperature of seawater varies from 1 - 4 °C in winter and 14 - 18 °C in summer. Closer to the equator, the average temperature increases and seasonal variation decreases. In Portugal, the sea temperature varies between 14 and 19 °C, and on the Banc d'Arguin in Mauritania from 18 - 26 °C. One thousand kilometres further south in Guinea-Bissau, the seawater temperature is nearly constant at 23 °C year-round.

Winters on the continental coast of Northwestern Europe are colder than they are in Great Britain and Ireland. The average seawater temperature of the Wadden Sea in winter is just 2 - 3 °C. The Irish Sea lies on the same latitude, but has average coastal seawater temperatures of 5 - 6 °C in winter, 3 °C warmer than the Wadden Sea. The Gulf Stream makes seawater temperatures in more northerly-situated Scotland as high as those along the Irish coast. On the south coast of Wales and England, winter seawater temperatures are 7 - 8 °C. These temperatures give a clear picture of how greatly average winter conditions differ. Average seawater temperature gives an indication of the likelihood that the tidal flats will freeze over, causing massive declines in prey, and leaving birds unable to forage through the ice. There is practically no chance of ice forming along the coast of the Irish Sea, but Wadden Sea tidal flats are covered in ice for at least a few days in one out of every three winters. The chance of ice forming in the Wadden Sea is, on average, slightly greater in the northeast (Denmark and Schleswig-Holstein) and a little less in the west (The Netherlands). The chances of ice forming further south are much less. In the harsh winter of 1984 to 1985, the Wadden Sea was covered in ice for almost two months, but only 200 km southwest in the Dutch delta area the ice only lasted one week.

## 1.2 Tidal movements

### 1.2.1 The tidal cycle

Most intertidal areas have two low tides a day, although much of the Pacific Ocean has only one low tide per day, or sometimes one on one day and two another. In fact, tidal cycles don't fit neatly within a 24-hour period. If there are two tides a day, a full tidal cycle takes an average of 12 hours and 25 minutes. This extra 25 minutes causes the time of high and low tide to shift slightly each day, so that after two weeks, high and low tide are back to the same time as before. Every 14 days, there is one spring and one neap tide. In spring tides, high water is extra high and low water extra low, creating a greater tidal difference than during a neap tide. This causes more of the lower tidal zone to be exposed during a spring tide than at neap tide. The reverse is true of the highest parts of the tidal zone, so salt marshes that are flooded at spring tide, aren't normally flooded at neap tide.

In addition to the 14-day cycle of spring and neap tides, there is a six monthly cycle, which causes spring tides to be even higher than normal in February and March, and again in August and September. Midway between these six monthly high tides, spring tide high water levels are lower and the low water higher. This six monthly cycle is particularly noticeable in flat coastal areas that have not been artificially retained (by dykes or seawalls), such as those in Mauritania. Here, areas many kilometres wide are flooded a few times in late winter and late summer, and remain dry for the rest of the year. During this dry period, the salt from the seawater concentrates forming a hard crust, which cars can drive across.

These variations between the neap and spring tides are most obvious in the lower and higher parts of the tidal zone. They make little difference to the areas that

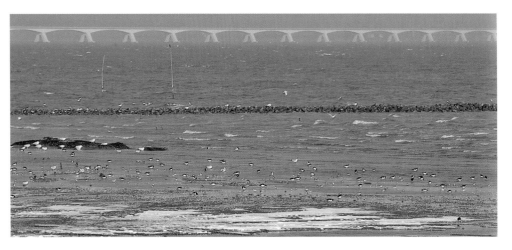

When ice covers the Wadden Sea in winter (top photo) most waders go to the Dutch Delta (lower photo) and tidal areas in France and Britain where there is less chance of ice forming.

Inlets and river mouths such as the Dollard (shown on page 15) tend to have very silty sediments. Their channels are narrow and deeply incised.

lie near average sea level. These areas are exposed for an average of 6 hours and flooded for 6 hours regardless of whether it's a neap or a spring tide.

### 1.2.2 Tidal range

The distance between high and low tide, which is called the tidal range, depends on sea currents and the extent to which water flow is constricted by the land or shallow seas and shoals. The tidal range in the middle of the North Sea is less than 1 m, but along the coast in spring tides it can be 1.5 - 6 m. Of these, tidal ranges are greatest along the English east coast, and least along the Danish and Dutch west coasts (Figure 1.3). In the Wadden Sea, the maximum tidal range varies from 1.5 m in the west to more than 4 m along the German coast, and down to 1.7 m further north. Enclosed bays, such as the Dollard and the Jadebussen in the Wadden Sea, experience greater tidal ranges. The English Channel is also enclosed, causing the tidal range to gradually increase from the southwest across to the Straits of Calais. Spring tides along the French coast can have spectacular tidal ranges of 6 - 12 m. Tidal ranges along the English west coast of the Irish Sea are equally huge and can vary greatly over small distances. In Southwest Wales the tidal range is 6 m, but further east along the Severn, between Wales and Cornwall, tides reach more than 12 m. However, this isn't the largest tide recorded. Tidal ranges of 15.6 m occur in the Bay of Minas, part of the Bay of Fundy in Nova Scotia.

The Mediterranean Sea is almost completely landlocked, and has little tidal movement. However, there is a large intertidal area on the east coast of Tunisia that has a 1.8 m tidal range at spring tide, and a much smaller range at neap tide. In most tidal areas, the tidal range at neap tide is a third smaller than at spring tide, but in the Tunisian intertidal area, neap tides are only 0.3 m high. This means that neap tides only expose a fraction of the 12 000 ha exposed during spring tides.[194]

It is very important for large ships to know when high and low tide will occur and what water depths they will encounter along different routes into harbours. Tidal movements can be predicted from astronomical data, but only with some difficulty as many factors affect the movement of the sun and the moon. Every year, the British Navy produces two fat books, the Admiralty Tables, which list the daily times and levels of high and low water at thousands of harbours across all the continents. Local times of high and low water depend on the speed with which the ebbing or flowing water can move through a coastal area. In deep water it can move quickly, but in a shallow or partially enclosed sea, incoming water is slowed causing high tide to occur a few hours later over a distance of less than 10 km. Times of high and low water can be predicted quite accurately, but tidal heights are more difficult to predict as many factors influence the water movement. For example, tidal heights in river mouths depend on the amount of water flowing down from the catchment, and will be greater when the

## TIDAL AREAS

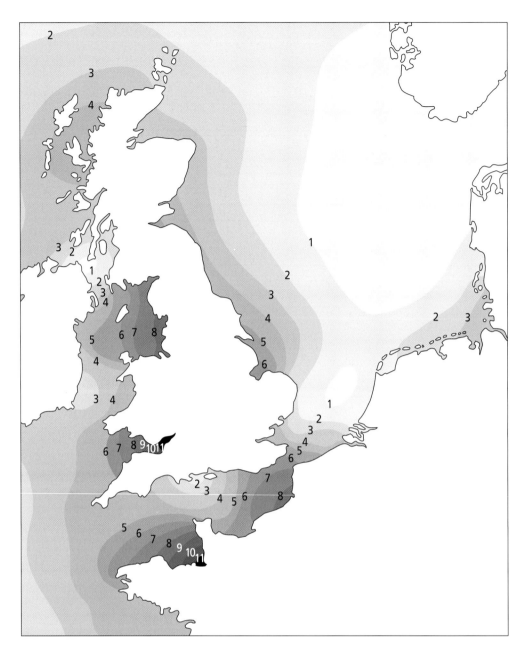

Figure 1.3 Average tidal differences (in metres) in Northwestern Europe during spring tides.[175]

river is in flood. Wind strength and direction can also have a large effect on tidal movements.

### 1.2.3 Wind and tidal movement

For birds that forage on exposed tidal flats, the movements of the tides are as important as the change from day to night is for others. Birds that forage during the day and rest at night can see when the sun rises and sets and adjust their daily activities around this. But it is far more difficult for birds that forage in the tidal zone to predict when the tide will fall or rise, as wind direction and speed can alter tidal movements. In the Wadden Sea, strong northwest winds can push the water 50 - 100 cm higher, causing large areas of the tidal flat to remain submerged at low tide. Conversely, a strong southeast wind lowers the sea level by some 50 cm, causing parts of the tidal flat to remain uncovered at high tide (Figure 1.4). Winds from the northeast and southwest have little effect on tidal water movements, and don't affect tidal flat exposure times in the Wadden Sea. Figure 1.4 shows the effect of wind on the exposure time of tidal flats near average sea level. Wind causes greater variation in the exposure times closer to the high and low tide lines.

The wind-caused variation in exposure times differs greatly between seasons. In Northwestern Europe, northwest storms are rare in May and June, but often occur in autumn and early winter. In the Wadden Sea, tidal areas near sea level are exposed for less than 10 hours on 15% of all days in April and May, compared to 50% of days from November to January. Fortunately for the Wadden Sea birds, periods of very short exposure times usually only last for one to two days. On one in 25 days, tidal flats at average sea level are not completely exposed, but such a situation lasts for two days only 1.5 times a year on average, and flats very rarely remain submerged for three days.[991]

The same northwest storm that causes shorter exposure times in the Wadden Sea causes longer exposure times on the other side of the North Sea, along the east coast of England and Scotland. Here, the wind effect is less pronounced, as this coast has twice the tidal range of the Wadden Sea. The chance of a strong storm damming up the water is even more important than the tidal range. Like the wind, big storms are most likely along the Atlantic Ocean coast of Northwestern Europe and increasingly less likely further south towards West Africa. The wind effect is greatest when the water is pushed into an enclosed space, which is why a northwest storm in the Western Wadden Sea may cause water to rise 0.5 m, but water levels will rise two to three times higher than this in the eastern Dutch and western German parts of the Wadden Sea.

## 1.3 Sediment characteristics

Most tidal flats are made up of sand and silt, sometimes in combination with rock and pebbles. There is a lot of terminology to describe sediment characteristics. Terms like coarse sand or soft silt seem clear enough, but even part of a sand flat may be described as silty if it contains slightly more silt than the rest of the flat. Because of this variation, geographical descriptions of the sea floor, such as those on charts, are not the most accurate indication of sediment characteristics.

We can objectively measure how sandy or silty an intertidal area is,[405] but unfortunately the different methods available do not always give comparable results. One often used technique is to dry sediment samples, then immerse them in hydrochloric acid and hydrogen peroxide to remove the calcium and organic matter. The sediment is then put in a bottle with plenty of water and shaken well. A sample of this water-mixed sediment is removed and the proportion of sediment remaining in suspension determined. Only silt, which consists of particles with a diameter of 0.016 mm or less, will stay suspended. The coarser particles that immediately sink are called sand. In the decantable fraction of the sediment a further distinction is made between clay (sediment particles smaller than 0.004 mm) and lutum (particles smaller than 0.002 mm). There are also other definitions and criteria for naming sediment types.

The method above is useful for very silty samples, but for sandy samples it's useful to determine the particle size of the sand grains. First, any shell fragments must be removed from the sample. Then, the dried and pre-treated sediment sample is shaken over a series of sieves, arranged from coarse to very fine mesh. Finally, the material on each sieve is weighed and a frequency distribution of the different sand grain sizes determined. Whatever remains on the 0.5 mm sieve is called 'coarse sand'. 'Medium fine sand' stays on the 0.25 mm sieve, 'fine sand' stays on the 0.125 mm sieve, and 'very fine sand' on the 0.062 mm sieve. The proportions of the four sieved fractions are used to describe how fine the sand or sediment is. Different laboratories use different types and numbers (4 - 11) of sieves, making it difficult to compare sediment samples analysed in different places. Luckily, it's common practice to determine the median grain size. The median grain size is the 'mid' grain size: half of the grains in the sample will be smaller than this value, the other half larger. This doesn't solve the problem completely, as a median grain size determined by using a large number of different sieves will be more accurate than one determined from just a few sieves. In addition, sometimes the median is calculated not just for the particles bigger than 0.016 mm (i.e. bigger than silt), but for the entire sediment sample, or for the fraction bigger than 0.050 or 0.062 mm. Nowadays, advanced and rather expensive machines called laser-particle sizers are increasingly used. Laser-particle sizers use light beams to measure the sediment and provide a number as the end result (actually, a series of numbers). If these machines are used incorrectly, the outcome is nonsensical. So it's a good idea to compare the output from the machine with the results obtained by sieving a sub-sample to ensure the results you receive are in the right order of magnitude.

The proportions of clay, lutum and silt are very closely related and therefore we will only consider the proportion of lutum. The amount of lutum is in turn, strongly correlated with the median grain size (Figure 1.5A). If sediment consists of coarse sand, there is almost no silt and clay in the sediment, but if the sand grains are smaller, the fraction of very fine sediment increases. The proportion of lutum is also strongly correlated with the amount of organic matter (Figure 1.5B) and calcium. However, all of these relationships differ locally. In the Dutch Wadden Sea, although the fraction of lutum may be the same, sand grains along the coast are finer than those near the open sea. This is due to local differences in sediment inflow, water currents, and how quickly different-sized particles sink in water.

Coarse sediment is more difficult to get into suspension than fine silt and settles out more quickly. This is why sand flats are usually found in exposed parts of the tidal zone and silt flats in more sheltered areas. In deeper water, currents are often stronger and the waves higher than in very shallow water. This causes sediment at the

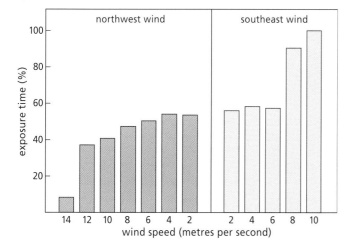

Figure 1.4 The effect of wind strength on the average exposure time of a tidal flat near average sea level. Only days with northwesterly or southeasterly winds were included in the analysis. Data were collected in the eastern part of the Dutch Wadden Sea.[968]

low tide line to be coarser and sandier than at the high tide line.[307] However, this is no longer always the case in the Wadden Sea. The Wadden Sea system has been put out of balance by numerous human-induced changes to the landscape over the ages. Physicists can't tell us precisely where the balance should lie, but there is no doubt that a natural balance existed. When human impacts disturb the balance, natural processes try to return the system to its previous levels. Closing down the Zuiderzee greatly reduced the number of sea trenches in the Western Wadden Sea, causing larger channels to start filling with sediment. This occurs slowly as the fine sand and silt can only remain in suspension to be transported at very low current speeds, and the sediment filling these channels is much siltier than would be expected in an undisturbed system.

As the proportion of lutum increases, the water-holding capacity of the tidal flat also increases (Figure 1.5C). A sample taken from a sandy tidal flat at low tide will consist of 20% water and 80% sediment. But a silty flat is made up of 70% water and 30% sediment. This is why the depth you sink to when walking across a tidal flat is a good measure of the flat's siltiness. For a human being, mud with less than 2% lutum is easy to walk on, at a lutum fraction of 3 - 4% you quickly sink to your ankles, and at 6 - 8% you sink down over your calves in the mud. These are only averages, because fat people with small feet will sink further than skinny people with big feet. The mud can no longer be walked across when there is more than 30% lutum. More lutum means the tidal flat has much less friction, so although it's not really possible to use a sled on mud with less than 3 - 4% lutum, it's a smooth ride from 6 - 8%.

The high correlation between the sediment water content and lutum content also has consequences for the structure of the mud surface. Silty intertidal areas are smooth, but sandy intertidal flats are often rippled (Figure 1.5D). This is because the motion of the waves and currents rearranges the sea floor every high tide, and during this time irregularities can appear in the sediment. When an irregularity appears in silt, the fine sediment flows smoothly over it and the surface stays flat. This, and the high water content, is why a silt flat shines at low tide. On sandy flats, any irregularities create sand ripples containing little shallow pools that are seldom deeper than 1 cm (Figure 1.5E).

Sediment size also has other consequences. Oxygen can penetrate several cm deep in sand flats, but only a few mm in silt (Figure 1.5F). Beneath this thin oxygenated layer, there is no oxygen in the sediment and the sulphate from seawater is reduced to sulphides, which smell like rotten eggs. Hydrogen sulphide doesn't just stink, it's also very poisonous. In fact, it's because it is so poisonous that we think it stinks: our ancestors who were put off by the horrible smell were more likely to survive. This poison makes it more difficult for benthic

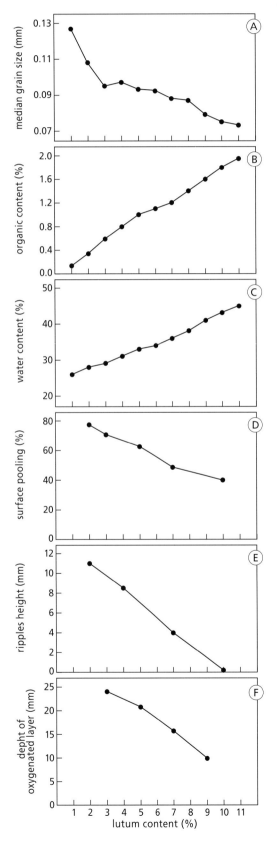

Figure 1.5 Sediment characteristics in relation to increasing lutum content: (A) Sand grain size (median size of grains larger than 0.016 mm), (B) organic content (% organic matter as determined by incinerating at 550 °C), (C) water content (water mass as % of water plus sediment), (D), percentage of surface covered in water pools, (E) soil relief (average difference in height as measured along a horizontally held ruler) and (F) depth of the oxygenated layer (measured with a redox-potential meter). Sediment data were collected from the top 30 cm of the sediment. All measurements were taken in summer at low tide in the eastern part of the Dutch Wadden Sea.[968]

animals to survive in black stinky mud than in sand. However, the sandy sea floor is more mobile than a sheltered silt flat, so animals living in sand flats are at greater risk of being flushed away with the sand.

## 1.4 Food for birds

### 1.4.1 The rich edges of the sea

Although most of the fish and birds that live in the tidal zone feed on animals, they are still indirectly dependent on plants. The marine food chain is based on single-celled algae. In the sea, these microscopic algae are food for very tiny animals called zooplankton, which in turn are eaten by small fish, that are themselves food for larger predatory fish. In shallow seas and particularly in tidal areas, benthic animals fill the role of the zooplankton. Benthic animals are animals that live in or on the sediment, for example, shellfish, worms and crabs. This makes benthic animals the most important link between the algae and the fish and waders. To complicate matters, fish and birds don't only eat benthic animals and benthic animals don't only eat plant material. However this doesn't diminish the crucial role benthic animals play in the intertidal system.

Food chains are rather like pyramids, with a predator at the top and many plants at the bottom. For example, one shark will eat many big fish. Each of these big fish eats a lot of little fish. These little fish might eat crustaceans that feed on zooplankton, which in turn graze on algae. This example of a food chain has six links, but most real food chains have fewer. In every food chain, 10% of the organic matter produced at one link is eaten by the next level. So for the sharks, you would expect that producing 1 kg of shark meat requires 10 kg of big fish, which requires 100 kg of little fish. Working back through the chain we can see it requires 100 000 kg of algae for a shark to gain 1 kg in mass.

Food chains on the Northwest European tidal flats usually consist of three levels, not including seals. Birds and fishes form one level that feeds on the next level, benthic animals, which feed on algae. The benthic animals transfer the nutrients and energy from algae into bite-sized food portions for the birds. This short food chain results in the great diversity of birds found on the tidal flats.[54] As we will discuss later, not all intertidal areas are equally rich in birds. This variation may be due to local differences in algae production and its effects on the density of the benthic animals.

Algae production is closely linked to the amounts of light and nutrients such as phosphates. These nutrients may be present locally or wash in from elsewhere. Most nutrients in the Wadden Sea originate from the North Sea. Incoming tidal currents are always stronger than the outgoing ones, causing intertidal areas to trap debris carried by the tide. This debris usually consists of dead plant material from nearby salt marshes or mangroves. Many of the world's intertidal areas lie in or near river mouths. These rivers are often polluted with animal waste, ferti-

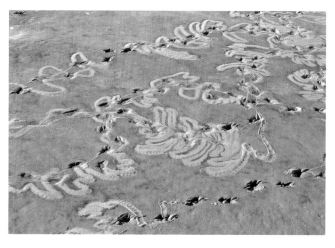

A redshank (top photo) and a whimbrel (middle photo) forage on the exposed tidal flat, on very soft silt and hard sand respectively. Each regularly makes shallow probes into the water and mud with its bill tip. Most probes don't yield anything, and even when the bird is successful it only gets to swallow one prey item per probe. In contrast, the shelduck can collect many prey items at once as it filters the surface sediment with its bill. Feeding shelducks leave distinctive tracks on the sediment (lower photo).

liser and effluent, which they carry to the sea. When such polluted water reaches the intertidal area, it provides extra fertiliser benefiting the local algae. This provides more food for the benthic animals and through this more food for the birds. However, a surplus of nutrients decreases the water clarity. Less light reduces algal growth and the amount of oxygen in the water. Benthic animals are fairly resilient, but they will die without enough oxygen. Although an inflow of nutrients can enrich the intertidal system, a surplus has the opposite effect.

In previous decades, the Rhine was a very polluted river, contaminated with toxic chemicals and heavy metals. Water from the Rhine mixes with water from the North Sea and flows into the Wadden Sea. This polluted the Wadden Sea, decimating the seals and coastal birds, and causing the disappearance of local dolphins. The Rhine also had very high nutrient levels from washing detergents, contaminated wastewater and agricultural runoff. In 1980, the phosphate outflow from the Rhine was 10 times higher than 30 years previously. Since then, effluent treatment has greatly reduced the levels of nitrogen and phosphate in the water.

Twenty years ago, the amount of organic material entering the western part of the Wadden Sea equalled the amount produced throughout the entire Wadden Sea.[701] This input of organic matter had been increasing since 1960 and peaked in 1980. During this period, the numbers of benthic animals also increased.[58] The effects of the nutrient-rich Rhine waters were hardly noticeable in the eastern part of the Dutch Wadden Sea, as they were greatly diluted, and had even less impact on the German and Danish Wadden Sea. There, the rivers Eems, Weser and Elbe meet the Wadden Sea and although they carry less pollutants than the Rhine, their contribution of nutrients and heavy metals is still considerable.

Could the decrease in phosphates since the end of the 1980s be limiting the abundance of benthic animals, fishes and shrimps in the Wadden Sea?[80] Long-term research on algae production in the Western Wadden Sea showed no signs of a decrease in 1993 and still doesn't.[117, 119]

Intertidal areas situated in river mouths are directly influenced by what is carried in the river water. As rivers often have high nutrient loads, tidal areas in river mouths are usually richer in nutrients than tidal flats that do not receive nutrients from outside sources. As rivers become cleaner through improved water treatment, tidal areas near the rivers can become poorer in nutrients. That this is not just a theory has been illustrated at the Dollard, an intertidal area in the Northeastern Netherlands. Until 1980, the potato processing and straw carton industries in the Northern Netherlands discharged huge amounts of effluent. These discharges reduced to almost zero in 1990. Prior to 1980, there was massive mortality of benthic animals at the discharge points, but across the Dollard as a whole, nutrients from the effluent led to an overall increase in benthic animals. Once the discharge stopped, the number of waterbirds in the Dollard halved.[260, 709] We can make three conclusions from this. First, bird numbers are limited by prey availability. When there was less food available, there were fewer birds. Secondly, the production of benthic animals is limited by the availability of algae. When there are less algae, the number of benthic animals decreases. Finally, algae production is limited by nutrient availability.

| | | |
|---|---|---|
| **Bivalves** | Mussel | *Mytilus edulis* |
| *Bivalvia* | Sand gaper | *Mya arenaria* |
| | Cockle | *Cerastoderma edule* |
| | Peppery furrow shell | *Scrobicularia plana* |
| | Baltic tellin | *Macoma balthica* |
| **Snails** | Mudsnail | *Peringia ulvae* |
| *Gastropoda* | | |
| **Crustaceans** | Amphipod | *Corophium volutator* |
| *Crustacea* | Shrimp | *Crangon crangon* |
| | Common shore crab | *Carcinus maenas* |
| **Polychaete worms** | Lugworm | *Arenicola marina* |
| *Polychaeta* | Ragworm | *Hediste diversicolor* |
| | King ragworm | *Hediste virens* |
| | Sandmason worm | *Lanice conchilega* |
| | Nephtyid polychaete | *Nephtys hombergii* |
| | Orbiniid polychaete | *Scoloplos armiger* |
| | Capitellid polychaete | *Heteromastus filiformis* |
| | Spionid polychaete | *Pygospio elegans* |
| **Fish** | Plaice | *Pleuronectus platessa* |
| *Pisces* | Flounder | *Paralycthys flesus* |
| | Common goby | *Pomatoschistus microps* |

Table 1.1 Common and scientific names of the 20 most common prey animals on the intertidal flats of Northwestern Europe

### 1.4.2 Most important prey species on Northwest European tidal flats

Most birds that forage on the intertidal flats are carnivorous, and can choose from a wide selection of prey. The only herbivorous waterbirds in Northwestern Europe are wigeon (*Anas penelope*) and brent geese (*Branta bernicla*). They feed on sea lettuce (*Ulva spp.*) and seagrass on tidal flats, and on red fescue (*Festuca rubra*), common salt marsh grass (*Puccinellia maritima*), common sea-lavender (*Limonium vulgare*) and other plants on the salt marsh.

If all the animal species birds encounter on the intertidal flats were listed, it would be a very long list. However, all these prey belong to just a few important groups. Fishes, crabs and shrimps live in the shallow water, while snails, shellfish, worms and crustaceans are found on the tidal flats. Around 20 species of animals form the main food for waterbirds in Northwestern Europe. Table 1.1 provides a list of the common and scientific names of these animals. The use of scientific or Latin names is often defended by the argument that these names are more precise. The irony is that most of the species listed in

Table 1.1 have been renamed by taxonomists in the last few years. So the Latin name of the cockle, *Cardium* changed into *Cerastoderma* and that of the mudsnail *Hydrobia* into *Peringia*. The polychaete worm *Hediste* changed into *Nereis* and once everybody got used to that, it officially changed back to *Hediste* again, although the old name *Nereis* is still the one most commonly used.

Birds can fish for young plaice, flounder and common gobies in the sea. In shallow water, they catch common shrimps and crabs. On the surface of the tidal flats they find periwinkles, mudsnails and a bivalve, the mussel. Other bivalves live buried in the sediment, the most common of which are: cockles, Baltic tellin, sand gapers, and locally the peppery furrow shell and thin tellin. An American newcomer, the American razor clam (*Ensis directus*), can also be included here. The amphipod *Corophium volutator* and many types of worm also live buried in the sediment. Most of the big worms have common names, but the small worms have to make do with their scientific names. The most common little worms are *Pygospio elegans* and *Heteromastus filiformis*.

A North American worm, *Marenzelleria cf. wireni*, was accidentally introduced to the Wadden Sea recently.[262] In the last few years, the likelihood of exotic species arriving has increased with the increase in shipping traffic. New species usually travel in modern ships as larvae in ballast water. In the past, when ships were not painted with nasty anti-fouling chemicals, the animals could latch onto the ship hulls. Before this, back when the ships were still made of wood, animals like the shipworm could burrow into the ship's hull. The sand gaper is rather a special case. This North American species crossed the Atlantic Ocean in the early Middle Ages, thanks to the Vikings.[637] The sand gaper provided an excellent food source for the Vikings as its meat could stay fresh during long voyages. Apparently not all of the sand gapers were eaten on the way back to Europe, and a single live one was thrown overboard or perhaps planted intentionally by a Viking.

Prey animals can be sorted simply according to where they live: swimming in water (fishes and shrimps), surface dwelling (common shore crabs, mudsnails, and some shellfish), or buried in sediment (worms and shellfish). In practice though there is some overlap. Shrimps live not only in the water but also on exposed intertidal flats where they bury themselves shallowly. Mudsnails don't only exist on the surface, but also often live just beneath it. Even benthic animals don't always live beneath the surface; cockles are regularly found on the surface, as are all sorts of worms, which crawl up their burrows to look for food.

Another way to sort species is by body size. Often, all animals larger than 0.5 mm are called macrofauna, and all animals smaller than 0.1 mm are called microfauna; meiofauna fit in between. Worms are long and thin and often try to actively crawl through sieves. Most worms smaller than 4 cm long can pass through a sieve with a 1 mm mesh.[978] Sometimes a 0.5 mm sieve is used to stop the worms from escaping. Little shellfish and snails are quite round so specimens bigger than 1 mm will stay on a 1 mm sieve. The benthic animals listed in Table 1.1 are all macrofauna, but in their first stages of life they may be too small to be detected in sieves and belong to the meiofauna, or even the microfauna.

On the tidal flats macrofauna are present in a density of around 1000 benthic animals per m², although this can vary by a factor of 10. The same variation in density is found in the meio and microfauna, whose average density is somewhere around 1 000 000 animals per m², a thousand times more than the macrofauna.[280] In spite of these very big numerical differences, by weight there are more macrofauna than micro and meiofauna. In total, macrofauna will have a dry flesh weight of 10 - 100 g per m², but the microorganisms will usually total around 1 g per m². For this reason alone, microfauna are of little importance as food for birds.

Species diversity (of both plants and animals) increases from the North and South Poles towards the equator. This is also true of the benthic animals, fish, crabs and shrimps found on the intertidal flats. If a random sample of around 200 cm² is taken on the intertidal flats of Northwestern Europe, and sieved over a 1 mm mesh, it would not usually contain more than 10 different species. A similar sample collected from a tropical tidal area would hold many, many more.[51, 946, 950]

### 1.4.3 Depth and sediment characteristics

Benthic animals can filter food from the water (suspension feeding) or scrape it off the sediment surface (deposit feeding). Some benthic animals actually eat the substrate, or are predators and feed on other benthic animals. A true filter feeder, like a cockle or mussel, can only feed when the tidal flat is submerged. These ani-

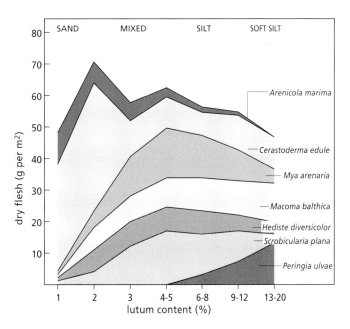

Figure 1.6 The relationship between the presence of various benthic animals, expressed in biomass (g dry flesh) per m², and sediment characteristics (lutum content, the percentage of soil particles smaller than 0.002 mm). Data collected during 11 years of sampling in August (1976 - 1986) in the eastern part of the Dutch Wadden Sea.[968]

mals are in trouble high in the intertidal zone, as there they can only feed for a short period at high tide. This is reflected in their decreasing growth rates as the average submersion time decreases.[25, 441] Deposit feeders can continue feeding at low tide, and could be expected to occur throughout the intertidal zone. A species like the Baltic tellin which is both a suspension feeder and a deposit feeder,[453] does indeed exist over the whole tidal zone. However Baltic tellins that live lower in the tidal zone have a higher growth rate.[63]

The common names of some benthic animals – sand gaper, sandmason worm, and mudsnail – can indicate which species are found more often in silty sediments and which in sand (Figure 1.6). Although silty sediments contain more food for benthic animals, oxygen cannot penetrate through silt as deeply as it does in sand. For this reason, benthic animals that get their oxygen from sediment water avoid silt. Animals such as mussels, that filter food from the water, also prefer sandy sediments. This is because there is more silt in suspension in the water above mud flats than above sand flats. Filter feeders that live in silt may find that as they inhale water their gills become blocked with silt.

Lugworms don't occur in the silt. These worms live in u-shaped tunnels and feed on sediment. Most of their food is from the surface, but if they crawled out to get it, they would immediately become someone else's next meal. Instead they pump the substrate down to them, a technique that only works in sediment firm enough to keep their almost funnel-shaped tunnels intact. A silty sediment is too soft, which is why lugworms are not found in silt. Conversely, they are also never found in coarse sand. Pure sand doesn't contain enough food to make it profitable for lugworms. The ideal tidal flat for a lugworm is a mix of sand and silt.

Sandy sediments are usually far less stable than silty ones. This means that shallow-living benthic animals are at more risk of being moved in sand than in silt. If such movements coincide with an increased predation risk, it's safer for the animal to settle on a sheltered sandy sediment. Some benthic animals (such as the lugworm and the Baltic tellin) settle in sheltered silty sediments for their first life stage, then crawl out of the sediment months or even years later, and let the outgoing tide take them to the more exposed sand flats.[63, 277, 278, 923] For this reason, these sheltered silty zones are called 'nursery flats'.

There are often low densities of benthic animals around the high and low tide lines. Instead, these animals reach their highest densities in the mid tidal zone, around average sea level and just below (Figure 1.7). Cockles are often less abundant at the low water line than further up the intertidal zone. This is not what you might expect. Cockles can only feed when they are submerged so you might think they'd be more abundant around the low tide line, where the tidal flats are submerged the longest. To understand the distribution of the cockle and other benthic animals we need to know more about the effects of predation pressure on survival.

## 1.4.4 Competition and survival

The distribution of benthic animals over the intertidal flats is only partly explained by each species' preference for specific areas within the tidal zone and certain sediment characteristics. Although the amphipod *Corophium volutator* is usually found in silty areas high in the tidal zone, it sometimes also occurs on lower areas, and is common throughout the intertidal zone in some estuaries. These differences exist because the distribution of benthic animals is largely determined by the presence of competitors and predators.

In summer, a new generation of mussels settles on an area of tidal flat, completely covering it in tiny mussels just millimetres long. A few weeks earlier these mussels were tiny larvae floating in the water; they only settle on the sediment once they have grown big enough. The newly settled mussel spat anchor onto each other, old mussels and other hard surfaces like shells or stones, with their long anchor threads or byssus threads to avoid being washed away. Within a few weeks of spatfall, common shore crabs and other predators have eaten most of the brood. Even so, young mussels that settle in late spring, still reach densities of 10 000 or more per m² in late summer. By then, they are about 1 cm long and, with more than one mussel per cm², completely cover the tidal flat. This makes life difficult for the animals that lived in the sediment before the mussels settled. If the young mussels survive their first winter, they will slowly

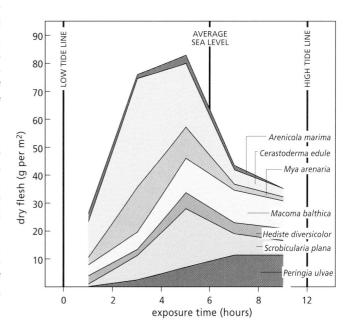

Figure 1.7 The relationship between the presence of various benthic animals, expressed in biomass (g dry flesh) per m², and the height of the tidal flat. Based on 11 years of sampling in August (1976 - 1986) in the eastern part of the Dutch Wadden Sea.[968]

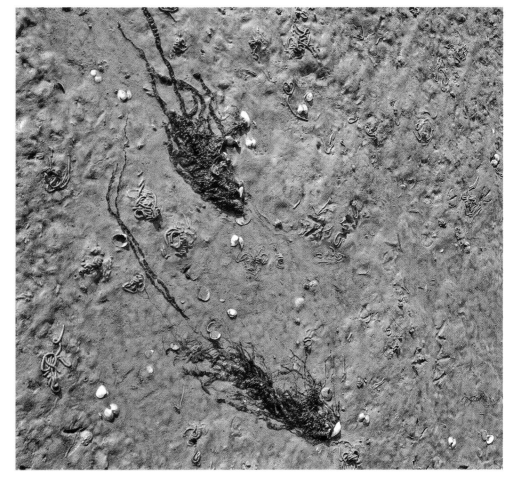

Mussel beds provide the most abundant food resources on tidal flats for waterbirds (photo top left), while shallow ponds and channels are attractive feeding grounds for birds that feed on shrimps and common crabs (photo top right). Waterbirds that forage on the exposed sediment at low tide find their food in the top few centimetres of sediment, not sitting out in the open. The benthic animals may be hidden in the mud, but they betray their presence by leaving all sorts of subtle signs on the surface. The toothpaste-like droppings of the lugworm are the most obvious, and are clearly visible in the adjacent photo, as are empty cockleshells and long strands of the seaweed *Enteromorpha*.

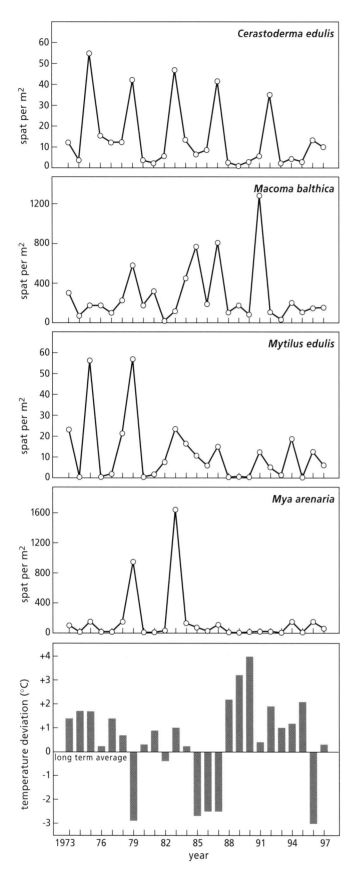

Figure 1.8 The number of juvenile benthic animals of only a few months of age on tidal flats in late summer from 1974 - 1997 and the annual variation in winter temperature over the same time period. Temperature data are expressed as the deviation from the long-term annual average. Based on a unique data set from the Balgzand in the Western Dutch Wadden Sea.[49, 56]

establish a stable mussel bed. Once this has happened, it is almost impossible for other benthic animals to settle. Not only is the sediment completely covered in mussels but small larval animals are likely to be sucked up by the filter-feeding mussels. However mussel beds are not just a monoculture of mussels. Mussels expel any inhaled silt from the seawater they filter, building up a layer of silt around the mussel bed.[897] Small pools of water remain between the built-up beds forming channels where fish, shrimps and crabs can live. Barnacles begin to grow on the mussels, periwinkles start to graze, and bladderwrack seaweeds attach themselves and provide a refuge for amphipods and common shore crabs.

If the mussel bed disappears and is not replaced by a new mussel spatfall the following summer, both the mussels and the other species that lived on the mussel bed will be lost. Other benthic animals can now colonise the bare sediment: which ones do depends on the fertility of the species nearby. One possibility is that cockles will settle here en masse. Like mussels, cockle spat can reach densities of 1000 – 10 000 per $m^2$. The presence of a cockle bed can make life very difficult for previous benthic residents. However, the cockles' effect is less dramatic than that of a mussel bed as cockles don't cover the sediment completely. This is why ragworms and Baltic tellins can still be abundant on a cockle bed. Cockles, like mussels, pump in and filter the surrounding water, making it risky for any larval benthic animals floating in the water to settle on a cockle bed.

Perhaps instead of cockles, peppery furrow shells (*Scrobicularia plana*) colonise the mud. These bivalves bury themselves deep in the sediment, extending two long tubes, the inhalent and the exhalent siphons, up into the water from their slightly gaping shells. The inhalent siphon sucks up little food particles off the sediment surface, while the exhalent siphon spits out any non-digested food particles. Peppery furrow shells vacuum the surface so intensely that they inhibit the settlement of any other benthic animals. Lugworms are similar. They feed on substrate from the top sediment layer so thoroughly that other benthic animals have little chance of settling where lugworms are. They have the same effect on their own offspring: young lugworms are much more likely to settle on tidal flats with few old lugworms. So at high adult densities few young lugworms will be recruited, but in years with few old lugworms more young can settle. As a consequence of this density dependent process there is little annual variation in the density of lugworms.[67]

So the early bird gets the worm, as far as settling on the substrate goes. Once a strong year-class of sand gapers or peppery furrow shells has settled, it could take five to ten years before this local generation is extinct and it becomes possible for other species to settle. As will be discussed in Section 4.3.6, the contribution of different species of benthic animals to total biomass varies strongly from year to year. This difference is largely due to competition between species. A species like the amphipod *Corophium volutator* is only able to settle somewhere that large benthic animals like the sand gaper, cockle

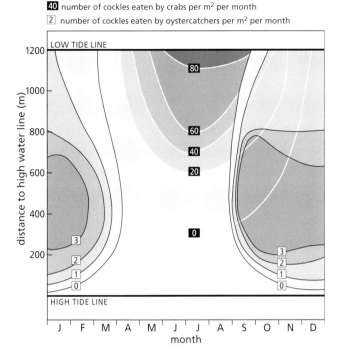

Figure 1.9 Few cockles living near low water survive the summer. This figure shows the number of cockles consumed by common shore crabs and oystercatchers monthly over the year. Predation pressure from oystercatchers is very small overall compared to that of common shore crabs, but can be significant at some sites and times. The few cockles taken in the winter or high in the tidal zone were undoubtedly the victims of oystercatchers, not common shore crabs.[764]

and lugworm are not present in large numbers.[283]

We know there is strong competition between the different benthic animals, but that's not all. As well as feeding on similar foods and making it impossible for others to settle, some benthic animals feed on others. The main benthic dwelling predators are the polychaetes *Nephtys hombergii*, king ragworms and sometimes ragworms. These aren't the main predators benthic animals must deal with though. The most important predators live in the water or on the sediment: common shore crabs, shrimp and fish. These predators are cold-blooded, and need less food than the warm-blooded birds, but they are so abundant that most benthic animals are at far greater risk of being caught by one of them than by a warm-blooded predator.

The effects of competition and predation on different species of benthic animals have been shown experimentally in America and Europe (see[722, 723, 724, 725]). In these experiments, benthic animals were removed from an area of tidal flat to see which species would resettle it. Sometimes part of the intertidal flat was fenced with thread to exclude birds or with fine mesh to exclude fish, crabs and shrimps as well as birds. If mesh was used, researchers could also introduce species, such as predatory worms. Excluding common shore crabs and shrimps was found to have the biggest effect on how the area was colonised. Crabs and shrimps exert strong predation pressure on very young benthic animals that have just settled. These predators are very abundant in spring and summer, unless there has been a very cold winter.

After a very harsh winter, shrimps and common shore crabs appear on the intertidal flats later in the season.[60, 61] This is probably the main reason why benthic animals in the Wadden Sea produce strong year-classes after harsh winters and very seldom after mild winters.[56] The spatfall of benthic animals has been carefully measured on the Balgzand, a tidal area in the Western Dutch Wadden Sea since 1973. The relationship between spatfall and the harshness of the previous winter is clearly visible in Figure 1.8. In the four winters with the highest winter temperatures (1988, 1989, 1990 and 1995) there was hardly any spatfall. But a harsh winter is no guarantee of spat the following summer. After the cold winters of 1979, 1985, 1987 there was a lot of spatfall but not after the equally cold winters of 1986 and 1995.

Predation pressure from common shore crabs varies in intensity across the tidal flat. Many crabs leave the channels and move onto tidal flats when the flats are submerged at high tide, and return to the channels a few hours later on the ebbing tide. This tidal migration doesn't occur in winter in Northwestern Europe, as the crabs remain in the channels with the fish and shrimps. In summer, small common shore crabs and shrimp stay behind on the tidal flats in shallow pools or bury themselves, rather than migrate with the tides. The large common shore crabs that migrate between channels and tidal flats stay below the high water line. As a consequence, cockles are more likely to survive if they live high in the tidal zone (Figure 1.9).[763, 764] So although cockles can feed longer and grow more quickly near low tide, they are unlikely to survive here. Predation risk explains why cockles are rarely present in their ideal growing conditions.

Some benthic animals avoid the problem of higher predation risk in the lower tidal zone by spending their first life phase higher in the tidal zone, and then allowing the ebbing tide to take them to a lower tidal zone the following winter, the Baltic tellin is a good example of this.[63] When Baltic tellins are older, they have better protection from predators: they are bigger which makes them more difficult to swallow, their shell is stronger so only big crabs can crush them, and they have a longer siphon, so they can live deeper in sediment[980] where they are out of reach of fish and crabs.

Portraits of an oystercatcher, a five-day old knot chick and a spoonbill (this page), plus a curlew, a turnstone and an avocet (page 27).

# 2. Portrait Gallery

## 2.1 Species and subspecies

### 2.1.1 A logical order

Birds come in different shapes and sizes. Usually, it's easy to sort them into categories based on their similarities, something we all, and even the birds themselves, do. We can categorise them with varying precision. The 18th century Swedish researcher Carl von Linné – better known by his Latin name Linnaeus – devoted his life to categorising the animal and plant kingdoms. He did this with the help of many assistants, who he sent all over the world to collect plants and animals (sometimes the collections

made it back to Sweden, but the assistants did not). In 1759 Linnaeus published the *Systema Naturae*, in which he named most of the waterbirds covered in our book. Only three small waders were not scientifically described until later: the curlew sandpiper in 1763, the purple sandpiper in 1764, and the little stint in 1812, by the researchers Pontopiddan, Brünnich and Leisler respectively.

The names given by Linneaus have two parts. First is the genus name (which starts with a capital letter, for example, *Charadrius*), then the species name (which starts with a lowercase letter, for example, *hiaticula*). *Charadrius hiaticula* is the scientific name used throughout the world for the ringed plover. Linnaeus grouped species that looked similar to him in one genus, for example, the ringed plover and the golden plover, which in 1759 was called *Charadrius apricaria*. Since then the golden plover has been put in a different genus and is now called *Pluvialis apricaria*.

In previous centuries, species names were based solely on inspections of museum specimens. These specimens were relatively scarce and not very well preserved. In spite of this, Linneaus sorted the waterbirds very well and the species names he gave are still in use.[414] But, apart from the ringed plover and the Kentish plover (both in the genus *Charadrius*), all of his genus names have been changed. On closer examination, biologists assigned four species that Linneaus put in the genus *Tringa*, the grey plover, red knot, common sandpiper and turnstone, into four different genera, *Pluvialis, Calidris,*

PORTRAIT GALLERY

| Common name | Scientific name | Biogeographic population | Population estimate | Estimated trend | 1% criterion |
|---|---|---|---|---|---|
| Great Cormorant | *Phalacrocorax carbo sinensis* | N & C Europe | 275 000-340 000 | Increase | 3100 |
| Eurasian Spoonbill | *Platalea leucorodia leucorodia* | East Atlantic | 9950 | Increase | 100 |
| Brent Goose | *Branta bernicla bernicla* | W Siberia (br) | 215 000 | Decrease | 2200 |
| Brent Goose | *Branta bernical hrota* | Svalbard (br) | 5000 | Increase | 50 |
| Barnacle Goose | *Branta leucopsis* | N Russia, E Baltic (br) | 360 000 | Increase | 3600 |
| Common Shelduck | *Tadorna tadorna* | NW Europe (br) | 300 000 | Stable | 3000 |
| Eider | *Somateria m. mollissima* | Baltic & Wadden Sea | 850 000-1 200 000 | Decrease | 10 300 |
| Mallard | *Anas p. platyrhynchos* | NW Europe (non-br) | 4 500 000 | Decrease | 20 000 |
| Wigeon | *Anas penelope* | NW Europe (non-br) | 1 500 000 | Increase? | 15 000 |
| Lesser Black-backed Gull | *Larus fuscus graellsii* | W Europe to W Africa (non-br) | 525 000 | Increase | 5300 |
| Great Black-backed Gull | *Larus marinus* | NE Atlantic | 420 000-510 000 | Stable | 4700 |
| Herring Gull | *Larus argentatus argentatus* | Baltic/Nordic (br) | 1 100 000-1 500 000 | Stable | 13 000 |
| Common Gull | *Larus canus canus* | N & W Europe (br) | 1 300 000-2 100 000 | Decrease | 17 000 |
| Black-headed Gull | *Larus ridibundus* | N & C Europe (br) | 5 600 000-7 300 000 | Increase | 20 000 |
| Sandwich Tern | *Sterna s. sandvicensis* | W & N Europe (br) | 159 000-171 000 | Increase | 1700 |
| Common Tern | *Sterna hirundo hirundo* | W & S Europe (br) | 170 000-200 000 | Stable | 1900 |
| Arctic Tern | *Sterna paradisaea* | N Eurasia (br) | >1 000 000 | | |
| Little Tern | *Sterna albifrons albifrons* | W Europe (br) | 31 000-37 000 | Stable | 340 |
| Eurasian Oystercatcher | *Haematopus o. ostralegus* | Europe (br) | 1 020 000 | Increase | 10 200 |
| Pied Avocet | *Recurvirostra avosetta* | W Europe (br) | 73 000 | Stable | 730 |
| Great Ringed Plover | *Charadrius hiaticula* | Europe, N Africa (non-br) | 73 000 | Increase | 730 |
| Great Ringed Plover | *Charadrius hiaticula* | W & S Africa (non-br) | 190 000 | Decrease? | 1900 |
| Kentish Plover | *Charadrius alexandrinus* | W Europe to W Mediterranean (br) | 62 000-70 000 | Decrease | 660 |
| Grey Plover | *Pluvialis squatarola* | E Atlantic (non-br) | 247 000 | Increase | 2500 |
| Eurasian Golden Plover | *Pluvialis apricaria apricaria* | NW Europe (br) | 69 000 | Decrease | 650 |
| Eurasian Golden Plover | *Pluvialis apricaria altifrons* | Iceland, Faeroes (br) | 930 000 | Stable | 9300 |
| Eurasian Golden Plover | *Pluvialis apricaria altifrons* | N Norway, N Russia | 645 000-954 000 | Stable | 8000 |
| Red Knot | *Calidris canutus canutus* | Taimyr Peninsula (br) | 340 000 | Decrease | 3400 |
| Red Knot | *Calidris canutus islandica* | NE Canadian islands (br) | 450 000 | Decrease | 4500 |
| Sanderling | *Calidris alba* | E Atlantic (non-br) | 123 000 | Increase | 1200 |
| Curlew Sandpiper | *Calidris ferruginea* | SW Europe, W Africa (non-br) | 740 000 | Increase | 7400 |
| Dunlin | *Calidris alpina alpina* | W Europe (non-br) | 1 330 000 | Stable | 13 300 |
| Dunlin | *C. alpina schinzii* | Iceland (br) | 940 000-960 000 | Stable | 9500 |
| Dunlin | *C. alpina schinzii* | Britain & Ireland (br) | 23 000-26 000 | Decrease | 250 |
| Dunlin | *C. alpina schinzii* | Baltic (br) | 3600-4700 | Decrease | 40 |
| Little Stint | *Calidris minuta* | Europe & W Africa (non-br) | 200 000 | Decrease? | 2000 |
| Purple Sandpiper | *Calidris maritima* | E Atlantic (non-br) | 50 000-100 000 | Stable | 750 |
| Bar-tailed Godwit | *Limosa lapponica lapponica* | W Europe, NW Africa (non-br) | 120 000 | Stable | 1200 |
| Bar-tailed Godwit | *Limosa lapponica taymyrensis* | W & S Africa (non-br) | 520 000 | Decrease | 5200 |
| Black-tailed Godwit | *Limosa limosa islandica* | Iceland, Faeroes, Shetland, Lofoten Is | 35 000 | Increase | 350 |
| Eurasian Curlew | *Numenius arquata arquata* | Europe (br) | 420 000 | Stable/Increase | 4200 |

| | | | | | |
|---|---|---|---|---|---|
| Whimbrel | *Numenius phaeopus phaeopus* | W Africa, W Europe (non-br) | 160 000-300 000 | Stable/Increase | 2300 |
| Whimbrel | *Numenius phaeopus islandicus* | Iceland, Faeroes, Scotland (br) | 610 000 | Stable | 6100 |
| Spotted Redshank | *Tringa erythropus* | Europe (br) | 77 000-131 000 | Stable | 1000 |
| Common Redshank | *Tringa totanus totanus* | E Atlantic (non-br) | 250 000 | Decrease | 2500 |
| Icelandic Redshank | *Tringa totanus robusta* | Iceland & Faeroes | 64 500 | Stable/Increase | 650 |
| Common Greenshank | *Tringa nebularia* | Europe (br) | 234 000-395 000 | Stable | 3100 |
| Common Sandpiper | *Actitis hypoleucos* | Europe (br) | 1 400 000-2 000 000 | Stable | 17 000 |
| Ruddy Turnstone | *Arenaria interpres interpres* | W Europe, NW Africa (non-br) | 94 000 | Increase | 1000 |

Table 2.1 The most recent 1% criteria for waterbirds discussed in this book.[1000] The 1% criterion is set at 20 000 for species with populations of 2 000 000 birds or more. Note that for some of the (sub)species the estimates and trends may already be out of date. In the biogeographic population column (br) denotes breeding populations and (non-br) denotes non-breeding populations; all other populations are present year-round.

*Actitis* and *Arenaria* respectively. Another change since Linnaeus is that species are grouped not just into genera, but are grouped within genera into families. So the genus *Pluvialis* is part of the plover family Charadriidae (the last part '-dae' shows that it is the name of a family), while *Calidris*, *Actitis* and *Arenaria* (and species now in the genus *Tringa* - such as the redshank *Tringa totanus*) are considered part of the sandpiper family Scolopacidae.[678, 691] The species is the clearest taxonomic unit. A species can be broken down into smaller groupings (such as subspecies and populations), but the further a species is broken down, the more subjective the grouping becomes and the more likely it is that experts will disagree over whether it is correct or not.

At the time when many waterbirds were first assigned scientific names, most biologists believed that God had created all the animals. After Darwin and Wallace published the theory of evolution in 1859,[150] this naming became more complicated. The categories of species, genera and families also needed to reflect the animal's ancestry. Species that belong to the same family must share a common ancestor, unique to that family. Apart from the fact that they do not even look similar, if grey plovers and red knots are not descended from the same ancestral birds then they don't belong in the same genus *Tringa*.[520] We can now be reasonably sure that the families we use for waterbird species do distinguish different evolutionary units and that all the species in such families stem from a common ancestor.[414]

Further problems are created by the way that evolutionary history usually branches like a tree. If you imagine that the distance from the ground to the treetop is a measure of time, then from the tree trunk (the common ancestor) all sorts of main and side branches split off, which we call families, genera and species. Many branches and twigs don't reach up to the treetop (the present) because they have already died out. How can the branches be sorted into a logical list? Something two-dimensional (like the spread of living species in the top branches) isn't easy to cram into a one-dimensional space (such as a species list). In this book, we have chosen the most commonly used species list, developed by the Dutch ornithologist Karel Voous,[904] as it gives the best overview.

Perhaps it's a little ironic that new developments in science have caused scientific names (which were meant as uniform and worldwide names for each species) to change more often than common names. So some sandpipers long described as belonging to the genera *Crocethia* (sanderlings) and *Erolia* (curlew sandpipers, little stints and purple sandpipers) are now considered part of the genus *Calidris*. Since every life form is the product of many evolutionary processes (which can only be reconstructed after very detailed research), it's understandable that as our knowledge grows, opinions on the boundaries between species change.[769]

Some name changes affect the waterbirds discussed in this book. In the Dutch Wadden Sea, the common black-bellied brent goose *Branta bernicla bernicla* has been split off from the easily distinguished white-bellied brent goose *Branta b. hrota* (which mainly occurs in the Danish part of the Wadden Sea) and the American black brent goose *Branta b. nigricans*.[572, 768] In 1998, on the basis of new biometric data, it was proposed that whimbrels be split into an American species *Numenius hudsonicus* (the Hudsonian whimbrel) and a Eurasian species *Numenius phaeopus* (the Eurasian whimbrel).[220] Of course, every proposal for a name change is critically evaluated by experts and it takes some time for any new names to come into common usage. In this book, we have taken a conservative approach and use the scientific names that have been in common use for the last 50 years.[904]

## 2.1.2 Variety within species: phylogenies and subspecies

The study of geographic variation within and between wader populations has become increasingly refined. Museum collections have expanded enormously, especially since the beginning of the 20th century. Powerful computers and specially developed statistical tools are now available to analyse data on body size, plumage and

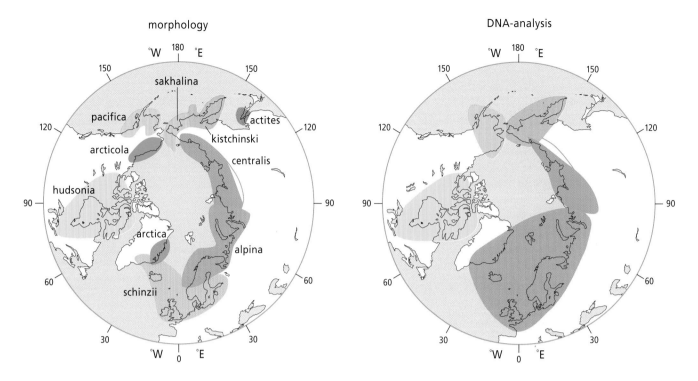

Figure 2.1 Geographical distribution of various dunlin subspecies based on morphometric and plumage differences (left) and on genetic variation (right).[220]

The red knot (in the mist net below and page 31) and the dunlin (below) are the two best-studied sandpiper species. They are often caught for ringing, which has greatly increased our knowledge of their migration routes.

moult from museum specimens and wild birds.[220] Laboratory-based research on genetics has been through revolutionary developments,[510] further influencing research on waterbirds. But the real key to our increased knowledge is the ability to catch large numbers of waders, and ring and measure them.

Two developments in catching techniques formed the basis of the recent knowledge explosion. First, the mist net (a fine-thread vertical net) that was developed in Japan proved useful for catching waterbirds on tidal flats. Secondly, cannon nets were developed in England to catch huge numbers of roosting waders. In cannon netting, a rectangular folded net is laid on the expected high water roost and when a flock of birds settles there, buried projectiles shoot the net over the birds. Using these techniques, large numbers of waterbirds have been caught and marked with individually numbered leg rings since 1965. These birds have been seen or caught at other places over the years.[679] As well as giving the birds individually numbered rings (with an inscription showing the administrative address), researchers record biometric data (such as the length of the bill, wing, leg and the body mass) and the stage of moult of most birds caught. These data can provide a lot of information. For example, if the size of the birds being caught changes over the winter, this could mean that birds from different breeding populations have successively moved through that area. Weight increases suggest the birds are building up fat reserves to make long-distance flights and can indicate when certain groups of birds are likely to leave the study area. The absence or presence of moult can also provide clues about the origin and identity of the birds. In this sort of international detective work, almost every piece of information is used.[178, 306, 679, 815]

Since large-scale ringing started, a detailed picture of the migration routes of European waders has emerged.[644, 645, 679] On the east coast of England, the Wash Wader Ringing Group, which developed the use of cannon nets, has been active for 35 years. They catch thousands of waders each year around the Wash, the most extensive tidal area in the United Kingdom. Through this, they keep a finger on the pulse of local wader populations. On several occasions they have been able to detect sudden periods of high mortality in redshanks and oystercatchers (due to food shortages and harsh winter weather) through the increased numbers of dead ringed birds found (known as recoveries).[131, 616] Similar progress has been made elsewhere in Europe. Extensive ringing work in the German Wadden Sea revealed how critical this area is for many waterbirds.[707] An intensive programme of catching, measuring and ringing waterbirds took place on the Dutch Wadden Sea island of Vlieland from 1971 until 1975 to investigate the Wadden Sea's international role for waders. Preliminary data from this research were presented in the report of the Wadden Sea Committee in 1974 and led to an about turn in thinking about the natural values of the Wadden Sea.[84] Extensive planned reclamations were cancelled.

This work wasn't restricted to Western Europe. Enthusiastic waterbird catchers quickly followed their beloved birds to Iceland, Lapland, Greenland and Northwest Africa. Usually, they returned with a few resightings of ringed individuals and fresh biometric data that provided further insights.[178, 505, 581] It quickly became apparent that birds from different wintering areas sometimes migrated along different routes to different breeding areas. Red knots are a good example of this: West European wintering knots breed in Northern Greenland and Northeastern Canada, while West African wintering birds breed in Siberia.[157, 178, 179, 684] The Siberian knots are slightly bigger than the knots from Greenland and Canada, which has led to their separation into different subspecies or races.[746] To distinguish between the two groups, each now has a subspecies name after the species name, making the full name of the European wintering knots *Calidris canutus islandica*, while the West African wintering knots are known as *Calidris canutus canutus*.

Similarly, ringed plovers that migrate through the Wadden Sea have been divided into different subspecies. Ringed plovers from Northern Europe and Siberia are called *Charadrius hiaticula tundrae*, while those from the rest of Europe, Greenland and Northeastern Canada are called *Charadrius hiaticula hiaticula*. A recent publication proposed that birds from Iceland, Greenland and Northeastern Canada be further distinguished using the name *Charadrius hiaticula psammodroma*.[220] The same publication also suggested that there are five subspecies of bar-tailed godwit. In winter, bar-tailed godwits from Northern Europe are found in the Wadden Sea (*Limosa lapponica lapponica*). Other bar-tailed godwits migrate through the Wadden Sea in large numbers every spring and autumn, winter in West Africa,[188] and breed on the Taimyr Peninsula. These West African wintering birds are slightly smaller than the birds that winter in the Wadden Sea and are now called *Limosa lapponica taymyrensis*. But, as is the case for the turnstone, it isn't always possible to find morphological differences between different breeding populations that can be used to differentiate subspecies. There is so little geographical variation in plumage and body size in turnstones that it doesn't seem useful to split them into different subspecies. However it's not inconceivable that turnstones that breed in Europe have completely different migration patterns from turnstones that breed in Greenland and Northeastern Canada. The former mainly migrate to West Africa,[788] while the latter winter mainly in Western Europe.

Over the last 30 years, there has been a real revolution in the techniques used to map the evolutionary histories of families and species. As well as using the body size, plumage and behaviour of birds, the characteristics of their genetic material (such as the order of certain pieces of DNA) are increasingly used to track phylogenies and verify the existence of (sub)species.[23, 510, 919] Amongst European waterbirds, the dunlin is certainly champion where the number of subspecies and the assortment of migration routes are concerned. Based on variations in plumage and body size, not less than 10 subspecies have been distinguished worldwide.[220] Laboratory research on variation in the DNA (isolated from tiny samples of blood) confirms much of the observed geographical structure.[920, 922] From the DNA work it appears that some populations were already following their own migration routes 200 000 years ago. However it isn't

## PORTRAIT GALLERY

Figure 2.2 The most important intertidal sites for waders on the East Atlantic flyway. The surface area of the various sites differs greatly. Tidal areas regularly used by more than 100 000 waders in winter are marked with an asterisk in the list of site names.

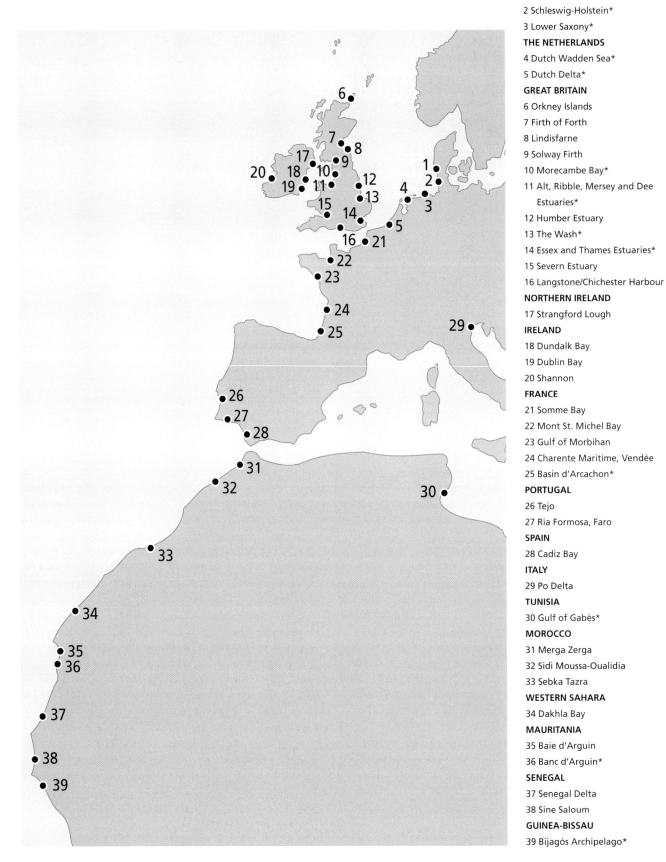

**DENMARK**
1 Danish Wadden Sea
**GERMANY**
2 Schleswig-Holstein*
3 Lower Saxony*
**THE NETHERLANDS**
4 Dutch Wadden Sea*
5 Dutch Delta*
**GREAT BRITAIN**
6 Orkney Islands
7 Firth of Forth
8 Lindisfarne
9 Solway Firth
10 Morecambe Bay*
11 Alt, Ribble, Mersey and Dee Estuaries*
12 Humber Estuary
13 The Wash*
14 Essex and Thames Estuaries*
15 Severn Estuary
16 Langstone/Chichester Harbour
**NORTHERN IRELAND**
17 Strangford Lough
**IRELAND**
18 Dundalk Bay
19 Dublin Bay
20 Shannon
**FRANCE**
21 Somme Bay
22 Mont St. Michel Bay
23 Gulf of Morbihan
24 Charente Maritime, Vendée
25 Basin d'Arcachon*
**PORTUGAL**
26 Tejo
27 Ria Formosa, Faro
**SPAIN**
28 Cadiz Bay
**ITALY**
29 Po Delta
**TUNISIA**
30 Gulf of Gabès*
**MOROCCO**
31 Merga Zerga
32 Sidi Moussa-Oualidia
33 Sebka Tazra
**WESTERN SAHARA**
34 Dakhla Bay
**MAURITANIA**
35 Baie d'Arguin
36 Banc d'Arguin*
**SENEGAL**
37 Senegal Delta
38 Sine Saloum
**GUINEA-BISSAU**
39 Bijagós Archipelago*

In general, waders are easiest to count when they are concentrated on high tide roosts at high tide. This isn't the case on the tropical tidal flats of Guinea-Bissau, where many waterbirds retreat into the mangrove forests at high tide.

always possible to find genetic differences between groups that behave very differently. For example, it has not been possible to distinguish between dunlins that breed in Northern and Western Europe, Iceland and Greenland on the basis of DNA, although the plumage, size and migration behaviours of these groups are very different (Figure 2.1). Since these behavioural differences have not yet led to differences in the DNA analysed, it's likely that they originated relatively recently (possibly after the height of the last Ice Age, some 16 000 years ago).

### 2.1.3 Geographically separate populations and the 1% criterion

Nature conservation isn't just concerned with saving species; we also need to conserve the variation within species. Within most species of waterbird, there are a variety of life forms and migration routes, and it is important that this variation is preserved long-term. The 1% criterion is a useful way of recognising the importance of an area for waterbirds.[15, 744, 788, 957] Any area that holds 1% of the individuals of any species at a time is considered internationally significant according to the Ramsar Convention (an international wetland conservation convention signed by many countries). Governments can nominate areas that host at least 20 000 waterbirds or 1% of a waterbird species to become Ramsar sites. All the sites shown in Figure 2.2 are important stopover areas for waders and fulfil the Ramsar criteria for one or more species.

To conserve diversity within species, the population structure of each species needs to be taken into account. This procedure is simple if different subspecies are recognised. For example, the importance of the Dutch Wadden Sea to red knots can be expressed as the percentage of the world population of the *islandica* subspecies (that are present in winter), separately from its importance to the *canutus* subspecies (present in May, July and August). When it is clear (as it is for turnstones) that the world population consists of separate breeding populations, even though these sub-populations are not recognised as subspecies, these 'biogeographical populations' are usually taken into account when population size is calculated and are allocated their own separate 1% criteria. In the case of the turnstones, five biogeographical populations are recognised for *Arenaria interpres interpres* worldwide.[744]

Over the years there can be all sorts of changes in the size of waterbird populations. The official 1% criteria currently considered valid for Dutch waterbirds (Table 2.1) are already at least partly out of date, although organisations such as Wetlands International try to keep them up to date. In at least eight (sub)species increases have levelled off or previously stable populations have started to decrease. For example, the population growth observed in the great cormorant, brent goose and herring gull over the 1970s and 1980s has stopped. The population of West African wintering knots may have halved since 1980.[996] The bar-tailed godwits that winter in West Africa are also likely to have decreased dramatically.

## 2.2 Waterbirds by species

On the following pages we introduce the main European waterbirds. Each bird's appearance is briefly described, and can be clearly seen in the accompanying pictures. Average body sizes and weights are given, but as you will read in Chapter 3, this doesn't reflect the large amount of variation found in the wild. Each species' distribution, migration, foods and reproduction are also briefly discussed; these areas are discussed more thoroughly in subsequent chapters.

The waterbirds in this book are known by the Dutch as 'wadvogels', which can be literally translated as the rather unglamorous sounding 'mudflat birds'. In English we don't have a specific name for this large group of birds that spend at least part of their lives on tidal flats, so throughout this book we refer to them simply as waterbirds. Different tidal areas offer different habitats, which waterbirds use in many different ways. As we cover the different species you may notice how much the use varies from (sub) species to (sub) species and from place to place. Although you sometimes see starlings out feeding on the tidal flats or crows opening mussels, it's a bit of a stretch to call them waterbirds. Here we limit ourselves to the bird species that are, for at least part of their lives, dependent on the saline tidal areas of Northwestern Europe.

Most birds out on the tidal flats are waders, birds with relatively long legs and (usually) long bills, which they put to good use walking around and foraging in shallow water or silt. But not all waders are found on the tidal flats. A number of species never go to coastal areas; others are more flexible and sometimes move via brackish tidal areas and river mouths to the saline tidal areas. Lapwings breed in large numbers inland, behind the dykes bordering the Wadden Sea. When it's time for their migration, they can be found on exposed tidal flats, often resting, but sometimes actively searching for food. Ruffs, common sandpipers and common snipe are also sometimes found along the edges of tidal flats, but are not counted as true intertidal birds. Before humans modified the European coastal landscapes, there would have been a slow gradation from freshwater to saline habitats. Back then, it would have been much more difficult to distinguish intertidal birds from the inland waders.

Sections 2.2.1 to 2.2.21 describe the 21 species of truly intertidal waders commonly found in Northwestern Europe. All of these waders are migratory, many of them breeding north of the Wadden Sea and wintering south of the Wadden Sea. There is wide variation in their migratory behaviour, but in general the waders use breeding grounds that extend into Siberia and Northern Canada, and have a stopover in the Wadden Sea before flying on further to wintering areas on southern, mostly African, coasts. This migration route is known interna-

Large numbers of shorebirds can be seen at high tide when they gather at permanent roosts, like the knots and dunlins shown here.

At low tide, birds foraging on the exposed tidal flats are usually very widely dispersed. The incoming tide often forces the birds to gather on higher lying areas, as has happened here on a tidal flat near the Danish island of Langli in the Northern Wadden Sea (shown above). A few golden plovers are foraging amongst the gulls and 'intertidal' waders. Other 'inland' waders such as the lapwing sometimes visit the edges of the tidal flat to make use of the abundant food supplies found there (right).

tionally as the East Atlantic flyway (Figure 3.1), and is discussed further in Section 3.1. In total, some six to seven million waders (and more than two million ducks and geese) use this flyway. Many of these birds pass through the Wadden Sea twice a year, stopping over for varying lengths of time.

Six wader species (oystercatcher, dunlin, red knot, bar-tailed godwit, redshank, and curlew), shelduck and herring gull are by far the most numerous species on the tidal flats. Together they make up more than 80% of all the birds found in Northwest European intertidal areas. Some of the less numerous species only visit particular parts of the tidal zone, while other more numerous species have their own peculiarities. Curlew sandpipers, for example, are very numerous on their high northern breeding grounds and African wintering quarters, but while in transit in early spring and in summer they visit only a few parts of the Wadden Sea, and are considered a real rarity at most sites.

International counts of the waders provide us with insights into the numbers of each species using the flyway. These numbers have been included with the species descriptions as a rough indication, but they can vary greatly from year to year. The data used here are based on the most recent Wetlands International publication,[1000] and have been supplemented with more recent counts where possible.[1001, 1002, 1003, 1004, 1005, 1006] The maps are also largely based on these publications. As well as each wader species' breeding and wintering areas (shown in yellow and blue respectively or in green where these areas overlap) the maps show the borders of the area used by the birds that visit the Wadden Sea. Some maps are more complicated, because the species have been divided into subspecies or populations that are kept separate for other reasons.

Sections 2.2.22 to 2.2.30 discuss other birds found in tidal areas such as gulls, terns, geese and ducks. Some of these species rely on tidal areas for food. Brent geese wintering in Western Europe depend on saline coastal vegetation and nearby pastures. The shelducks that spend their whole lives in Western Europe are also very strongly connected to the tidal landscape. Other species are less dependent on tidal areas. Great cormorants can live on freshwater fish, and herring gulls get some of their food from the North Sea. These birds are opportunists that make use of the rich food resources provided by the tidal flats and while doing so often compete with the waders. Coastal habitats such as dunes and salt marshes can also be very important breeding grounds for many of these non-wader species. There are plenty of reasons to include these non-waders in this book. These birds pop up regularly throughout this book, as they are an important part of the intertidal community.

## 2.2.1 Oystercatcher (*Haematopus ostralegus*)

*Dutch:* Scholekster. *German:* Austernfischer.
*Danish:* Strandskade

*Appearance and behaviour*

Oystercatchers are certainly very conspicuous. At 44 cm in length and weighing 425 - 820 g, they are one of the largest and heaviest waders in the intertidal area.

The oystercatcher has striking black and white plumage year-round. Its head, breast, and the upperparts of the wings are glossy black, while its underparts and rump are white. In flight, its broad white wing-bar is visible. Its red eyes, solid pinky-red legs and yellow tipped orange-red bill (which is up to 10 cm long) are in striking contrast to its black and white feathers.

It is very difficult to determine an oystercatcher's sex visually, but on average females are a little larger, have a longer and thinner bill and slightly brown rather than black plumage. Juveniles are easily distinguished from adults. Their bills and legs are less colourful than adults' and the dark parts of the plumage are a dull brown black. In their first years of life juveniles have a white neck ring, which adults only have in winter. Oystercatchers aren't conspicuous on tidal flats simply because of their striking plumage, they are also very abundant year-round and rather noisy. Out on the tidal flats the birds argue amongst themselves, and in the breeding season they greet every approaching threat to the nest with a penetrating alarm call.

*Distribution, subspecies and migration*

Three subspecies can be distinguished, one of which breeds in central Russia, and a second in Eastern Asia. Western Europe is home to the third subspecies, *ostralegus*, which breeds in coastal areas from the White Sea to Brittany, as well as Iceland and a few sites in the Mediterranean. As the map shows, oystercatchers are mostly restricted to coastal areas and only breed well inland in Iceland, Great Britain, The Netherlands and Northwestern Germany. After the breeding season, the birds that bred inland migrate out to the coast. Some coastal breeding oystercatchers are resident year-round, but most spend the winter in Southern Europe. In general, oystercatchers don't make long migratory flights. In Britain, the local breeding birds are joined by birds from Iceland and Norway for the winter, making a total of more than 300 000 oystercatchers. Four hundred thousand oystercatchers winter in the Wadden Sea, mostly in the west. Some of these are local birds, while others breed around the Baltic Sea and Northern Russia. In very cold winters, some oystercatchers leave the Wadden Sea for the Dutch Delta or for Northern France. Of the one million oystercatchers that use the East Atlantic flyway, few fly direct to Africa. Around 7000 oystercatchers used to winter in Mauritania.

*Food*

Inland, oystercatchers feed mostly on earthworms, crane fly larvae, caterpillars and adult insects. In coastal areas,

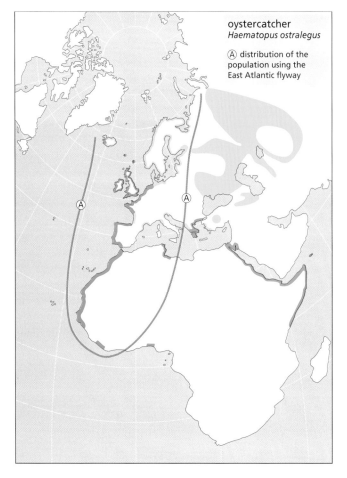

oystercatcher
*Haematopus ostralegus*

Ⓐ distribution of the population using the East Atlantic flyway

PORTRAIT GALLERY

oystercatchers eat a wide variety of animals. They use their long robust bills to find prey living in or on the sediment and detach prey fastened to rocks. They can use their sensitive bill tip to track hidden prey and feed in the dark by touch, but the bill is also strong enough to hammer open or pierce thick bivalve shells. In tidal areas oystercatchers often specialise on a particular prey type, such as large mussels or cockles.

*Reproduction*
The Wadden Sea is a very important breeding area for oystercatchers and the number breeding there increased greatly during the 20th century. In 1996, 46 000 pairs were counted in the Wadden Sea. Breeding oystercatchers are particularly numerous in the Wadden Sea salt marshes. The nest contains three to four eggs and is usually made either amongst dense vegetation in salt marshes and dunes, or out in the open on beaches, sand and gravel banks or along the sides of dykes. They can also breed on rocky areas, on roofs of houses, or inland on agricultural land. Inland nest sites are often on bare arable fields, where the birds can forage on the short, richly fertilised meadows nearby. Oystercatchers are extremely conspicuous during the breeding season, when, as part of their territorial behaviour, small groups of flying or running oystercatchers carry out the very noisy "tepiet" ceremonies, described in Chapter 5.
Both parents take an equal share of parenting, both in incubating and caring for the young. This parenting is

Oystercatchers are always very noisy, especially when danger threatens their chicks and they raise the alarm loudly (as shown opposite). Others stay alert with their chicks (above). The juveniles amongst this group of roosting oystercatchers (below) can be distinguished by their white neck rings.

far more intensive than that of most other waders, whose chicks usually have to find their own food from the first day of life. In contrast, oystercatchers provide food for their young for the first six weeks. It takes months for the young to become fully independent.

### 2.2.2 Avocet (*Recurvirostra avosetta*)

*Dutch:* Kluut. *German:* Säbelschnäbler. *Danish:* Klyde.

*Appearance and behaviour*

The avocet is around the same length as the oystercatcher (44 cm) but is much more slender, weighing around 225 - 400 g. Like the oystercatcher, the plumage of the avocet is completely black and white. Avocets are mainly white, but the cap and hindneck are black, as are three bands on the scapulars, wing coverts and primaries. These bands are most visible in the butterfly-like flight. The most obvious features of this graceful wader are its improbably thin, long blue-grey legs, and long (7 - 9 cm) dark bill, which is clearly upcurved at the end. The bill is most strongly curved in females, and is the only visible difference between the sexes. The penetrating call "kluut, kluut, kluut" is very distinctive, but can vary depending on the situation.

*Distribution, subspecies and migration*

Avocets have a curious distribution. Some breed and are resident at inland sites in East and Southern Africa. Others breed inland along a corridor stretching through Southeastern Europe and Turkey up to China; these birds migrate to southern coasts. There is also a breeding population (there are no distinct subspecies) on the coast of Western Europe. Approximately 17 000 pairs breed between Spain and the southern tip of Sweden. Most of these birds, 10 000 breeding pairs, breed in the Wadden Sea. Strangely enough, although the avocet was absent as a breeding bird in Britain for many years, more than 400 pairs now breed there.

After the breeding season, many avocets gather at a few sites in the Wadden Sea where they moult and fatten up before migrating south. Their favourite moulting sites are sheltered enclosed harbours where they can find plenty of food, such as the Dollard in The Netherlands, Leybucht and Jadebusen in Germany and around the Rømø-dam in Denmark. Every autumn, tens of thousands of avocets use these moulting grounds.

Many of these avocets winter in North Africa (Morocco and Tunisia) and West Africa (Senegambia and Guinea-Conakry). Most take a coastal migration route, but ringed avocets have been observed in Italy, indicating that some birds use an inland route. There are also important wintering areas along the coasts of France, Spain and Portugal. In mild winters some thousands of avocets remain in the Wadden Sea.

*Food*

Avocets use their uniquely shaped bill for a unique way of feeding. While walking through shallow water or soft sediment, the avocet quickly sweeps its bill through the top layer of silt. With a scything movement, worms and other benthic animals are scooped up and consumed in small gulps. This method is used to catch chironomid larvae in inland muddy areas. Their long legs allow avocets to forage in relatively deep water. The sediment must be very soft to feed in this way, which is why avocets are usually found in very silty tidal areas. Avocets have webbed feet, useful for walking across the silt and swimming short distances.

Avocets sometimes forage together in tight groups, occasionally in the presence of other species, as in summer when they catch shrimp in shallow tidal pools. An avocet can also hunt visually, walking quickly and pecking at prey visible in clear water, on the sediment, or amongst vegetation on the water's edge.

*Reproduction*

Avocets often breed in small colonies, nesting within a few metres of other pairs. Inland nests are often near silt flats where the birds can forage. In the Wadden Sea, avocets usually breed in salt marshes but will sometimes breed on agricultural fields or meadows behind dykes where the nests are safe from high tides. These nests are usually near tidal areas where the adults can forage. Later, the adults take the chicks to these feeding grounds and, for the still flightless chicks, travelling even short distances appears an almost insurmountable problem.

Both parents share the incubation of the four eggs (24 days) and caring for the chicks. Like most young waders, avocet chicks can forage from their first day of life. However the parents must still help to keep them warm and protect them against any threats. Parents dive-bomb predators or pretend to have a broken wing to lure potential predators away.

Elegant avocets, with their distinctive appearance and way of feeding, would be difficult to confuse with any other waterbird. Their butterfly-like flight is also very characteristic.

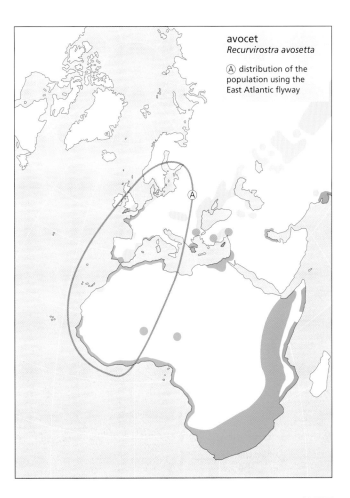

### 2.2.3 Ringed Plover (*Charadrius hiaticula*)

*Dutch:* Bontbekplevier. *German:* Sandregenpfeifer. *Danish:* Stor Præstekrave.

*Appearance and behaviour*
The ringed plover has a black-tipped yellow-orange bill, a distinctive black and white pattern on the head and a black breast-band. The underparts are white; the upperparts and cap are a dull grey-brown to olive-brown. The legs are bright orange-yellow like the bill. After the breeding season, the black on the head and breast becomes dark brown, making the birds less conspicuous. The ringed plover is similar in size to the dunlin (19 cm and 56 g). Unlike a dunlin, it shows the typical plover feeding behaviour; step, step, look, take a few more steps, run, peck, step...

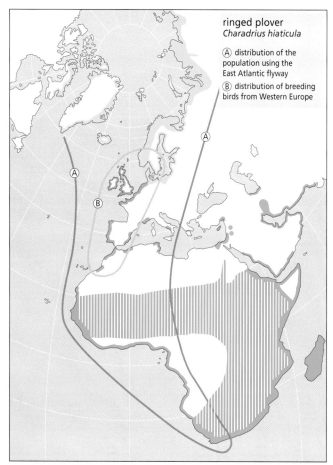

ringed plover
*Charadrius hiaticula*

Ⓐ distribution of the population using the East Atlantic flyway
Ⓑ distribution of breeding birds from Western Europe

*Distribution, subspecies and migration*
The ringed plover's breeding grounds extend from the coast of Brittany via Scandinavia along the extreme north of Siberia and up to the Bering Strait. Ringed plovers also breed in Iceland, Greenland, the Eastern Canadian Arctic and almost all coastal areas of Western Europe. In the northern regions they also breed far inland. Ringed plovers that breed in Lapland and the Arctic areas to the east belong to the subspecies *tundrae*, while those breeding in Iceland, Greenland, Canada and Western Europe belong to the subspecies *hiaticula*. Although it takes practise to

identify any external differences between the two subspecies in the field, there are obvious differences in their migratory behaviour. In May, when the local breeding birds already have young, more than 10 000 *tundrae* stopover in the Wadden Sea on migration.

Almost all ringed plovers winter in either Africa or Europe; few visit other parts of the world. The more than 200 000 ringed plovers that winter in West Africa use the East Atlantic flyway, while many of the Siberian breeding birds take a more easterly route overland to Eastern and Southern Africa. Birds of both subspecies that breed in the Arctic tend to fly to the most southerly wintering grounds. Breeding birds from Western Europe also migrate, but only as far as their wintering grounds in Portugal and Morocco. It's not yet clear if they all follow the same migratory routes. The wintering sites of ringed plovers from Iceland and Canada are still unknown.

*Food*

Ringed plovers spend much of the year in tidal areas, where they forage on visible prey such as worms, small crustaceans, and insects. The bill is used to peck at what they can see, not to search for prey hidden in the sediment. They make good use of their feet while foraging; standing on one foot and softly tapping the sediment with the other makes some benthic animals more active and therefore more visible.

*Reproduction*

A total of around 1400 pairs of ringed plovers breed in the Wadden Sea, although this number varies greatly from year to year. Most birds breed in the Schleswig-Holstein area. Ringed plovers lay their four eggs in a shallow nest scrape, usually on a barren shell or gravel beach. Eggs are laid from late March onwards in the Wadden Sea and if conditions are suitable, two clutches can be raised.

Ringed plovers usually nest on exposed shelly beaches. The pair regularly swaps shifts when incubating (shown on page 40). Other birds that approach the nest are chased away with a threatening display (above). The Kentish plover (right) is much paler and appears more slender than the ringed plover.

## 2.2.4 Kentish Plover (*Charadrius alexandrinus*)

*Dutch:* Strandplevier. *German:* Seeregenpfeifer. *Danish:* Hvidbrystet Præstekrave.

The Kentish plover is slightly smaller, more slender and less colourful than the ringed plover. Instead of a broad black breast-band, Kentish plovers have two dark side patches on the breast. Black feathering on the head is limited to a small eye stripe (and a small black cap in breeding males). The bill and legs are black.

Kentish plovers breed over much of central Europe and Asia. They also breed in coastal areas as far north as Jutland. The Wadden Sea is near the edge of their range, and has a relatively small breeding population. Nevertheless, the 500 pairs that breed there (mainly in the Schleswig-Holstein region), makes the Wadden Sea an important stronghold for the species. The number of breeding pairs is very variable, however, and has decreased recently. The decline is mainly due to their preference for nesting in sparsely vegetated coastal areas. Nesting birds have managed to survive the natural hazard of spring floods for many years, but now they must also deal with increasing human disturbance as recreational use of these sites increases.

Most Kentish plovers breed south of the Wadden Sea, and only a limited number pass through. Kentish plovers aren't solely intertidal birds and although they sometimes forage on sandy tidal flats, they prefer beaches and shoals.

### 2.2.5 Grey Plover (*Pluvialis squatarola*)

*Dutch:* Zilverplevier. *German:* Kiebitzregenpfeifer. *Danish:* Strandhjejle

*Appearance and general behaviour*

The grey plover is the largest (29 cm) and heaviest (175 - 320 g) plover found in the Wadden Sea. Like all plovers they hunt visually and have a distinctive way of searching for food. They stand motionless, intently watching, before taking a few steps and looking around again then snatching a worm or other prey item from the sediment with an accurate peck.

In spring, grey plovers have a very distinctive black and white breeding plumage. The black belly, breast, throat and cheeks contrast with a broad white band extending from the forehead to the shoulder. The feathers on the upperparts are dark with white fringes. Males have glossy pure black underparts; in females the black feathers are edged with white. Outside the breeding season, pure black or white feathers are mostly replaced by flecked feathers and the underparts are often completely white. Only the axillaries or 'armpits' remain black year-round. These black axillaries are clearly visible in flight and are a useful identifying character, as is their distinctive call of "pieee-uuu-wieee".

The grey plover's breeding plumage (as seen in this bird near its nest on the Canadian tundra; photo top left) is very different from its winter plumage (top right). In late summer and autumn grey plovers in the Wadden Sea can be seen wearing all sorts of patchy intermediate plumages (lower right). The brownish individuals are juveniles.

*Distribution, subspecies and migration*
Grey plovers breed in tundra along most of the Arctic Ocean coast, from the White Sea to Canada. No subspecies are described. After breeding, grey plovers fly south along many different flyways and are found on the coasts of every continent (except Antarctica). Some Siberian-breeding birds pass through the Wadden Sea on migration, as may some that breed in Canada. Many winter on the West African coast. Up to 20 000 birds winter in the Wadden Sea, depending on the severity of the winter. The Wadden Sea is a particularly important site for grey plovers in spring, with some 140 000 grey plovers counted in May.

*Food*
On the breeding grounds, grey plovers feed mainly on insects and other invertebrates. Outside the breeding season they live in intertidal areas where they feed on worms and, to a lesser extent, other small animals at low tide. Grey plovers have excellent vision, which they need to detect the tiny movements that betray the presence of their prey, and can even hunt visually at night.

*Reproduction*
Grey plovers breed out on the open tundra, particularly on dry stony areas with little vegetation. The female lays four large eggs in a shallow nest scrape. Both parents incubate the eggs and care for the chicks.

## 2.2.6 Golden Plover (*Pluvialis apricaria*)

*Dutch:* Goudplevier. *German:* Goldregenpfeifer. *Danish:* Hjejle

Golden plovers are slightly smaller than grey plovers, with less black on the belly, and have golden yellow and brown flecks on the upperparts. Most breed in Great Britain, Iceland, Scandinavia and Siberia; very few breeding pairs remain in Germany and Denmark. While on migration, golden plovers are very numerous in Western Europe, particularly in inland areas. They usually feed on earthworms in meadows and agricultural land. Although they only occasionally forage in tidal areas, they frequently roost on the edges of tidal flats. Flocks of sleeping golden plovers are often seen on the tidal flats on days near full moon. At this time of the month moonlight enables the birds to find earthworms so successfully at night that they can afford to sleep during the day.

A golden plover in breeding plumage on the breeding grounds in Lapland.

grey plover
*Pluvialis squatarola*

Ⓐ distribution of the population using the East Atlantic flyway

Knots in spring, with their brick-red plumage (shown on page 45), are easily distinguished from the other sandpipers. In the winter their plumage is pale grey (above), just like the other sandpiper species. They can then be identified by their larger size and plump appearance.

## 2.2.7 Red knot (*Calidris canutus*)

*Dutch:* Kanoetstrandloper. *German:* Knutt. *Danish:* Islandsk Ryle

*Appearance and general behaviour*

The red knot is the largest of the Wadden Sea sandpipers (24 cm and 140 g). They may have been named after King Canute, who was known to be particularly fond of them for culinary reasons. Alternatively, their name may refer to the soft hoarse "gnut-gnut" call that they sometimes make.

The red knot is rather dumpy and sturdy in appearance. It has rather short legs and a straight bill that is comparatively short and wide for a sandpiper. The bill is almost black and the legs are a light-grey green. In breeding plumage, knots have beautiful rusty red underparts, which vary from brick red to chestnut in different individuals. The upper feathers are black flecked with different shades of red and brown. These markings form an attractive pattern, perfectly camouflaging the bird on the tundra. In autumn, grey winter plumage replaces the red feathers on the underparts and the dark feathers on the upperparts. The upperparts then appear grey, while the rump and underparts are paler with fine, somewhat darker, streaks.

Knots in flight are recognisable by the thin white wing-bar and pale rump and tail. The underside of the wing is also pale.

*Distribution, subspecies and migration*

Knots only breed in the High Arctic, the Wadden Sea birds in the north of Canada, Greenland and the Taimyr Peninsula in Siberia. Two of the six subspecies recognised use the East Atlantic flyway: *canutus* that breed on the Taimyr peninsula and *islandica* that breed in Greenland and neighbouring parts of the Canadian Arctic.

Knots are true long-distance migrants and can travel from their breeding grounds to Northwestern Europe in a single flight. One such flight is apparently enough for the 345 000 *islandica* knots that remain in Western Europe for the winter. Many of these birds move on from the Wadden Sea to winter in the Dutch Delta, France, or the British Isles. In mild winters some 60 000 to 100 000 *islandica* knots remain in the Wadden Sea, but in harsh winters they migrate further south or west. The *canutus* knots stopover in the Wadden Sea on the way to their wintering grounds in West Africa (mainly Mauritania). Most return to the Wadden Sea in late spring, where they spend a few weeks rebuilding their fat reserves for the final flight back to the breeding grounds. International counts show that the number of knots in the Wadden Sea peaks in May, at more than 400 000. These are mainly *islandica* knots, as most *canutus* are still arriving in May. Throughout the month, the number of knots in the Wadden Sea remains high. Most of the knots are in Schleswig-Holstein, which is thought to provide the good feeding grounds required for their long flight north.

*Food*

There is little food available in the High Arctic at the start of the breeding season. As the snow recedes, knots are able to search for insects and other invertebrates and will even eat vegetation when necessary. Later in the season, more insects are available for the birds and their young. As soon as possible, knots move back to the coast. Outside the breeding season, they are only found in tidal areas. There they peck visible prey off the sediment surface and use their sensitive bills to detect prey hidden in the upper few centimetres of sediment. They prefer to feed in areas with high densities of small shellfish such as cockles and mussels. These shellfish must be less than 2 cm long so that they can be swallowed whole. The shell is then crushed in the knot's stomach.

Foraging knots roam the tidal flats in dense groups. This may allow the birds to find the areas with the highest prey density more quickly.

*Reproduction*

What a knot wants when it searches for a nest site is not completely clear to researchers. The nests are often in dry places, but can also be in swampy areas. Usually, they are in barren stony sites, but sometimes they are surrounded by dense vegetation, at least by High Arctic standards. Perhaps knots simply chose the first places to become snow-free for their nest sites. The thaw only starts in early June and they must start breeding as early as possible to benefit from the short Arctic summer.

Parents share the 22 day incubation of the four eggs, on a 17 hours on, 17 hours off schedule. The female leaves the breeding area as soon as the chicks hatch. The male cares for the chicks for the next three weeks. He will only leave the breeding grounds once the chicks have fledged (and are therefore able to fly).

### 2.2.8 Sanderling (*Calidris alba*)

*Dutch:* Drieteenstrandloper. *German:* Sanderling. *Danish:* Sandløber.

*Appearance and general behaviour*

The sanderling, at 20 cm and 55 g, is one of the smallest sandpipers in the Wadden Sea. During the breeding season, sanderlings are easily distinguished from other similar-sized sandpipers by their plumage. The head, throat, breast and upperparts become a deep reddish-brown to warm orange-red flecked with black and white. The underparts are white, and its rather short bill and legs are dark.

After the breeding season, the upperparts become very pale grey and in winter appear almost completely white. Sanderlings also look very pale in flight, but the primaries remain dark and contrast with a distinct broad white wing-bar. The underside of the wing appears white, while the dark central tail feathers contrast with an otherwise pale tail.

By far the easiest way to distinguish sanderlings from other sandpipers is by their behaviour and preference for sandy habitats. As their name suggests, sanderlings don't like silty areas and in the Wadden Sea they are only found on beaches or sand flats, including those with waves breaking on them. They search for food along the water's edge in small groups, scuttling along behind a retreating wave, then speedily running in front of the incoming water, sometimes flying to keep ahead of a breaking wave.

*Distribution, subspecies and migration*

Sanderlings breed almost exclusively in the High Arctic, usually near the coast. Their main breeding grounds are in Greenland and islands in the Canadian Arctic. They also breed along the coast of the Taimyr Peninsula, the Lena Delta and a number of islands in Siberia. No subspecies have been described. After the breeding season, hundreds of thousands of birds migrate to wintering grounds in Australia, Southern Africa and South America. They are rarely seen inland.

The sanderlings seen in Western Europe breed in Greenland and Siberia. Most sanderlings make long nonstop flights to Africa. Some 123 000 sanderlings have been counted along the East Atlantic flyway, of which 20 000 winter in the Wadden Sea. The actual number may be much higher, as beaches aren't searched systematically during counts and sanderlings tend to spread out along coastlines rather than congregating in large numbers. By late May, when birds are fuelling up for the final flight to the breeding grounds, up to 50 000 sanderlings may be present in the Wadden Sea. The biggest numbers are found on the sand flats of Trischen, an island in Schleswig-Holstein.

*Food*

On the breeding grounds, sanderlings find little food

Some of these sanderlings running from a breaking wave still have white winter plumage (page 46), but most are in brown breeding plumage, as is this bird with its chick on a nest in the Canadian tundra (above).

along the edges of the melting snow at first, and must feed on vegetation as well as insects and larvae to survive. By the time their chicks hatch, there are plenty of insects available on the tundra. Outside the breeding season, sanderlings are almost exclusively coastal birds. They are a regular feature of sandy tidal areas, where they forage alongside other sandpipers. They find their prey mainly by probing the sediment surface with their bills. Sanderlings are usually found on the exposed North Sea side of the Wadden Sea islands, where they forage along the tide line, searching for food left by the waves. This might be small marine animals carried in on the currents, drowned insects, or sandhoppers and small shellfish dislodged from their refuges by the waves.

*Reproduction*

Sanderlings nest on the Arctic tundra in dry stony places with little vegetation. The four eggs are incubated for 24 days. Sometimes both parents take care of the eggs and chicks, but often the female produces two clutches in nests not far from each other at the same time. The first nest is then cared for by the male, the second by the female.

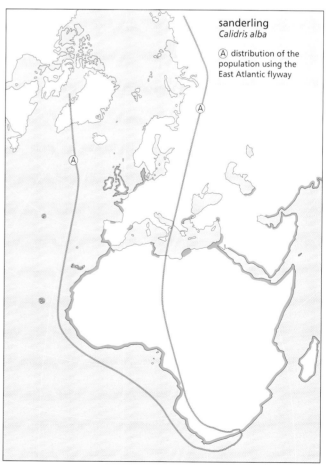

sanderling
*Calidris alba*

Ⓐ distribution of the population using the East Atlantic flyway

## 2.2.9 Curlew Sandpiper (*Calidris ferruginea*)

*Dutch:* Krombekstrandloper. *German:* Sichelstrandläufer.
*Danish:* Krumnæbbet Ryle.

*Appearance and general behaviour*

The curlew sandpiper is 21 cm from tip to tail, 32 - 44 mm of which is due to its bill. Their average weight is 63 g. Outside the breeding season, the species can be difficult to distinguish from its close relative the dunlin. A curlew sandpiper usually appears more slender than a dunlin, due to its longer neck, legs and bill. The down-curved bill, to which the species owes its name, isn't as obvious as might be expected.

In the breeding season, the curlew sandpiper is immediately recognisable by its brick red underparts and black, white and brown-flecked upperparts. The breeding plumage of the curlew sandpiper can be even more colourful than that of the red knot (which is bigger and heavier billed). Female curlew sandpipers are usually less brightly coloured than males and have more white feathers amongst the red. Over the summer, more and more pale grey spots appear amongst the red in both sexes. In winter, the curlew sandpiper is light grey with almost completely white underparts.

Throughout the year, the curlew sandpiper is recognisable in flight by its white wing-bar, white rump and darker tail.

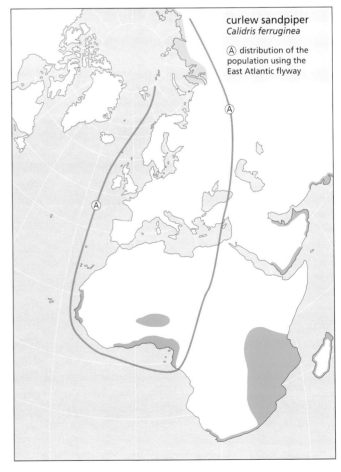

curlew sandpiper
*Calidris ferruginea*

Ⓐ distribution of the population using the East Atlantic flyway

*Distribution and migration*

Their breeding grounds are restricted to the far north of Siberia well within the Arctic Circle, from the mouth of the River Ob almost up to the Bering Strait. No subspecies have been described. In winter, the population of one million birds is distributed across Australia, the south coasts of Asia, and Africa. Many winter in East and West Africa both in coastal and inland areas. They are also found in Southern Africa.

The wintering areas in West Africa (such as the tidal areas of Guinea-Bissau where 340 000 have been observed) lie on the East Atlantic flyway, but most curlew sandpipers do not use the flyway route. Every spring and autumn, large numbers (up to 21 000) are seen in tidal areas between the Rivers Elbe and Eider in the Schleswig-Holstein area. Elsewhere in the Wadden Sea only tens or in some cases hundreds of migrants are seen. It is believed that most curlew sandpipers take a more easterly route over inland Africa and Europe and observations of birds inland agree with this. This explains why the numbers of curlew sandpipers in the Wadden Sea differ so much from year to year. Prevailing easterly winds can cause the number of migrants passing through the Wadden Sea to temporarily increase. Curlew sandpipers only remain in North Sea tidal areas for a short time. In spring, from late April to early June, small numbers of birds pass through quickly. Then, from the second half of July, the post-breeding birds appear. After a successful breeding season, juvenile birds are also seen in the Wadden Sea from mid August to mid September.

*Food*

On the breeding grounds, curlew sandpipers search the tundra vegetation for small insects and other small animals. During migration and on the wintering grounds they prefer open muddy inland areas. In tidal areas they are often accompanied by dunlins and feed on similar prey, mainly worms and other benthic animals. These they peck from the mud surface or detect using their sensitive bills.

*Reproduction*

The curlew sandpiper's High Arctic breeding grounds consist of sparsely vegetated tundra that is largely covered with pools of water early in the breeding season (second half of June). The nests are often in quite damp places. Although breeding birds often appear to have sought out neighbours, the distances between the nests are too great to describe them as colonial breeders.

The males do not have a well-defined territory or a strong bond with their mate and leave the breeding grounds after just a few weeks. The females take sole responsibility for the incubation of their four eggs (which takes around 20 days) and caring for the growing chicks during their first two weeks of life. Most waders that breed in the High Arctic appear very faithful to their breeding area. The curlew sandpiper is an exception to this rule. Their breeding areas and breeding success seem to fluctuate with the weather; if spring is late they breed at (somewhat) lower latitudes.

In the breeding season, the curlew sandpiper is the same brick-red as a red knot, but is easy identified by its long slightly down-curved bill (shown on page 48). In winter its bill looks even longer, because the birds are usually more slender at this time of year. Here a group of curlew sandpipers forages along the edge of the incoming tide on the mudflats of Guinea-Bissau (right).

### 2.2.10 Dunlin (*Calidris alpina*)

*Dutch:* Bonte strandloper. *German:* Alpenstrandläufer. *Danish:* Almindelig Ryle

*Appearance and general behaviour*

The dunlin is the smallest sandpiper regularly seen in the Wadden Sea, with a length of 19 cm (23 - 44 mm of which is the bill) and an average weight of 48 g. The fairly long legs and slightly down-curved bill are dark. During the breeding season, the dunlin is easily recognised by its black belly patch. At this time the dunlin's breast and throat are white with fine dark streaking and the upperparts are reddish brown with fine black and white markings.

In winter the black patch disappears, the underparts are mainly white and the upperparts a light brown-grey, making dunlins difficult to distinguish from other small pale sandpipers. The species is then most easily identified in flight by its white wingbar and dark-centred rump, which are useful characteristics year-round. The underwings are almost completely white.

Outside the breeding season, dunlins are usually quiet. When flying they occasionally make a soft "treeeaa" call. A group of dunlins can often be heard making satisfied soft chattering calls.

*Distribution, subspecies and migration*

Dunlins are one of the most abundant waders on European tidal flats and more than one million have been counted in the Wadden Sea in spring and autumn. Dunlins are very visible as they often gather in large groups of many thousands of birds. The world population, comprising around three million dunlins, breeds in a small area along the north coasts of Europe and Asia, Alaska, Northern Canada, Greenland and Iceland. Most breeding grounds are within the Arctic Circle, but dunlins also breed in Denmark, Northern Germany and the British Isles.

Nine different subspecies have been described based on differences in plumage, body size, and bill shape and length. Two subspecies (*alpina* and *schinzii*) are found in the tidal areas around the North Sea. The breeding grounds of the subspecies *alpina* (population 1.4 million) stretch from the north of Scandinavia up to the Ural region. The subspecies *schinzii* breeds in Southern Sweden, Denmark (600 pairs) Northern Germany (some pairs in Schleswig-Holstein and some hundreds along the Baltic Sea coast), the British Isles (4000 – 8000 pairs) and Iceland (more than 350 000 pairs). The *schinzii* dunlins migrate with the *arctica* subspecies (which breed along the east coast of Greenland) through the British Isles to wintering grounds on the coast of West Africa. Some winter around the North Sea and Southern Europe. In almost every season, dunlins are the most abundant waders in the Wadden Sea. They only move to the milder Dutch Delta area and the south of England in very harsh winters, but even then tens of thousands usually remain in the Wadden Sea.

*Food*

On the inland breeding grounds dunlins feed on a variety of invertebrates including insects, spiders, snails and worms. Outside the breeding season, most dunlins stay in coastal, preferably intertidal, areas. Here, they feed on worms, small shellfish and crustaceans that they peck from the surface or, more commonly, detect with their sensitive bill in the upper sediment. Often, the prey are so tiny that it is practically impossible to see the birds swallowing. Dunlins enjoy the company of other dunlins and also other wader species while foraging. Sometimes they continue feeding at high tide on the water's edge on marshes or beside dykes or in swampy places inland. They are active day and night.

*Reproduction*

On the breeding grounds, males perform singing display flights above their territories, alternating short gliding flights with quick fluttering movements to regain height. These flights can last for some minutes. The song consists of a continuous "wggga" that becomes a long chattering trill, then fades away slowly. When the male lands after a display, he pauses for a moment with his wings stretched in a V-shape. Males also sing from high lookout posts where the singer makes himself more conspicuous by periodically lifting one wing to reveal its bright underside. The High Arctic breeding grounds are usually in peat or other wet parts of the tundra. The less northerly subspecies *schinzii* breeds in coastal meadows. The four eggs and the breeding bird are usually hidden in a clump of vegetation. Incubation takes 20 - 24 days. Both parents share the incubation, but the male takes most of the responsibility in caring for the chicks.

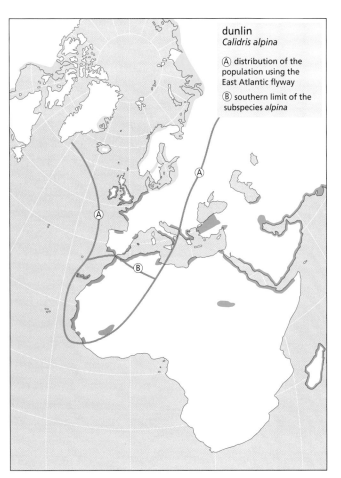

dunlin
*Calidris alpina*

Ⓐ distribution of the population using the East Atlantic flyway

Ⓑ southern limit of the subspecies *alpina*

PORTRAIT GALLERY

With its black breast and chestnut upper parts, the dunlin is at its most beautiful in the breeding season, as is shown by this breeding bird calling from a peat mound near its nest in a marshy part of the Northern Norwegian tundra (top photo). Dunlins seen in tidal areas in spring and summer still have this beautiful plumage. In the winter, however, they are the same dull grey as all the other sandpipers (right).

## 2.2.11 Little Stint (*Calidris minuta*)

*Dutch:* Kleine strandloper. *German:* Zwergstrandläufer. *Danish:* Dværgryle.

*Appearance and general behaviour*

Little stints are only 12 - 14 cm long and weigh just 23 g, far smaller and lighter than other intertidal sandpipers. The only relative of a similar size is Temminck's stint, a predominantly freshwater species. In spring, little stints are very colourful with rusty brown, dark-flecked neck and cheeks and neat black stripes on the crown. The brown upperparts are attractively flecked with black and white and a light-coloured V is visible across the scapulars. Little stints can be distinguished from other sandpipers not only by their size, but also because they are more active.

*Distribution*

The little stint's breeding grounds are restricted to Northern Siberia and the north of Lapland. Nothing is known of the population size. However it is clear that the numbers vary greatly depending on their breeding success.
Little stints spend the northern winter in coastal areas and inland wetlands of Southern Asia and Africa. More than 200 000 little stints have been counted in West Af-

The little stint broods four eggs on the tundra (top photo). A newly hatched chick weighs less than 5 g (photo page 53 below), but fledges after 15 days, and a few months later can be seen out foraging on the African coast or at an inland site, such as the river Niger in Mali (photo right).

At low tide, the purple sandpiper searches for its dinner on rocky areas amongst the newly exposed seaweed.

rican tidal areas along the East Atlantic flyway. Here, little stints are very much intertidal birds but not all of them use the flyway. Very few little stints follow the coast during migration. Most migrate over inland areas and take a more easterly route. This is why so few migrating little stints are seen in the Wadden Sea. In spring, only a few tens are seen and even in the summer when their numbers on the tidal flats are higher, they never exceed 1000. Most summer sightings are from inland areas.

*Reproduction*
Like some other sandpipers, little stints often lay double clutches. The female pairs with a male, lays four eggs and leaves these eggs in his care. She then goes to another area and lays another four eggs that she incubates and cares for on her own.

## 2.2.12 Purple Sandpiper (*Calidris maritima*)

*Dutch:* Paarse strandloper. *German:* Meerstrandläufer. *Danish:* Sortgrå Ryle.

*Appearance and general behaviour*
The purple sandpiper is 20 - 22 cm long, similar to the dunlin, but because of its dumpy shape it looks bigger and at 64 g it is much heavier.
Its name refers to the colour of the upperparts in breeding plumage, which are mostly dark brown but have a purple sheen in certain lights. In winter the upperparts and breast are grey, but much darker than those of other waders. The legs and base of the bill are yellow-orange. Purple sandpipers are typically found on rocky shores and gravel beaches where they scurry around amongst the seaweed. The purple sandpiper is not really an intertidal bird.

*Distribution, subspecies and migration*
The High Arctic breeding grounds of the purple sandpiper lie on both sides of the Atlantic Ocean and include the islands of Eastern Canada, Greenland, Iceland, Spitsbergen, and the north coast and associated islands of Siberia, up to 110° longitude. They also breed throughout Norway.
In winter purple sandpipers move south to the east coast of North America and the west coast of Europe, but never south of the Bay of Biscay. Little is known of their migration. Some 10 000 hardy individuals remain on the west and north coasts of Scandinavia throughout winter. It's not clear if all the birds undertake a short migration south, or if only the most northerly breeding birds move south. Most of the birds wintering in England come from Scandinavia; those that winter in Scotland are mainly from Canada (which is probably where the Wadden Sea purple sandpipers also come from). Some purple sandpipers that breed in Greenland winter in Southern Greenland, others in Iceland. A total of 50 000 purple sandpipers is estimated to use the northern part of the East Atlantic flyway. Only some hundreds remain for any time in the rock-poor Wadden Sea.

*Food*
Purple sandpipers find their prey amongst rocks, especially in seaweed. They are especially active in the wave zone and make good use of the waves that repeatedly expose different sides of the seaweed. They mainly feed on snails, bivalves and crustaceans but also eat insects that live in rotting seaweed. Because their food sources are available year-round, purple sandpipers can survive along rocky coasts that are kept ice-free in winter by the warm Gulf Stream.

*Reproduction*
The breeding grounds are in areas of barren or scarcely vegetated tundra. The nest and four eggs often lie on stony ground where the dark bird is beautifully camouflaged. The parents share incubation. Once the chicks hatch, the female leaves the breeding grounds and the male raises the chicks.

### 2.2.13 Bar-tailed Godwit (*Limosa lapponica*)

*Dutch:* Rosse grutto. *German:* Pfuhlschepfe.
*Danish:* Lille Kobbersneppe.

*Appearance and general behaviour*
The bar-tailed godwit is 38 cm long, somewhat smaller than an oystercatcher, and is also more slender, weighing just 320 g. They have long black legs and a distinctive long bill that is slightly upcurved at the end. The head and underparts of the male can be deep brick red in spring, although this varies somewhat between individuals. Outside the breeding season they are much less colourful; in winter the upperparts are pale grey-brown and the underparts are pale beige. Females are larger and have duller breeding plumage than males with only a light red-brown tinge to their beige underparts. In flight the bar-tailed godwit is easily distinguished from the black-tailed godwit, as the bar-tailed godwit's pale barred rump extends up its back and it has less obvious wingbars.

*Distribution, subspecies and migration*
The breeding area of the bar-tailed godwit covers a narrow discontinuous strip from Lapland, along the north coast of Siberia and up to Alaska. Only the subspecies *lapponica* uses the tidal areas of the North Sea. The other subspecies migrate along different routes to the coasts of the Pacific Ocean. Two distinct groups of bar-tailed

godwits use the East Atlantic flyway. One population of 125 000 birds (subspecies *lapponica*) winters exclusively in Europe mainly around the North Sea. Tens of thousands of these godwits winter in the Wadden Sea, although in cold winters, many (up to 80 000) stay in the British Isles. The British birds return to the Wadden Sea in early spring so that by April almost the entire European wintering population is found in the Wadden Sea.

The other, much bigger, migratory population (subspecies *taymyrensis*, 700 000 birds) winters in tidal areas of West Africa, passing through the Wadden Sea in late spring on their way to the breeding grounds. In May, birds from both populations are found on the intertidal flats of the Wadden Sea making a total of more than

As well as the difference in colour, there is also a large size difference between male and female bar-tailed godwits. This is especially obvious in the breeding season (upper photo on page 54; male on the left, female on the right). In winter both sexes are beige grey (lower photo on page 54).

300 000 bar-tailed godwits. In spring, more than 10 000 non-breeding bar-tailed godwits remain in the Wadden Sea. Bar-tailed godwits are present in the Wadden Sea year-round and are abundant in most seasons.

*Food*
Away from the breeding grounds, the bar-tailed godwit is usually found in silty or sandy parts of the intertidal area. Its long bill and legs enable it to find prey even when the sediment is covered with 10 cm of water. All types of benthic animals are eaten, both small ones that are pecked off the mud surface and larger ones retrieved from deep in the sediment. In spring, groups of bar-tailed godwits are often seen in meadows near the coast where they feed on crane fly larvae.

*Reproduction*
The breeding grounds closest to the Wadden Sea are in Northern Norway, where bar-tailed godwits breed in the tundra just above the tree line. The nest is usually made amongst low vegetation, but can also be out in the open up on a dry spot in a swampy area. Parents share incubation duties and caring for the chicks.

When you compare this black-tailed godwit with the bar-tailed godwits on page 54, there are clear differences in plumage and bill shape between the species.

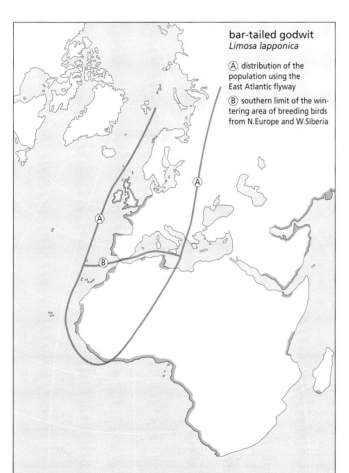

### 2.2.14 Black-tailed godwit (*Limosa limosa*)

*Dutch:* Grutto. *German:* Uferschnepfe.
*Danish:* Stor Kobbersneppe

The black-tailed godwit, a well-known meadow bird in The Netherlands (home to 45 000 pairs), is seldom seen on tidal flats. When it is, it only forages in places where the sediment surface is very soft. In winter, these black-tailed godwits migrate to West Africa. There most stay in inland areas, but some forage on the silty tidal flats.
But this subspecies is not the reason black-tailed godwits have been included in this book. Another subspecies of black-tailed godwit, *islandica*, breeds in Iceland. The 56 000 Icelandic birds winter in Western Europe as far south as Portugal and Morocco. Some prefer inland areas but most are intertidal birds in the winter. Every winter, some 4 200 can be found in British silt-rich tidal areas. Increasing numbers of these Icelandic godwits (more than 1500) pass through the Dutch Wadden Sea on migration.

### 2.2.15 Curlew (*Numenius arquata*)

*Dutch:* Wulp. *German:* Grosser Brachvogel.
*Danish:* Stor Regnspove.

*Appearance and general behaviour*
The curlew is the largest intertidal wader. Females, at 60 cm and 1 kg in weight, are usually larger than males. From a distance curlews appear grey-brown but viewed closely the buff feathers have variable black-brown bars and stripes. These form a regular pattern of fine lengthwise stripes on the head and breast. The curlew's long down-curved bill is very distinctive. The males' bills are 11 - 13 cm long and those of females 14 - 16 cm.
It is not always easy to observe these details at close range as curlews are still hunted in some countries. As a result, curlews are very shy and often their call, a very loud 'kuur-liep', is heard before they are seen.

*Distribution, subspecies and migration*
Curlews breed throughout much of Europe and Asia. In Norway they breed in a small strip of coast almost up to the North Cape. Further east, the breeding grounds don't extend as far north. Eurasian curlews are divided into two subspecies based on small differences in their size and plumage: *arquata* in the west and *orientalis* in the east. The border between these two subspecies lies somewhere near the Ural region, but exactly where is not known.

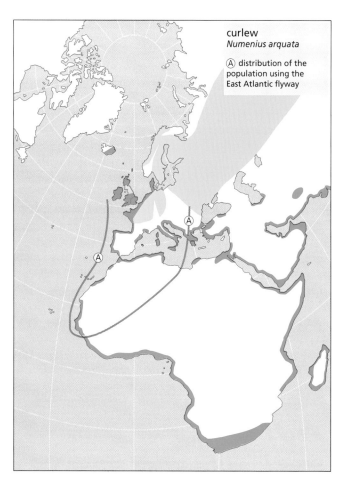

After the breeding season, most curlews migrate south. Those from mid and Northern Europe tend to initially migrate west, and many winter around the North Sea and Ireland. Although they often winter in tidal areas, large numbers are also found inland.
Many other curlews winter in the southern part of the East Atlantic flyway. It is estimated that more than 400 000 curlews in total are found along the flyway. Most curlews in West Africa seem to have come from breeding areas further east and to have followed a different migration route.
Curlews are present in the Wadden Sea throughout the year. They are most numerous in late summer when there are usually more than 200 000 present, but numbers are also high in winter. Even in harsh winters, when many birds migrate south, there are never fewer than 100 000 curlews. Those birds mainly stay in the western part of the Wadden Sea, but some curlews are found inland in winter.

*Food*
On the inland breeding grounds, curlews tend to use damp sites where they feed on a variety of soil invertebrates and berries. In tidal areas they prefer bigger prey such as crabs, worms and large bivalves. Curlews find their prey by quietly walking along the tidal flats and delving into likely looking places with their bills. The curlew's long bill is well suited to extracting buried prey. Curlews are not entirely dependent on tidal areas. Although some individuals will usually forage only in tidal areas, others spend the winter feeding in inland meadows where they find worms and crane fly larvae. Other birds forage on tidal flats at low tide and use inland grasslands at high tide. Males in particular appear to prefer foraging inland.

*Reproduction*
During the breeding season, curlews do not maintain a connection with the coast. They prefer to nest in swamps, peat areas, dry heather or dunes and have only recently become meadow birds. Most curlews that breed in the Wadden Sea use the dunes on the Dutch islands. Curlews make beautiful display flights at their breeding sites. With rapid wing beats they fly some tens of metres up into the air, and then make a long gliding flight down with a loud yodelling trill. The four large eggs are usually laid very early in spring and incubated by both parents. The eggs hatch after 29 days and the chicks are cared for by both parents. After five weeks the young are able to fly.

PORTRAIT GALLERY

Even in winter, curlews are still numerous on the tidal flats. Tidal flats covered with snow or ice can be defrosted beneath their icy cover by the tidal influx of seawater. The curlew can then use its long bill to extract worms from under the ice (top photo). The lower photo shows that curlews are similar in size to oystercatchers but considerably bigger than bar-tailed godwits, which, like the curlew, have grey-brown striped plumage in winter.

### 2.2.16 Whimbrel (*Numenius phaeopus*)

*Dutch:* Regenwulp. *German:* Regenbrachvogel.
*Danish:* Lille Regnspove.

*Appearance and general behaviour*
At 43 cm long and 500 g in weight, the whimbrel is slightly smaller than the curlew. The stance, bill shape and plumage of the two species are very similar and differences are most easily distinguished when the two are seen together, preferably side-by-side. The most obvious difference is in the head markings. The whimbrel has a distinct dark eye-stripe, which contrasts sharply with its pale supercilium, dark lateral crown-stripes and white median crown-stripe. Other differences are that the whimbrel's dark bill is shorter than that of a curlew, whimbrels have a less stately posture, and they appear more energetic while foraging.
The flight call of the whimbrel is very distinctive, a high, clear trill "bi-bi-bi-bi-bi".

*Distribution, subspecies and migration*
The whimbrel breeding sites nearest the Wadden Sea are in Iceland, Norway and Northern Scotland. The breeding grounds extend further north across Scandinavia and Northern Russia up to Siberia. The whimbrels breeding here belong to the subspecies *phaeopus*. There are also isolated breeding grounds further into Siberia, in Northern Alaska, and in Canada used by another three subspecies.

The subspecies *phaeopus* uses the East Atlantic flyway and winters mainly in coastal areas in West Africa and further south. There is still some uncertainty about the numbers that winter in Africa. For most waterbirds, estimates of the numbers on the wintering grounds and on the breeding grounds are similar, but this isn't the case for the whimbrel. Only 100 000 whimbrels have been counted in Africa, yet estimates from the breeding grounds suggest that there should be seven times as many. More intensive surveys are required at both the wintering and breeding grounds. Many whimbrels in Africa are found along beaches and amongst mangroves where they are more widely distributed than most other waterbirds (which are usually concentrated at a few sites at high tide). This may cause many whimbrels to be missed during counts.

Large numbers of whimbrels are seldom observed in the Wadden Sea. Whimbrels apparently migrate in a broad front without any special interest in coastal areas. Although they have a strong preference for tidal areas in Africa, in Western Europe they often forage inland.

*Food*
Whimbrels seem to make do with whatever food is available. Inland they feed on a variety of invertebrates found in the soil and on plants, such as crane fly larvae in grasslands. In spring on the northern breeding grounds, their main foods are berries that appear from beneath the melting snow. In summer, they feed on berries, seeds and

Whimbrels breed in open areas but prefer to hide their nests amongst scrubby vegetation or low bushes, so that they are more difficult for predators to find (shown on page 58). Whimbrels that migrate through Western Europe are usually seen inland and seldom stop in the Wadden Sea. When they do, they prefer places where they can easily find crabs (right).

buds. On the tidal flats, they are particularly fond of crabs. In West Africa they have specialised on the most numerous species, the fiddler crab. This is discussed in more detail in Section 4.5.4.

*Reproduction*

The whimbrel breeds both on marshy and dry places, preferring to nest in the cover of low vegetation. The four eggs are incubated by both parents, and the non-incubating bird often guards the nest from a nearby rock or a treetop. The chicks hatch after 28 days, and fledge five weeks later. Until then, they are guarded by both parents.

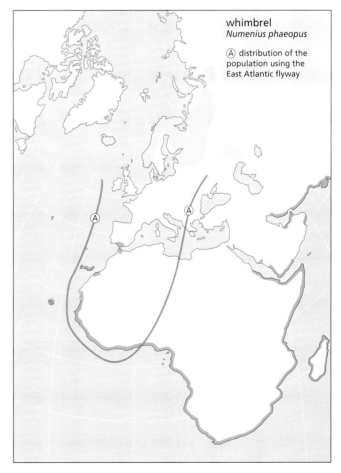

whimbrel
*Numenius phaeopus*

Ⓐ distribution of the population using the East Atlantic flyway

### 2.2.17 Spotted Redshank (*Tringa erythropus*)

*Dutch:* Zwarte ruiter. *German:* Dunkler Wasserläufer. *Danish:* Sortklire.

*Appearance and general behaviour*

Waders belonging to the genus *Tringa* are slender and medium-sized with long thin legs and a relatively long bill. At 31 cm and 164 g, the spotted redshank is slightly larger than the more abundant (common) redshank. In spring the male is completely black with small white flecks on the upperparts and flanks. The female is more heavily flecked below and often appears lightly spotted. Males don't retain their black breeding plumage for long and appear lightly spotted by the end of spring. Birds about to depart for the breeding grounds are sometimes seen in the Wadden Sea in late May. When they return in July, their black plumage has faded to a dappled grey, passing from slate-grey into the pale grey winter plumage. Their winter plumage is much paler and less spotted than that of the redshank. Apart from that the species are similar, both having a red bill-base and striking red legs. The legs of the spotted redshank are slightly longer, making them appear more slender.

*Distribution, subspecies and migration*

The spotted redshank breeds in Northern Scandinavia and across a broad zone of Northern Siberia from the Pechora River to the Bering Strait. There are estimated to be 30 000 breeding pairs in Europe, but numbers in Siberia are unknown. Since the species almost completely disappears from Europe in winter, it is assumed that most birds winter in Africa. Their known wintering grounds lie south of the Sahel, in Mali, Nigeria and Chad. Little is known of routes used to reach these sites, but most appear to migrate via inland areas. Only a very small proportion of migrants take a coastal route. Counts in the Wadden Sea peak in spring and autumn at up to 15 000 birds. In the Wadden Sea spotted redshanks favour silt-rich areas such as the mouth of the Dollard and the Elbe. The birds remain at these sites for some time in autumn. The Wadden Sea provides a good moulting and fattening site, allowing the birds to complete their journey to tropical Africa with fresh feathers and the necessary body reserves. No major coastal stopover sites are known to be used on the migration south from the Wadden Sea. Similarly, the vast majority of spotted redshanks wintering in West Africa uses inland areas.

*Food*

Inland, spotted redshanks forage in waterways or marshes, feeding on everything from small insects to fishes 6 cm long. The birds usually walk in the water but will also swim. In tidal areas they search for worms, crabs and smaller prey in silty sediments. In summer, spotted redshanks hunt young fish and shrimps trapped in pools. The redshank forcefully sweeps its bill through the water, disturbing prey sitting on the sediment, which the bird can then pursue. When a bird is very successful at this foraging method, other redshanks are quick to join in. This 'social foraging' is discussed at greater length in Chapter 4.

*Reproduction*

The spotted redshank breeds in open areas in the northern forests and the taiga. The nest is made in a dry area, not far from suitably wet foraging sites. The nest and four eggs are usually well hidden amongst vegetation or rocks. It is not yet known if the female assists at all with incubation. In most cases she appears to leave the breeding grounds well before the eggs hatch. The chicks are cared for exclusively by the male.

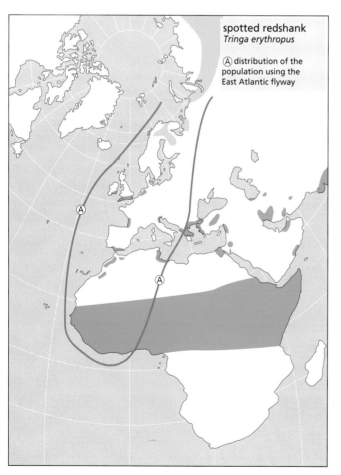

spotted redshank
*Tringa erythropus*

Ⓐ distribution of the population using the East Atlantic flyway

Throughout the year the spotted redshank has white edges to its back feathers and a white underwing. In the breeding season, the rest of the plumage is pitch black (as is beautifully shown on the photograph on page 119 of a bird in the breeding grounds). The spotted redshanks that pass through the Wadden Sea in May are usually more or less spotted (top photo). When they return later in the summer not much black remains on the upperparts (lower photo). Later in the year the underparts become completely white.

# PORTRAIT GALLERY

PORTRAIT GALLERY

## 2.2.18 Redshank (*Tringa totanus*)

*Dutch:* Tureluur. *German:* Rotschenkel. *Danish:* Rødben.

*Appearance and general behaviour*

The redshank has fairly uniform grey-brown plumage that is somewhat speckled or streaked, and is darker above than below. The winter plumage is somewhat paler, particularly on the underparts. The bill-base is red and the long legs are orange-red. The redshank is 28 cm from bill tip to tail, weighs 120 g, and is found on tidal areas year-round.
It often appears rather nervous and noisy. As well as the three note "tu–re-luur" call, it has a repertoire of "tuut" sounds that reflect the bird's mood in frequency and pitch.

*Distribution, subspecies and migration*

The redshank's breeding grounds cover much of Europe and temperate Asia. The birds seen in the Wadden Sea breed on Iceland, the British Isles, the Northwest European mainland and Scandinavia. These redshanks are officially divided into two subspecies: *totanus*, which breeds on the mainland, and *robusta*, which breeds in Iceland. The two subspecies are indistinguishable in the field. Some researchers make a further distinction between *totanus* from northern and southern regions. To make it even more confusing for bird watchers the 35 000 breeding pairs found on the British Isles are mostly viewed as an intermediate form between *totanus* and *robusta*. Most of these British birds remain on their islands through the winter, unlike the Icelanders that spend the winter on the coast of Western Europe. In total, 64 500 *robusta* redshanks are estimated to migrate to Western Europe (as far south as Portugal) for the winter.

In addition 250 000 *totanus* redshanks are estimated to use the East Atlantic flyway. These are mainly birds from (Northern) Scandinavia and the Wadden Sea countries, many of which winter on West African coasts. Most take a coastal migration route, but there is also some evidence for inland flights over the Sahara.

Redshanks are numerous in the Wadden Sea throughout the year. More than 12 000 pairs breed here, particularly in marshes where they can forage on nearby mudflats. In spring they are often accompanied by redshanks from Scandinavia and Iceland. However these birds have only a short stopover in the Wadden Sea, causing little increase in total numbers. Peak numbers are reached in late summer when there are around 60 000 redshanks in the Wadden Sea. Even in winter when temperatures drop below freezing, many redshanks remain in the Wadden Sea, particularly in the western part. The 10 000 or more birds that remain for the winter probably breed in Iceland.

*Food*

Redshanks forage in wetlands and drier areas inland, where they feed on a variety of insects found amongst vegetation and larger prey such as earthworms and crane fly larvae. In the tidal areas, they feed on worms, crustaceans, small bivalves and snails. Redshanks mainly hunt by sight, but can also use their bills to detect hidden prey by touch. This allows redshanks to forage at night, especially in the winter months. They often work together in a group. When they hunt for shrimps and small fish in shallow water they are often accompanied by other *Tringa* species.

*Reproduction*

Most inland breeding sites are in damp grassland, but around the Wadden Sea redshanks usually nest in salt marshes and other areas seaward of the dykes. The female lays four eggs in early May, but pairs are present on the breeding grounds much earlier than this and enthusiastically perform aerial displays there.

Redshanks nest in dense vegetation and will often use almost exactly the same nest site for many years. Incubating birds often hide by pulling grass and other vegetation over themselves while they are on the nest. Redshanks will remain camouflaged in this way until a person or other threat approaches within one metre of the nest. Once disturbed and as long as the threat remains nearby, the birds will continuously circle the nest making alarm calls. Their defensive behaviour is more intense when chicks are present and predators will often be dive-bombed, sometimes with the assistance of a neighbouring pair.

The parents share incubation and caring for the chicks. Some records suggest that the female leaves the breeding area first, just before the chicks fledge.

Redshanks are found in the Wadden Sea throughout the whole year. Their plumage changes little during the course of the seasons.

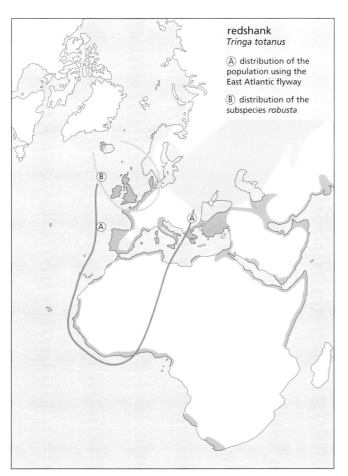

redshank
*Tringa totanus*

Ⓐ distribution of the population using the East Atlantic flyway

Ⓑ distribution of the subspecies *robusta*

### 2.2.19 Greenshank (*Tringa nebularia*)

*Dutch:* Groenpootruiter. *German:* Grünschenkel. *Danish:* Hvidklire.

*Appearance and general behaviour*

The greenshank is the largest of the *Tringa* sandpipers, 30 - 35 cm long and weighing 210 g. Its grey-green legs are conspicuously long and its bill is slightly upcurved. The upperparts are grey-brown with distinct white fringes to the feathers. The head and neck feathers are very finely patterned with long dark grey streaks. The underparts are almost completely white.

The calls, a far-carrying "tew-tew-tew" and a snapping "krieup-krieup" distress call, are very distinctive.

*Distribution and migration*

Greenshanks breed throughout much of Scandinavia and across a broad zone of taiga and pine forest in Siberia and further east. 1500 pairs breed in Scotland, some of which winter in England and Ireland. Other greenshanks winter further south: south of the Sahel in inland Africa, in Southern Asia, and in Australia. These birds probably migrate across a wide corridor over inland areas. Birds that breed in Scandinavia are known to fly directly over Europe and the Sahara to Mali.

There are also some greenshanks that migrate along the coast and winter in African tidal areas. A total of 19 000 individuals has been counted at these sites but there are likely to be many more, as the species is widely distributed making it difficult to count. In winter almost none remain in the Wadden Sea. Migrating greenshanks pass through the Wadden Sea rapidly in spring. The return trip in late summer doesn't pass as quickly and the number of greenshanks in the Wadden Sea is then highest, with a peak of around 15 000 birds in August. Greenshanks are evenly distributed across the Wadden Sea, seldom forming large flocks or mixing with other waders on roosts.

*Food*

Inland, greenshanks usually forage in or beside waterways, where they eat a variety of insects, worms, frogs and small fish. On the tidal flats they take all sorts of small prey from the surface and search with their bills in the silt. They will often actively hunt fish and shrimps in shallow water. A hunting greenshank walks quickly, moving its bill across the sediment surface to disturb prey, which it then chases with a slightly open bill. Greenshanks often use this fishing technique in collaboration with other *Tringa* species or fishing birds.

*Reproduction*

Greenshanks usually nest on dry ground in a marshy area. The nest sites often have trees nearby that are used as lookout posts. The same nest site is used for many years. Although the nest scrape is sometimes made in an exposed area, it is usually hidden between rocks, woody debris or clumps of vegetation. Incubation takes 24 days. Both parents incubate the four eggs and care for the chicks.

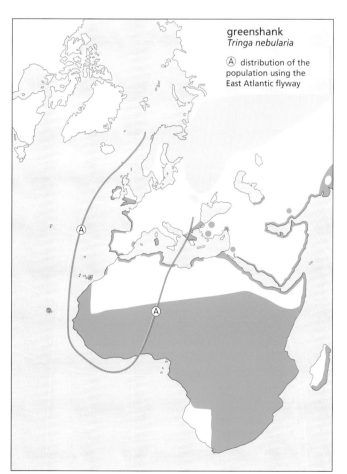

greenshank
*Tringa nebularia*

Ⓐ distribution of the population using the East Atlantic flyway

Greenshanks can be seen on migration in the Wadden Sea, and are especially numerous in late summer (page 64 top). They usually breed in the northern pine forest zone where they like to use a tree near the nest as a lookout post (lower photo on page 64). The common sandpiper (below) seldom visits the tidal flats and mainly uses inland areas when on migration.

## 2.2.20 Common Sandpiper (*Actitis hypoleucos*)

*Dutch:* Oeverloper. *German:* Flussuferläufer.
*Danish:* Mudderklire

Common sandpipers are closely related to the *Tringa* sandpipers, and are almost as big as dunlins. The common sandpiper usually holds its body horizontal, regularly bobbing its tail. They are very mobile and often fly close to the ground or water, alternating a rapid wing beat with short gliding flight. They repeatedly make a high call "hididididididi" in flight.

The common sandpiper is mainly an inland bird, and breeds in large numbers over much of Europe and Asia along rivers and lakes. Most winter in Africa, remaining near the water's edge, preferably freshwater, on migration.

They are not abundant in the Wadden Sea and counts never reach more than hundreds of birds. In the Wadden Sea, they mainly occur along the (silt rich) edges of intertidal areas where they feed on insects, small crustaceans and other benthic animals. Common sandpipers probably feel more at home in tropical tidal areas, where mangroves provide cover and small fiddler crabs are easy prey. The number of wintering common sandpipers on the tidal flats of Guinea-Bissau is estimated to be 9000 and the number in the mangroves at tens of thousands.

### 2.2.21 Turnstone (*Arenaria interpres*)

*Dutch:* Steenloper. *German:* Steinwälzer.
*Danish:* Stenvender.

*Appearance and general behaviour*
At 23 cm, the turnstone is robustly built, with short orange legs and a short dark bill. The breeding plumage, seen in most migrants passing through the Wadden Sea in May, is strikingly patterned with black and white on the head and breast and orange-brown and black on the upperparts. The head patterns differ between the sexes; the black and white contrast boldly in males but coloration is slightly duller in females. After the breeding season, both sexes lose their distinctive head markings and their upperparts become darker and greyer. Turnstones in flight are easily recognised by their broad white wingbars and diamond-shaped white patch on the back. They have a distinctive way of foraging that has given rise to their name.

*Distribution, subspecies and migration*
Turnstones breed in a narrow coastal zone, well inside the Arctic Circle. Their main breeding grounds only extend further south in Scandinavia. A total of 20 000 pairs breeds in Scandinavia, some tens in Denmark and a few around the Wadden Sea. This distribution is reversed in winter, when turnstones are found along the coasts of all continents, south of 40° north. They only winter further north than this in the Wadden Sea and on British Isles.

Their movements between the breeding and wintering grounds are very complicated and not fully understood. Turnstones from two areas use the East Atlantic flyway, but their migratory behaviour is so different that they are considered separate bio-geographic populations.
The 67 000 turnstones that winter along the coasts of Western Europe mainly breed in Greenland and North-eastern Canada. However, turnstones ringed in Mauritania have been recaptured in Greenland, so presumably some of the Greenland birds prefer to winter on tropical beaches. Of the second group, comprising breeding birds from Scandinavia and neighbouring Russian coasts, some winter in Western Europe. But most migrate to North and West Africa, where a total of 317 000 turnstones has been counted. The turnstones that winter in Southern and East Africa originate from breeding grounds in Siberia and follow a completely different migratory route.
The number of turnstones in the Wadden Sea peaks at around 8000 in spring. Up to 4000 remain in winter, and will stay even through very harsh winters. However it is apparent that the Wadden Sea isn't an optimal site for this species when these numbers are compared to the 44 000 wintering birds found on rocky British coastlines.

*Food*
Although turnstones breed near the sea in the high north, the coast is often still frozen during the breeding season, so the birds must feed on terrestrial insects and vegetation. However they feed in coastal areas as soon as possi-

# PORTRAIT GALLERY

Male turnstones have a very prominent black and white pattern on the head and breast. This male (page 66) is incubating eggs that are just starting to hatch on the tundra of the Taimyr Peninsula. The turnstone's facial markings can be surprising when viewed head-on. Here, three juvenile birds rest alongside adults at the tide line of a sandy flat (above). At high tide turnstones often congregate in small flocks, avoiding the larger flocks formed by most other waders.

**turnstone**
*Arenaria interpres*

Ⓐ distribution of the population using the East Atlantic flyway

ble, both during and after the breeding season. The opportunistic turnstone can find food in tidal areas, on sandy or gravel beaches, on rocky shores, and amongst mangroves. Their foods include worms and sediment-dwelling bivalves, sandhoppers, remains of prey left by bigger waders, and parts of dead animals washed in on the tide. The turnstone uses a very simple and effective trick to find food, which is unique in the wader world. It uses its bill to toss stones, shells and other items over. Once such an item has been turned over it becomes the equivalent of a richly laid table for the turnstone, especially if the item was something big like a dead bird or fish. Turnstones sometimes even roll up seaweed like a carpet in their search for food.

The strong bill can also be used to break open shells and crabs, and to peck barnacles from rocks. This is why turnstones are often seen in 'harder' locations in the Wadden Sea. Sometimes these sites are mussel beds, but more often they're the edges of tidal flats, beaches and piers, along the high tide line, or the stone bases of dykes and harbours (where the turnstones can eat leftovers from the fishermen and any nearby fish and chip shops).

*Reproduction*

On the High Arctic breeding grounds, the turnstone makes its nest scrape on a dry and often barren part of the tundra. Its pied plumage provides good camouflage in these exposed sites. In Scandinavia turnstones hide their nests among dense vegetation or between rocks, sometimes even nesting completely underneath a rock, so that the incubating bird is hardly visible.

The four eggs are incubated for 24 days, and the chicks, which can forage for themselves from the first day of life, can fly after three weeks. Both parents take an equal share in incubating and caring for the chicks. Predators and other threats are zealously attacked.

### 2.2.22 Great Cormorant (*Phalacrocorax carbo*)

*Dutch:* Aalscholver. *German:* Kormoran. *Danish:* Skarv.

*Appearance and general behaviour*

The cormorant is nearly 1 m in length from bill tip to tail and weighs around 2400 g. Its plumage is mainly black. In spring, the cormorant has varying amounts of white on the head and neck and a white thigh patch, and its black feathers acquire a beautiful blue-purple metallic sheen. For the rest of the year its plumage is somewhat duller. The young non-breeding birds have browner plumage with varying amounts of pale grey or white on the underparts. The posture and behaviour of the cormorant are very distinctive. Cormorants sit deep in the water when swimming with their tail spread out on the surface, taking flight with great effort. When the cormorant is not swimming or flying, it rests by the water's edge in an inelegant slouch. They will often sit on a sandbank on the tidal flat but prefer to be higher up, on a pier, pile or beacon. They then often assume their characteristic pose for drying their feathers by extending or flapping their wings.

*Distribution, subspecies and migration*

European cormorants belong to two subspecies. One subspecies, *carbo*, breeds only on the coast of Norway and the British Isles, mainly on rocky shores. This subspecies is resident in these areas, usually remaining there through the winter. The second subspecies, *sinensis*,

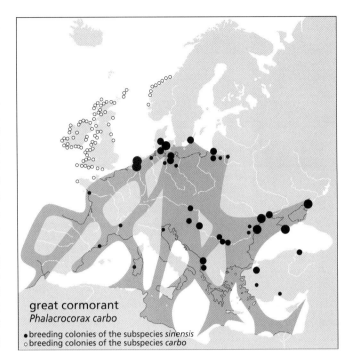

Breeding colonies and possible migration routes of European cormorants. The *carbo* subspecies only travels short distances and is not found in the Wadden Sea. The *sinensis* subspecies breeds on the mainland and makes long migratory flights across Europe and to Northern Africa.[1012]

mainly breeds in the Wadden Sea countries and south of the Baltic Sea but also breeds in inland Great Britain, the catchment of the Danube, and the Black Sea. All members of this second subspecies are migratory. Cormorants from The Netherlands mainly migrate to Western France and the Western Mediterranean. Most appear to take an inland route. Breeding birds from Denmark have been observed in The Netherlands, Italy and Tunisia.

For many years, fishing communities have viewed the cormorant as a competitor and cormorants were greatly persecuted which kept their numbers very low. Conservation measures have allowed their numbers to increase sharply since 1975. Now, cormorants that were previously only present in freshwater and around river mouths are appearing more and more in tidal areas. For many years they were present in tidal areas only as non-breeding summer visitors, and in 1950 only one small tidal colony existed, in the mouth of the river Weser. Now more than 1400 pairs breed in the Dutch Wadden Sea.

Most cormorants found in the Wadden Sea in spring are birds that breed locally. In summer and autumn the number increases to almost 10 000 birds, most of which remain in the Wadden Sea to moult. In the winter, fish are relatively scarce in the Wadden Sea and cormorants are seldom seen, although the number of wintering birds seems to have increased in recent years. Most cormorants migrate south in winter, or remain in freshwater areas, as long as these are not frozen over.

*Food*

Cormorants can dive down to 20 m and catch a variety of fish (which they swallow whole), everything from

sticklebacks weighing just tenths of a gram to pikeperch of more than 1 kg. When they can catch these large fish, they only need to feed for less than half an hour per day to meet their daily requirements, but often it takes them much more time than this.

The species and size of fish eaten reflect what is available in the feeding grounds. In the Wadden Sea, flatfish make up more than half of the cormorants' diet. In late summer this proportion can increase to up to 80%.

The cormorants that breed in the Wadden Sea arrive very early in the year when little food is available locally. So, early in the breeding season, these birds commute back and forth to freshwater areas for the necessary top up.

Cormorants often fish in groups. When one cormorant finds a school of fish, its conspecifics can immediately tell what has happened by the bird's behaviour and quickly join in. Gulls are also quick to take advantage of the discovery (lower photo). Cormorants that aren't fishing search for a place to roost. They prefer to use beacons and other high structures in the Wadden Sea (as shown on page 68), but will also roost in large numbers on sand banks (top photo).

### 2.2.23 Spoonbill (*Platalea leucorodia*)

*Dutch:* Lepelaar. *German:* Löffler. *Danish:* Skestork.

*Appearance and general behaviour*
Spoonbills, with their all-white plumage and long, spatulate bill, are unmistakable. In spring they sport a flamboyant crest that waves in the breeze and a yellow breast patch. The spoonbill's long legs make it look rather like a heron, but it is most closely related to storks and ibises.

*Distribution, subspecies and migration*
Spoonbills are divided into four subspecies that breed mainly in Asia, but also in Africa and Europe. European birds all belong to the subspecies *leucorodia*. Some members of this subspecies breed in Southeastern Europe. The spoonbills in Western Europe are completely separate from these southeastern ones, during both the breeding season and migration. Because of this, the 3000 Western European spoonbills are considered a distinct population that uses the East Atlantic flyway. Some of these birds breed in France, but most breed in Spain and The Netherlands. The breeding colonies in the Wadden Sea are the species' most northerly breeding grounds.

Spoonbills have bred for many, many years in the Western Dutch Wadden Sea. In recent years, colonies have also established on islands further east. There are around 1000 pairs breeding in the international Wadden Sea. In autumn these birds migrate with their young in small

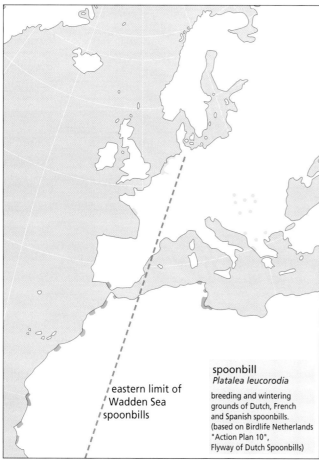

eastern limit of Wadden Sea spoonbills

**spoonbill**
*Platalea leucorodia*
breeding and wintering grounds of Dutch, French and Spanish spoonbills.
(based on Birdlife Netherlands "Action Plan 10", Flyway of Dutch Spoonbills)

groups via various stopover points along the French coast to Southern Spain and Morocco. Some spend the winter there, but most spoonbills from Western Europe winter with the Spanish birds in coastal areas of Senegal and Mauritania. They are sometimes joined there by the local spoonbill subspecies that breeds in Africa. These spoonbills from Mauritania are residents that have already bred by the time the European spoonbills start to return to their breeding grounds.

*Food*

The spoonbills that breed in the Wadden Sea often feed in freshwater areas in early spring, when sticklebacks are their main prey. Later in the season there is enough food for them to forage on the intertidal flats. There they wade alone or in small groups, taking large steps through the shallow water and scything backwards and forwards with a partially open bill. They also use this technique to forage at night. They catch everything that their bills encounter, and any shrimps and small fish are quickly gulped down. They also catch larger flatfishes that are sometimes too large to fit down the spoonbill's throat.

*Reproduction*

Spoonbills often breed in large reed beds inland. On the Wadden Sea islands, they also have colonies in the sand dunes and in salt marshes, where they make a simple nest on the ground. Increasing numbers of foxes steal the eggs and chicks of the inland breeding birds. This has made the spoonbills relocate to nature reserves that are difficult for predators to access, and the fox-free Wadden Sea islands. Perhaps these spoonbills will eventually start to nest in trees, as they do in Spain.

The female lays three to four eggs in the messy twiggy nest, from early March onwards. Both parents incubate the eggs and feed the chicks for more than seven weeks. The parents regurgitate half-digested food, which the chicks peck from their bills. Later, when the young birds are out on the tidal flats with their parents, they continue to be regularly fed.

Spoonbills form breeding colonies on various Wadden Sea islands. In early spring, they forage in ditches inland of the dykes. Later in the season they often fish in groups on the mudflats. In this way, prey such as fish and shrimps are hunted more efficiently and there is a greater chance that they will end up within reach of a spoonbill (as shown on page 70).

### 2.2.24 Brent Goose (*Branta bernicla*)

*Dutch:* Rotgans. *German:* Ringelgans. *Danish:* Knortegås.

*Appearance and general behaviour*
The brent goose is the smallest of the European geese and, at 58 cm and 1400 g, is only slightly larger than a mallard. Its bill, head, neck, breast and legs are black, while the upperparts are dark grey. Its white rear end is particularly conspicuous in flight. Adult brent geese have a small white crescent on each side of the neck. They make a distinctive soft rolling call "rrok".

*Distribution, subspecies and migration*
Brent geese breed in Arctic areas in Northern Siberia, Northern Greenland and Northern Canada, and on isles like Spitsbergen. Four subspecies are recognised, two of which live around the Wadden Sea. The most numerous is the subspecies *bernicla* that breeds in Taimyr and mainly winters around the Southern North Sea. Some of these geese migrate a little further south, as far as the mouth of the Gironde. In total, some 200 000 brent geese migrate through the Wadden Sea and are dependent on this area for much of the year. In May, almost the whole population is found in the Wadden Sea.

The other subspecies, *hrota*, the white-bellied brent goose, breeds around Spitsbergen and winters in Northeast England and Denmark. Several thousand of the Danish wintering birds are found on the Danish tidal flats in autumn.

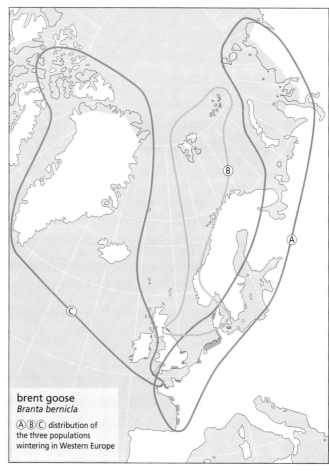

**brent goose**
*Branta bernicla*
Ⓐ Ⓑ Ⓒ distribution of the three populations wintering in Western Europe

Brent geese (shown on page 72) feed on seaweed and seagrass on the tidal flats, a food resource that is quickly exhausted. So in winter they often graze on inland grasslands. They can also be found grazing on young sprouting plants on the salt marshes in spring.

Barnacle geese (right) are not dependent on tidal areas. They mainly graze on inland grasslands, but are regularly seen on salt marshes.

*Food*

Brent Geese have a stronger association with the sea coasts than other goose species. They are often found in tidal areas where they feed on seagrass and algae such as sea lettuce. In the past, seagrass covered much of the Wadden Sea tidal flats, forming the main autumn food for brent geese. But seagrass has become very scarce and the geese have turned to the grasslands behind the dykes as an alternative food source. Thanks to artificial fertilisers this grass now has a much higher protein content, making it relatively easy to digest in winter. In spring brent geese are usually found on the grasslands beyond the dykes and on salt marshes, feeding on the fresh young sprouts of the salt marsh plants.

*Reproduction*

Brent geese chose a partner for life, and they remain together day and night. They arrive together on the High Arctic breeding grounds and often reuse their old nest sites. Brent geese usually breed in a colony, close to the sea or fresh water. The nest is sometimes made in the shelter of a few rocks, but is more often on an exposed area of tundra. The three to five eggs are kept warm in a bed of down. Only the female incubates and she must leave the nest every few hours to feed. The male always stays nearby, guarding the nest. Both parents take an active role in caring for the chicks. They migrate together from the breeding area and arrive as a family in the Wadden Sea, making it easy to determine how successful the breeding season has been.

### 2.2.25 Barnacle Goose (*Branta leucopsis*)

*Dutch:* Brandgans. *German:* Nonnengans.
*Danish:* Bramgås.

The barnacle goose is bigger than the brent goose, and is easily distinguished by its pale underparts and yellow-tinged or white head. These geese often cluster in a tight group, continuously making a high yelping call. The hundreds of thousands of barnacle geese that winter in The Netherlands, Germany and Denmark breed on the Novaya Zemlya archipelago, adjacent coastal parts of Siberia, Northern Russia and islands in the Baltic Sea. Originally, nearly all barnacle geese wintered on coastal grasslands. Although barnacle geese are still often found on salt marshes, many now use the good grazing provided by agricultural grasslands. Barnacle geese only use tidal flats for roost sites. In spring and autumn many thousands graze on salt marsh vegetation on the Wadden Sea islands and along the coast of the mainland.

PORTRAIT GALLERY

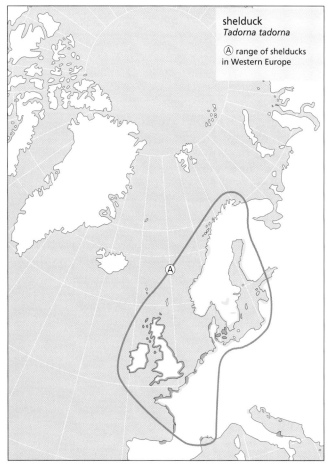

### 2.2.26 Common Shelduck (*Tadorna tadorna*)

*Dutch:* Bergeend. *German:* Brandgans. *Danish:* Gravand.

*Appearance and general behaviour*
Common shelducks are somewhere between geese and ducks in their appearance and behaviour. They are easily recognised by their bold plumage and large size, 58 - 70 cm long and 1 kg in weight. Their plumage is mainly white, with a glossy black head and neck, a broad chestnut breast-band, and a broad black stripe over each wing and along the belly. The male has a prominent bulge at the base of its bright red bill and is larger than the female. In spring shelducks are mostly seen in pairs, with the male displaying conspicuously. He dips his head and whistles to try and impress other males. If this doesn't work, he tries to chase them away. Outside the breeding season shelducks are usually found in large groups on tidal flats.

*Distribution and migration*
Common shelducks breed in Western and Southern Europe, and in a broad band from the Middle East to Northeastern China. On the Asiatic breeding grounds they are usually found in inland areas of brackish water and salt lakes. In Europe, the common shelduck is usually found in coastal areas, but has penetrated inland over recent decades.
The Western European breeding grounds extend from the North Cape of Norway south to Brittany. Within this

area, the most important breeding sites are the British Isles with 50 000 pairs, the Wadden Sea with almost 5000 breeding pairs, and the Baltic Sea.

After the breeding season, an estimated 300 000 Northwest European shelducks remain on the British Isles and in the Wadden Sea. More than 250 000 have been counted in the Wadden Sea in autumn. Many of these birds will remain in the Wadden Sea during mild winters. When the temperature falls below zero they migrate to the Dutch Wadden Sea, to the Dutch Delta area and further afield to England and Northern France. Some travel as far as Spain and North Africa, but most of the shelducks that winter at these sites are from more easterly breeding grounds.

The Wadden Sea is also a very important moulting site for the common shelduck. When ducks and geese moult they lose all their flight feathers simultaneously, making them flightless for some weeks. In July, after the breeding season, nearly all Northwest European shelducks fly to two quiet shallow coastal areas, the mouth of the River Weser and mouth of the Elbe, for the duration of this risky moulting period. Although there are small moulting areas along the English coast, most English shelducks also use these Wadden Sea moult sites.

*Food*

The shelduck mainly feeds on small shellfish and snails, but will also eat a variety of other small benthic animals and vegetation, which it sieves from the soft silty sediment through its bill. Shelducks are usually found in sheltered silt-rich areas of the Wadden Sea. They can feed while walking, wading, or swimming and dipping their heads under the water, which allows them to forage at different water depths and so make use of much of the tidal cycle. They often continue foraging at high tide.

*Reproduction*

Shelducks are real cave dwellers in the breeding season. They like to nest in rabbit burrows in the sand dunes, but will also use other holes provided by tidal debris, dense vegetation, or purpose-built artificial burrows. Eight to twelve eggs are laid in the down-filled cup. The female incubates the eggs for 28 days, while the male stands guard nearby or on the adjacent feeding grounds. The chicks take at least eight weeks to fledge. Initially, both parents accompany the chicks, but later the family bond weakens. The young of several clutches, which are by then self sufficient, often end up in a sort of crèche under the care of a few adult birds.

Shelducks are almost always seen in pairs in spring (shown on page 74). The male, identified by the knob on its bill, chases away any rivals that come too close (below). Shelduck chicks are jaunty little black and white fluff balls (photo top left). After a month they look like rather odd pied ducks, and will only acquire the colourful adult plumage in autumn (photo top right).

### 2.2.27 Eider Duck (*Somateria mollissima*)

*Dutch:* Eidereend. *German:* Eiderente. *Danish:* Ederfugl.

*Appearance and general behaviour*
The eider duck is a sturdy duck, 60 cm from bill tip to tail, weighing up to 2100 g. Its head appears streamlined in profile, with the bill joining the forehead in a continuous line. As is usual for ducks, the two sexes look very different. In winter and spring, the male is mostly white apart from a black cap, tail, sides and belly. The white breast has a rosy glow, and the nape is adorned with lime-green feathers. After the breeding season, the male's plumage is mostly patchy black. The female's plumage is grey to reddish brown with black-brown barring and is the same year-round. Eider ducks are true marine birds, and are nearly always found in large flocks on the water. Outside the breeding season they only come ashore to roost, and even that is often done on the water.

*Distribution and migration*
Eider ducks breed on the northern coasts of Europe, Asia and North America, almost anywhere where the coasts are regularly free of ice in spring. The Wadden Sea is one of their southernmost breeding grounds. Different subspecies are recognised, but the more than two million

Thanks to her beautifully camouflaged plumage, the female eider duck can hardly be seen, hidden amongst the vegetation on her nest (left). Males in their spring plumage are much more conspicuous, especially when, as shown here, they are displaying (below).
The male leaves the breeding area once his ducklings hatch. Often the females take their chicks and seek out the company of other breeding females, forming a sort of duckling crèche (shown on page 77).

eider ducks that winter in Northwest European and Scandinavian waters all belong the subspecies *mollissima*.

Six thousand pairs breed in the Wadden Sea, mainly on the islands. These birds are local residents, as are most eiders in England, Ireland and Norway. In winter, the Wadden Sea breeding birds are joined by some of the eiders that breed in the Baltic Sea, as much of the Baltic Sea freezes over.

In winter some 100 000 eider ducks are counted in the Wadden Sea area annually, but this only includes the birds that can be counted from the shore. Many more stay in deeper waters, further south of the islands. Another 200 000 wintering birds have been counted offshore during aerial surveys. In summer, non-breeding eiders outnumber the local breeding birds. Many of these non-breeders appear to be young eiders from the Baltic Sea. The Wadden Sea also provides many thousands of eiders with a safe moulting area in July and August as they renew their primaries and are temporarily flightless.

*Food*

Like true seabirds, eider ducks find their food underwater. Using their feet and wings they can dive down to 10 m in search of benthic animals. They mainly feed on shellfish, but will also take crabs, starfish and occasionally fish. Bivalves such as mussels are preferred. They can swallow shells up to 4 cm long whole, but prefer to take smaller ones, which are crushed inside the strong stomach.

In the Wadden Sea, eider ducks use the tidal cycles differently from other waterbirds. At high tide they dive for food on the intertidal flats. They will still swim and dip their heads under the water while foraging in shallow water, but are almost never seen feeding on exposed tidal flats. At low tide they often roost on sandbanks.

*Reproduction*

Eider ducks are almost always seen in flocks and their display behaviour in spring seems rather like a social game. During this time, females can often be seen swimming, surrounded by males that try to impress them by dipping their heads and making loud cooing noises "ahuua". Eventually, pairs are formed and remain faithful for at least one season. Together they go ashore to start their nest; the female chooses and builds the nest, while the male remains in the general vicinity.

The nest is usually near the coast. Nests around the Wadden Sea are often hidden in dense sand dune vegetation. The four to six eggs are surrounded by a thick, warm layer of down from the female's belly. The female is also responsible for incubating the eggs for 26 days, while the male stays nearby and keeps an eye on the nest. She occasionally leaves the nest to drink something and will sometimes feed early in the breeding season, but mostly relies on her reserves. By the time the chicks hatch, her weight has dropped by one third. Even so, she still must take care of the chicks, as the males leave the breeding site when the chicks hatch.

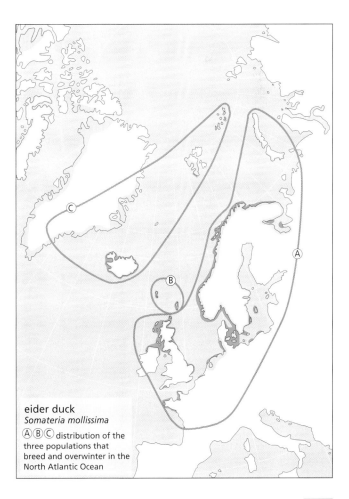

eider duck
*Somateria mollissima*
Ⓐ Ⓑ Ⓒ distribution of the three populations that breed and overwinter in the North Atlantic Ocean

### 2.2.28 Ducks

Tidal areas with their transition from wet to dry and salt to fresh are naturally very attractive to a variety of waterbirds. Many duck species are seen more or less regularly around the Wadden Sea, but it's difficult to determine if they deserve to be included in this book. Here, we discuss the species that are most numerous on the tidal flats and describe what is known of their relationships with intertidal areas. Unfortunately, our knowledge is often limited as, amongst other reasons, ducks tend to be more active at night, making it difficult to determine what they are up to on the tidal flats.

**Wigeon** *(Anas penelope)*
*Dutch:* Smient. *German:* Pfeifente. *Danish:* Pibeand.

The more than one million wigeon that come from Northern Russia and Siberia to Western Europe almost exclusively use coastal areas. In the Wadden Sea they mainly graze on the saline grasslands outside the dykes. In autumn more than 300 000 have been counted out on the tidal flats feeding on seagrass and seaweed at low tide. In cold winters many wigeon migrate to the British Isles and France, but even then more than 100 000 remain in the Wadden Sea. Wigeon depart in March and early April for their breeding grounds. Vegetation sea-

ward of the dykes grows far slower than that inland, so wigeon that pass through tend to stay inside the dykes.

**Mallard** *(Anas platyrhynchos)*
*Dutch:* Wilde eend. *German:* Stockente. *Danish:* Gråand.

Many millions of mallard are present in Western Europe year-round. Their abundance is due to their ability to adapt to different environments, as is shown by the large numbers found throughout the Wadden Sea. For some, the Wadden Sea provides suitable breeding grounds. Others only forage there for part of the year. As well as large numbers of small benthic animals, mallards find many seeds in the silty sediments on the edge of the tidal flats. In summer, mallards will sometimes join waders and gulls socially foraging on numerous shrimps and small fish stranded in small tidal pools. The number of mallards in the Wadden Sea peaks in autumn and summer, and can reach up to 165 000. Most mallards seem to use quiet tidal areas to roost and as a base for nightly foraging trips inland.

**Teal** *(Anas crecca)*
*Dutch:* Wintertaling. *German:* Krickente. *Danish:* Krikand.

This little duck is often seen in silty sheltered areas of the Wadden Sea, especially when on migration. In autumn the number of teal in the Wadden Sea can reach 56 000. They seem to prefer freshwater and brackish tidal areas such as the Dollard and the mouth of the Elbe. Many teal roost in these areas during the day, but they also forage around salt-tolerant plants and on the exposed sediment. It's not easy to observe what the teal are sieving from the silt, but analysis of stomach contents shows that they feed almost exclusively on seeds.

**Pintail** *(Anas acuta)*
*Dutch:* Pijlstaart. *German:* Spießente. *Danish:* Spidsand.
and **Shoveler** *(Anas clypeata)*
*Dutch:* Slobeend. *German:* Löffelente. *Danish:* Skeand.

More than 50 000 of the 70 000 pintails that winter in Western Europe stay in the Wadden Sea. They are often seen on the edges of tidal flats and in brackish tidal areas. These sites are also used by many of the 4000 shovelers that spend autumn and winter in the Wadden Sea (10% of all the shovelers that winter in Western Europe). Sometimes both species can be seen foraging together on the tidal flats. It's known that both feed on a variety of small animals as well as seeds, but it's not clear how important the tidal flats are as a food source.

Wigeons are real vegetarians, living on seaweed and seagrass on the tidal flats, and grazing on the salt marshes (shown on page 78). Other ducks like the teal (above) are only found on the siltiest edges of tidal flats. Mallards are real opportunists, fishing here alongside spoonbills and greenshanks in a shallow pool on the tidal flat (right).

PORTRAIT GALLERY

Gulls are outspoken opportunists. A herring gull flying by while searching for food (top photo) is a characteristic sight in all coastal areas. Gulls sometimes try to eat large prey on the tidal flats, like this common gull trying to eat a mussel (photo middle left). But they are also content with more modest mouthfuls. These black-headed gulls track small but numerous benthic animals, like the amphipod *Corophium volutator*, beside the incoming tide (photo middle right). These three gull species are abundant on tidal flats year-round, unlike the lesser black-backed gulls (lower right), which winter in coastal areas further south.

## 2.2.29 Gulls

The following five species of gulls are regularly present in the Wadden Sea:
- **Black-headed Gull** *(Larus ridibundus)*
*Dutch:* Kokmeeuw. *German:* Lachmöwe. *Danish:* Hættemåge.
- **Common Gull** *(Larus canus)*
*Dutch:* Stormmeeuw. *German:* Sturmmöwe. *Danish:* Stormmåge.
- **Lesser Black-backed Gull** *(Larus fuscus)*
*Dutch:* Kleine mantelmeeuw. *German:* Heringmöwe. *Danish:* Sildemåge.
- **Herring Gull** *(Larus argentatus)*
*Dutch:* Zilvermeeuw. *German:* Silbermöwe. *Danish:* Sølvmåge.
- **Great Black-backed Gull** *(Larus marinus)*
*Dutch:* Grote mantelmeeuw. *German:* Mantelmöwe. *Danish:* Svartbag.

*Appearance and general behaviour*

Even though other less common gull species are regularly seen in the Wadden Sea, these five species are the most numerous in all seasons. The plumage of all five species is mainly white, although in the breeding season the black-headed gull has a dark brown head. In three species, the upperpart of the wing is pale grey with black wing tips. The upperparts of the lesser black-backed gull are slate-grey, while those of the great black-backed gull are almost black. These larger species take some years to mature, and the plumage of the juveniles is brown-flecked and slowly gets paler with age. It can then be difficult to distinguish these two species.

The great black-backed gull, with a length of 74 cm, is clearly the largest. The black-headed gull (38 cm) and the common gull (41 cm) are the smallest. The other two are intermediate at 55 cm.

*Distribution and migration*

The great black-backed gull has quite a limited distribution, and mainly breeds around the North Atlantic Ocean, where it prefers to nest on rocky shores. In the last few years, a few pairs have bred around the Wadden Sea. The numbers in the Wadden Sea peak at more than 15 000 in autumn.

The other gulls have a much wider distribution and are also much more abundant in the Wadden Sea. They form large breeding colonies in the dunes and salt marshes on almost all the islands in spring. The black-headed gulls and herring gulls are particularly abundant and many thousands of pairs nest; in total around 130 000 pairs of black-headed gulls and some 80 000 pairs of herring gulls. There are also 37 000 breeding pairs of lesser black-backed gulls and 10 000 pairs of common gulls. In spring these four very noisy gull species occur in large numbers almost everywhere in the Wadden Sea.

Lesser black-backed gulls migrate to Africa and winter there, but the other gulls do not migrate far. The breeding birds from the Wadden Sea don't fly more than hundreds of kilometres from their breeding colonies and mostly stay in the vicinity of the North Sea. There they are joined by breeding birds from Scandinavia and further east. These eastern gulls often migrate through the Wadden Sea and on further to the west and south. Others use the Wadden Sea as a wintering area.

This inflow of northern birds can create huge numbers of gulls on the tidal flats in autumn. There can be 240 000 black-headed gulls plus more than 100 000 common gulls and 330 000 herring gulls present. In winter these numbers decline to 40 000, 58 000, and 157 000 respectively and when temperatures drop well below zero, they halve again, as many gulls fly further south and west, mainly to England and Ireland.

*Food*

Originally, black-headed gulls and common gulls lived inland and foraged in marshes and grasslands. There they feed on earthworms by trampling the grass with their feet, and they are also agile enough to catch flying insects. Over the last century these opportunists have settled increasingly in coastal areas and, like the herring gulls, learnt to feed on the abundant benthic animals such as worms, shellfish and crabs, and to hunt fish and shrimps in shallow water. Herring gulls and lesser black-backed gulls will go further into the North Sea where they can catch fish from the surface waters. But the number of prey caught in this way is very limited. Herring gulls in particular are very dependent on the waste from fishing boats.

In harbours, cities and on rubbish dumps gulls can find even more garbage. When bad weather makes it difficult to feed on tidal flats and in the North Sea, most gulls will temporarily return to inland food sources.

Gulls use a whole arsenal of methods to collect food on the tidal flats. Benthic animals are sometimes tracked by walking gulls, or flushed out by the gulls quickly trampling the sediment with their feet. The shallow trampled pits seen everywhere on the tidal flats bear witness to this. Small shellfish are swallowed whole, shell and all. Larger shells are dropped from a great height onto a hard surface to break them open. Gulls will make shallow dives to catch swimming prey, and regularly reveal themselves to be successful thieves. Gulls often steal prey from waders or terns, and in spring they specialise on stealing the eggs and chicks of other birds, including the smaller gulls.

*Reproduction*

Gulls are also very opportunistic when it comes to choosing a nest site. In the Wadden Sea area they nest in marshes or sand dunes. Gulls prefer to use dunes over marshes, particularly on islands where there is no risk of ground predators such as foxes, as there is less risk of nests being flooded by an overly high tide. All gull species nest in colonies. In black-headed gull colonies the nests are sometimes only 1 m apart. Herring gulls and lesser black-backed gulls sometimes form mixed colonies. Black-headed gulls often nest with terns.

The nest is sometimes little more than a shallow scrape, but usually both parents enthusiastically collect nesting materials, mostly vegetation.

### 2.2.30 Terns

Four species of terns are common in the Wadden Sea:
- **Sandwich tern** *(Sterna sandvicensis)*
*Dutch:* Grote stern. *German:* Brandseeschwalbe.
*Danish:* Splitterne.
- **Common tern** *(Sterna hirundo)*
*Dutch:* Visdief. *German:* Flussseeschwalbe.
*Danish:* Fjordterne.
- **Arctic tern** *(Sterna paradisaea)*
*Dutch:* Noordse stern. *German:* Küstenseeschwalbe.
*Danish:* Havterne.
- **Little tern** *(Sterna albifrons)*
*Dutch:* Dwergstern. *German:* Zwergseeschwalbe.
*Danish:* Dværgterne.

*Appearance and general behaviour*

Terns, like gulls, are largely white, with pale grey on the upper wing. They are much smaller (24 - 40 cm) and more slender than gulls, have longer, narrower wings, and are more graceful in flight.
All terns have a distinct black cap. Some species can be distinguished by their bill colour: the sandwich tern has a black bill with a yellow tip, while the little tern has a yellow bill with a black tip. The Arctic tern and the common tern both have a red bill. The bill tip of the common tern is dark, although this is not particularly useful as a distinguishing characteristic. Apart from their bills, the two species can only be distinguished by very subtle differences in their size and colour, and are difficult to differentiate.

*Distribution and migration*

Within Western Europe, three of the four tern species are found only in coastal areas. The common tern is the exception; breeding mainly in freshwater areas, it also makes itself at home on the sea coasts of Western Europe. Its breeding range extends up to the north of Scandinavia. Little terns and sandwich terns have a limited distribution in Western Europe, extending north to Denmark and Scotland. The northernmost breeder is the Arctic tern, which breeds in a broad zone around the North Pole. The Wadden Sea is one of the southernmost breeding grounds for this species. In the far north Arctic terns use coastal and inland areas, both of which are still largely frozen over at the start of the breeding season. Like waders that breed high in the Arctic, Arctic terns have only a small window of time in which to rear their young. In August they leave the breeding area and migrate south along the coast and spend the winter in the waters around Antarctica. Arctic terns must complete this 50 000 km journey twice a year. The other terns also migrate south, but not as far. Many terns that breed in Western Europe winter on the coasts of West Africa.
The Wadden Sea is important to all four tern species, especially the sandwich tern: 17 000 pairs of sandwich tern nest in just a few very large colonies in the Wadden Sea. Common terns and Arctic terns are somewhat more

abundant overall. Their colonies are widely distributed and can be found on every island and along the coast of the mainland. In contrast the little tern is something of a rarity, with only 1000 pairs breeding in the Wadden Sea. It's only possible to admire the terns in the Wadden Sea in spring. In late summer they start to migrate slowly south with their young, and most have left by the end of September.

*Food*

All terns feed on small fish, which they catch by diving into the water from a reasonable height. The sandwich tern is the heaviest and can dive the deepest; they consequently can catch the largest fish. Sandwich terns prefer to fish in the coastal areas of the North Sea where they feed on small herrings in the surf. When this is not possible because of high winds, they use the more sheltered waters of the Wadden Sea. The little tern can only catch small fish from shallow water and can always find suitable feeding grounds in the Wadden Sea. The common tern feeds in both the Wadden Sea and in neighbouring freshwater areas. The Arctic tern feeds on shrimps, crabs, worms and squid as well as fish in the Wadden Sea. On their northern breeding grounds Arctic terns feed on crane flies and other small insects.

*Reproduction*

All terns nest on exposed coastal areas, mainly on open beaches and high sand flats, but also in salt marshes. The terns are safe from most land predators at these isolated breeding sites, but risk flooding during high tides. Terns nest in colonies, often seeking the company of other species. Sandwich terns are the best example of a colonial breeding bird. They nest in their thousands, just tens of centimetres from their neighbours and almost all start to nest on the same date.

The nests of both common and Arctic terns are usually at least several metres apart. Both species actively defend their nests and chicks, repeatedly dive-bombing while aggressively using their bills and claws to scratch the intruder. Little terns make their nests on large beaches, where they must face not only the risk of high tide flooding but also disturbance from human tourists on sunny days. Increased disturbance in recent decades has caused this small graceful tern to disappear from many beaches. Little terns now breed only in protected reserves. The terns usually lay two to three eggs in a simple scrape in the sand. The young remain in the nest for their first days of life, guarded by one parent, while the other brings food. Later the chicks hide amongst vegetation so that both parents can go foraging. Even when the young can fly, they continue to be accompanied and fed by their parents.

The Arctic tern is the only truly intertidal tern species. Arctic terns do catch fish, but they also eat benthic animals such as crabs, shrimps and worms. Here, a worm is fed to a newly fledged juvenile (shown on page 82).

In the Wadden Sea, sandwich terns nest in few very large breeding colonies, like this one on the Dutch island Griend (photo right). These colonies are often kilometres away from the deep water where these terns prefer to fish. They must fly large distances to gather fish for their young.

## 2.3 Characteristic attributes of waterbirds

Waterbirds can only exist on the tidal flats if they are able to survive in a saline and windswept environment with strong fluctuations in temperature. Each species has become specialised in catching and processing certain prey, requiring further external and internal adaptations. Many waterbirds migrate across vast distances between their Arctic breeding grounds and the wintering grounds. To be able to fly more than 5000 km in one burst requires pointed wings, trained muscles and a strong heart. In the following paragraphs we discuss some of the waterbirds' special adaptations, especially those found in waders.

### 2.3.1 Feet

Have you ever walked across a mudflat and stopped to take a look around as others struggle on? Did you notice that the long-legged lanky people tend to sink just a few centimetres into the mud, while fatter people with small feet sink over their ankles in the sludge? This is an important phenomenon for waterbirds. Avocets forage on silty tidal flats but have webbed feet that spread their weight and stop them from sinking too deeply. Sanderlings, as their name suggests, are usually found on sandy flats and beaches. They are actually missing a (fourth) hind toe, which gives them a high mass-to-footprint ratio (Figure 2.3) but can dash between the waves very quickly, searching for highly mobile prey on their tiny feet. When we look across all the Wadden Sea waterbird species, it's clear that species that weigh more have larger feet. (Figure 2.3A). Waterbirds such as curlews, sanderlings and golden plovers that spend most of their time on hard substrates have relatively small feet and short toes, and so have a relatively heavy footprint (Figure 2.3B). Species that use soft sediments have a light (*Tringa* species) or very light footprint (avocets and little sandpipers). Like the avocets, geese, ducks and gulls have webbed feet, but these latter birds probably mainly use them to propel themselves through the water.

### 2.3.2 Bills

Bills must meet very specific and often rather incompatible demands. Waterbirds have to move their bills through sand and silt, so the bills must be very strong and not get worn down. But at the same time the birds must use this bill to find prey, so they need delicate sensory organs on the bill surface. Once they find prey, they need to process it, at least to the point where it can be quickly swallowed. The design of a bill is an intricately designed compromise between all these demands,[998] something many engineers would drool over. Biologists go into ecstasies about bills because their design can teach us so much about the life and behaviour of the birds carrying them.[110, 160, 293, 423, 424, 840]

A bird's bill is built around its upper and lower jaw bones. Shallow pits containing the sensory organs are hidden on the bone surface. All of this is covered in a horny layer that resists wear by growing continuously, rather like our nails, only faster. The hardness of this horny layer in oystercatcher bills makes their growth easier to measure than that of any other waterbird species. The average growth is almost half a millimetre per day, three times the average nail growth of humans.[418] If an oystercatcher didn't continuously wear its bill on the sand and mud, after a year it would appear to be wearing a long orange carrot (Figure 2.4).

The bill of an oystercatcher is straight, as are the bills of plovers, red knots, sanderlings and redshanks. However the bills of many waders are slightly up- or down-curved (Figure 2.5). A straight bill is the best option me-

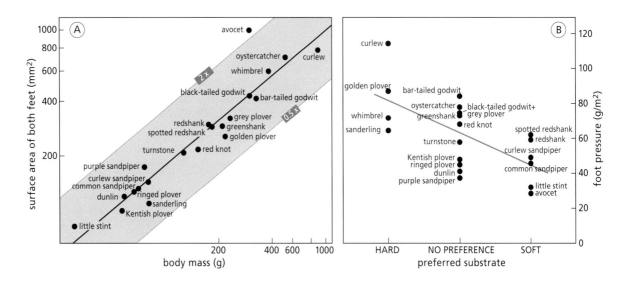

Figure 2.3 Waterbirds with 'small feet' mainly forage on hard substrates; waterbirds with 'large' or webbed feet mainly forage on soft silt. (A) The surface area of both feet in relation to body weight. The avocet stands out as its webbed feet give it more than twice the surface area expected based on its body weight. (B) The 'footprint' (mass divided by the surface area of both feet including the hind toe) in relation to habitat preference. Species on the left prefer harder substrates, those on the right softer. Species in the middle do not show a clear preference. These data were collected by P. Duiven.

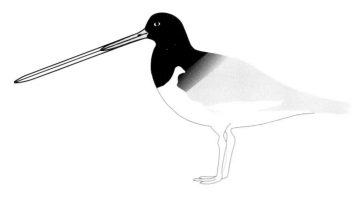

Figure 2.4 If an oystercatcher's quickly growing orange-red bill wasn't worn down by continual use, after a year the oystercatcher would look like this! (Based on a drawing by Jan B. Hulscher).

as its bill probes in the wet sand (Box 2.1).[696]

If researchers want to know which animals are hiding in the sediment, they scoop it up and wash it through a sieve. The silt and sand are washed through but the benthic animals remain on the sieve (Chapter 4). The shelduck can also use this technique because of the system of small ridges (called lamellae) on the underside of its upper bill. The shelduck scoops up the top layer of sediment and water and presses this against these ridges, retaining any bigger food particles in its 'sieve'. It then scrapes any foods, such as tiny mudsnails, off the lamellae with its tongue. The spoonbill doesn't have a sieve in its spatula-shaped bill tip, but the tip is stuffed full of very sensitive tactile organs. At the tiniest sign that a fish or shrimp is within its gaping bill, the spoonbill snaps its bill shut and, with a quick toss of its head, throws the prey down its throat.

chanically if you want to be able to probe deep and hard in the substrate. A bird wants to be able to penetrate the substrate with the least effort per unit 'firmness', while investing the least in the bill's 'structural material'. By minimising the structural size of the bill, space becomes available in the bill for a long tongue that can be used to transport prey quickly and easily from the bill tip to the mouth. The common snipe, woodcock and sanderling all use this strategy as, to a lesser extent, do the *Tringa* species. A curved bill is more useful for retrieving prey from around corners and down burrows and is how the whimbrel extracts small crabs from their holes. Curved bills are also useful if, like the curlew, you want to pull long soft prey such as ragworms from the mud or, like the avocet, need a large surface area to increase your chance of finding prey in soft silt. However, curved bills often require extra reinforcement, which doesn't leave much room for a long tongue. So birds with curved bills can often be seen tipping up their bills to get the prey to their throats. This makes it easy to tell when curlews and whimbrels are swallowing prey. Even species like the bar-tailed godwit have a reinforced bill and only a short tongue, so must tip up their bills a little to get the prey to their throats. In relatively straight-billed species such as godwits and most sandpipers the lower end of the upper mandible is hinged, which allows them to open the very tip of the bill while keeping the rest closed. This is rather handy for grabbing deeply buried prey.

Plovers and turnstones have short, strong bills which are rarely used to probe in the substrate. Turnstones use their robust bills to push aside seaweed and stones in search of any small crustaceans that might have been hiding beneath them. No one has been able to come up with a good explanation for the thickened shape of the plover bill. Species with a soft horny layer on the bill (i.e. not oystercatchers or gulls) have many sensory organs on both the outside and inside of the bill (hard-billed species only have sensory organs inside the bill). These small organs are always the same type, but can be used for very different purposes. Many species use them to detect prey by the vibrations that the prey make in the sediment.[295] The knot uses these organs to detect buried shellfish by measuring pressure fields generated

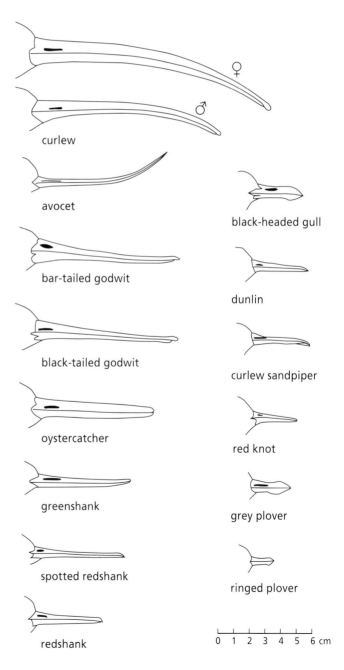

Figure 2.5 The bill shapes of waders and gulls found on the tidal flats.[961]

### Box 2.1 How knots can detect shellfish from a distance

Red knots can distinguish between trays of wet sand that contain small stones hidden 5 cm below the surface and trays without any stones. They do this by inserting their bill tips 0.5 cm into the sand for just a few seconds.[696]

This ability to detect stones hidden in the sand centimetres away is extraordinary, as the stones do not move, are the same temperature as the sand, have no electromagnetic field and no smell. This means that the knots aren't using any of the mecha-

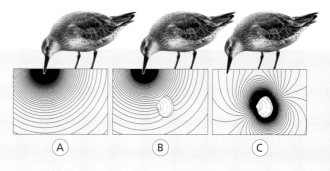

Figure B2.1.2 Hypothetical cross section through a tidal flat showing pressure fields created when a red knot probes the upper surface of the wet sediment. In (A) there are no hard objects present to disrupt the pressure field. In (B) a buried shellfish prevents the displaced interstitial water from rapidly moving away, resulting in increased water pressure around the shellfish that allows the knot to detect it from several centimetres away (C).

nisms for prey detection known in other birds and mammals. Knots were unable to detect stones buried in dry sand or in liquid mud, which suggests that the detection mechanism is based on an interaction between water, sand grains, buried items such as stones and shells, and of course the bill itself. Increased pressure created by the probing bill causes the water between the sand grains (called the interstitial water) to try to flow away. The buried stones simply obstruct this water movement.

Beneath the protective horny layer covering the knot's bill tip are numerous rows of forward-pointing tiny pits in the bone. Each of these pits contains clusters of 10 to 20 'Herbst corpuscles', small organs able to measure pressure differences (Figure B2.1.1). When knots probe with their bills in the sediment a spherical pressure field builds up around the bill tip, due to the inertia of the interstitial water. This field is disturbed by compact buried objects such as stones or shellfish, which prevent the water in the pressure field from dispersing (Figure B2.12). The knots appear able to interpret the information this disturbed pressure field provides and even amplify it by rapidly moving their bills up and down.

This very specialised sensory organ explains why knots can detect cockles and Baltic tellins much more efficiently than oystercatchers, which have to touch their prey in order to find them.[417, 690] This also explains why knots don't mind tidal flats that have low shellfish densities and why they only forage on moderately wet sediments. In dry mudflats the pores between the sand grains are filled with air rather than water. This air is easily dispersed, so even if stones or shells are present, no discernible pressure disturbance can be created. As soon as the retreating tide causes an area to dry out, knots fly to lower-lying flats. It also explains why the knots do not like flats that are too wet. Sand grains are more buoyant and float in very wet sediments. This means that in very wet sand repeated probing will not lead to an increased pressure field and any stones or shells will remain undetected. As knots cannot distinguish between buried stones and shells, it is logical that they never forage on 'unsorted sediments' (sediments containing a mixture of different-sized sand grains and stones), even if suitable shellfish are living amongst the stones. As the knots cannot distinguish between living and inert buried objects, they would waste too much time hauling inedible items out of the sediment at such sites. They would be better off going somewhere else.

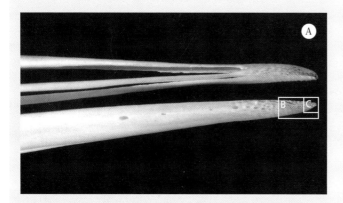
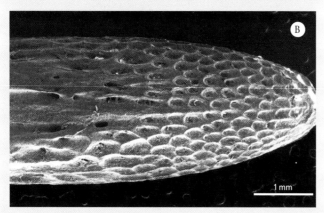
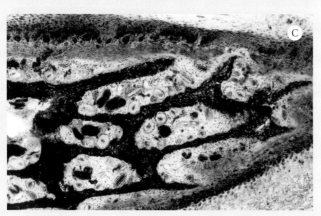

Figure B2.1.1 The red knot's sensitive bill tip. (A) The bill tip after removal of the horny layer and soft parts. (B) The Herbst corpuscles lie inside these tiny, tear-shaped pits in the bone, covered by a protective horny layer. Herbst corpuscles register subtle water pressure differences in damp sand, allowing the knots to detect buried shellfish from a few centimetres away. (C) Every tiny pit contains a group of onion-like Herbst corpuscles encircling a central nerve (coloured black).

These knots probe the top layer of the mudflat with a very sensitive measuring device (upper photo), specially designed for finding buried shellfish – their bill tips (see Box 2.1). Turnstones (middle photo) mainly use their stout bills to turn over small stones and seaweed. They watch for any amphipods, shrimps and crabs revealed by their delving. Oystercatchers can use their chisel-like bills to skilfully wedge open mussels (lower photo). They can also cut with their bills to neatly detach bivalve flesh from shells.

### 2.3.3 Eyes and ears

Plovers have bigger eyes than sandpipers, geese and gulls. This makes plovers better than other waterbirds at visually searching for food at night.[547] Bigger eyes enable them to catch more of whatever light is present. In addition, plovers have a greater absolute density of rod cells in their retinas than sandpipers do. These rod cells allow the plover to register shapes and movement even when the light is poor. There is also a stronger relationship between the numbers of rods and cones, (cones are used for seeing colours and seeing clearly in very bright daylight) in plovers than in other waterbird species.[740]

Plovers probably also use their ears to find their prey. You may have looked at ostriches in the zoo and noticed the holes on both sides of their scarcely feathered heads. These holes are the ostriches' ears. Waterbirds' ears are rather similar, but they are kept well hidden under their head feathers. Instead of an external ear, waterbirds have skin folds which seem to perform a similar function.

Waterbirds can use their 'hidden' ears to hear better than we humans can with our obvious ears. It's very likely that waterbirds can hear very low sound frequencies (below 10 Hz).[5] With this sort of hearing ability, waterbirds in the Wadden Sea should be able to hear approaching storms west of Ireland!

### 2.3.4 Intelligence

It's truly amazing that knots can detect the presence of buried prey from very subtle pressure differences in the wet sand. This requires not only very accurate pressure sensors but also the mathematical ability to quickly and appropriately process the information received by the bill tip. Bar-tailed godwits, sanderlings and other sandpipers interpret and react on vibrations made by quickly moving or buried prey. This indicates that, like the knots, their brains are well developed for processing stimuli. It's rather satisfying for biologists that the brains of the touch-hunting sandpipers and visually-hunting plovers are very different in structure. It is especially noticeable that the very front part of the cerebrum is much bigger in sandpipers than in plovers.[638] This 'trigeminal expansion' is the main area where the nerve stimuli from the bill tip are processed. In this way the information received from the bill can lead to the appropriate behaviour that allows the bird to capture the prey it has just detected. Waterbirds' foraging behaviour depends not only on their bills but also on their brains.

### 2.3.5 Digestion

When a bird specialises in a particular habitat on a particular food source, it doesn't just need the right sort of feet. It also requires suitable eyes, ears and brain for finding its food and a suitable bill to enable it to consume the food it finds. However these are not the only adaptations required; birds also need their digestive systems to be tuned to the foods they consume. Some birds, like the brent goose, consume vegetation (seagrass, seaweed and salt marsh grasses) that must then be converted to commonly-used nutrients (sugars, proteins and fats). Other birds, like eider ducks, purple sandpipers and red knots, feed on whole shellfish or, like the oystercatcher, only consume shellfish flesh. Still others, like the avocet, little stint and curlew sandpiper, feed on soft worms.[454]

Let's start by looking at the shellfish eaters. An oystercatcher can use its strong bill to break through the prey's shell or cleverly prise the valves apart, so that it can remove the flesh. The diet of eider ducks, purple sandpipers and red knots consists mainly of shellfish, which they swallow whole. These birds must crush the shells in their stomach to reach the juicy flesh. Perhaps it's not so surprising that purple sandpipers and red knots have heavily muscled stomachs, twice as big as those of other waders studied (Figure 2.7). The oystercatcher's stomach is somewhat smaller than average, but the stomach masses of most other waders lie reasonably close to the expected values. An obvious exception is the avocet's stomach, which is only half the expected size. This is because avocets in the Wadden Sea only eat soft prey. Species that eat soft prey or shellfish meat are able to start digestion in the first part of the stomach (the proventriculus). In contrast, birds like the knots and sandpipers that use their muscular stomachs to crush shells[687] must leave all the digestive work to their intestines. This may explain why the intestine mass of purple sandpipers and red knots is higher than that of other species, and also explain the strong relationship between relative stomach and intestine masses (Figure 2.7). As well as their increased digestive capacity, the intestines of purple sandpipers and knots need to be tough, which may also make them heavier. The crushed and sometimes sharp shell fragments must move through the gut before being passed out. The compact, calcium-rich droppings of these birds are easily found and contain an invaluable archive of dietary information.[171, 686] The birds pay a high price for eating whole shellfish: they must carry relatively heavy digestive organs.

### 2.3.6 Salt glands

A shellfish diet may be even more costly.[471] The salt content of snails and bivalve flesh is as high as that of seawater. This means that as well as consuming many indigestible shell fragments, eider ducks and red knots must take in a lot of water and salt. All the shell fragments and the saltwater are quickly removed via the intestine, but it is not possible to absorb large fat, sugar and amino-acid molecules through the intestine wall, while simultaneously excreting all the tiny salt molecules. This means that salt levels in the blood of shellfish-eating birds could get dangerously high after a meal, if the birds hadn't found a cunning solution to this problem too.[626,

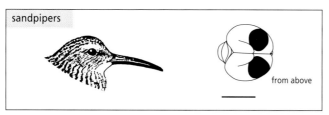

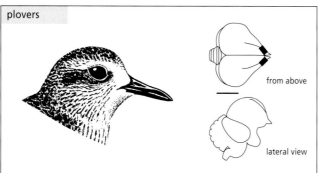

Figure 2.6 The brain structure of a sandpiper (upper figure) and a plover (lower figure). Viewed from above, it is possible to compare the size of the areas associated with foraging. The black shading represents the parts of the brain directly connected to the bill tip. This area is much larger in the tactile-foraging sandpipers than in the visually-hunting plovers.[638, 658]

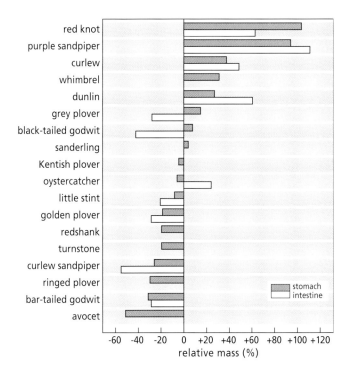

Figure 2.7 Relative fresh masses of the stomach and intestine of the (carnivorous) waders of the tidal flats ordered by increasing relative stomach mass (those at the bottom have the lowest relative mass). Relative mass is the difference between predicted and actual organ masses. Both stomach mass and intestine mass increase in proportion to body weight (in statistical terms: the allometric exponent = 1), enabling each species' predicted stomach and intestine mass to be easily calculated from its average body weight.

[800] On top of the bird's skull, on the ridge between the eye sockets are two small glands. These 'salt glands' filter salt molecules from the blood and produce a thick, very salty liquid that is excreted through the nostrils. Have you ever noticed a drip hanging off a herring gull's bill? It's the gull's way of excreting the salt-rich liquid produced by the salt glands.

Like the digestive organs, the mass of the salt glands increases in proportion to body mass (Figure 2.8), and just like the digestive organs, the salt glands of purple sandpipers and knots are (almost) twice as heavy as average. Oystercatchers and brent geese also have champion salt glands: their salt glands are more than twice as large as would be expected based on their body mass. Oystercatchers probably take in a lot of salt with shellfish flesh, as would brent geese when they forage on seagrass, seaweed or salt marsh vegetation. A comparison with species that live mainly in freshwater habitats (in this case golden plovers and black-tailed godwits from inland areas) shows that all intertidal birds have relatively large salt glands,[800] even though they are, by necessity, somewhat larger in some species than in others.

## 2.3.7 Plumage

They say that clothes maketh the man, and this is even more the case for waterbirds' feathers. Without wing feathers, waterbirds could not fly. Without body feathers, they would die of cold. The dense feather and down layer of eider ducks and brent geese insulates them so well, that only when temperatures drop below zero do they need to use extra energy to maintain their body temperature (at 41 °C).[437, 936] For an oystercatcher this limit lies at around 10 °C, but the smaller waders will need to increase their energy expenditure once temperatures drop below 20 - 22 °C.[460, 466] This means that all of these small species find it very cold on the Dutch tidal flats, where the mercury seldom rises beyond 20 °C! Purple sandpipers are near the southern limit of their winter distribution in The Netherlands and have heavier and more downy plumage than other sandpipers.[817] They seem able to withstand colder conditions.[818]

Plumage is necessary to keep the bird's warm body insulated, but what determines the colour and pattern of its feathers? In many cases the bird wants to be inconspicuous and has feathers that provide camouflage. This is why most birds on tidal flats in winter are dull grey. But if sex and violence come into play, it often pays to be conspicuous. Waterbirds that are searching for a partner, or have to defend their partner against rivals, often have an area of coloured feathers. Male eider ducks start to adorn themselves before winter has even arrived, and are only an inconspicuous grey-brown in summer when the females are incubating the eggs and taking care of the young. From late summer onwards, noisy males can be seen dipping their heads to the females, in their beautiful black and white plumage with green neck patches.

The males and females of most waders that breed on the taiga or the tundra, only put on their striking rusty-red or black plumage in late spring (Box 2.2). This breeding plumage is probably not meant to provide camouflage up in the Arctic[447, 660] as the best way to find a nesting red knot on the tundra (as in the picture) is to search for its rusty red breast. Individual birds try to impress others with their beautifully coloured feathers. By producing such beautiful plumage, they show other birds

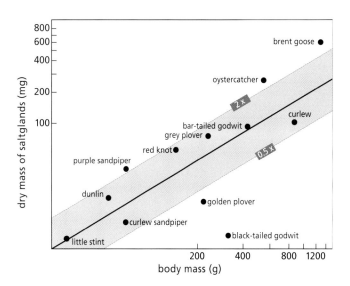

Figure 2.8 Salt gland mass in relation to body mass for brent geese[810] and waders. The area where organs are between half and twice the average mass is show in grey.[650]

how 'healthy' they are, in spite of the demands made by the long migration north.⁶⁷⁰ Large, aggressive species like the shelduck, oystercatcher and avocet don't need colourful plumage. In Western European tidal areas, they are a conspicuous black and white year-round. In gulls, the need for camouflage seems to disappear with age; only the plumage of young birds is brown or grey; the older birds are predominantly white.

The horn-like material that feathers are made of doesn't last forever, so feathers, and particularly parts that are white, wear down with use. This can cause the colour of a bird to slowly change; for example, as the white edges wear off their back feathers, young sandpipers lose their speckled look and change into more uniform plumage. However to change into a really different colour the bird has to replace its feathers, which is called moult. A new feather (the pin) grows from a small gland that lies in the skin at the base of the old feather. As the new feather grows bigger, it pushes the old one from the skin.³⁰¹ Such a new feather is rather like a butterfly emerging from a cocoon: beautiful, fragile and initially very vulnerable. This is why the moult, and especially that of the very important primaries, usually takes place in safe, food-rich areas, so that the risk of 'misuse' (for example, trying to escape from a predator with half-grown primaries) and damage is as small as possible. Like all waterfowl, shelduck replace all their primaries at once, making them temporarily flightless. Most Western European shelducks seem to agree that the quiet tidal flats around the mouth of the River Elbe are the best places to moult.⁵⁹⁶, ⁸³⁸ Many thousands aggregate there for a few weeks in late summer. In September, as soon as they can fly they spread out over the whole Wadden Sea.

Waders usually only replace their primaries once they've arrived on the wintering grounds (Box 2.2), which can be either the Wadden Sea or elsewhere in Western Europe or in West Africa.⁸⁴ Even though they must replace all their feathers including the primaries, they don't do it all at once and become temporarily flightless as shelducks and eider ducks do.⁶⁰⁶ The moult into the breeding plumage only involves contour feathers and also usually takes place on the wintering grounds just before migration to the northern breeding areas.⁶⁷⁰ Waterbirds do not moult on migration. Clearly, it's not very sensible to fly long distances on new feathers.

Every day the feathers are preened to keep them in good condition, and coated with a fatty substance from a gland at the base of the bird's tail. This 'preen gland' produces wax-like fats that are squeezed out through a sort of little nipple.⁴³² Birds quickly wipe their heads over the preen gland to coat their head feathers. They also use their bill to get wax from the gland which is then smear-

Red knots have conspicuously rusty-red underparts in the breeding season. This bright colour makes it possible to find the knot breeding here on a stony part of the Canadian tundra.

It really is a miracle that small and very warm-blooded creatures like this ringed plover (top right), grey plover (middle), and dunlin (lower right) are able to withstand the extreme weather conditions they encounter through the year. Their secret lies in their highly water-resistant plumage that they replace once or twice a year. The grey plover's breast feathers are in very heavy moult and the pins of its newly developing black feathers are very obvious. The dunlin is moulting its primaries. A gap is visible in its right wing where the growth of new primaries is just starting.

ed over the rest of their plumage, including the primaries which are coated one by one.[728] A waterbird in the rain is someone to be envied, as the water-resistant wax from its preen gland keeps its plumage completely dry.

## 2.3.8 Flight equipment: wings, flight muscles and heart

A bird that has to cover large distances at high speed is best off with long pointed wings.[111, 609] This means that the outermost primary forms the wing tip. This wing shape is found in most waterbirds regardless of whether they are waders, gulls, terns or waterfowl. It has the disadvantage that the bird is not very manoeuvrable, but this is not usually a big problem in the open tidal landscape.

## Box 2.2 Feather moult in waders

During winter (October to March), most adult waders in the Wadden Sea have grey plumage, but in summer (May to August) these birds are adorned with beautiful black, white and rusty-red feathers. The birds gradually moult their body feathers between March and May, temporarily causing them to wear rather spotty transitional plumage. In autumn, the birds replace both their body feathers and their flight feathers. The flight feathers of the wing (10 large primaries on the outside and 10 smaller secondaries closer to the body) moult in two waves, starting from where the primaries and secondaries meet. One by one new primaries start to grow, until finally the tenth primary (the wing tip) is replaced. As soon as the first (innermost) primaries are full-grown, the first (outermost) secondaries begin their moult, until finally the secondaries closest to the bird's body are moulted. It takes 80 to 100 days until both moult waves reach the last feathers and moult is complete. Wing moult occurs in such a way that the gaps in the wing are never large enough to seriously reduce the bird's flight abilities and manoeuvrability. Both wings are usually moulted in parallel to prevent the birds being unbalanced in flight.

In moult studies, the primaries are numbered 1 to 10 from the inside to the outside, the secondaries numbered from the outside to the inside, and the age and relative length of each flight feather is recorded. An old feather is scored as a 0, a recently lost feather or a small pin is scored as 1, a feather that has reached a quarter of its final length as 2, a feather that is half grown as 3, a three-quarters grown feather as 4, and a fully grown new feather as 5. The scores for all of the primaries on one wing are added up to produce the primary moult score, which increases from 0 in September to 50 in November.

Figure B2.2 Seasonal changes in the plumage of a male grey plover moulting and wintering in the Wadden Sea (right), and an illustration of the wing feather moult that occurs from September to November (left).[788]

The common sandpiper is a clear exception to the rule, but its inclusion as a waterbird is rather marginal anyway. Common sandpipers have rounder wings and, like lapwings and woodcock, fly more slowly and in a different way than waders of the open tidal flats. Common sandpipers have a rather slow discontinuous wing beat and whenever they stop flapping their wings they start to glide downwards. Their body shape and way of flying makes common sandpipers, unlike other waterbirds, very manoeuvrable. This is especially useful if they are surprised by an owl or bird of prey while in their more densely vegetated breeding and wintering habitats, which are usually along rivers and streams. Common sandpipers then throw themselves into the water and try to escape their predator by swimming underwater.[940] Waterbirds are usually vulnerable to surprise attacks, which might be one of the reasons why they prefer open habitats.[686]

As well as wings, waterbirds need powerful flight muscles and the machinery to provide good blood circulation to keep muscles going. The heart pumps the blood around, making it an important part of that machinery. Waterbirds usually have relatively large flight muscles and a big heart. The size of the wings is unchangeable, but by adapting the size of the flight muscles and heart to the mass that must be carried in flight, individual waterbirds can keep their 'flight costs' as low as possible.[673] Red knots from the Wadden Sea show a large variation in the size of their flight muscles depending on their body mass. These muscles can weigh as little as 10 g, or at other times (and masses) up to three times as much (Figure 2.9A). Reducing the required muscle power by having a lower body mass also means that the amount of blood that has to be pumped around can be reduced. And indeed, the size of the heart also varies by a factor of three (Figure 2.9B).

## 2.3.9 Reproductive organs

None of the bird's organs varies so greatly through the year as its reproductive organs.[586] The mating season is very short so there is no reason to keep the testicles and follicles in a state of readiness. The bar-tailed godwit is one of the few species whose changes in reproductive organs are reasonably well known (Figure 2.10). Research has shown that the males don't start to enlarge their testicles until the very last moment, some 10 days before their departure for the Siberian tundra on the 1st June. Even though the mass of the testicles at the time of de-

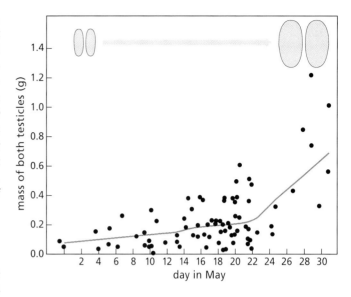

Figure 2.10 Rapid changes occur in testicle size of bar-tailed godwits after 20 May, around 10 days before they leave the Wadden Sea for the Siberian breeding grounds. The line shows the average testicle mass through May. Based on data from catching casualties collected over a 15-year period.

parture is already four times the winter mass, they still aren't able to function yet. Bar-tailed godwits are already beautifully coloured and will sometimes farewell the Wadden Sea singing display songs, but at the time of departure from the Wadden Sea they are not yet fertile. It's generally accepted that other northerly breeding waders also try to minimise weight and flying costs by postponing the ripening of their reproduction organs for as long as possible.

Of course they quickly make up for this 'backlog' as soon as they arrive on the breeding grounds, and the ripening of their reproductive organs isn't the only thing happening at full speed. The sperm of avocets, oystercatchers, gulls and terns consist of neat straight rods with a little tail,[729] but those of bar-tailed godwits and other sandpipers are spiral-shaped. It's possible that long spiral-shaped sperm can reach higher speeds that straight sperm. This suggests that there is sperm competition in some sandpipers; male sandpipers try to use fast, spiral-shaped sperm to overtake their rival's sperm in the race to the egg. You can find out more about this and other fascinating aspects of waterbird reproduction in Chapter 5.

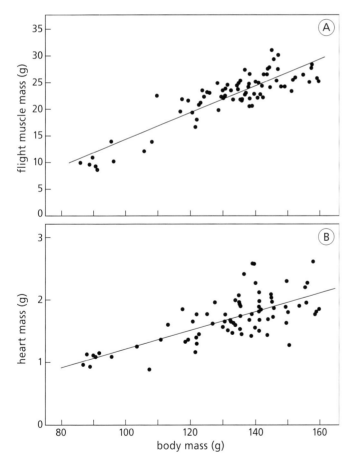

Figure 2.9 (A) Flight muscle mass and (B) heart mass in relation to body mass of red knots in Western Europe. Based on data from frost victims from the Dutch Wadden Sea and the Eastern Schelde, and lighthouse victims from Westerhever in Schleswig-Holstein (the German Wadden Sea).

# 3. Migration

## 3.1 Wadden Sea: migration crossroads

### 3.1.1 Central Station Wadden Sea

When we plot the flight paths of migratory birds, spun like a web around the world, (Figure 3.1), we see that many paths cross in the Wadden Sea.[83] This isn't surprising, as the Wadden Sea is a large, strategically positioned area. The extensive tidal areas of the Wadden Sea lie midway between the vast northern breeding grounds and the southern wintering grounds. Not far to the north are the taiga and tundra areas of North America and Eurasia (Figure 3.2). Less than 5000 km to the south are two of the main West African wintering sites for European birds, the Banc d'Arguin in Mauritania and the Bijagós Archipelago in Guinea-Bissau. Because of this, the Wadden Sea acts like a funnel, through which millions of birds pass on northward and southward migrations.[788, 789, 831] It is like nature's version of 'grand central station'.

### 3.1.2 Flyways

The flyway in which the Wadden Sea plays such a crucial role is called the 'East Atlantic flyway'.[17, 788] A flyway is a general term describing a collection of species-specific migration routes. Flyways are not precisely defined, be-

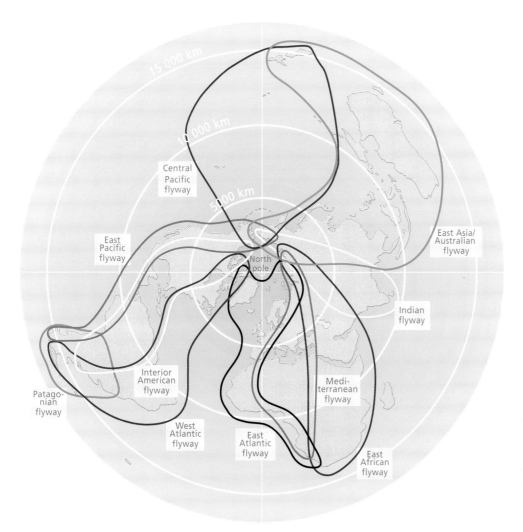

Figure 3.1 The routes of the 10 main flyways worldwide, as seen from above the arctic and boreal breeding grounds. A flyway is rather an abstract concept, comprising many overlapping species-specific flight paths used by waterbirds on their southward and northward migrations. This map projection displays the rest of the world's position relative to the North Pole and from this perspective all directions shown and distances displayed are accurate.

cause migration routes vary between species, and the borders are incompletely known. The 10 large flyways currently recognized show greatest overlap in the 'source' area (Figure 3.1).[582, 583, 624, 643, 679, 813] All of these flyways 'start' in the boreal and arctic breeding areas around the Northern Arctic Sea. At these sites, it's sometimes unclear which flyway local breeding birds use (Figure 3.1). In the eastern part of the Taimyr Peninsula in Russia, you can find curlew sandpipers that winter in Australia sitting beside sanderlings that winter in Southern Africa[886] and brent geese that spend the winter in the Wadden Sea.

There are large differences in the mix of species and the number of birds that use the different flyways.[154, 158, 299] But in spite of the overlapping nature of flyways, and the vagueness that sometimes accompanies this, flyways are important and useful concepts. Researchers who handle the same individual birds at different sites along a flyway exchange information, gaining far more than bird ringers working in isolation. These recaptures of ringed birds have led to many long-term professional relationships between researchers.

Being able to identify the flyways is very important for conserving these natural areas. In North and South America, protection of wetlands through the Western Hemisphere Shorebird Reserve Network (WHSRN) is based largely on their use by migratory shorebirds.[151, 377, 593] Similar reserve networks are now being created in Asia and Australasia (East Asian-Australasian Shorebird Site Network),[912] and in Europe and Africa (Africa-Eurasia Waterbird Agreement, AEWA). The aim of the latter agreement is to facilitate species conservation and management throughout the flyway.[85, 779] Under the framework of this agreement it will hopefully be possible to monitor and protect the wader and waterfowl populations of the East Atlantic, Mediterranean and East African flyways.

### 3.1.3 Cryptic travellers

Outside of the breeding season, most waterbird species gather in large flocks in tidal areas. However in the Wadden Sea some species only occur in low densities, which make them much less conspicuous. The common sandpiper is one such 'cryptic traveller'. As their name suggests, they are very common (and are among the most numerous breeding waders in Europe[653]), but are seldom seen, and never in large numbers. Common sandpipers, and related *Tringa* species such as green and wood sandpipers, hop from one small wetland to another. Like migrating songbirds, they only remain at each site for a short time. Their cryptic way of life means they often escape the attention of researchers,[738] and in doing so, conservationists. 'True' intertidal birds and species such as spoonbills depend on just a few, very specific, coastal areas for their survival. Most of these coastal areas are

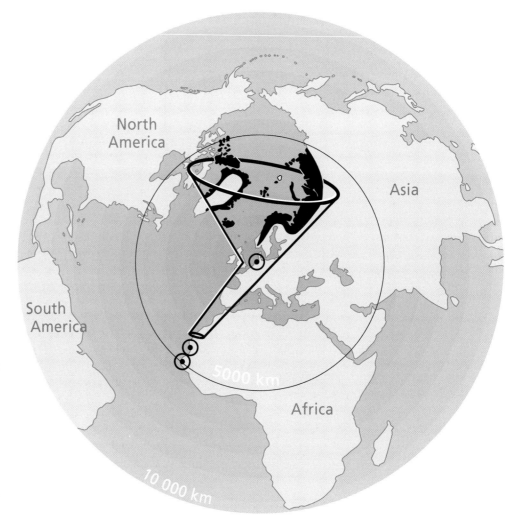

Figure 3.2 The Wadden Sea's strategic location is rather like the neck of a funnel running from the extensive arctic breeding grounds of North America and Eurasia down to the two large tidal areas of West Africa. This map shows the position of other countries relative to the Wadden Sea, and from this perspective the directions shown are accurate. The distances displayed on the map are distances to that point from the Wadden Sea following the great circle route.[364]

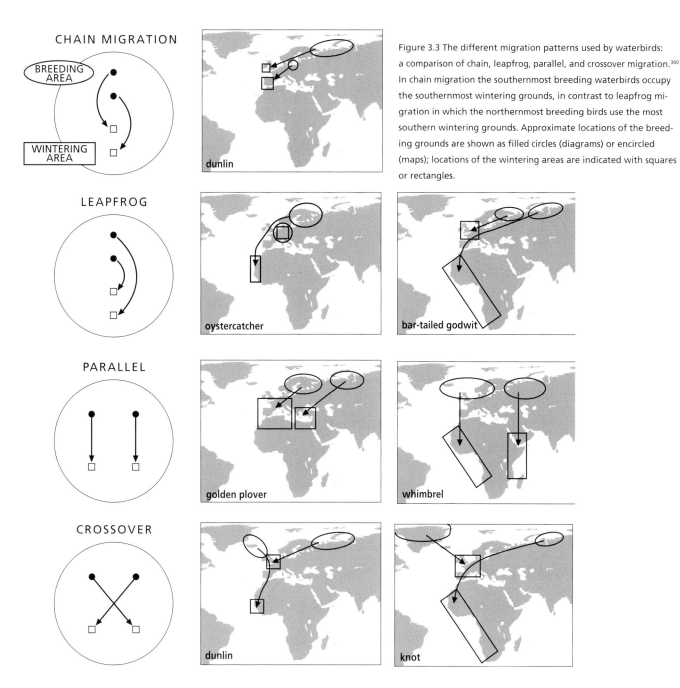

Figure 3.3 The different migration patterns used by waterbirds: a comparison of chain, leapfrog, parallel, and crossover migration.[360] In chain migration the southernmost breeding waterbirds occupy the southernmost wintering grounds, in contrast to leapfrog migration in which the northernmost breeding birds use the most southern wintering grounds. Approximate locations of the breeding grounds are shown as filled circles (diagrams) or encircled (maps); locations of the wintering areas are indicated with squares or rectangles.

threatened to some degree. The scarcity of suitable habitats makes these species easy to study, but also means they are most in need of this attention. This is discussed further in Chapter 6.

### 3.1.4 Migration patterns: Chain or leapfrogging patterns

Seen on a large scale, flyways are the routes northerly breeding birds use to move south as winter approaches, but it's not always a simple trip south.[269] Such 'simple' or 'chain' migrations are actually very rare in waterbirds.[188, 360] The only birds on the East Atlantic flyway that make that sort of migration are dunlins from Northwestern Europe and Western Siberia (Figure 3.3). In many species, the most northerly breeding populations winter further south than more southerly breeding ones. Oystercatchers, ringed plovers, redshanks, curlews, and bar-tailed godwits all show this 'leapfrog' pattern (Figure 3.3).[188, 360, 425, 643, 761]

After a stopover in the Wadden Sea, Siberian-breeding bar-tailed godwits overwinter in West Africa, while godwits that bred in Lapland and Northern Russia spend the winter in Western Europe.[188, 220] Other species have parallel migration patterns, in which birds that breed at similar latitudes but different longitudes each migrate to their own southerly wintering areas; golden plovers and whimbrels are good examples of this (Figure 3.3). In some species, populations from different breeding areas cross over each others' flyways en route to the wintering grounds. In the East Atlantic flyway, red knots from the northeast (Siberia) migrate the furthest south to West Africa, and the red knots from the northwest (Canada and Greenland) winter in Western Europe (Figure 3.3). Dunlins do the opposite, as northeasterly breeding birds winter in Western Europe, while northwesterly breeders (Iceland and Eastern Greenland) continue to migrate south to West Africa.

Bar-tailed godwits are champion long-distance migrants. They possess not only the long pointed wings of true 'athletes', but also a great capacity to store fat, the fuel for flight, both subcutaneously and in the abdomen.

At first glance, 'leapfrogging' seems a very illogical migration strategy. Why should birds from the far north migrate all the way to the tropics? Why should birds that live in the coldest and most energetically 'expensive' sites for the breeding season also undertake the longest flights and so have the highest flight costs? To answer these questions we need to understand more about migratory behaviour, energy consumption, and other physiological and ecological limits to the existence of waterbirds. In the second to last section of this chapter we will return to this.

## 3.2 Flight travels

### 3.2.1 Flight techniques

Species like brent geese, waders and terns that make long flights between their breeding and wintering grounds, have narrow, pointed wings. The technical term for this is having a high 'aspect ratio' which is the square of the wingspan divided by the wing area.[609, 628, 630] Birds with these type of wings fly both fast (usually 50 - 70 km per hour and sometimes more)[604, 695, 985] and relatively efficiently.[537, 609, 719] This makes them more than capable of reaching their destinations on time. Other waterbirds (like the spoonbill and shelduck) have much lower aspect ratios and so migrate shorter distances.

A more efficient way to cover long distances is to use thermals (the relatively warm air which rises from sun-heated land).[384, 455] Thermals are often used by large birds such as birds of prey, cranes and storks, which have relatively short and broad wings.[384] These birds have low aspect ratios, but use the thermals to gain height. By rising to the top of a thermal, then slowly soaring down and forwards while searching for the next thermal, they transform the energy gained from their elevation into distance travelled. The disadvantage of soaring is that you can only migrate at midday (once the sun has heated the land to create thermals), so journeys tend to take a long time.

Even though the Wadden Sea waders are not shaped for soaring, they can still do it, particularly on descent from high altitudes (such as on arrival from a long migratory flight) or just before landing at a high tide roost. In these situations they hold their wings stiff and use gravity to spiral down. On rare occasions, waders soar rather than actively flapping their wings on the ascent of a long-distance flight. Red knots departing from a fjord in Southwest Iceland (en route to the tundras of Canada and Greenland), must cross a 700 m mountain range. In 1987 and 1988 groups of 50 to 100 knots were seen mak-

Mid April in Mauritania, most of these bar-tailed godwits, red knots and grey plovers are ready to fly north.

Late May in the Wadden Sea, these bar-tailed godwits have beautifully coloured plumage. Males and females can be easily distinguished in flight.

ing good use of the local sea breeze that blew up over the mountains.[14] The knots circled at high speeds alongside the mountains, until they could soar over them. In 1992, the weather conditions were completely different, and the birds did not do this.[386] Although waders that set off on northward migration from the Banc d'Arguin in Mauritania sometimes seem to be using the thermals above the warm Saharan sand[695] they must still flap their wings to gain altitude.

### 3.2.2 Flight costs

To understand the distances that migrating waterbirds cover and how long it takes them, we need to examine their flight costs. Flying faster means the destination can be reached sooner, but it's more energetically expensive per unit time than flying slowly. A heavier bird (with larger fat stores) might be able to fly further, but the flight will cost more per time unit than for a lighter bird

**Box 3.1 Do shorebirds make decisions?**

Throughout their lives, birds and other animals must continuously respond to their environment. Their behaviour has far-reaching consequences and is of vital importance during migration and reproduction. Biologists researching this area tend to speak of 'decisions being made' rather than 'observed behaviours', and like to determine whether the choice that their study animal has made is optimal or not.

This does not mean that they believe that the bird has an overview of all possible alternatives and consequences, which it uses to make a conscious decision. The biologist is only attempting to rank the alternatives and see whether the bird makes the 'right decision'. For him or her it is simply a way of understanding why the bird is doing what it does. Evolutionary theory determines what the correct decision is, or in other words, natural selection is the most important force shaping animal behaviour: only animals that made the right choices have survived and successfully raised offspring. This theory gives researchers better insights into, for example, the migration strategies, food choices, and love lives of shorebirds.

Some real-life examples should help clarify this way of thinking. Imagine a curlew flying up when a peregrine falcon flies over. Why does the curlew do this? The first answer is that the curlew sees the peregrine falcon and flees. Right! The second answer is that the curlew flees because it does not want to be eaten. Also right! The first answer focuses on the direct causes of the curlew's behaviour: the image on the curlew's retina leads, via all sorts of pathways in its brain, to the avoidance behaviour. The second answer refers to the behaviour's function, or in other

words, the predictable consequences. Curlews that do not take flight are much more likely to be caught by the peregrine than curlews that fly up. This sort of continual natural selection against curlews that react inappropriately has shaped curlews' escape behaviour through the process of evolution. Their brain capacity determines what the birds can and cannot do. For example, it isn't necessary for the curlew to make complicated calculations about its chances of being caught, but it is necessary for it to recognise in time that the peregrine falcon is a dangerous bird. As well as clear thinking, this requires good vision.

In the following example we shall assume, for simplicity's sake, that female curlews play a decisive role in the formation of a pair bond. Imagine that a behavioural ecologist observes a female curlew form a pair bond with a male Arnold, when another male, Bernard, was also available. In this case the behavioural ecologist records that the female has chosen Arnold. The female had three possible behaviours: (1) pair bond with Arnold, (2) pair bond with Bernard, (3) no pair bond. Of these three behaviours the first (pair bond with Arnold) was displayed. If by choosing Arnold the female produces the most offspring possible, then she made the optimal decision. It is very possible, and even very probable, that our bird fell deeply in love with Arnold, but as yet there is no commonly accepted methodology to determine whether this occurred. Conversely it seems unlikely that the female had made numerous complicated calculations in her head about the expected number of offspring produced for every possible behaviour before she decided.

# MIGRATION

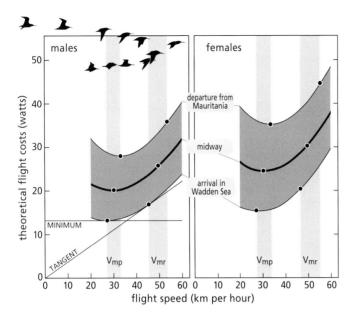

Figure 3.4 Theoretical flight costs of male (left) and female (right) bar-tailed godwits flying from Mauritania to the Dutch Wadden Sea in relation to flight speed (the air speed) and changing body mass.[669] On the left figure, the horizontal line intercepts the power curve at the male's minimum power speed ($V_{mp}$), the most efficient flight speed. The point where the tangent from the origin touches the same curve is the flight speed that costs least per distance covered, the maximum range speed ($V_{mr}$). Flight costs were calculated based on the bird's mass on departure from Mauritania (350 g for males and 430 g for females; upper curve), midway through the journey (middle curve), and on arrival in the Wadden Sea (214 g for males and 252 g for females; lower curve).

carrying less fat. To migrate as cheaply as possible, a bird must make the right 'choices' (see Box 3.1). To understand these choices we need to know about the factors influencing flight costs.

Let's start with the standard 'power curve'.[849] This curve describes the relationship between an individual's flight speed and the amount of energy that the individual needs to generate per unit time to remain at this speed. Note that the ground speed (which we see as a bird flies past) is not the same as the flight speed (which the bird achieves). If a bird is flying in windy conditions, it may travel faster or slower across the ground than it flies in the air, depending on whether it encounters tailwinds or headwinds. The bird's 'own' air speed has to be subtracted from or added to the wind vector in the flight direction, to determine its speed relative to the earth's surface - the ground speed.[5] Once the true airspeed is known, the 'rate of energy use', or the power, can be plotted. This is expressed in watts (W), the number of joules used per second. To get an appreciation of this unit, it's useful to know that a bicycle lamp uses 3 W and a small light bulb uses 25 W. Sleeping human beings use around 80 W; sleeping knots use around 1 W.[936] As flight speed increases, the power required decreases slightly at first, but then rises increasingly steeply (Figure 3.4). This is because birds flying at very low speeds need to work hard to remain airborne, while at higher speeds they need to overcome the rapidly increasing air resistance.

Using basic aerodynamic equations[631] we can calculate and graph the power curves for 'Afro-Siberian' bar-tailed godwits (Figure 3.4). This is done separately for the sexes, as female bar-tailed godwits are much larger than males.[669] We also give the curves for three different weights, as the birds are much heavier on departure than on arrival (because they have used their fat reserves flying thousands of kilometres). It's clear that with increasing body weight, the power curve is higher. Even though the female bar-tailed godwits are larger than the males, their curve on arrival in the Wadden Sea is below the curves of males departing from Mauritania because females on arrival are lighter than males on departure. En route to the Wadden Sea, the most efficient flight speed (symbolised by $V_{mp}$, the minimum power speed) decreases with decreasing body weight (Figure 3.4). This means that to keep flight costs to a minimum, birds should fly more slowly as the journey proceeds.

But since waterbirds cover such large distances it is probably more useful to know which speed will cost least per distance covered, which is known as the maximum range speed, $V_{mr}$. The maximum range speed can be easily read from the graph. A line drawn from the zero on the graph will touch the power curve at the point that the ratio of flight costs (y-axis: joules per second, J/s) to flight speed (x-axis: metres per second, m/s) is smallest. (You can calculate this yourself: J/s divided by m/s is J/m, which is the energy required per distance covered!). From this theoretical approach you can see that the cheapest flight speed for bar-tailed godwits from Mauritania to the Dutch Wadden Sea is 12.5 - 15.3 metres per second (or 45 - 55 km per hour). The average flight speed of 50 km per hour measured at departure[695] fits the predicted speed exactly.

A bird's size and weight are not the only factors determining flight costs and the shape of the power curve.

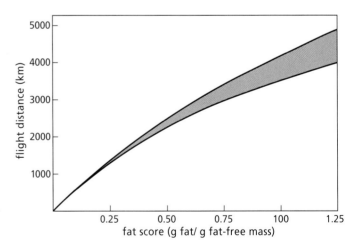

Figure 3.5 The law of diminishing returns as demonstrated here by the effects of increased fat storage on distance flown. Calculations are made for red knots with a fat-free mass of 130 g flying continuously at sea level, in calm conditions, at the optimal speed ($V_{mr}$). As the amount of fat deposited increases (represented here as the ratio of fat: non-fat tissue), flight range also increases, but not in a straight line. The shaded area represents the range of two different prediction methods.[914]

Flight costs decrease with increasing altitude, because air resistance decreases as the atmosphere gets thinner – so on long-distance flights it makes sense to fly high. By flying in a V-formation birds at the rear can profit from the lift (uprising air) generated by the birds flying at the front.[24, 144, 427, 521] This phenomenon is a bit like the slipstream effect experienced by cyclists and ice-skaters. By varying the intersects and the tangents of power curves (like Figure 3.4) you can show that smart waterbirds should fly faster in headwinds and slower in tailwinds. Not only can flight conditions be taken into account, but you can also consider the conditions during the preparation for migration when calculating the optimal flight speed.[11, 387] In most cases, both time and energy are important to a migrating bird, so waterbirds need to fly faster than predicted on the basis of the simple power curves. The fast flight of red knots above Southern Sweden in early June (en route to Siberia[362]) of on average 21 metres per second (75 km per hour) reflects their being very heavy, flying high (1 km or more), and possibly being in a hurry to arrive on the breeding area on time.

Do these models also tell us how far a waterbird can fly if it has a certain body weight?[153] Well, yes and no. Yes, because the models can clearly demonstrate how far a bird can fly with a certain amount of extra weight (assuming that this weight consists of fat; Figure 3.5).[11, 914] But it's obvious that not every extra gram of fat is directly proportional to extra kilometres. The law of diminishing returns comes into play, so as a bird gets heavier, flying becomes more expensive and the extra distance travelled per gram of weight decreases. Also, there are few trustworthy calibrations of these theoretical flight cost models,[632] and the additional costs of carrying more fuel have only recently been measured for the first time (see Box 3.2). Furthermore, a bird that adapts and reduces the size of its muscles and organs, as well as using its fat, can fly half as far again or even twice as far as a bird that only uses fat under the same conditions.[631]

Finally, even with accurate measurements of flight costs, it is still impossible to predict precisely how far a waterbird can fly, until we know the speed of the medium that it's flying in, the wind speed! Since wind speeds can be as high as the bird's flight speeds, a bird can make virtually no progress in strong headwinds.[214] But with a

**Box 3.2 Flight costs of knots: measurements in a wind tunnel**

It's preferable that the bird you're studying doesn't fly away when you are trying to measure its flight costs. This makes a wind tunnel the obvious place to take these measurements, as here the birds can fly normally but are unable to escape. Over the past 35 years,[875] a number of research groups have used wind tunnels to try to measure birds' flight costs. Many of these studies used homing pigeons or other domestic pigeons bred specifically to fly in wind tunnels.[595, 627, 749] To measure energy consumption, each flying bird had to wear a small mask over its head connected to a tube that carried the exhaled air to the analysing equipment. Of course, experimental birds that have to wear this sort of equipment when they fly are less relaxed than birds flying in natural conditions. There were also problems with turbulence in the wind tunnels: the whirling air made it very difficult for the birds to find a good flight rhythm.

Nowadays, it isn't necessary to make the birds wear 'facemasks' when they fly in wind tunnels as energy expenditure can be calculated from the bird's loss of 'doubly labelled water'.[503, 537] In this method, the bird is injected with a small amount of 'heavy' water ($D_2O^{18}$ - water made from the stable hydrogen isotope deuterium and the stable oxygen isotope oxygen-18) and the decrease in the concentration of heavy water in the bird's blood over time allows the researchers to calculate how much energy the bird has used. At the University of Lund in Sweden a wind tunnel has been purpose-built to research flight in migratory birds.[632, 633] The tunnel's creators have managed to reduce the turbulence in the part of the wind tunnel where the birds fly (the test-section) to the point where it is no longer detectable. Furthermore, the tunnel has been designed so that the air pressure in the test-section is the same as the air pressure outside the tunnel, which makes it possible for a bird-trainer to simply walk in and out of the wind stream. This means that it is very easy to release a bird into the tunnel and to continuously stay in contact with the bird; this is handy if the trainer needs to encourage it to fly a bit higher or lower or in a slightly different way.

In autumn 1998, the University of Lund and the Royal Netherlands Institute for Sea Research started a two-year co-operative project to measure the flight costs of red knots using doubly labelled water in the Lund wind tunnel.[519] They wanted to know what extra flight costs were incurred by birds carrying a higher body mass on long-distance flights. Red knots from the Dutch Wadden Sea were taken into captivity and, through their carers spending a lot of time with them and being as friendly as possible, the birds became tame. These birds were then flown in the Lund tunnel. Some individuals had no difficulties flying in the wind tunnel, but others just didn't seem to like it very much. By continuing the experiments with the more enthusiastic flyers, the researchers managed to record data from five different birds that flew for 10 hours at a speed of around 55 km per hour, almost as fast as knots fly in the wild. Even though knots may fly for more than 70 hours in one stretch on migration, a 10-hour flight can probably still give researchers a good indication of what is happening in the wild. In any case, those 10-hour flights rank very high on a worldwide list of wind tunnel flight times. Until recently the maximum flight record of 3 hours was held by specially bred domestic pigeons.[595] Since then a thrush nightingale has reached a flight time of 17 hours in the Lund wind tunnel.[473, 488]

These preliminary measurements confirm that the heavier a flying knot is, the more energy it uses. But the increase in energy use was less than that predicted by aerodynamic theory.[999] This information can now be used to improve models of flight costs. Through this we will be able to predict the maximum flight distances different waterbirds are capable of, dependent on their species, weight, and body composition. This will help clarify limits to flight capabilities of migrating waterbirds.[153]

strong tailwind, its speed and the distance covered are easily doubled. So let's look at the effects of wind in more detail.

### 3.2.3 Wind subsidies

Some people are lucky enough to visit Iouik on the Banc d'Arguin in Mauritania in late April and early May, and see the waders depart for the Wadden Sea. When the flocks of birds arranged in lines or V-formations gain height and slowly disappear from view in a north-north-east direction, in spite of the prevailing northerly winds, you can really appreciate the importance of wind conditions for migrating birds. It's hard to imagine that these small birds will fly continuously for two to three days and nights before they settle in the Wadden Sea, let alone them fighting the strong trade winds along the northwest coast of Africa! However, the wind conditions the birds encounter change with altitude. Birds that fly at more than 3 km above sea level after departing Mauritania usually encounter a strong tailwind of on average 15 km per hour.[677, 680] These tailwinds continue for the next 2000 km or so at altitudes of 3 - 6 km (Figure 3.6) providing substantial assistance to the birds. The wind

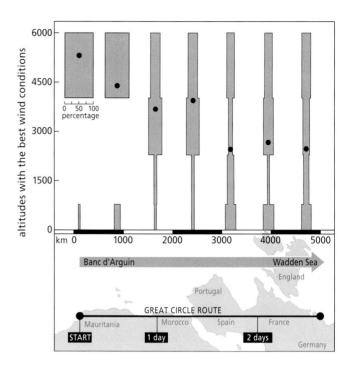

Figure 3.6 Relative distribution of optimal flight altitudes for waders en route from the Banc d'Arguin in Mauritania to the Wadden Sea, based on wind data from 40 departure dates in the first week of May.[677] Dots represent the average optimal flight altitudes at different stages of the journey.

conditions usually change at the latitude of Northern Morocco, and often the birds have to fly lower (around 2 - 3 km above sea level) to find the best wind conditions. Closer to the Wadden Sea, the best winds are often found at even lower altitudes. If the birds had flown at sea level from Africa to Europe they would have encountered headwinds averaging 8 km per hour. When migrating from the Wadden Sea to the Taimyr Peninsula in Siberia in late May and early June, if the birds consistently choose the best flight altitude they can count on average tailwinds of 22 km per hour. At sea level they would have had headwinds of 2 km per hour to contend with.[677]

Do birds fly high enough to encounter the favourable winds during migration? And how do they find the best flight altitude? As they depart from Banc d'Arguin for Western Europe, groups of waders disappear from view at altitudes of 500 – 1500 m.[681] Birds were still ascending at that point, and it's assumed that they continued to gain height. In late April and early May, thousands of birds fly the final kilometres of their journey from West Africa to the Wadden Sea at sea level, providing good viewing opportunities along the Dutch coast.[123] But in some years, coastal bird watchers in North Holland (a province in the central Netherlands) see almost no passing waders. In these years, it's probably more profitable for the birds to continue flying at higher altitudes and descend only when they're above the Wadden Sea. Radar observations along migratory routes confirm that waders often migrate at high altitudes. Red knots en route to Siberia from the Wadden Sea fly over Southern Sweden and Southern Finland at altitudes of 500 m to 4 km.[179, 362] Waders migrating over the West Atlantic Ocean also travel at thou-

This flock of turnstones tries to quickly gain altitude just after departure from the Banc d'Arguin in Mauritania.
In the Wadden Sea, flocks of departing shorebirds are often seen after high tide (photo on page 103).

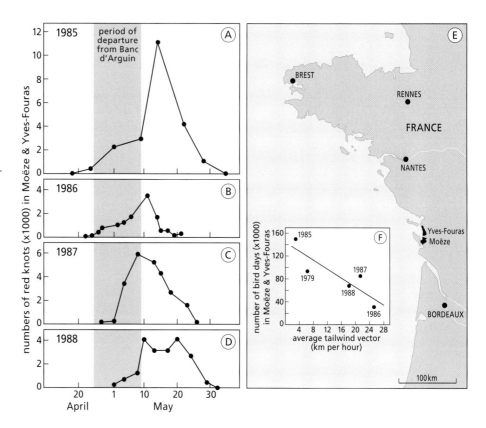

Figure 3.7 The variation in the number of knots observed in May 1985 to 1988 (A-D) in two coastal reserves in the Vendée in France (Moëze and Yves-Fouras, E) can be largely explained by the average tailwinds the knots encountered each year en route from the Banc d'Arguin to France (F). In these calculations it was assumed that the knots were always able to find the optimal flight altitude (up to 6 km). Figure based on[187] and additional data from T. Piersma, D. Bredin and P. Prokosch.

sands of metres above sea level.[731, 732, 939] Large, heavy migrants like brent geese and eider ducks, which have difficulties gaining altitude,[385] migrate at much lower altitudes than waders and usually fly not far above the sea or land.[13, 366] Birds probably determine which altitude provides the most favourable winds through trial and error. Researchers using tracking radar found that common terns and Arctic terns crossing Southern Sweden first climbed to high altitudes, then descended to the optimal flight altitude.[4]

When assessing the strength of tail or headwinds, researchers must assume that migrating birds can regularly, or even continuously, determine their position. The birds could then calculate their ground speed and determine whether they could be blown off course by crosswinds and by how much. It is not impossible that this assumption is correct, but caution is necessary. At higher altitudes winds are stronger because the friction caused by the earth decreases. This means that at altitudes over 3 km winds can easily reach hurricane speed (9 - 12 on the Beaufort scale). Migrating birds capable of reaching flight speeds of 60 - 75 km per hour (7 - 8 Beaufort), cannot stay on course in these winds.[677] At altitudes of 5 - 6 km, the wind is so strong that on a quarter to a third of all days birds would be unable to avoid being blown off course. So even though flying high has some advantages because of the tailwinds that can be encountered, it can also be risky. Red knots that have been blown off course are regularly found in Switzerland; these birds are completely exhausted and will almost certainly die.

In spite of this, tailwinds are probably crucial for many long-distance flights.[806] The knots that fly from the Banc d'Arguin to the Wadden Sea are a good example of this. Most knots complete the journey in one go, but some descend on route in Western France.[179] The knots that stopover in France are always very thin, which has led to the assumption that knots prefer to fly direct to the Wadden Sea, and stop in France only if they run out of fuel.[179] From 1985 to 1988, the numbers of migrating knots were carefully counted during the May migration in two nature reserves in the Vendée: Moëze and Yves-Fouras (Figure 3.7). Every year, the number of knots started to increase from the time that the first departure was observed on the Banc d'Arguin.[680] The knots only remained in the Vendée for a short time, and their numbers declined steeply after or by 20 May. In 1985, observations on knots dye-marked in Mauritania and France confirmed that the French birds came from Mauritania,[655] and that they continued from France to the Wadden Sea, especially to Schleswig-Holstein in Germany.[90] The numbers of knots at the French sites were very high in 1985, very low in 1986, and intermediate in 1987 and 1988 (Figure 3.7).[90] There was no difference in the body weights of knots about to leave the Banc d'Arguin in 1985 and 1986,[656] so differing fuel supplies couldn't explain why more knots stopped in France in 1985 than in 1986. However, wind conditions could provide an explanation (Figure 3.7). In 1986, strong tailwinds were available; but in 1985 there was little wind assistance. Data from 1979,[179] 1987 and 1988 reinforced this pattern. Furthermore, the weights of knots caught in Schleswig-Holstein around 22 May were also correlated with the tailwinds. In years with little tailwind, when many knots made a stopover in France (such as 1985) the average weights of knots in the German Wadden Sea were significantly lower than in years with higher tailwinds.

All of these data point towards the fact that knots which winter in Mauritania rely on tailwinds during their migration north. Without this wind subsidy they have to make an extra stopover. Such stopovers expose the birds to many risks (unfamiliar locations, French hunters!) and delay refuelling for their final flight to the breeding grounds. This raises the question, why don't knots accumulate bigger fat reserves on the Banc d'Arguin before migrating north?[654, 985] At the end of this chapter, we will return to this.

## 3.2.4 Adaptable flying machines: sports physiology of the tidal flats

Like an aircraft fuelling up before takeoff, migratory birds prepare for long flights by storing subcutaneous fat. Fat, the fuel for flying birds, is made of simple triglycerides: glycerol molecules with three long fatty-acid tails. This fat builds up on the bird's body in a thick white layer around the breast, belly and back. Bird fat looks very much like farm butter, and a very fat bird can feel rather like a big lump of butter in the hand. People who harvest migratory birds on the point of departure (as many people in Western Europe and North America did until early in the 20th century, and some Southeast Asians still do) found that these birds could be perfectly fried or baked in their own fat. The now almost-extinct Eskimo curlew was nick-named the 'doughbird' by 19th century American hunters, because of the thick white fat layer it deposited before flying to South America.[81, 300, 540] Waders departing from West Africa have fat loads (percent of the total body mass that is fat) of up to 30%.[665, 985] Bar-tailed godwits leaving the Dutch Wadden Sea for the Western Siberian breeding grounds can also reach up to 30% fat.[662, 669] However 30% isn't the maximum amount of fat migratory waders can put on.

Over half of the body weight (55%) of Alaskan bar-tailed godwits, ready for the 11 000 km flight to New Zealand, consists of fat.[668] This is highest proportion of fat that has been measured in birds, and is achieved because of the very small abdominal organs of the Alaskan bar-tailed godwits.[668] The birds studied had died shortly after departing on migration, when they flew into a lighted radar beacon on the Alaskan peninsula during an autumn storm. They had small livers, stomachs and intestines. These organs had important roles when fat was being stored but are less important during the long flights.[673] A reduction in abdominal organs just prior to departure has also been observed in some other wader species,[687, 697] usually those that make long, relatively dangerous flights.[662] These flights, such as over the polar ice-cap, provide few possibilities for safe stopovers and if an exhausted wader had to stop flying it would have little hope of survival.[81]

A bird's flight muscles and heart are a completely different story. They are very much in demand during flight. Flight muscles keep the birds airborne and propel them, and the heart pumps blood to supply the muscles with fuel and oxygen. The fatter the bird, the more power the muscles and heart need to provide. The flight muscles and hearts of bar-tailed godwits from Alaska were just as expected based on their body weights.[668] To

| *Body part* and function | Changes during fattening up | Changes during flight |
| --- | --- | --- |
| *Body fat*, stores energy | Continual increase | Continual decrease |
| *Breast muscles*, generate flight power | Increase, especially just before departure | Continual decrease |
| *Leg muscles*, generate walking power | Increase during fattening up, decrease before departure | Decrease |
| *Heart*, pumps blood around the body | Increase, especially just before departure | Continual decrease |
| *Liver*, biochemically converts nutrients | Initial increase, decreases just before departure | Decrease |
| *Stomach*, breaks down and predigests food | Initial increase, decreases just before departure | Decrease |
| *Intestine*, absorbs nutrients | Initial increase, decreases just before departure | Decrease |
| *Kidneys*, remove waste products | Sometimes decrease slightly before departure | Small decrease |
| *Lungs*, extract oxygen from the air and excrete carbon dioxide | No measurable change | No measurable change |

Table 3.1 Functional changes in the size of organs during migration.

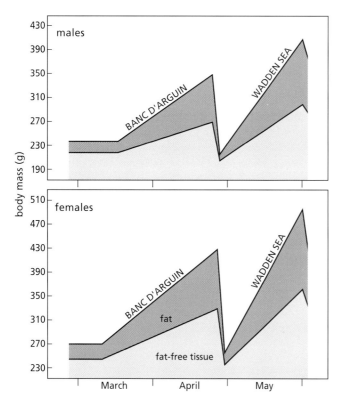

Figure 3.8 Changes in the total body mass, fat storage and fat-free mass of bar-tailed godwits that fuel up and then migrate from the Banc d'Arguin to the Siberian breeding grounds via the Wadden Sea.[669]

understand what happens before and during the extreme endurance event of a long-distance flight, it's necessary to examine individual organs.[673] The bad news is that birds often have to be dissected for this. Researchers make as much use as possible of accidental deaths; birds that fly into obstructions such as lighthouses and radar beacons or die during freak weather events, capture casualties, and birds killed by human hunters (as in Southeast Asia). But recently, new techniques, such as ultrasound, have become available making it possible to study organ size in living birds. Using these techniques we are on the edge of making exciting advances in understanding how and why organ sizes change in migratory birds.[182, 671]

First, let's review what we know now about the changes in the size of organs. While fattening up, a migratory bird needs large abdominal organs: a big stomach to process food, a long and active intestine to extract nutrients, and a big liver to convert nutrients into the substances the body requires (Table 3.1). The liver also helps to remove toxic substances ingested with the food, while the kidneys excrete the other waste products. Leg muscles are very important at this time, as fattening birds generally walk around a lot when feeding. During long-distance flights, the abdominal organs and leg muscles are functionally less important. By reducing those organs before departure, they can get rid of surplus weight making it possible to fly further for the same

A red knot arriving in the Wadden Sea after flying many thousands of kilometres often weighs just 100 - 120 g. A recent arrival (upper photo) can be distinguished by its shrunken abdomen and seemingly extended cloaca. Three to four weeks later the bird's shape has changed greatly. It has rebuilt its subcutaneous fat layer and filled its abdominal cavity with fat (middle photo). Its weight has now doubled to more than 200 g (lower photo).

The sort of photo that is shown on page 106 can only be taken in the Wadden Sea at the end of May, when the bar-tailed godwits are round and very heavy. Their tubby shape is particularly visible as they fly straight towards the camera. With their newly acquired rusty-red breeding plumage, they are now ready to depart for the Siberian breeding grounds, more than 4000 km away.

amount of fat. However there's no point in downsizing the flight motors (the heart and breast muscles). At departure these organs have to be big, otherwise the fat birds cannot even become airborne,[273, 385] which can be a real problem. Sometimes waders have to slim down before departure![433] As the birds become lighter during the flight they have less need for pumping and flying power, and the economical bird should gradually break down its heart and breast muscles.[629, 631] There is evidence that breast and heart muscles are indeed broken down during the flight,[70, 669] and that the digestive machinery and liver are also broken down further during long flights.[426] After flying for many thousands of kilometres, the birds are light-weight and vulnerable when they finally touch down.

Blood composition also changes dramatically as birds prepare for migration. Just before departure, red blood cells increase and the blood gets 'thicker'.[693] Since every blood cell with a certain amount of haemoglobin can transport a certain amount of oxygen from the lungs to the organs, thicker blood can carry more oxygen around the body per unit time. It can also remove more waste products such as carbon dioxide (via the lungs) and uric acid (via the kidneys). But it takes more effort to pump thick blood around,[902] so the heart enlarges just before departure.[693] Lungs play an important role in endurance events, but show little change in size as they are important both while fattening and during the flight.[554, 692]

Birds must also prepare for arrival on the breeding grounds. Birds need a lot of calcium to produce eggs.[351] But on the tundra, where many of waterbirds breed, calcium is limited. So it's not surprising that female waders have been recorded eating lemming bones selectively during the laying period.[529] Females may also store calcium in their own skeletons: knot females (but not knot males) make their skeletons 10 - 20% heavier during the May fattening period in Iceland.[694] They are probably building up a calcium store for egg production thousands of kilometres away and some weeks later.

By the early 1960s, American researchers had developed a theory that migratory birds simply gain fat and use this as flight fuel.[617] The birds were thought to be like aeroplanes: both fuel up, empty the tanks during flight, then refuel (and so on). This comparison, however, is out of date. Birds (the flight machines) must gain their own crude oil (food) and refine this to kerosene (fat). Their machinery must continuously adapt to local conditions and the precise stage of the migration.[34, 435, 517, 657, 673] Migratory birds are more like top athletes, whose organs work together in a very precise and coordinated way. The biomedical world could learn a lot from migratory birds, which, for example, possess very efficient enzymes for quick fat breakdown.[434] In human beings, weight loss is far more difficult as we can't burn pure fat very well.[121, 934, 952]

In spite of all this detailed knowledge, there are few good descriptions of seasonal changes in body composition of migratory birds. One reason for this is that it is difficult to determine the weight changes of birds as they fly halfway across the world; another is that we have insufficient information to determine these weights for each organ.[179, 209, 581] But researchers have had some success at describing changes in the body composition of bar-tailed godwits that fly from Mauritania via the Wadden Sea to the Taimyr peninsula in Siberia (Figure 3.8). When they are fuelling up in Mauritania and the Wadden Sea, bar-tailed godwits build up both fat and fat-free tissue. Just like the knots discussed above, they leave the Banc d'Arguin lighter than when they leave the Wadden Sea a month later (even though the distances to be travelled to the Wadden Sea and then on to the Taimyr are similar, around 4500 km). En route from West Africa to the Wadden Sea they not only burn nearly all of their fat, they also lose a lot of fat-free tissue. Although they have greatly reduced digestive organs on arrival in the Wadden Sea, they don't seem to have any problems fuelling up. Their daily mass increases are constant throughout their stopover in the Wadden Sea.[669] This is probably because fat-free tissue is much 'easier' to deposit than fat (being eight times less energetically expensive). Bar-tailed godwits can probably quickly restore their abdominal organs to their old state, and then start on the real refuelling.[697]

### 3.2.5 Moulting waterbirds don't migrate

Intercontinental flights might seem exhausting to humans, but they are nothing compared to the travels of migratory birds. The preparation and the actual flight push the birds to their physical limits. There is no scope for further endeavours, such as moult (feather replacement).[301] Moult starts deep in the skin with the forma-

---

**Box 3.3 'Smoking' tidal flats: the secrets of three-dimensional formation flight**

The co-ordinated flight of a large flock of waterbirds swooping above a tidal flat is an impressive and beautiful sight. Often the flock appears to continuously change colour as the birds alternately turn their light bellies and dark backs to the observer, and it seems that every bird knows exactly what he or she has to do. At lectures about waterbirds, the most frequently asked question is "how it is possible for such dense groups of birds to make such co-ordinated flight movements without bumping into each other?" When many thousands of shorebirds simultaneously whirl through the air, quickly changing direction,[377] it's hard to believe that they can successfully perform these manoeuvres without ever colliding. And if collisions never occur, how the birds manage to avoid them really is a very interesting question. A number of researchers have put their minds to this problem and produced some rather nice hypotheses.[168, 398, 533, 700, 702, 780] In 1931, the English behavioural ecologist Selous put forward the rather far-fetched idea that telepathy was involved: a fast, invisible and telepathic thought transfer made it possible for hundreds or thousands of birds to make the same turns at the same time.[780] When Potts filmed groups of dunlin using high speed cameras, he discovered that the movement is always started by one or a few birds that swoop towards the group's centre of gravity.[702] The nearest individuals react to this in just 67 milliseconds. After this the movement transfer speeds up and the wave takes only 15 milliseconds to jump from neighbour to neighbour. Since this transfer time is much shorter than the 'alarm' reaction time measured for dunlins in the laboratory (38 milliseconds), it was concluded that the birds can anticipate the group movement because they can see it approaching. Exactly the same happens during the co-ordinated movements of dancers performing a ballet or a cancan, or football supporters making a 'Mexican wave' in a sports stadium. However, with a reaction time of 67 milliseconds dunlins are much faster than humans. In a stadium a transfer speed of 108 milliseconds has been measured for a Mexican wave, half the average human alarm reaction time of 194 milliseconds. The individual shorebirds that initiate these group movements are in a way the group's 'leaders', even though the leadership changes from second to second.

So it's easy to explain how these wave-like movements can arise in a dense flock of waders, but we haven't really done justice to each individual's contribution. In a flock all the birds are flying at full speed. Even a drawing of the movements of a group of three birds flying in a two-dimensional plane (based on the assumption that every bird makes the same turns and maintains about the same speed), shows how incredibly complicated these movements are. The birds must continuously change their positions relative to each other.[398, 700] Yet it was possible to show with the help of a computer simulation that objects that (1) keep a certain minimum distance from their nearest neighbours, (2) adapt their speed to that of the neighbours, and (3) move themselves towards wherever most of the others are, behave almost like a flock of birds. These simulated flying objects were unable to make errors, but nature is not so perfect. When neighbours' mistakes are intercepted and corrected across the group and using the three behavioural rules, it is possible to make small error-making 'objects' behave like a flock of dunlin.[45]

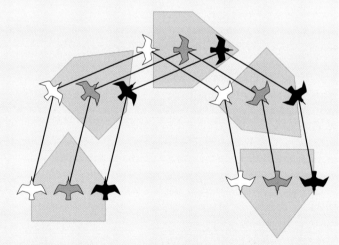

Figure B.3.3 As a flock of waders wheels around in the air, each of the participants must move in the same way, continuously changing position relative to the other birds. This diagram demonstrates how complex this is for just three birds (one black, one grey and one white) moving in a two dimensional (flat) plane, along one turning flight path (the grey arrows). The three dimensional reality, involving many, many more birds, is far more complicated.[700]

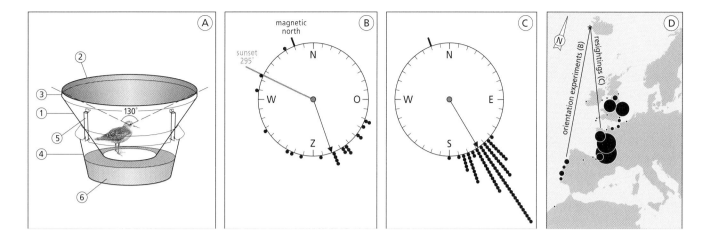

Figure 3.9 An orientation cage (A) used for research on dunlins in Iceland in September 1995, and the directions (B) in which the dunlins tried to escape from the cages. These escape directions overlapped nicely with the directions expected for dunlin about to depart on migration (C), based on what is known from ringing data (D). The ringing data comprise dunlins ringed in Iceland and resighted elsewhere in July to October, and dunlins ringed outside of Iceland in July to October and resighted in Iceland during the breeding season. In (A), (1) shows the aluminium funnel, (2) the fine mesh net that closes off the funnel, (3) the aluminium band that holds the net, (4) the plastic container to which the funnel is fastened with elastic bands (5) that is filled with beach sand (6). In (B) and (C) the arrows show the average departure directions, and every dot represents one observation. In (D) the size of the filled circle is proportional to the number of resightings; a resighting from Mauritania is not shown.[765]

tion of a gland with good blood circulation (the feather follicle). Here, the feather protein keratin is secreted and starts to form the new feather. The growing feather pushes the old feather out of the skin. The growing feather is initially surrounded by a protective skin or sheath and takes two to three weeks to reach its final length. When the feather is almost ready, the protective skin crumbles and the feather unfolds. The new feather follicle and blood vessels that transport nutrients and wastes are no longer necessary. This need to mobilise high quality proteins and to supply extra blood just under the skin surface, apparently make it impossible to moult and migrate simultaneously.

But this isn't the whole story. Once a year waders replace their flight feathers. While this is happening, they fly around for two to three months with 'holes' in their wings (see Box 2.2). The only way to avoid this is to replace all the feathers simultaneously, as waterfowl do (which leaves them completely flightless for a few weeks). Flight feather moult can affect waders' manoeuvrability and ability to fly for weeks. During moult not only is flight more energetically expensive, but there is also a greater risk of being caught by a peregrine falcon. To reduce these problems, waders carry little fat during the moult, so they are as light as possible.[84] This is another reason why moulting waterbirds cannot undertake migratory flights.

### 3.2.6 Orientation and navigation

The meanings of the words orientation and navigation are often confused, but if you're talking about bird migration, it's worth knowing the difference. Orientation is the ability to find and keep to a certain direction, and doesn't directly involve finding a particular place on the globe. In contrast, navigation is the ability to find a particular place while crossing unfamiliar terrain.[2, 29, 911] To navigate you use orientation mechanisms.

Let's start with orientation, the ability to know where north, south, east and west are. Over the last 55 years there has been a lot of research on orientation mechanisms.[46, 857] Much of this work has been on songbirds and racing pigeons. The results are probably relevant for understanding the orientation ability of waders and waterfowl since the only clear differences are in small details between species and groups of species.[394, 765] Racing pigeons are resident rather than migrating birds, but are easy to breed and to handle. They are ideal subjects for studies of orientation and navigation, because of their eagerness to fly back to their own roost. The directions taken by pigeons upon release can be followed until the birds disappear from the observer's view.[485, 905, 907]

Northerly-breeding songbirds are also quite easy to study during migration, as they are active at night and attempt to move in their preferred migratory direction. If kept in cages, they flutter around and seek to escape in the direction of the next stopover or wintering site.[484] Researchers have made good use of this behaviour by building funnel-shaped cages with an inkpad in the centre and absorbent paper around the walls. The birds' feet become inky and if, just after sunset and during the night, they attempt to fly, their feet mark the absorbent paper in the direction of the migration route. The density of the marks in different directions is easy to quantify.[218] In later experiments, researchers used large sheets of Tipp-Ex typing correction paper, which the restless birds scratched with their claws in the escape direction (Figure 3.9).[765] In the latest design of orientation cages, the walls of the funnel are a circular row of panels attached to electronic recorders. If a bird flies or jumps against a panel, the panel is pushed backwards and triggers the recorder for that direction.[767] There are two published studies on the use of orientation cages to record migration directions in waders, one on Pacific golden plovers,[771] the other on dunlins in Iceland (Figure 3.9).[765]

The dunlins were kept overnight in orientation cages in late summer (the departure period). When the evening sun was visible, birds oriented themselves in a similar direction to that expected from ringing data, southeast (Figure 3.9). On overcast evenings they generally showed no migratory restlessness; the only dunlin that did oriented itself northwards.[765] Another striking result was that the dunlins in the cages kept magnetic north directly behind them. Such a course would lead them to the Portuguese coast (Figure 3.9), an important fuelling spot on the way to the West African wintering grounds. The hand-reared Pacific golden plovers could only orient properly if they were able to see stars in the sky.[771] These studies on dunlins and golden plovers suggest that both the setting sun, the starry night sky, and the earth's magnetic field are important orientation tools for migrating waders. In fact, this covers all known compasses: the sun compass, the polarisation compass, the star compass, and the earth magnetic compass.[5]

*Sun compass* – At midday in the Northern Hemisphere the sun is due south. If you know the time of day, you can also calculate where south is from the sun's position in the morning and afternoon. Extensive research on pigeons and songbirds has shown that birds have an internal clock which they use to determine the time of day.[71, 370] With the aid of this clock they can determine which way is south, based on the sun's position at a certain time of day.[5, 47, 217] But this isn't all that birds can do.

*Polarisation compass* – In a clear sky (or in patches of clear sky) polarised light follows an orbit that moves during the course of the day along a more or less north-south orientation. People can see these stripes of polarised light in the sky only if they look through a polarisation filter, but birds can observe polarisation patterns without artificial tools. The polarisation pattern is not informative at midday (but then the sun is an easy orientation point); it is most useful at sunrise and sunset.[579, 906, 911, 916] At high latitudes, daily changes in the polarisation pattern may be too complicated to be useful for orientation, but closer to the equator birds can use this extra compass, because they can see and interpret polarisation patterns when the sky is visible at sunset and sunrise.

*Star compass* – Photographing stars in the night sky usually requires an exposure time of several hours. These slow exposure photos show that the stars appear to move in a circular pattern, as the earth turns on its axis. In the Northern Hemisphere, the stars move in semi-circles around the North Star, which points exactly north. In the Southern Hemisphere, stars describe a circle near the Southern Cross. On and near the equator, neither of these turning points is visible. Experiments using orientation cages in planetaria have shown that night-migrating songbirds learn the position of this turning point in the night sky during their first months of life. This might be the North Star, but in a planetarium the birds can be tricked by 'moving the sky' around another star, for example, Betelgeuse. The birds then orient themselves on that turning point.[5, 217] Once they know this star compass, birds do not need to see the whole night sky to orient themselves. A small piece of the night sky will usually show ample stars or constellations. However, the closer the constellations are to the North Star, the more useful they are for orientation. Over the centuries, both the northern and southern celestial turning points change location in relation to the stars. But since migratory birds learn these turning points in their youth, they are quite capable of adjusting to such gradual changes.

*Magnetic compass* – Even though it has nearly been replaced by the pocket-sized GPS, the old-fashioned compass remains a useful tool for the explorers amongst us. A compass can be used to determine wind direction, but you must take into account the fact that magnetic north is not necessarily the same as true north. With a small twist, you can adjust your compass to account for this 'declination', the angle between magnetic and true north. The 'magnetic field lines' also create another angle, the angle relative to the earth's surface, called the 'inclination'. At the equator these field lines are horizontal, but at the magnetic poles they rise steeply upwards or downwards (Figure 3.12). Merkel and Wiltschko's spectacular discovery in 1965 that robins can orient themselves on the basis of the earth's magnetic field[570]

Shorebirds tend to set off for far lands when the sky is clear, in the hours just before or just after sunset. As the sun sets, the birds have the maximum number of orientation techniques available for their use.

Dunlins that breed in Iceland (photo on page 111) have been used in experiments to find out how migratory shorebirds orient themselves.

MIGRATION

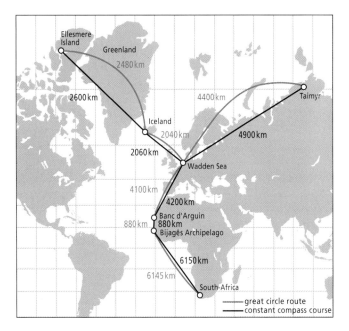

Figure 3.10 The difference between a constant compass course (the loxodrome) and a great circle route (the orthodrome) for the migration routes of red knots that breed in northeast Canada and Siberia. The two types of migration routes are drawn on a Mercator projection, which allows the compass course to be accurately plotted but causes increasing distortion of the relative surface area nearer the North Pole.

was at first greeted with disbelief. Since then, the evidence for a magnetic compass in birds has accumulated.[2, 217, 765, 945] It has been clearly shown that birds can only orient themselves if the magnetic field lines form an angle relative to the earth's surface. In other words, in places where the inclination is bigger or smaller than zero. The field lines that lie horizontally at the (magnetic) equator are no problem for a hand-held compass, but a bird can't use them for orientation. Because of this, the magnetic compass of migratory birds is also called an inclination compass.

Now that we have a basic understanding of the different compasses, we can consider how birds use these compasses to navigate. To start with, we must realise that a migratory bird can chose different sorts of migratory routes.[3, 6, 364] For example, migrating from point A to point B, birds can follow a constant compass direction. On the mercator projection of the globe, this is a straight line; in the technical literature it is also called the 'rhumbline' or the 'loxodrome' (Figure 3.10). This is not the shortest route, however. Around a sphere, the great circle or 'orthodrome' gives the shortest route, and is followed by aircraft. This explains why when you fly from Amsterdam to Tokyo or Vancouver you almost circle the North Pole. For orientation purposes, the great circle is a difficult route, since the compass direction that has to be followed changes continuously through the journey. If we draw the two types of routes to the breeding and wintering areas from the Wadden Sea (Figure 3.10), it is obvious that when migration routes lie along a north–south axis, there is hardly any difference between a compass course and the great circle. The distance and the locations that are flown over are almost the same, as is the angle relative to north on which the bird must depart.[681]

However the distances are very different for large east-west movements in northern regions, including all migratory routes between the Wadden Sea and the arctic breeding areas (Figure 3.10). When a bird flies along a constant compass course to the Taimyr Peninsula instead of via the great circle route, the bird spends an extra half day travelling (10% of the journey time), covering an additional 500 km. Migration distances from Iceland to Northeastern Canada are similarly affected. If we assume that birds know exactly where they are heading, the main problem with following the great circle route is that the birds would have to continually adapt the compass bearing being followed. A flight from the Wadden Sea to Taimyr would require departing in a northeastern direction and arriving on an east-southeast course (Figure 3.10). This requires both being able to read your compass well enough to know which course to take, and knowing where you are all the time. Migratory birds may not need to possess all this information *per se*. Biologists have shown that simple behavioural rules in conjunction with deviations in the declination pattern could enable birds to fly in the polar regions on more-or-less great circle routes, whether or not they make use of a magnetic compass or an internal clock.[7, 12, 468]

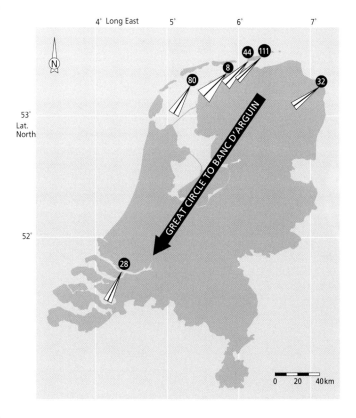

Figure 3.11 Departure directions of different wader species (grey plover, red knot, dunlin, bar-tailed godwit, redshank, greenshank and turnstone) leaving The Netherlands in August to winter in West Africa. The numbers represent the total numbers of flocks observed. After[681] complemented with new data.

In the Wadden Sea, careful observers can see the orientation abilities of migratory waterbirds for themselves. It's even possible to use these observations to get an idea of the migratory route being followed. Waders departing from the Wadden Sea in late summer always do so in the late afternoon and fly in a V-formation calling loudly, while steeply gaining altitude and heading southwest. When one such group flies over, you can use a compass to determine the direction the flock were heading when they disappeared from view.[681] With enough observations, the directions can be plotted (Figure 3.11). In July to September waders depart from the Dutch Wadden Sea in a southwesterly direction (about 225°). Since this is west of the great circle course to the most important wintering area in West Africa, the Banc d'Arguin (216°), and much further west than the constant compass direction (208°) this suggests that the waders keep more to a great circle than a constant compass route. But as we have already seen, the differences between the two routes are not very big on this flyway (Figure 3.10).

On flights like these, that closely follow the coast, experienced migrants can also orient themselves by large-scale geographical landmarks, particularly coastlines. Brent geese and knots departing in late May and early June for breeding grounds in Siberia, Greenland and Northeastern Canada mainly fly on a constant compass course (Figure 3.10).[14, 179, 209, 361] This course roughly follows the coast.[362]

However, there are many indications that waders also use the great circle route, especially when they return from the arctic breeding grounds to the wintering areas.[7] We, the authors, have observed red knots depart from the Taimyr Peninsula in late July and early August flying northwest over the pack ice. This is what is expected from a great circle route (Figure 3.10). Radar observations from an icebreaker along the Eastern Siberian coast provided further evidence for migration on a northeasterly great circle route in the direction of American wintering grounds.[7] Depending on the location and time (and, probably, opportunities for orientation) migrating waders and waterfowl can follow both the constant compass and the great circle routes.

Juvenile waders face a different challenge. Completely inexperienced, they migrate without assistance from their parents. They probably make their way to the wintering grounds solely by using an innate compass direction and a clock that tells them how long they have to continue flying in a certain direction.[371, 395, 634, 635] Over time, all species (including those such as brent geese that migrate in family groups) build up a cognitive map in which all possible odour, sound, visual and magnetic beacons have a place in a two or three dimensional perspective.[43, 944] Research on the significance of all these possible orientation beacons and the use of different types of migration routes is still in its infancy.[6]

It is now clear that whereas sanderlings and curlew sandpipers fly diagonally across Africa and Asia,[215, 814, 943] red knots in Southern Africa fly via West Africa and Western Europe to Siberia.[178] This means that the knots travel more than 2000 km extra compared to the most direct route. When they leave in mid-April, they probably fly obliquely across the Southeast Atlantic Ocean to Guinea-Bissau (Figure 3.12). If they followed the African coastline they would have to fly an additional 200 km.[766] Red knots are almost never observed along the rest of the West African coast during northward migration, apart from some thousands that were seen in Ghana in April 1991.[614]

During the 6500 km journey from Southern Africa to Guinea-Bissau they must fly for more than three days and three nights over the open ocean (Figure 3.12A). So it's very important that they don't miss the coast of West Africa on approach to Guinea-Bissau! On departure from Southern Africa, red knots can use all four compasses,

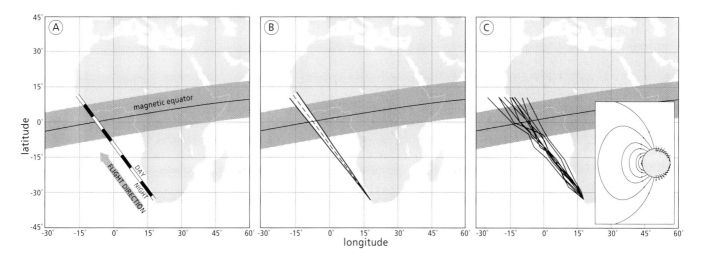

Figure 3.12 An overview of red knot migration from South Africa to Guinea-Bissau across the East Atlantic Ocean. The knots are temporarily unable to use the star compass while crossing the true equator or the magnetic compass while crossing the magnetic equator (shaded). (A) shows the route and parts of the journey covered by day and by night at an average flight speed of 65 km per hour, (B) shows that red knots that depart with a deviation of 2° or less arrive within 200 km of the Guinea-Bissau coast, while (C) shows the possible migration routes for red knots that determine their flight direction eight times en route allowing for an 'inaccuracy' of 10° (see text). The insert in (C) shows how the magnetic fields radiate from the earth; near the equator these field lines lie parallel to the earth's surface. After[5] and[766] complemented with author's data.

including using the turning point of the Southern Cross as a star compass. At that stage they have fairly predictable southeasterly trade winds as tailwinds. However, as the knots approach the equator both wind assistance and orientation options decrease. Unless birds are flying above the clouds, the sun and the polarisation compasses don't work. Around the magnetic equator (which is obliquely situated over the true equator; Figure 3.12) the magnetic field lines are almost completely horizontal, so a bird's magnetic compass is useless. For 3° around the true equator, neither the Southern Cross nor the North Star can be seen, so birds can't use a star compass either. This means that the birds have very limited orientation sources around the equator, especially in bad weather.

These limited orientation sources make it very important for red knots to choose the proper course on departure from Southern Africa. If birds determine their direction only once (at the start of their flight) and hold that course strictly, they would have to have a precision of 2° in order to make landfall in West Africa (Figure 3.12B). This requires greater accuracy than orientation researchers think birds are capable of. If birds determine flight direction eight times during the journey but with a more realistic 'unreliability' (a sort of standard deviation[766]) of 10°, then only 64% of birds would arrive within 200 km (the distance the birds can see) of the West African coast, and in this way survive the trans-oceanic flight (Figure

Red knots that commute annually between the edge of the Sahara in Mauritania (top photo) and their breeding grounds on the Siberian tundra (lower photo) must be able to deliver a winning performance in a variety of arenas.

3.12C). Since red knots have an annual survival of around 80%, their real success has to be much higher. A flight from Southern Africa to Guinea-Bissau across the East Atlantic Ocean is a remarkable achievement; waders and other waterbirds are thought to achieve even greater feats than those already known from their better studied passerine relatives.[766] Perhaps red knots, like pigeons, are capable of hearing infrasound. These are long wave, low frequency sounds that develop where sea waves break onto the shore, especially during storms and in thunder, and where strong winds noisily rush over mountain ranges. Perhaps red knots fly in the direction of the strongest infrasound when they approach the equator, and in this way finally reach the Guinea-Bissau intertidal area.

It's obvious that flying over the equator is not easy. There are no celestial or magnetic beacons for orientation, and the winds are unreliable. Sailors have always known this; the lack of wind and oppressive humidity of the doldrums made sailing across the equator an unpleasant experience.

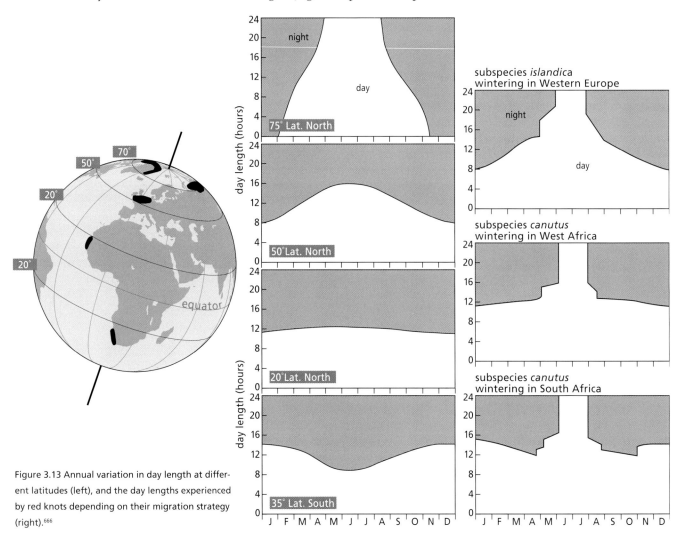

Figure 3.13 Annual variation in day length at different latitudes (left), and the day lengths experienced by red knots depending on their migration strategy (right).[666]

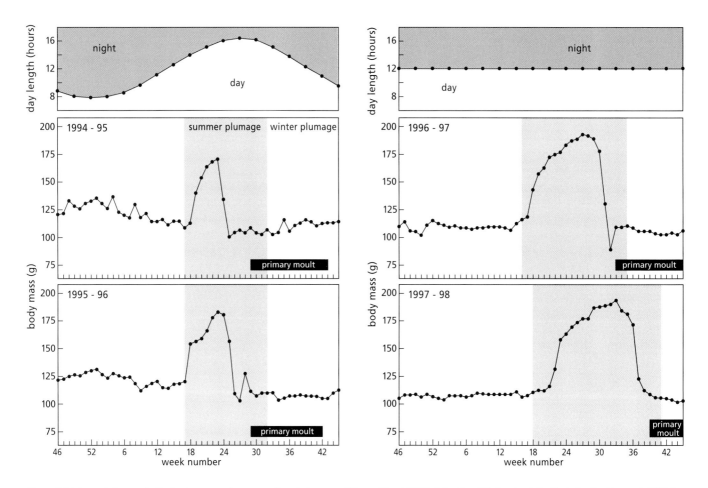

Figure 3.14 Annual changes in the body mass, primary moult and plumage of the red knot K218 at the Royal Netherlands Institute for Sea Research (NIOZ) on Texel. K218 was monitored for four years. Initially he was kept in an outdoor aviary with natural light with a group of his conspecifics. He and his group then spent two years in an indoor aviary, where day length was kept constant with 12 hours of artificial light and 12 hours of darkness. The period when he was in summer plumage is shaded grey, and times when he was moulting his primaries are designated by the black bars.

### 3.2.7 Travel planning: the role of the internal clock

Spending part of the year near the equator also has implications for the birds' ability to determine the time of year. Most waterbirds are real 'calendar birds'; their presence in the Wadden Sea area is more dependent on the time of year than weather conditions (except during exceptionally harsh winters). Curlew sandpipers only visit the Wadden Sea in July and August (and the mouth of the River Elbe in May). Spotted redshanks are only seen in May, July and early August. Purple sandpipers are only seen in the winter months on dykes and piers. Around 30 April large numbers of bar-tailed godwits and grey plovers arrive in the Wadden Sea from the south, regardless of the weather.[123, 654, 669, 680] Brent geese depart from the Wadden Sea between 22 and 25 May,[209] bar-tailed godwits from 28 May till 1 June,[669] and knots in the first days of June.[179]

How do these birds know so precisely what time to go? Many bird species seem capable of precisely estimating the time of the year by measuring changes in day length. If the days become longer then summer is on the way; if the days are getting shorter winter is approaching.[367, 368, 369, 370] However, such a system only works if a bird always stays in the Northern (or Southern) Hemisphere. For birds in the Banc d'Arguin in Mauritania, just north of the equator, the increases in day length in spring are very small (Figure 3.13C), perhaps too small to indicate that it is spring and will soon be summer on the breeding grounds. Day length changes are even smaller closer to the equator in Guinea-Bissau. Birds that fly to the Southern Hemisphere must deal with the reverse pattern before migrating north, decreasing rather than increasing day length (Figure 3.13D).

A simple solution to this problem would be an internal calendar, one that doesn't depend on differences in day length.[368, 370] The existence of such internal calendars has been proven, mostly in song birds,[368, 369] but also in waders.[120, 680] To demonstrate the existence of an internal calendar you must remove all external cues about the time of year, by keeping birds in indoor aviaries with constant temperature and day length.[368] If birds continue to moult (into and out of breeding plumage, as well as the annual flight feather moult) and put on migratory fat reserves at the right time, even though they lack any information about the seasons, then there are signs that an internal clock is involved. Of course, even in a closed aviary we cannot be completely sure that there is nothing in the environment that changes through the year in a way that the birds could detect. The most convincing way to prove an internal calendar is when birds in a constant environment continue to moult and fatten up over intervals that are shorter or longer than a year. An internal clock is only visible when it is running ahead of or behind schedule.

The latter is exactly what was observed in seven red knots housed together for four years (see Figure 3.14 for the patterns in moult and weight of one of the seven; the other birds were similar). When the knots were in an outdoor aviary their 'annual' cycles took exactly one year. After being brought into an internal aviary with constant conditions, the moult and weight cycling took almost two months longer (cycles of 13 and 14 months). There was apparently nothing in their surroundings that they could set their clocks by. How free-living knots set their clocks is unknown. For the captive knots, we can plot moult and weight patterns against the change in day length (Figure 3.14). Is 21 December, the point at which days become longer instead of shorter (in the Northern Hemisphere), important for their biological clocks? In the Southern Hemisphere, 21 December is the longest day of the year; this could be the calibration point for knots in Southern Africa.[120] Knots that winter near the equator would have to measure day lengths very precisely. Alternatively, knots could calibrate their internal clocks in the (northern) summer period, on the longest day. The problem here is that in the polar regions it's always light in summer and the days are all equally long. That (some) waterbirds have an internal calendar is obvious, but how they set their yearly clock is still a mystery.

Birds that winter thousands of kilometres from their breeding grounds, and have only a small window of time when breeding is possible need to arrive on the breeding grounds at precisely the right time. Arriving on time means departing on time.[680] In addition, successful migration and breeding require the bird to have built up its energy stores and moulted into attractive plumage for a potential partner at the right time. Waterbirds cannot manage this without a good calendar.

### 3.2.8 When do we leave?

Some days out on the tidal flat there is a sense of restlessness in the air: it's time for the birds to begin their long journey. They are nervous and make short flights in ever increasing groups. Then suddenly, usually as evening approaches and on an incoming tide, they launch into the open air, calling loudly. At first they fly in a cluster, as they do for most of the year (see Box 3.3), but the group soon stretches into a V or line formation (Figure 3.15). The formation quickly gains height and sets on a consistent course, the departure direction. Once they are in formation and the course is straight, they stop calling.[14, 33, 681, 878] Sometimes the V formation falls apart and the birds return to the tidal flats. Perhaps one individual can't keep up with the group and returns alone to the mudflats. This bird has probably decided that it is not ready for such a long and dangerous flight.

The moment of departure is crucial in many ways. It has to be the right time of year, and the departing birds must be physically able to complete the migratory flight successfully (Figure 3.16). Have they laid down enough fat and prepared their body organs? How are the wind conditions at the time of departure? Are there enough flockmates available to fly in a proper formation? Do they have enough orientation cues available to choose the correct migration course?

And why do waders and other migratory birds tend to depart in the hours before sunset?[681] Is this so they can see the setting sun and associated polarisation patterns? Or is it so they can consecutively use and compare the sun, polarisation and star compasses? Maybe at night the air streams are quieter and more predictable than in the day.[456] Leaving in the evening could mean they can capitalise on a final profitable and safe feeding day, or arrive in time for breakfast at the next stopover site. Perhaps there is less chance of overheating and dehydrating during the flight because of the lower temperatures at night?[470] Or is the time before sunset just an arbitrary appointment to ensure that birds planning to migrate do so together? It's likely that all of these reasons are important; long-distance flights are so life changing and dangerous that all circumstances must be optimal. Like a pilot who carefully checks the weather and plane before take-off, a departing migratory bird will evaluate all the conditions that can influence the success of its flight. A bird that doesn't take the checklist seriously is, to all intents and purposes, a dead duck.

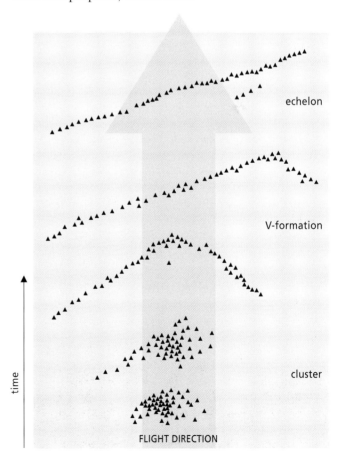

Figure 3.15 Diagram showing how departing flocks of waders arrange themselves as they start a journey of many thousands of kilometres.[681]

# MIGRATION

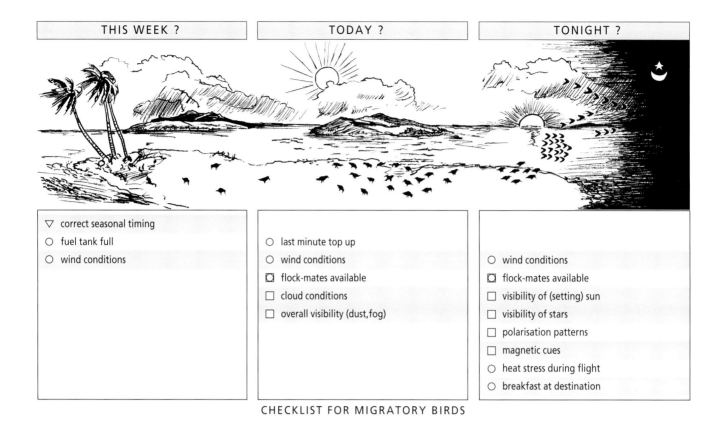

Figure 3.16 A checklist for migratory birds. Only when each item has been 'ticked off' will a bird decide to migrate at that time of year, on that particular day, at that particular moment, with that particular group, in that particular direction. Triangles denote items relating to the time of year, squares denote items relating to orientation, and circles denote energy-related items.[681]

## 3.3 Havens in a hostile world

### 3.3.1 Adequate habitats

You would get a surprise if you saw a curlew sandpiper scurrying down a small path in an inland pine forest. In fact, curlew sandpipers have probably never been seen in forest, and neither have brent goose, red knots or Arctic terns. Maybe once in a while an oystercatcher will breed on the forest margin, but this looks so strange that it's hard to take that oystercatcher seriously.

Every individual bird has very specific environmental

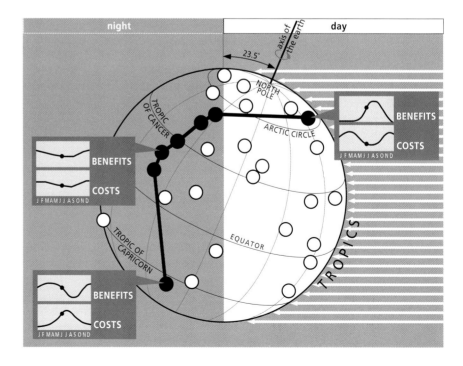

Figure 3.17 The globe with a number of theoretical wader habitats (shown as circles). A potential migration route for a species of wader is shown (filled circles), along with seasonal costs and benefits at three places along the route (insert graphs). As the earth turns on its axis, day and night are experienced every 24 hours. But because the earth is tilted at an angle of 23.5° relative to the sun, the day and night lengths vary with latitude and time of year. On 21 June, as shown here, day lengths are longer in the Northern Hemisphere. At this time of year (indicated as black dots on the graphs) birds in the Northern Hemisphere experience the greatest advantages. In the Southern Hemisphere the patterns are reversed, and waders in the south encounter the best conditions during the northern winter. Conditions in the tropics change much less through the year than in higher or lower latitudes.

118

requirements[858] that vary between species and seasons, but birds can only survive in their species-specific habitat. These habitats must provide suitable food, low risks of predation and disease, and in some seasons, nesting opportunities. Adequate habitats are those that manage to meet all of the bird's requirements.

## 3.3.2 Seasons, day length, and climate

Seasons exist because the earth's axis between the North and South Poles is not perpendicular to the sun (which provides our warmth). In winter the hemisphere (either northern or southern) faces away from the sun; in summer it faces towards it (Figure 3.17). In the Northern Hemisphere (where most waterbirds breed), winter days are short and nights are long and cold, but in summer the days are long and nights are comparatively warm. In the high arctic the sun may shine for the complete rotation of the earth (the whole day) in summer, making it attractive to many lifeforms. But once summer is over and the opposite hemisphere is angling towards the sun, you must be able to either survive the winter or move away. To profit from the polar summer a bird must arrive on time, which requires the ability to move quickly and relatively cheaply. Of more than 9000 bird species in the world, it is predominantly the tens of waterbird species that make use of seasonal opportunities on a worldwide scale.[112, 551]

Different species are specialised in completely different ways, but all must deal with the seasons and the opportunities they offer. Almost all phenomena relating to the migration, foraging behaviour and reproduction of waterbirds are a direct result of the 'tilting of the earth' and the resulting seasons. The seasons ecologically constrain not only the birds, but also their predators, parasites and prey. Let's now have a look at food, parasites and predators; followed by short ecological descriptions of the most important 'adequate' habitats along the East Atlantic flyway.

## 3.3.3 Food, parasites and predators

It is obvious that birds need food, but how much and of what type is more difficult to determine. First, each bird's precise food requirements depend on many varying external and internal conditions; secondly, prey animals are almost never easy prey. Later in this chapter, we'll discuss which factors determine the energy demands, and hence the daily food requirements of waterbirds. In Chapter 4, waterbird food and intertidal foraging will be covered in depth.

There is more than just food to consider, however. Waterbirds ingest a variety of viruses and bacteria with their food, not all of which are necessarily benign. They also consume all sorts of parasitic worms: round worms, tapeworms, flukes and hookworms. These parasites have different life stages, just like butterflies and other insects (which exist as eggs, caterpillars, pupae and finally butterflies). Parasites use different host species at the different stages, and in the final stage can live in the body of a

Some waders, like the spotted redshank, breed in or near pine forests (see also page 250).

bird (which is the so-called end host). In this phase the parasites usually start to produce eggs.

Prey such as mudsnails, bivalves, amphipods and crabs are suitable intermediate hosts for many parasites.[501, 832, 836, 839, 850] When birds eat infected prey they can become the next host for the parasites. Waterbirds can usually live with most parasites (the parasites don't benefit from a dead host), but in some cases parasites can lead to death,[832, 891] as is the case for eider ducks in the Wadden Sea. In years where few mussels and cockles are available, eiders have to eat shore crabs. Shore crabs are the intermediate host of a type of hookworm,[850] so the eider ducks that are already in poor condition from food shortages are repeatedly infected with hookworms. This frequently leads to the death of the eider duck. There is one known case in which hundreds of migrating red knots died from a heavy infestation of hookworms.[663] Bar-tailed godwits infected with tapeworms are usually unable to completely moult into their beautiful rusty-red breeding plumage. Infested godwits are often unable to depart for the breeding grounds in time and with adequate energy reserves.

Intertidal areas are usually unsuitable for micro-organisms and insects, and are therefore relatively disease-free. This is partly because intertidal areas are saline, and it may also be because of the high-intensity solar radiation. In a freshwater marsh, both birds and humans have much higher chances of contracting parasites, particularly if the marsh lies in a warm climatic zone. Malaria has only been eradicated from The Netherlands in the last 50 years, but on the intertidal flats the chances of catching malaria were always small. Is it partly the low parasite numbers that make tidal areas so attractive to birds? Is it a coincidence that some of the 'true' waterbirds that breed exclusively on the high arctic tundra are never seen in freshwater wetlands? Parasites are also scarce on the northern tundra.[661, 736] The image of a tundra with clouds of mosquitoes (which can be intermediate hosts of blood parasites such as malaria) is only true of southern tundra areas such as in Lapland. In the far north where red knots, sanderlings and grey plovers breed, mosquitoes are mostly absent. By breeding in the high Arctic and wintering in tidal areas, it's possible these waders can avoid parasitic infections year-round.[661]

These parasite-free environments may even make it possible for the birds to make their enormous migratory journeys. In parasite-free habitats the birds do not have to invest in costly immune defence mechanisms, and can spend all their energy on energy-consuming migration.[661] People have a higher likelihood of getting ill after doing physical work. Well-trained sportspeople have much bigger chance of falling sick with a cold or the flu than the average human.[282, 404, 602] Tidal areas may be a healthy environment for both people and waterbirds.

### 3.3.4 Safety

So suitable habitats must provide ample food and be free of disease-causing organisms, but waterbirds also want to reduce their risks of being attacked by predators. The tendency for sandpipers to group in large flocks is often explained as a way of decreasing an individual's chance of being caught by a bird of prey.[373, 859] This seemed unlikely in Europe in the 1970s, when the peregrine falcon, a very dangerous bird of prey for waders, was hardly ever seen. However, in the last 30 years, peregrine falcon numbers have increased enormously (Figure 3.18). The risk of being caught by a peregrine is now much higher than it was in the 1970s, particularly if a sandpiper is out on its own or somewhere where it can be surprised by a raptor (such as beside a sea dyke or low dunes). With the return of the peregrine, this explanation for sandpipers' strong flocking behaviour seems very plausible.[141, 686]

### 3.3.5 Tundra

Tundras are seemingly endless expanses of treeless areas in the polar regions (Figure 3.19). The scarce vegetation consists of lichens, mosses, some sedges, and a single species of Arctic willow.[357, 708, 770, 827] The soil is permanently frozen (permafrost) and covered by snow for most of the year (except for June, July, August and sometimes early September). For a waterbird, the short arctic summer is as cold as the winter in the Wadden Sea,[935] and they must deal with food shortages caused by the snow and frozen ground. As soon as the snow melts, tundra

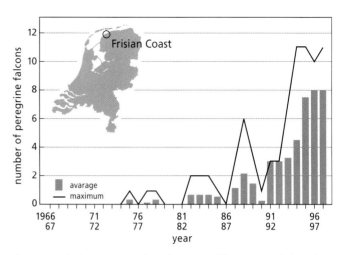

Figure 3.18 The increasing number of peregrine falcons counted along the Frisian Wadden Sea coast during monthly shorebird counts carried out from September to February between 1966 and 1998.[222, 481]

plants commence their short growing and flowering season. This is followed by a short peak in the abundance of spiders and some insects, especially flies, beetles and mosquitoes. Often this insect peak is over by the time the wader chicks hatch. There are no indications that the tundra provides more invertebrate prey than the temperate zone.[879] Predators of eggs (Arctic foxes, gulls and skuas), chicks (Arctic foxes, stoats, skuas, rough-legged buzzards, and snowy owls) and adult birds (stoats, gyr and peregrine falcons, and snowy owls) are present, but their densities vary greatly between places and years.[798] In some years lemmings are numerous and provide the predators with attractive alternative prey, so the waders and their clutches are relatively safe.[745, 812, 890] The intense cold and dryness of the arctic winter, combined with the continual solar radiation of the arctic summer, greatly reduce the numbers of pathogenic organisms and parasites. This means that on the tundra there is little chance of illness.

### 3.3.6 Taiga

The taiga are extensive, damp pine forests, interspersed with peat moors, bogs and heathlands south of the tundra (Figure 3.19). Few humans live there, apart from in Scandinavia where people may occur in relatively high densities. In winter the days are short, and it's as cold, or colder, than on the tundra, but with an even thicker snow layer. On the Siberian taiga the frozen soil, or permafrost, may extend much deeper than on the tundra. Summer lasts longer in the taiga than the tundra, which allows more organisms to complete their life cycles. This applies to both the birds' prey and their parasites. In summer, at least from the longest day until August, the taiga is humming with mosquitoes. Predators are relatively abundant, including mustelids, foxes, and many species of owls and raptors. Waterbirds breeding in the taiga are relatively scarce (mainly the *Tringa* sandpipers) and lead a cryptic life. Even though some *Tringa* sandpipers nest in or between trees, they usually feed in the peat marshes, far from their nest sites. They also take their chicks to feed in these peat marshes.

### 3.3.7 Steppes and artificial steppes

South of the taiga are the deciduous forests (which are by definition unsuitable for waterbirds), and the grass steppes.[41] Although some open steppe areas are of natural origin, nowadays many are the product of modern agriculture. Steppes are the habitat of large grazers. Where once wild bison and aurochs, woolly rhinoceros and mammoths fed, cows, horse and sheep can now be seen. In summer, these grasslands are rich in insects, spiders and worms, in densities comparable to or greater than those on the tundra, providing suitable breeding sites for many species of waders. However, there are also abundant predators and birds of prey on the steppes that create problems for eggs, chicks and sometimes even adult birds. Parasites and pathogenic organisms are probably much more common here than on the breeding grounds further north.

In winter, when the ground is frozen or covered in snow, ice or water, there is nowhere for waders to go. The heavily fertilised grasslands near the coast with their high densities of crane fly (Tipulid) larvae and earthworms can be suitable foraging grounds for species such as the curlew and the oystercatcher. Grassland use is particularly important on days when strong northwesterly storms cover the tidal flats in water, when ice covers the flats, or if human shellfisheries have over-exploited feeding areas. Coastal grasslands are very important foraging sites for brent geese.

### 3.3.8 Dunes, marshes and tidal flats

Waterbirds also breed in dunes and marshes in temperate areas. Ground predators may be absent from some islands, although raptors are usually present. Although, spiders and insects are abundant in the short vegetation over summer, birds such as oystercatchers, avocets and redshanks that breed in the marshes, often feed on the tidal flats. Not only are pathogens and parasites relatively scarce in saline environments, tidal flats can also provide abundant food. The numbers and availability of prey fluctuate strongly with the time of year. Prey availability is usually highest in summer when the animals reproduce and have the fastest growth rates. In winter, benthic animals bury themselves deeper in the sediment and food availability is low, yet the birds' energy requirements are highest due to the low temperatures and strong winds.

Figure 3.19 Circumpolar map showing the tundra and taiga areas of the world, as depicted on the CAFF-website of the World Conservation Monitoring Centre in Cambridge.

## 3.3.9 Fuel stops along the Atlantic coast

To the south of the Wadden Sea and the Dutch delta, are the coasts of France, Spain, Portugal and Morocco and their numerous small intertidal areas (Figure 2.2). These areas are used as small fuel stops by migrating birds travelling between the Wadden Sea and West Africa.[18, 225, 461, 464, 664] Most consist of lagoons behind sand dunes, or estuaries of rivers running into the Atlantic Ocean. They usually have at least a small area of tidal flat, containing a similar benthic fauna to the Wadden Sea.[685, 751] There are small differences in the species composition (differences that increase as you head further south) but here, as in the Wadden Sea, bivalves, crabs and polychaete worms provide suitable food for the birds. In many places, people have created saltpans or fishponds on the tidal flats, sometimes enhancing the feeding opportunities for the hungry migrant birds.[636] The parasites are similar to those in the Wadden Sea. However, as these estuaries and lagoons are small, edge effects are relatively larger. Refuelling migrants find these places much more dangerous than large intertidal areas.[143, 928] Raptors and other predators can easily attack from the landward side. Human hunters made many of these areas unsafe for migrating birds for years. In France, this hunting continues.

## 3.3.10 Banc d'Arguin (Mauritania)

The Banc d'Arguin is a large tidal area on the edge of the Sahara Desert and the Atlantic Ocean, just south of the Tropic of Capricorn (Figure 3.20). Even though the area lies in the tropics, a large cold coastal upwelling means that its climate is not truly hot and tropical.[17, 948] The richness of the feeding grounds may be partly a consequence of this nutrient-rich deep upwelling.[951]

The Banc d'Arguin comprises a complex of tidal flats and islands, and covers around a third the area of the Dutch Wadden Sea. Here, the majority of the tidal flats are seaward of the islands, rather than between the islands and the mainland as they are in the Wadden Sea. No rivers flow into the sea, so the area is entirely marine. Sea grasses cover much of the tidal flats. The benthic fauna of the flats is similar to that in the Wadden Sea (shellfish, worms and crustaceans),[17, 221, 950] but there are more species, the individual organisms are generally smaller, and the total biomass is lower. The availability of small prey, however, is much higher than in the Wadden Sea.[685] Fiddler crabs, which are an important prey here, are rarely found in northern areas. These crabs are mainly eaten by large species such as whimbrels, curlews and gull-billed terns.[245, 975] It is energetically cheaper to winter

There are two very important sites for waterbirds on the coast of tropical West Africa. The Banc d'Arguin in Mauritania (below) is probably the best site for waterbirds as it's not too hot and is relatively open. In Guinea-Bissau the tidal flats are surrounded by dense mangroves and tropical forests (facing page) that conceal a variety of predators and disease-carrying insects.

on the Banc d'Arguin rather than in the Wadden Sea.[935] There are no mosquitoes and the clean saline environment probably results in low numbers of pathogens and parasites. But there are some fearsome predators. Peregrine falcons, lanner falcons, barbary falcons, marsh harriers and short-eared owls hunt waders from the air.[17, 113, 651] Roosting waders must also be wary of ground attacks from jackals, hyenas, desert foxes and sand adders. The Banc d'Arguin has a small human population; some hundreds of fisherfolk, the Imraguen, live there and they leave the birds in peace.

### 3.3.11 Bijagós Archipelago (Guinea-Bissau)

The islands and tidal flats of the Bijagós Archipelago lie in an estuary formed by a complex of large rivers (Figure 3.21). These river mouths go through a large transition from freshwater, through brackish, to saline waters. Most local islands contain some fresh water, are densely covered in rainforest, and encircled by mangrove forests.[947, 965, 967] The climate is tropical with little seasonal change. At midday, temperatures on the beaches and the barren saltflats between the mangroves, the *tannes*, can reach over 50 °C. The tidal flats are partially covered in sea grass with large numbers of fiddler crabs along their edges, just as in the Banc d'Arguin.[965, 967] The benthic fauna is also comparable with that of the Banc d'Arguin in other ways (composition, species diversity, prey size and availability),[685, 967] but physically the river mouths are very different with their brackish, turbid water. Guinea-Bissau lies in the part of Africa historically known as the 'green hell', because of the high risk of white colonists catching tropical diseases and chronic parasitic diseases. Migrating waterbirds may also risk infection near freshwater and on land. The coasts are rich in raptors, which use the mangroves and the forest fringes as hunting perches. There are many ground predators, including mongooses, monkeys and many species of snakes. The edges of the tidal flats in Guinea-Bissau are fairly dangerous for waterbirds. The human population is quite high, but there are few records of people hunting the migrant birds.

### 3.3.12 Coastal lagoons along the African Gold Coast

The weather near the coastal lagoons of the African Gold Coast, from Liberia to Nigeria, is very similar to that in Guinea-Bissau, but other conditions are very different. There are no real tidal flats, apart from the edges of mudflats where mangrove forests reach the sea.[289, 601, 614, 615, 675] Most lagoons are separated from the sea by a sandbar or narrow row of sand dunes. Once every few years, these sandbars are breached and there is direct tidal contact between the sea and lagoon. In some lagoons, people have made this opening to the sea more permanent, so that there is always contact between the lagoon and the sea. In other places, people try to dam the outflow. In all sites the lagoons fill up with freshwater during the rainy season (April to October).[675] Then, in the dry season, the water slowly evaporates and the salinity increases. Over this time, which usually coincides with the northern winter and the presence of many migratory birds, large parts of the lagoon dry out. A lot of waders, particularly curlew sandpipers, make use of the benthic fauna in these slowly drying lagoons for a number of weeks or months. The shallow edges of the lagoons are even more important, and large numbers of *Tringa* sandpipers, herons and terns gather to feed on the numerous small fish (tilapia).[615] The lagoons are also used intensively by people, who harvest the larger tilapia in the wet season. In the dry season, locals take advantage of the high salinity and lowering water levels and harvest the salt. Hunting pressure on waders and waterfowl varies greatly from site to site, but is low in countries such as Ghana. Birds of prey, owls and ground predators are relatively scarce because of a high human presence. Due to the saline nature of the coastal lagoons the number of parasites is low, but more parasite-infested freshwater habitats are always nearby.

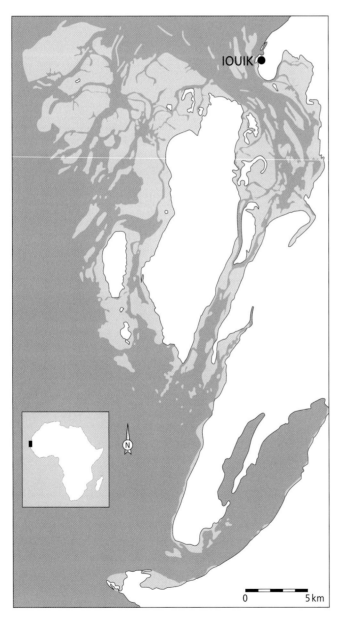

Figure 3.20 Islands and intertidal flats of the Banc d'Arguin in Mauritania.

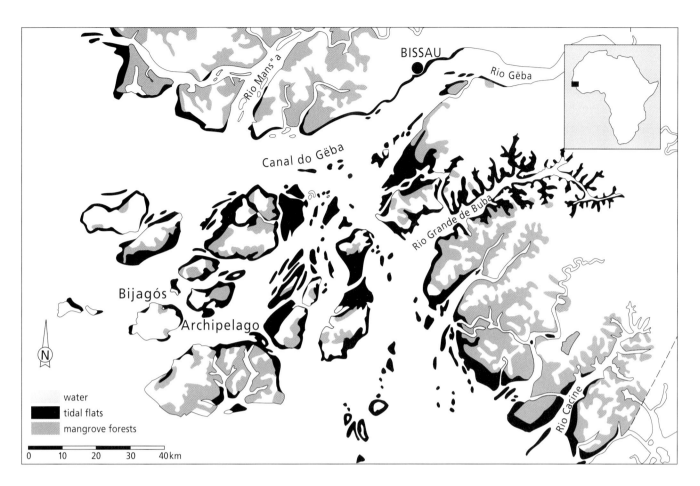

Figure 3.21 Islands and intertidal flats of the Bijagós Archipelago in Guinea-Bissau. One large city lies within this area (Bissau).

### 3.3.13 Tidal flats in Southern Africa

There are a number of temperate inlets and estuaries in Namibia and South Africa.[380, 885, 887] They are similar in appearance and size to the larger of the 'fuel stations' along the Atlantic coast of France, Portugal and Morocco. The most important areas in Southern Africa are Walvis Bay, Sandwich Harbour, the Berg River estuaries and Langebaan Lagoon.[813] These tidal areas have similar benthic fauna to European tidal flats.[685, 714] As they are in the Southern Hemisphere, the seasons are reversed so waders from the north spend the southern summer there. According to some researchers[402] the reversal of the seasons makes these areas especially attractive for northerly breeding waders, as food availability increases enormously during the southern summer. Birds of prey are abundant in Southern Africa, so the predation pressure on waterbirds can be quite high.[380] The chances of parasitic infection are probably similar to those Southern Europe and Northwest Africa. In Southern Africa waterbirds are not hunted.

## 3.4 Balancing the books

### 3.4.1 Time and energy budgets

You can only spend your money once, and the same applies to time and energy. A year has 12 months, or 365 days – apart from leap years in which humans have an extra day. Because time is limited, a breakdown of how waterbirds spend their time and energy through the year is a very useful tool if we want to understand migration. In one year, a bird has to complete a reproductive cycle, moult, and migrate between the breeding and wintering grounds. Let's look at the two populations of bar-tailed godwits that use the East Atlantic flyway as an example (Figure 3.22).[188] One population breeds in the Northern European tundra around the White Sea and winters in Western Europe. The other breeds in Western Siberia, winters in West Africa and must migrate more than four times further than the first population. This is nicely mirrored in how the two populations spend the year. The Afro-Siberians spend almost half of their time (and almost half of their energy) on migration, either in preparation or in actual flying. In contrast, the European birds spend only a quarter of their annual time and energy budget on migration. Instead, they spend more than half of their annual energy budget withstanding the relatively cold Western European winter.

Although the time available is finite, there is no strict limit to the energy budget, as long as demands are covered. Regardless of whether we're talking about a blue whale, an armadillo, or a migrating bird, an animal that spends more than it earns is done for. The bigger the animal, the longer that shortages can be covered. Waterbirds with large fat stores can survive for one to three

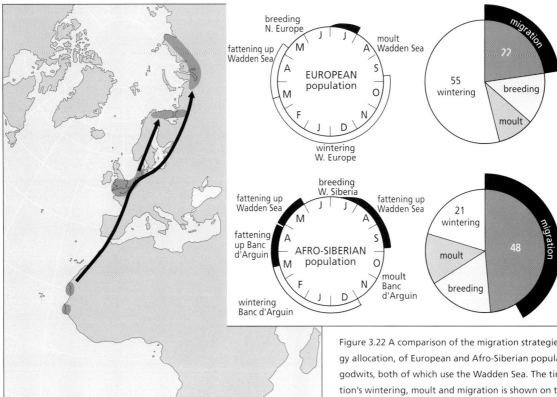

Figure 3.22 A comparison of the migration strategies, and time and energy allocation, of European and Afro-Siberian populations of bar-tailed godwits, both of which use the Wadden Sea. The timing of each population's wintering, moult and migration is shown on the left hand circles. This is then 'translated' into the total energy budget, shown as pie charts on the right, which provide a breakdown of the relative amounts of energy spent on migration, breeding, moult and wintering. The outer dark bands represent the percentage of time each population spends on migration in one way or another.[188]

When red knots arrive on the Siberian tundra in mid June (facing page) it is still very cold and much of the tundra is still covered in snow. It can also be rather unpleasant when they arrive in the Wadden Sea in August (photo below). These dunlins and red knots hold their bills up slightly as they wait for a summer shower to pass.

weeks,[676] as long as they take it easy and it's not too cold.[658, 659, 936]

The good thing about time and energy is that researchers can collect hard data on it. We can make calculations to quantify numerous ecological factors in different places along the East Atlantic flyway. We can determine whether the warm conditions in Southern Africa offer an energetic advantage to birds making the long migration there compared to birds wintering in Europe.[815] The food supplies of these localities can be compared in terms of the ultimate energy income. Using this approach it's possible to see whether the higher costs of living in the cold are compensated for by higher food availability. If the assumption that an animal 'wears down' slower when it spends less energy is true,[514] then a comparison between the energy expenditure of different populations could tell us which migration strategy gives the best chance of survival.[658] For a better understanding of waterbird migration, some knowledge of energy budgets is essential.[265, 682]

## 3.4.2 Balance sheets

An energy budget has both income and expenditure, and each can be broken down into different components. These components need to be defined in a way that is measurable in practice (Figure 3.23). Let's look at the sources of energy expenditure and income.

*Basal metabolism* – The basal metabolism represents the energy used when a bird is asleep, is not digesting food, and doesn't need to produce any heat to keep its body temperature constant. To measure this, a bird has to be placed in a thermally neutral environment, it has to be resting and have an empty stomach. Basal metabolism is mostly determined by the abdominal organs and brain, and only partially by the muscles. To see whether there are systematic differences between species and birds from different climate zones, basal metabolism has been measured in a large number of different waders.[460, 466, 515] Differences do exist between birds from different places and are probably the result of local adaptations to climate and 'work load'. The colder the environment and the harder the birds have to work, the higher the basal metabolic rate.[692] The basal metabolism reflects the size of the metabolically active organs of a bird.[692, 913] These organs change dramatically under different circumstances, and the basal metabolism of migratory birds changes accordingly.[689]

*Thermoregulation* – When it is too cold for basal metabolism to maintain constant body temperature, birds must 'thermoregulate' to prevent their body temperatures dropping. On average, birds have a body temperature of 40 - 43 °C,[706] 3 - 6 °C higher than the 37 °C body temperature of most mammals (including humans). Cold winter weather means the birds require a lot of energy to compensate for the loss of heat to the environment and maintain body temperature at 41 °C.[936] How can they restrict their heat loss?

### Box 3.4 Using copper knots to estimate energy consumption/use

Copper models of knots were placed outside in different natural habitats to derive formulae (*f*) to predict the maintenance metabolism of free-living knots in the wild, based on local air temperature, wind speed and global solar radiation (*f*{temp, wind, radiation}). These copper knots were hollow knot-shaped models made of two thin layers of copper with a heating wire between the copper layers. Each copper model was wrapped in the skin and feathers of a real knot. The copper knots were placed on site and heated to 41 °C (like most birds, knots are quite warm) with a 12 V car battery. Energy use/consumption increases with lower temperature, less radiation, and increasing wind speed. The energy consumption of the model provides a measure of the 'windchill' of real knots. Using a mathematical model we were able to calculate the energy consumption of the copper knots (upper panel). By calibrating these copper knots with living knots in the laboratory (lower panel), we obtained the final formulae that can then be used to estimate the maintenance metabolism of free-living knots on the basis of their location, behaviour and microhabitat (middle panel).

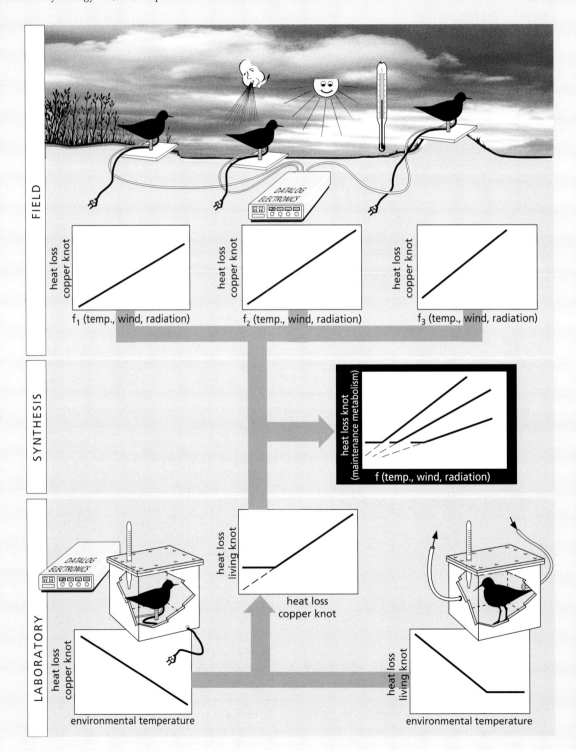

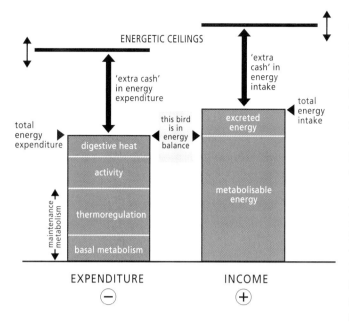

Figure 3.23 Diagram of a daily energy budget with the various components of energy expenditure on the left, and energy income on the right. The presence of possible energetic ceilings is shown, in relation to both energy expenditure and energy intake. In this example the bird has a neutral energy balance: its total energy expenditure equals the amount of metabolically available energy taken in.

First of all they can adjust their insulating layer, mainly through their plumage. By erecting feathers, they trap more air between their feathers and their bodies, increasing the insulation. They can also change their blood circulation. Decreasing the flow of blood to certain parts of the skin will decrease their surface temperature. This means less heat is lost, since the heat exchange between the skin and the air layer depends on the temperature difference between them.

Secondly, changes in body posture can conserve heat. Poorly insulated parts can be covered by body feathers: the head tucked between the shoulders, the wing bend covered by breast feathers and one leg raised into the belly plumage.

Finally, there are behavioural adaptations. Birds can take shelter from the wind amongst vegetation, flockmates, or other waders. There are many ways birds can lose or gain energy to the environment (Figure 3.24). To know how much heat waterbirds exchange in the field, it's not enough to simply measure temperature. Instead, the influences of the weather can be integrated simultaneously by using a heated model of a bird's body (Box 3.4). Using such a 'copper bird' we can determine the influence of the direct environment, such as vegetation or flock mates, on the energy flow.[99, 935, 936]

*Maintenance metabolism* – What we measure in the heated copper bird models represents the basal metabolism plus the additional costs of thermoregulation. This is what we call the maintenance metabolism. With weather data and information about the use of different habitats, good estimates can be made of the maintenance metabolism of different species in a variety of situations.[674, 936] Open intertidal flats turn out to be expensive places to stay. Amongst thick vegetation and in dense groups of conspecifics less heat is lost. If a bird turns its back to the wind its feathers are blown up by the wind and its insulation decreases. In the field, waterbirds usually stand with their heads facing into the wind. Standing side on to the wind increases heat loss by 9% in red knots.[935, 936] The absolute amount of energy that can be saved by flocking and orientation depends on the weather. The colder it is (low temperature, high wind speed, little sunshine) the higher the possible savings. So we expect that birds should make more of an effort to save energy in the cold. They do; foraging knots walk closer together at lower ambient temperatures, and walk more with their heads into the wind.[935] Since the maintenance metabolism may constitute more than half of the total energy expenditure, maintenance metabolism provides a good approximation of the relative energy costs.[99, 935] However, waterbirds are also very active; they spend a lot of time on the go.

*Activity* – Flying, walking and foraging are energetically expensive. Activity costs depend on the time-budget of the bird, but are very difficult to measure in practice.[658] Instead, coarse estimates are usually made. It's

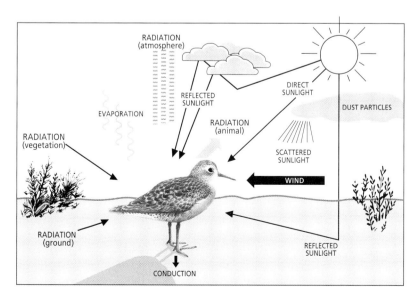

Figure 3.24 Methods of heat transfer between a shorebird, in this case a red knot, and its natural environment.[935]

This turnstone (top photo) does not seem to be bothered by the cold on a sunny, calm day near the North Pole. However these knots wintering in an ice-covered salt marsh in The Netherlands do seem to be feeling the cold (lower photo). They try to reduce their heat loss by standing close together with their feathers fluffed up.

generally assumed that walking and foraging costs three to five times as much energy as the basal metabolism. This is much cheaper than flight, which costs at least ten times more than basal metabolism. A further complication is that birds may use part of the heat generated during activity to cover the costs of thermoregulation.[98] The estimated costs of thermoregulation and activity can't just be added up. Activity is an expense column that requires more research.

*Digestive heat* – A feeding waterbird uses heat to warm its cold-blooded food up to the bird's own body temperature of 41 °C. Heat is produced during digestion, which could in principle be used for thermoregulation (just like activity generated heat).[539] Digestive heat depends on the source and amount of food ingested, and cannot be kept completely separate from thermoregulation costs.

*Energetic ceilings* – Obviously, there must be limits to the energy coming in and being spent.[187, 376, 917] The highest possible level of energy expenditure may depend on the maximum amount of oxygen that the lungs can extract from the air and the blood can circulate to the working organs. When oxygen limits energy expenditure the ceiling can be reached in just a few minutes, as it is with athletes suffering cramp. Cramp occurs when the athletes' muscles receive insufficient oxygen and begin to use anaerobic respiration. But as well as releasing energy, anaerobic respiration releases an acid into the muscles causing serious pain, which is only removed once the muscle activity decreases.[22] Energy expenditure may also be limited by the speed at which the liver and kidneys can get rid of the waste products produced by cell activity. Here, the timescale is somewhat longer; it can take a whole day before the build up of waste products in the liver leads to problems. Energy intake can also be limited by the digestive tract, the liver, or the ability of the blood to circulate nutrients to where they need to be used or stored. There are good indications that nutrient intake can limit human athletes. Participants in the Tour de France and comparable marathons (endurance experiments that may last for weeks) can only manage to continue if they are on a very light and sugar-rich diet.[925, 926] They reach an energetic ceiling at about five times their basal metabolism. This is also the level of energy expenditure of waders that winter in the Wadden Sea.[658, 935]

*Extra cash* – The difference between actual energy expenditure and the energetic expenditure ceiling can be, but is not always, used for energetically expensive activities such as display and fights. If birds have to be as economical as possible (as they often do), then this energy store is seldom used.

*Energy available for metabolism* – Not all of the energy present in food is available to the animals eating it, as their digestive tracts are unable to extract all the nutrients present.[454] Some energy-rich compounds can't pass through the bird's intestinal wall as their breakdown is not completed during the limited time the food is in the intestine. Plants are often hard to digest, so herbivorous birds and mammals have a rich bacterial flora in their intestines to help them with this. Herbivores are careful to feed on only the most digestible plant species and parts. Animal food is generally much easier to break down, although some waterbird foods have very hard parts that can pass through the bird's intestine intact (such as the shells and chitin of molluscs, worms and crabs).[171, 973] In fact, some mudsnails swallowed by shelducks, survived passing through the ducks' entire digestive tract.[118] The snails managed this by sealing off their very hard shells with a little lid of indigestible horny material.

*Excreted energy* – This cost column is very easily defined: all the energy that leaves the body in droppings. But even this isn't as straightforward as it first seems. All

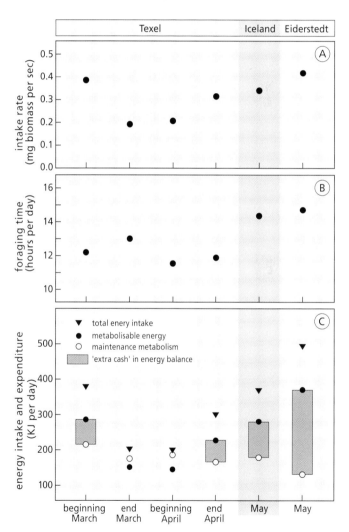

Figure 3.25 The energy intake rate (A), amount of time spent foraging (B) and a reconstruction of the total energy intake and expenditure (C) of red knots at three different sites (Texel in the Dutch Wadden Sea, Eiderstedt in the German Wadden Sea, and Iceland) in spring. The knots studied in Texel and Iceland belonged to the subspecies *islandica* and were en route to their breeding grounds in Greenland and northeast Canada. The Eiderstedt knots belonged to the subspecies *canutus* and were en route to their breeding grounds in Siberia.[688]

# MIGRATION

organisms continuously repair themselves and excrete broken down body proteins.[454] This excretion increases when a bird isn't taking in enough energy and is forced to burn it's own body tissues.

*Flexible energy intake* – This is a very important part of the energy budget, as it must cover the costs of fat storage prior to long-distance migration and plumage replacement during moult. Fattening and moulting waterbirds must eat more than is required for their own maintenance and activity and store the surplus energy and nutrients as fat, larger body organs, and feathers. Body moult can also add to thermoregulation costs, as feather growth in a cold environment requires extra blood circulation near the skin surface.[538]

## 3.4.3 Cost-benefit analysis

Just as researchers can calculate energy budgets on timescales varying from one day to one year (compare Figure 3.22 with Figure 3.23), the timeframes of cost-benefit analyses can also vary. Let's start by comparing estimates of energy income and expenditure for knots on the spring migration through the Wadden Sea and Iceland, and then compare the energetic consequences of wintering in the cold (the Wadden Sea) with wintering in the tropics (on the Banc d'Arguin).

Estimates of the food intake of knots were made at three sites, six times between early March and late April (Figure 3.25).[688] The intake rates varied by a factor of two, more than the variation in the daily foraging time (which is limited by the times of low and high water). Multiplying the intake rate by the daily foraging time and the energetic content of the prey gives the daily energy intake for each of the six time periods (Figure 3.25). Subtracting the excreted energy from this gives us the daily energy intake available for metabolism. These intake estimates can then be compared with cost estimates of the maintenance metabolism. The daily intake of

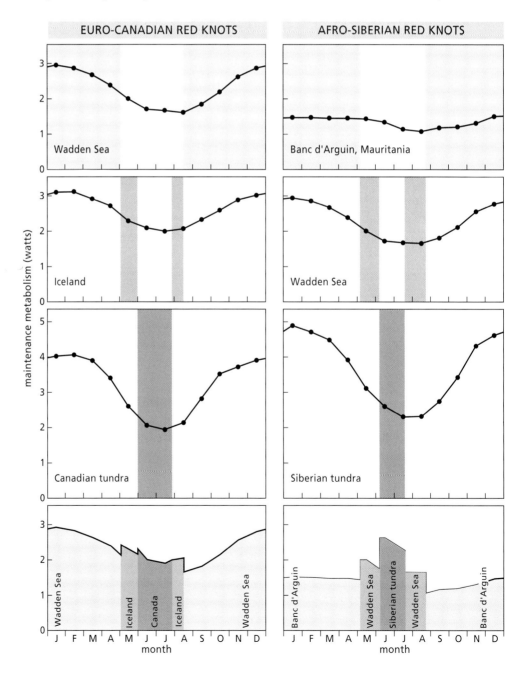

Figure 3.26 Comparison of the maintenance metabolism of two subspecies of red knots, one of which (*islandica*) breeds in Greenland and northeast Canada and winters in Western Europe, while the other (*canutus*) breeds in Siberia and winters in West Africa. The upper three graphs show the maintenance metabolism requirements during the year at each of the three main sites used by the two subspecies. The bottom graphs show the annual maintenance metabolism experienced by the two subspecies.

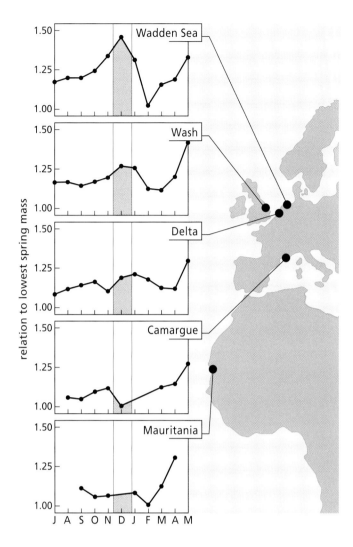

Figure 3.27 The mass cycles of dunlins on different wintering grounds.[177, 288, 439, 789, 658] Only dunlins in the north put on a reasonable fat store in winter.

available energy should always be more than the maintenance metabolism, and we can see this is the case (Figure 3.25). The difference between these two estimates, the 'spending money', can be used to cover activity costs and for fattening up. In some cases this difference is small. If we look at some average estimates of activity costs, knots in Iceland should be able to store around 0.5 g fat per day, and knots in the German Wadden Sea around 4 g per day. The first estimate is on the low side, the second on the high.[688] Nevertheless, these calculations show that creating energy budgets can provide realistic estimates of fattening rates.[308, 877]

So what about comparing the energetic consequences for one species of using two different flyways? Let's look at two populations of red knots, both of which use the Wadden Sea, but at different times of the year. Breeding birds from Greenland and Northeastern Canada winter in the Wadden Sea, while the Siberian breeding knots winter in West Africa. Most Siberian knots stay in Mauritania (which will be discussed in the following calculations) or in Guinea-Bissau. We'll restrict ourselves to comparisons of the maintenance metabolism, as this the only part of the budget where reliable results have been gathered.[935] Using weather data from all the sites where the knots stay,[585] and knowledge of their habitat use at these sites, we can calculate the annual cycles of maintenance metabolism for each site from the predicting equations (Box 3.4, Figure 3.26). From these annual patterns, we can then cut and paste relevant 'items' for the different knot populations. It's immediately obvious that costs are very different for the two populations: the Siberians' costs are far higher in the summer, the Canadians' higher in the winter (Figure 3.26).

There is a big difference in costs between wintering in the Wadden Sea or on the tropical Banc d'Arguin (Figure 3.26). In West Africa the maintenance metabolism costs are half of that in Western Europe and the bird can profit from these savings for the whole eight-month wintering period. Knots in West Africa usually spend less time foraging and flying than their conspecifics wintering in the Wadden Sea, making it even cheaper to live in the tropics. But to be able to winter in Mauritania, a knot has to fly an extra 4 – 5,000 km. This extra flight time is more than made up for by the cost savings due to the favourable West African climate. It's estimated that knots that winter in West Africa annually use half as much energy as knots that winter in Western Europe (including the spring and autumn migration and breeding season).[658] This knot research corroborates some rough estimates of the energetic consequences of wintering in the tropics instead of the Wadden Sea for bar-tailed godwits, for which fewer data were available (Figure 3.22).[188]

### 3.4.4 Freeze or fly

In some winters, the Wadden Sea gets very cold and large areas freeze over. In these harsh conditions some waterbirds migrate further south on the spur of the moment.[211] Such 'frost flights' are especially characteristic of oystercatchers.[124, 420, 421, 425] They remain in the Wadden Sea for as long as they can, until they have only just enough fat left to make it to the French coast. The French coast is a far from ideal travel destination, since even if the coast is free of ice, it's never free of hunters. Many birds are killed when they reach France.

Other oystercatchers, knots and dunlins hang on as long as possible, even once most of the tidal flats have disappeared under ice. These birds must survive on their fat reserves.[161, 646, 992] Winter fat reserves are only found in waders that winter in northern latitudes (Figure 3.27). In West and Southern Africa waders do not have a mass peak in the middle of the northern winter.[177, 811] Strangely enough, purple sandpipers also don't show this sort of mass peak, even when they winter north of the Arctic Circle.[817, 818] Purple sandpipers forage amongst seaweed on rocky shores for a variety of small prey. Unlike tidal flats, the steep and wave-swept rocky shores are never covered in snow and ice. This means purple sandpipers are not at risk of suddenly being left without food and don't need the 'life insurance' fat reserves provide. The chance of African tidal flats being covered by snow and ice is also nil. Since their foraging areas are always available, African wintering birds have no need to take out life insurance either.

By dividing the size of their energy store by estimates of maintenance metabolism, it's possible to see how long waders can survive if they are without food in midwinter i.e. if they just wait patiently till conditions improve (or till death arrives). Knots can survive for four to five days on their fat reserves.[658] Larger waders like curlews and oystercatchers, can hang on for more than a week.[420, 425] It's so much colder in Western Europe than in Africa, that any differences in the energy stores of knots and other waders between the two sites are completely consumed by their greater energy use in Europe. The higher risk of ending up in a Wadden Sea winter without food is not compensated for by proportionately bigger fat stores: they don't store enough energy to fully compensate.[658] This shows how 'expensive' such life insurance is, and that knots and other waterbirds prefer to carry as little weight as possible. They are only prepared to get fat if they really have to, either in very risky wintering sites or just before migration.

## 3.5 Migration strategies

### 3.5.1 Migration strategies?

Waterbirds use the Wadden Sea for many different reasons. Some species breed here and try to remain year-round. Oystercatchers will only head south when starvation threatens in a frozen Wadden Sea.[420] The Wadden Sea is the southern limit of some species' wintering areas

Some avocets from the Wadden Sea winter along the coasts of France and Portugal. Only in mild winters do small numbers of avocets remain in the Wadden Sea (photo above).

There are often mangrove forests on tropical tidal flats, like those shown here in Guinea-Bissau (page 135, lower photo). Even though mangroves are not the preferred habitat of most waterbirds, birds like these curlew sandpipers can still make use of them by roosting on aerial roots at high tide. However the abundant predators often cause the birds to fly off in a panic (page 135, top photo).

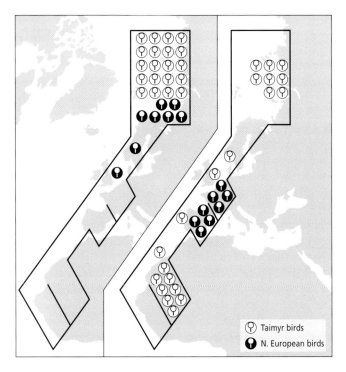

Figure 3.28 Those who are first to leave from the more southerly breeding grounds get first choice of the most northerly wintering grounds (left). As these northerly wintering grounds fill up, the birds that had to depart later from the breeding grounds must fly on to more southerly intertidal areas (right). This is how the leapfrog migration pattern of the bar-tailed godwit (also discussed in Figure 3.3), which inspired this diagram, has been explained.[5]

(such as eider ducks), while for others it's their northern limit. Species such as curlew sandpipers and little stints only forage in the Wadden Sea whilst travelling between the far south and the high north. For others, like the dunlin[306] and common sandpiper, Western Europe is just one link in a chain of stopover sites. Yet others are completely dependent on the tidal flats neighbouring the North Sea outside of the breeding season (Canadian knots). Each of these reasons could be called a migration strategy, in the sense that alternative ways of migrating don't appear to have worked. Even though Western Europe is relatively close to their breeding grounds, curlew sandpipers never winter in Western Europe.

### 3.5.2 Why leapfrog?

One of the most intriguing migration patterns known is the way some northerly breeding populations fly past southerly breeding populations to reach their wintering

grounds (Figure 3.3). This 'leapfrog' pattern[761] in which northernmost breeding individuals use the most southern wintering grounds, has given rise to much debate amongst migratory bird biologists. All participants in the debate assume that there is competition for the best feeding grounds, mates and nest sites within species. One theory is that birds try to winter as near as possible to their breeding grounds, reducing their energetic costs and the risks of migration.[643, 644, 647] The early bird (from the breeding areas) gets the worm (in the wintering areas). The southernmost breeders finish breeding first and can therefore occupy the closest wintering areas (Figure 3.28). The northern breeders arrive later and find the closest wintering areas are already full. If they are unable to chase off the occupants (for example, by being big and strong), they must fly further south to wintering areas where there is still space available. In support of this theory, there are indications that in various wader species the individuals wintering the furthest north are the largest birds.[647] Researchers working on grey plovers have found that the biggest individuals are indeed dominant.[271, 867]

The pattern in which the largest members of species use the most northerly wintering grounds also fits with the idea that the smaller a bird is, the more it feels the cold, and the further south it must winter.[467, 587, 588, 658] Southern wintering grounds have better weather (lowering the birds' maintenance metabolism costs) than northern areas, and may also have more food available.[402, 568] This explanation also fits with another theory, that wintering close to the breeding grounds is the best solution for southern breeders as it assists them in estimating the local weather conditions on the breeding grounds in spring. This way, they can be sure that they arrive on the breeding grounds just as the weather becomes suitable.[8, 10] It's important to arrive on the breeding grounds early so you can secure the most attractive partner (Chapter 5). But if you are too early and future territories are still covered in snow and ice, you risk starving to death.[575]

These discussions about leapfrogging don't lead to a clear conclusion. However, they do make the 'biological ingredients' for a universal explanation clear. Competition (for foraging areas, food, breeding areas or mates) is important. The costs of migration both in terms of energy and the risk of death, also play a role. This also applies to the relative energetic costs of staying in different climate zones, dependent on body size. The importance of arriving on the breeding grounds on time, and the possibility of planning this arrival time well, also play a role. Maybe latitude and body-weight dependent predation risks are also important, and perhaps the risk of disease, and of navigation difficulties once en route. Knowing the importance of all these factors would help elucidate the different migration patterns. It would explain not just leapfrog migration, but also stepwise, and crossover migration (Figure 3.3). This knowledge could also clarify why the females of some species winter further south than the males, while for other species the reverse is true.[467, 587, 588]

**Intertidal areas in Western Europe**
*Advantages*
- Closer to the breeding grounds
- Greater density of food organisms
- Individual prey items are often larger than those on the West African tidal flats

*Disadvantages*
- Energy use is high and borders the maximum that is physiologically attainable
- Periodically, large areas of the tidal flats are covered in snow and ice
- Large fat reserves are required to survive periods when no food can be found; the extra fat reduces manoeuvrability making the birds more vulnerable to predators
- Prey bury deep in the sediment making them (temporarily) unavailable
- Low temperatures and the reduction in plumage insulation during moult make the times suitable for moulting very short

*Consequences*
- Periodic high mortality
- Need to search for food over a much larger area than in West Africa
- May be able to make use of the higher food availability in spring with relatively few competitors

**Intertidal areas in West Africa**
*Advantages*
- Much lower energy consumption due to warmer weather
- More stable and predictable food supply
- Despite the longer journey, travel costs have not increased proportionally due to the relatively favourable prevailing winds

*Disadvantages*
- Individual prey items are often very small
- Heat stress can restrict the ability to 'work hard'

*Consequences*
- Lower annual mortality?
- Much higher densities of foraging birds on exposed tidal flats
- Competition for food during the build-up of energy stores preceding the migration north

Table 3.2 The ecological advantages and disadvantages of wintering in different intertidal areas.[568, 658]

### 3.5.3 Wintering in West Africa as opposed to Western Europe

When the first wader researchers returned from Mauritania in 1973, with stories of the birds of the Banc d'Arguin,[176] they were not only deeply impressed by the huge numbers of birds, but also by the abundance of food. The tidal flats were packed full of small shells, completely different from what people were used to seeing on English tidal flats. In 1980, a follow-up survey of the Banc d'Arguin waterbirds and their food confirmed that the birds were indeed very numerous, but found that there wasn't that much food.[17] The 1973 researchers had been mistaken. The huge numbers of shells that they found on tidal flat sediments had been dead for hundreds or thousands of years. As the Banc d'Arguin lacks big sand eaters like lugworms (Chapter 4), and has low rates of sedimentation, shells remain on the surface instead of being worked into the sediment. The preliminary conclusions in 1980 were that were huge numbers of birds, but not so much food.[17, 221] This has been confirmed by more recent 'in depth' research.[950]

The expectation that Banc d'Arguin was a 'land of milk and honey' for waterbirds could also be dismissed by observations that by far the majority of species spent almost all their available time foraging.[17, 221] In spring, during the fattening period, there was even greater foraging activity, especially at night.[984] It was discovered that the fattening rates on the Banc d'Arguin were relatively low.[985] And as we saw previously, it's only possible for bar-tailed godwits and knots to reach the Western European tidal flats in a single flight with the help of tailwinds which blow at high altitudes over the Sahara.[669] The Banc d'Arguin, a relatively small area on the edge of the Sahara, is certainly not a land of milk and honey. So why do a quarter of all coastal waders on the East Atlantic flyway go to the trouble of flying there?

Maybe we've made another mistake. Maybe the intertidal flats of West Africa actually are lands of milk and honey and that's why they become so densely populated by birds. So densely, that in spring, when all the birds want to eat more so that they can reach the arctic breeding grounds in good condition and with beautiful plumage, the Banc d'Arguin ecosystem is unable to fulfil their needs. There are arguments that support this latter theory.[243, 568] In late summer, many West African wintering waders migrate south through Western European tidal areas when these areas are still unoccupied by the northerly wintering birds. The dominant adult birds migrate from Western Europe to West Africa while the juveniles are yet to arrive in the Wadden Sea. If these adult birds really didn't want to travel to the West African tidal flats, wouldn't they chase the weaker and later arriving young birds to West Africa? And why do they fly to West Africa straight after breeding, instead of spending more time in Western Europe and beginning primary moult as ruffs and black-tailed godwits do? These two species winter in inland African wetlands, areas that have relatively unpredictable ecological conditions. Birds arriving on the Banc d'Arguin or on the Bijagós Archipelago can at least be sure that the foraging areas are available half of the time (dependent on the tides), and that there is something to eat. The prey might be a bit small, but it's warm, so the birds don't use so much energy (Table 3.2).

The difficulties waders in West Africa face in building up fat reserves could result from strong competition for food. This competition occurs because these areas are so attractive to many waterbirds.[568] The journey to West Africa might be long and may make it difficult to return to breeding areas on time, but West African tidal flats certainly have many advantages over Western European ones (Table 3.2).

### 3.5.4 Hop, skip or jump?

Some of the fat shortages that waterbirds experience before departing from West Africa could be compensated for if they didn't find it necessary to fly from Mauritania

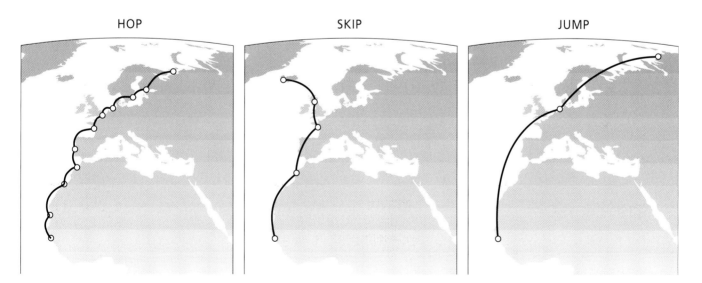

Figure 3.29 Hop, skip or jump? The different ways that shorebirds can travel from West Africa to the arctic breeding grounds via the Wadden Sea and other intertidal areas in Western Europe.[654]

to the Wadden Sea in a single long flight.[448, 985] A single flight 'jumper' uses much more energy than a multiple flight 'hopper' (Figure 3.29),[654] since the jumper faces greater costs under the law of diminishing returns (Figure 3.5). To remain in the air for a long time and fly further in one go, a bird needs to store a lot of extra fat so it has enough power to keep all the rest of the required fat airborne. A 'skipper' taking medium-length journeys has to store less fat each time, and this is even more the case for a short-length 'hopper'. By the time they arrive in Western Europe, hoppers have used much less total energy than jumpers.[654] Of course, each time they arrive in a new area they must find food and safe places. Most passerines, which also migrate over very large distances, use the hop strategy, so it is surprising that of the waterbirds only redshanks and curlews fly the 4 - 5,000 km to Western Europe in short stages.[654, 985] Some dunlins could be included as skippers. Why do so few waterbird species use hop or skip strategies en route from the West African tidal areas to the arctic breeding grounds?

To be a hopper or a skipper, and to be able to fuel up en route, requires suitable fuel stops! Many tidal areas lie exactly on the migratory route between the Banc d'Arguin and the Wadden Sea (Figure 2.2). Most areas are small, but the tidal flats are usually rich in benthic prey.[652, 685] Redshanks and dunlins make good use of these.[461] You hardly ever encounter knots and bar-tailed godwits there, even though there is quite a lot of food for them. Maybe the areas are just not attractive enough, in spite of the food.[654] Maybe there are dangerous parasites that we don't know about,[661] or perhaps species like knots and bar-tailed godwits think the small enclosed tidal areas are too dangerous. The surface area of open tidal flats is quite small, which means that the chance of surprise attacks by raptors (and human hunters) is greater than on the open tidal flats of Mauritania or the Wadden Sea. To fatten up quickly, turnstones, preparing for a long flight from the Scottish coast to the arctic breeding grounds, spend less time watching for predators than they do in winter.[567] Fattening up is best done in a very safe place.

It might be even more subtle than this. If both time and energy are important,[11] West African wintering birds would want to fly as cheaply and quickly as possible from Mauritania through Western Europe to the arctic breeding grounds. This makes it worthwhile to store more fat than is needed to just reach the nearest suitable habitat.[365] So, on the Banc d'Arguin, birds overload themselves so that they can skip the Moroccan and French coastal areas, and they do this again in the Wadden Sea so they can fly to Siberia in one go. Achievable fattening rates and the time it takes to find good foraging areas are traded against the additional flight costs of carrying extra energy reserves.

It is intriguing to think that of all waterbirds, the typical intertidal waders don't use the option of hopping from one small tidal area to another as they travel between their wintering and breeding grounds. The *Tringa*

Grey plovers and ringed plovers that winter on the edge of the Sahara (facing page, top) and bar-tailed godwits that winter on a Senegalese beach (photo right) must all make it back to the northern tundras, 10 000 km away, in time to breed. This male grey plover with his first chick (facing page, bottom) has been very successful at this. The journey from West Africa to Siberia is usually completed in two long flights.

Figure 3.30 The most important 'decisions' (see Box 3.1) a waterbird must make during its annual migratory journeys so that it can achieve (biological) success. By making the right choice at exactly the right time the bird increases its chance of surviving and producing offspring.[236, 248]

sandpipers that breed in the taiga, such the common sandpipers seen in the Wadden Sea, are typical hoppers. Black-tailed godwits and ruffs that use inland wetlands are not jumpers either. The true waterbirds of this book are jumpers and only make stopovers during their long migrations when it is absolutely necessary because of a serious mishap.

### 3.5.5 Optimal travel plans

To survive, find a mate, and produce offspring, migrating waterbirds must do many things well. During the year, they have to make all sorts of important subconscious and conscious decisions (Figure 3.30). They have to combine the necessary with the possible to the best effect. For every species and every individual the decisions and considerations are very different, but in all cases we as observers are dealing with the birds whose ancestors made the right choices in biological terms; their ancestors produced the most offspring. We will discuss this in more detail in Chapter 5, but to understand migration strategies we need to know how different decisions affect the number of offspring.

We only know some of the possible consequences of different decisions for a few species. It seems to be just enough for us to build computer models which can calculate the migration decision trees[236, 915] using the mathematical technique of 'stochastic-dynamic programming'.[534] In this method, a problem is solved backwards. A waterbird that arrives on the breeding grounds too early may starve and freeze to death before it can even find a mate. But if the bird arrives too late, the chicks miss out on the food peak and too little of the arctic summer remains to raise healthy offspring. A lightweight waterbird that arrives on the breeding grounds with low energy reserves is also not good (it has a higher chance of starving, needs more time to forage, less time for romance etc.). The model shows that the heavier a female brent goose is when she departs from the Wadden Sea, the better the chance that she will return with chicks.[208, 209] There are costs to arriving early or late, or being too skinny on arrival (Figure 3.31). Even though such mathematical curves have not been measured for any species, quite realistic calculations can be made which have the usual form (flowing downwards like a row of playground slides).

So how are these calculations made? It starts with a chain of suitable habitats whose ecological conditions (food richness, climatic conditions, predator density etc.) and flight distances are known, and with the optimal weight and time of arrival at the breeding area. From these, a researcher can calculate what the best fattening speed at the last fuelling station should be, and when the birds should start this fattening. This is the first calculation. You can repeat this calculation, and work backwards to the wintering grounds. From these calculations you end up with a large table showing each site's best fattening rate and departure time. You can then use this to go through the series of decisions in the normal order. This is necessary as in practise, chance events often mean that the back-calculated optimal weight and distance trajectory on the migration from wintering to breeding grounds is not reached. A certain decision doesn't lead to a precisely described next state with certainty, but to a random normal distribution of future states. On the basis of the body weight that an individual bird has at the time of departure, all the possible decisions following from this, and the bad and good luck the bird encounters along the way, can be simulated. The average of thousands of individual simulations gives the predicted group of optimal decision rules.[236, 915] These decision rules consist mainly of (time and site dependent) weight levels at which the bird has to stop foraging and depart for the next stopover. Until such a weight is reached, the bird has to continue feeding. By following these rules, this virtual migrant reduces its chances of ending up in situations in which further migrations are no longer useful, since it's physically impossible for it to arrive on the breeding grounds on time.

When finally a model has been produced that simulates waterbird behaviour accurately and has been validated by observations, researchers can start believing that the information put into the model is not only necessary but is an accurate description of the real situation. Once this happens, the model can be used to determine what the effects on the number of offspring would be if you remove one or more fuelling stations, lower the food density in the intertidal areas used, increase predator density, add different types of predation risks, and so on. However, the model isn't that well developed yet, since even simulations of the migration of one of the best-researched migratory intertidal birds, the red knot,[236, 915] don't really reflect what we know from the field.[666]

It is not possible to adequately simulate the timing of waterbird migration with the ambitious stochastic dynamic programmes and through this predict the effects of ecological changes. However, such modelling does help put different phenomena into context.[248] It was possible, using combinations of fattening speeds and predation pressures at different fuelling stations, and uncertainties about fattening speeds and wind conditions en route, to simulate migration situations in which birds could store so much fuel at some locations that they could skip intermediate fuel stations according to the optimal decision rules.[915] This suggests that, in principle, collecting the necessary field data plus improving the dynamic models, could lead to very realistic simulations of bird migration.[230] Time will tell.

## 3.6 The origin of waterbird migration

### 3.6.1 Different questions

At first glance, it seems logical for bird migration researchers to be interested in the question: why exactly do birds migrate? But when you think about it, you realise that it would be rather strange if birds didn't. After all, flying birds can easily move, and since they may find the best conditions at different sites depending on the season, why not make use of their abilities to fly from one good place to another! A better question might be:[5, 6] why don't all flying birds migrate? But answering this doesn't really help us understand waterbird migration, nor the migration patterns that we can now map using various observation techniques.

When we ask questions about bird migration, we need to distinguish between questions about how the different migration patterns work now, and how these migration patterns originated.[658, 699] There are three aspects of the current workings of migratory behaviour that we can investigate: (1) how the migration patterns of individual waterbirds develop during their life cycle,[180, 747] (2) the numerous interactions between the environment and physiology that make migratory behaviour possible,[47] and (3) the results, in terms of survival and number of offspring produced, of various migration strategies.[493, 592, 761] Until now, this book has focused mainly on the physiological aspects of migration (Section 3.2). Only in the previous paragraph did we try to investigate the function of the migratory behaviour in waterbirds (Section 3.3).

Questions about the origin of migration patterns have rather a historical nature, and can only be answered by trying to reconstruct previous migration patterns in the world as it used to be. Unfortunately these histories can only be judged on consistency and logic; they can never be verified experimentally. Still, let's have a look at what information there is on the origin of waterbird migration.

### 3.6.2 Subspecies formation and ice ages

To ensure that birds can fly well and as cheaply as possible, their bones are lightly built and hollow. This is why they don't make good fossils. Every week, when the Dutch fishing fleet returns to harbour with 'new' fossilised mammoth molars, fished up from areas of the North Sea that were once tundra and steppes, the fishermen never bring back half a bucket of sanderling skulls or curlew sandpiper sternums. To begin to understand the history of waterbird populations and migration, we must search for clues in the genetic material, the DNA.[23, 573, 919] This search has been going on for twenty years.

Molecular research on waders has focussed on pieces of genetic material from the energy factories of the cells, the mitochondria.[918, 919] Mitochondrial DNA has two advantages: one, it's only inherited through the mother's line, so it's not rearranged during reproduction, and two, it has parts that are 'selection neutral'. This means that some parts of the DNA are not used, and are just 'filling' or 'junk' DNA. This in turn means that changes in the base pairs making up the 'junk' DNA, only occur through chance mutations (so over a long time they occur with some regularity). If you know how different the base pairs of different groups of birds are, and if you know the

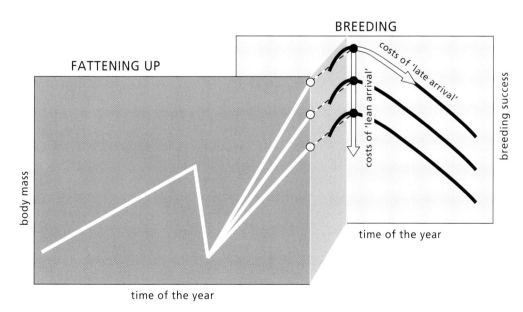

Figure 3.31 Stylised diagram of the relationship between fattening speed in a southern intertidal area and breeding success in the Arctic.[248, 667] Breeding success may be higher or lower depending on the time of departure and the weight at that time. There are costs associated with not arriving on the tundra at the right time or having a low arrival weight.

To keep up to date with the status of different shorebird populations, researchers must do more than just count birds. The birds must also be caught, ringed (photo left: bar-tailed godwit female in Guinea-Bissau), and measured (facing page, lower photo: newly hatched curlew sandpiper in Siberia). This is the only way that changes in migration route, body condition and survival can be detected. Shorebirds are caught at night using the practically invisible mist nets (top photo). When shorebirds gather on a sandspit or the edge of a salt marsh during the day, they can be caught by shooting a large net over them (lower photo). A large cloth is quickly put over the captured birds, in this case dunlins and red knots, which quickly quieten and are then easy to extract from the net.

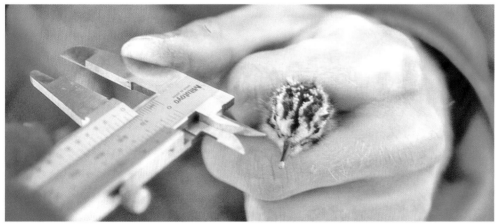

Detailed data on the distribution of shorebirds can be gathered by following the birds with the help of tiny radio transmitters. In the photo above, a transmitter weighing 1.5 g is glued to the back of a red knot. The bird preens the 10 cm long antenna into its tail feathers perfectly. It is then possible to accurately map the presence and movements of the radio-marked birds using hand held and automatic radio-receivers. The radio-transmitter falls off after around two months due to natural process of skin cells sloughing off and regrowing.

rate of random mutations, then you can calculate how long ago these groups had a common ancestor. In this way, you can calculate how long ago species, subspecies and populations diverged. So you can go back in time, even without fossil records!

Currently, three wader species have been investigated in detail: dunlins,[920, 921, 922] turnstones,[921] and red knots.[28] Even though all three are sandpipers, there are large differences in genetic variation between the three species both within and between birds from each different subpopulation.[27] With red knots, and to a lesser extent, turnstones, not only was there little genetic variation within each subspecies, but there were hardly any differences between birds from the different breeding areas. This shows that all of these birds are descended from one breeding population that recently went through a demographic bottleneck. The numbers of knots and turnstones that reproduced successfully must have been so small for a while, that by chance much of the genetic variation in their mitochondrial DNA was lost. For red knots, it is obvious that such a demographic bottleneck occurred around 10 000 years ago and that during this time only a few hundred female knots remained.[28]

Even though all three species are distributed worldwide and abundant, there is much more genetic variation in dunlins than there is in knots and turnstones.[920, 921, 922] This is the case not only for the variation within populations, but also the variation between populations.

Dunlin populations have never been very small; there were no demographic bottlenecks. The large genetic differences between dunlin subspecies show that these subspecies diverged much earlier than 10 000 years ago. In fact, most dunlin subspecies are much older than the 25 000 years since the last inter-glacial period. Previously, people concluded that the world population of dunlin diverged into subpopulations around that time.[354] But the dunlins that now breed in central Canada (subspecies *hudsonia*) had already diverged some 200 000 years ago.[922] The dunlins that breed in Greenland, Northern Europe and Western Siberia (subspecies *alpina*) diverged from the dunlins which breed in central and Eastern Siberia and Alaska around 117 000 years ago. The latter group diverged again around 70 – 80 000 years ago into three populations. One now breeds in Alaska (subspecies *pacifica*), one in Eastern Siberia (subspecies *sakhalina*), and one in central Siberia (subspecies *centralis*). In spite of the inherent inaccuracy of these time estimates, it is striking that all divergences of the dunlin populations correlate with phenomena from the ice ages. The Canadian *hudsonia* population seems to have diverged from the other dunlins at the beginning of the Wolstonian Ice Age,[149] while the European group diverged from the easterly breeding dunlins at the beginning of the early Devensian Ice Age. The divergence of the dunlins in Eastern Siberia and Alaska coincides with the height of the most recent ice age, the Late Devensian, when there were a few isolated tundra areas around the Bering Strait.

Genetic research suggests that populations breeding in different ice-free areas or refugia could indeed become genetically different. These concepts of ice age refugia and subspecies formation in arctic breeding birds have fascinated evolutionary researchers in museums for many years, although they drew their conclusions from external morphological differences.[354, 372, 545, 699, 729] In Europe, there are three subspecies of dunlin (*alpina, schinzii* and *arctica*) easily distinguishable by their external morphology, which seem very closely related genetically. It's obvious that the dunlins of Iceland (*schinzii*) and Eastern Greenland (*arctica*) have diverged only 9000 years ago from the other *alpina*-dunlins, after Iceland's icecap had melted away, around 10 000 years ago.[27] The relationship between climate change and the origin of flyways is particularly well illustrated by the history of the red knot.

### 3.6.3 The last ice age and red knot flyways

Since the knots that winter in Africa and breed in Siberia (subspecies *canutus*) and the knots that winter in Europe and breed in Greenland and Northeastern Canada (subspecies *islandica*) are likely to have diverged less than 10 000 years ago,[28] this species' flyway history can be easily reconstructed from what we know of changes in climate, geomorphology and sea level.[658] The knots' limited genetic variation shows that they were repeatedly threatened with extinction. They must have had a very difficult time 10 – 15 000 years ago, when tundra still extended across The Netherlands and into the Southern North Sea (Figure 3.32). The icecaps started to melt,[174] the tree line moved north to the edge of the icecaps, replacing the tundra, and the sea level rose so fast[276] that temporarily no exposed tidal flats remained. This left the knots, with their dependence on tundras for breeding and on extensive shellfish-rich tidal flats for wintering, in a tight spot. We know that only a few hundred female knots may have survived.[28] When the ice cap covering Northern Europe started to retreat, knots couldn't winter in Europe as there were no tidal flats and it was too cold. All knots had to spend the winter in West Africa. Without the Wadden Sea as a fuel stop, the distances to the tundra in Northern Canada were probably too great, so knots had to breed in Western Europe. Breeding on the Canadian tundra only started around 8000 years ago, once Western Europe could provide suitable wintering areas in the form of the tidal flats around the North and Irish Seas. The moderating influence of the Atlantic Ocean causes spring, and therefore the breeding season, to start earlier in the Iceland and Greenland, than in Northern Russia and Siberia. To arrive in time on the tundra, knots that bred in Greenland and Northeastern Canada were better off wintering in Western Europe, and the same applies today.

Even though the two knot subspecies that currently exist in the Wadden Sea have only 'recently separated' and still look very similar, they now face very different limiting factors. Spring migration is the critical time of year for the tropical winterers that breed in Siberia. For the breeding birds from Greenland and Northeastern Canada, winter is critical. Although individual prey are

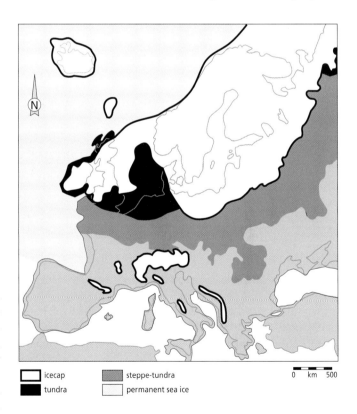

Figure 3.32 During the last ice age, around 15 000 years ago, The Netherlands (and the neighbouring part of the continental shelf of the North Sea) was covered in rough tundra where red knots, sanderlings, grey plovers and brent geese may have bred.

relatively large, and so the birds have a high possible food intake, food availability in the Wadden Sea and British estuaries is rather variable in winter. The prey year-classes are quite different sizes[68, 552] the shellfish can bury too deep in the sediment to be caught[982] and ice can stop birds from accessing the tidal flats. Knots which winter in Western Europe put on extra fat reserves in winter, to carry them through times of scarcity. Birds in West Africa don't do this. The Wadden Sea is critical for the survival of both subspecies. Without the Wadden Sea, Siberian knots could never reach Africa. Without the Wadden Sea, the Greenlanders and Canadians could not fatten up for their migration to Iceland and the breeding areas, nor could they survive the winter.

The evolution of waterbird flyways is largely determined by climate change, and the opportunities and threats provided by the various ice ages, warmer interglacials and sometimes short transition periods. Over a geological time scale, migration patterns show harmonium-like movements. With these movements, both populations and the flight paths change enormously in size and structure. Even though most waterbird species are likely to be millions of years old,[26] and many of the characteristic qualities inherited unchanged from their ancestors even older, the migration patterns we see today are relatively new. They are less than 10 000 years old.

### 3.6.4 New migratory traditions: learning and/or genes?

Changes in flyways can happen in much less than 10 000 years.[824] Over the last hundred years, there have been big changes in the use of Western European wintering and breeding grounds, particularly by geese and ducks.[172, 212] The food rich IJsselmeer has become an important wintering area for diving ducks, and increasingly lush agricultural pastures can harbour more and more geese in winter. In 1970, some pairs of breeding barnacle geese settled on small isles in the Baltic Sea; until then barnacle geese had been assumed to only breed on the tundra.[500] These Baltic barnacle geese, which mainly winter in The Netherlands, have diverged very successfully from the North Russian breeding population. Just thirty years after the colonisation, there are now more than 2000 breeding pairs[499] and many thousands of winterers in The Netherlands. In the breeding areas, they are now so numerous that they are getting in each other's way, and density dependent processes (such as food shortages, parasites and epidemics) are starting to halt the increase.[438]

An overview of historical changes in bird migration behaviour,[824] found no examples of apparently suboptimal migration routes in species whose young migrate with their parents from the breeding grounds to the wintering areas (such as geese and swans). (Migration strategies were judged as suboptimal if they could have been much shorter, as is the case for Alaskan wheatears that could overwinter somewhere in America, rather than flying all the way to East Africa.) The overview found seventeen cases where the migration route had recently changed. In species that leave their young to find their wintering areas themselves, there were 26 cases of changed migration routes, but also another 14 cases (35%) whose present migration strategy appeared suboptimal.[184, 824] The blackcap was one of the species with independent young that had recently changed its migration route. For this passerine, researchers could show by cross-breeding and selection experiments that the migration direction, and the fast changes in this, have a genetic basis.[48] It's also likely that migration direction and distance are genetically based for most waterbird species (with the exception of the barnacle and brent geese of course) and are not easily influenced by speedy cultural differences.

Migration strategies that are primarily genetic in origin (rather than being culturally transmitted or learnt) are difficult to change, unless (1) there are very few migration genes involved, (2) pairs form on the basis of these migration genes, (3) changes in the dominant inherited migration genes will arise from mutations, or (4) the population sizes are very small.[183] These in-built constants might explain why, within the well-studied group of waders, there is not a single example of large changes in migratory routes in historical times. However, there are examples of species or populations that have become extinct, or are on the point of extinction. The Netherlands has lost a (probable intertidal) wintering population of Asiatic golden plover.[444, 445, 446] In Europe, the slender-billed curlew is or was on the point of extinction,[356, 663] and the Eskimo curlew in America is in similar trouble.[81, 300] If these three wader species had had a goose-like evolutionary flexibility and had been able to change their migration route over time, then there would have been a chance that we would have encountered them as waterbirds in this book.

Large mussels are profitable prey for oystercatchers and can be eaten quickly when handled correctly. Experienced oystercatchers insert their bills between the two valves to cut the mussel's adductor muscle in a single snip, enabling them to eat the meat from the gaping shellfish.

# 4. Food

## 4.1 Decisions

A bird foraging on the tidal flats must continually make 'decisions'. A feeding bird has already decided to try one particular spot rather than go elsewhere, and while foraging, it must make further choices with every step and peck. The bird can search for its prey by touch or sight. If it searches by touch, it must decide how deeply to probe its bill into the mud. If the bird restricts itself to prey visible on the surface, it must look around and decide if it will pursue all the prey it can see or only those that are nearby. It's very easy to think of more examples of decisions, but can we relate all these decisions to a common denominator? In other words, are there common rules that can explain foraging behaviour?

### 4.1.1 Optimisation

The 'optimal foraging theory' is based on the assumption that the choices a foraging bird makes are aimed to maximise its food intake rate. This theory was first formulated in 1966[216, 527] and has given rise to thousands of

---

**Box 4.1 'Handling time', 'handling efficiency' and other definitions**

Some phrases used to describe and analyse foraging behaviour have quite specific meanings in the context of 'optimal foraging theory'. Here are the definitions of some of them.

- The time that the bird spends searching for prey, but is not busy eating them, is called the *searching time*.
- The time that the bird spends actually eating the prey is called *handling time*. This term is used instead of 'eating time', because handling time also includes the time it takes to kill the prey, open it, clean it and/or break it up into manageable portions. It is assumed that when the bird is handling one prey item, it's not already searching for another. This means that searching time and handling time by definition exclude each other. When a bird spends time on a prey item that is eventually not eaten, this is still considered handling time. In this event the expression *negative handling time* is used.
- The combination of search and handling times is called the *foraging time*. Time spent resting and preening is not included as foraging time.
- The total time that a bird spends in its feeding area is called the *foraging period*. This includes any rests and preening pauses while in the feeding area.
- The expression *foraging activity* means that part of the total foraging period in which the bird is actively foraging; this is the foraging time as a percentage of the foraging period.
- The total amount of food that a bird consumes, the *food consumption*, is mostly expressed in milligrams or grams of dry meat.
- The term *intake rate* refers to the speed at which food is eaten during foraging, mostly expressed in milligrams dry meat per second foraging time.
- While foraging the bird eats only during handling time and not during searching time, by definition. The intake rate during handling is called the *handling efficiency*, mostly expressed in mg dry meat per second handling. This is also called the 'profitability'. When the handling efficiency is larger than the intake rate during foraging the prey is said to be *profitable*.

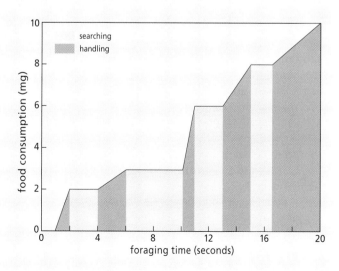

Figure B4.1 Birds alternate searching and handling times while foraging. In this example the total food consumption (V) is 10 mg, the foraging time (F) 20 seconds, and the summed handling time (ΣH) 10 seconds. The intake rate is V/F = 0.5 mg per second and the average handling efficiency of the five prey eaten V/ΣH = 1 mg per second.

publications, especially in the 1970s and 1980s. A number of attempts have been made to summarise all these data, insights and assumptions, for example.[415, 486] The underlying belief of the optimal foraging theory is that if food can be collected quickly, less time needs to be spent foraging, which leaves more time for other important matters affecting the survival of the individual and its offspring. Like every theory, optimal foraging theory is an over-simplification of reality. This is not necessarily bad, and can even be rather useful. The only question is, can this theory accurately predict the behaviour of a foraging bird? There is ample literature available that shows how good birds are at (almost) making the 'optimal choices' invented for them by researchers. But let's start with an example that provides useful insights into the foraging decisions of a waterbird, even though the predictions of the model were not met.

### 4.1.2 How curlews eat common shore crabs

Curlews walking along tidal flats in late summer often encounter small (5 - 15 mm) common shore crabs. Even when the curlews are actually searching for other prey, they still eat many of these crabs. Although the little crab has eight legs and two pincers, it is too small to withstand the strength of a large bird like a curlew, which can swallow it in a single gulp. When the crabs are larger they can struggle more effectively, which means the curlew must gulp a few more times. This is why it takes the curlew longer to eat a large crab than a small one. A curlew can handle a small crab in just a few seconds, but a large crab will take over a minute. Large crabs cannot be swallowed whole, so most of the handling time is spent pulling the crabs apart. Large crabs are shaken, until their legs and pincers fall off. The curlew then has to search for the pincers and legs that have been dropped, before swallowing them one by one. The larger the crab, the less often it is swallowed whole (Figure 4.1A).

Crabs that are swallowed whole provide the most energy when they are around 2 cm in size. If they are larger, they are less profitable. Crabs larger than 3 cm are never swallowed whole. It takes a lot of effort for a curlew to swallow a 2.5 cm crab, but the curlew could choke if it tried to eat whole prey larger than 3 cm. The curlew also risks choking when it eats a crab larger than 4 cm, even when the legs and pincers have been removed. Curlews usually only eat the pincers and legs of these very large crabs, leaving the crab defenceless on the tidal flats. In doing so, the curlew misses out on two thirds of the meat.

Can the decision of which crabs to eat whole and which in pieces be explained by the optimal foraging theory? If the theory is correct, the bird should always choose the fastest way to eat its prey. To determine this, you must first measure how long it takes on average to eat different-sized crabs. You also need to know the relationship between the crab's size and its meat weight, so that you can calculate the amount of energy produced (expressed as mg meat per second of handling). This allows you to take into account the fact that when crabs are shaken sometimes all of their legs and pincers aren't

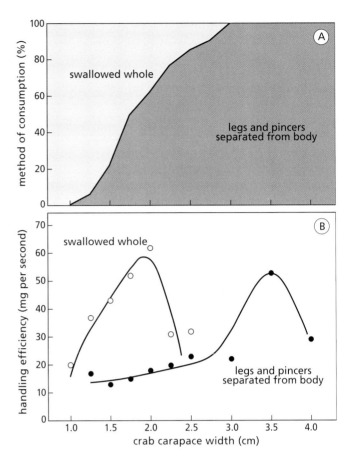

Figure 4.1 (A) Small common shore crabs (*Carcinus maenas*) are swallowed whole. Larger individuals tend to be shaken until their legs and pincers break off and can be eaten separately from their bodies. (B) Handling efficiency (mg dry meat eaten per second handling) relative to carapace size for a curlew feeding on common shore crabs, eaten whole (hollow circles) and in pieces (filled circles). All observations are of an individually colour-ringed curlew feeding on tidal flats.[73]

eaten (Figure 4.1B). Looking at Figure 4.1A, you would expect that when crabs are smaller than about 2 cm swallowing them whole would yield a greater reward per unit time than shaking them, and that shaking would be the most profitable option for larger crabs. But this isn't the case. Figure 4.1B shows that shaking is less profitable for all size-classes smaller than 2.5 cm, so shaking is less profitable than swallowing a crab whole.

From this observation you could conclude that the curlew's decision to eat the legs and pincers separate from the body isn't due to it being faster to eat crabs this way. If it were faster, you would expect that only crabs larger than 2.5 cm would be shaken (Figure 4.1B). But curlews shake crabs as small as 1.5 - 2 cm (Figure 4.1A). Perhaps this is because a curlew eating a large intact crab risks being pinched by the crab. It also risks damaging its oesophagus or even choking if it swallows prey that are too large. There is less of a risk with small prey, but the risks increase with increasing crab size. These observations, inspired by optimal foraging theory, show that although a curlew could theoretically eat faster, it is prudent not to. We really need to know what the curlew's risks of oesophagus damage are when eating different-sized shore crabs in different ways. Only then will we know if curlews have made the right choice.

When making foraging decisions birds must consider more than simply maximising their food intake. Oystercatchers for example, take a certain amount of time to open and eat a cockle, even though it has been shown experimentally that they can do this faster when they have a limited time available.[841] In an experiment, captive oystercatchers were forced to meet their daily food requirements in just one third of the normal foraging time. While they managed to do so, during the experiment one of the oystercatchers broke its bill. So although birds could feed faster, if doing so puts them at greater risk of bill damage they're better off feeding more slowly. The birds also need to consider other factors such as parasites and predators. Perhaps some prey were infected with a parasite harmful to birds and the birds could determine which prey were infected after a short inspection. By refusing infected prey, the bird's intake rate is not as high as is theoretically possible, but it's much better for the bird to opt for a long-term advantage (staying healthy) than a short-term advantage (a higher intake rate). A bird that can choose between a poor and a rich feeding ground would be expected to forage in the rich area. But if a peregrine falcon visits the rich area more frequently, the bird is probably better off choosing the safer area even though it results in a lower intake rate. Although the world is more complicated than optimal foraging theory would suggest, the theory can still be a useful tool for researchers trying to understand a bird's foraging behaviour.

### 4.1.3 Why birds often refuse small prey

For a given feeding method, eating large rather than small crabs gives a curlew a higher prey yield (Figure 4.1B) The prey yield is expressed as the amount of meat consumed per second handling time, the handling efficiency. Larger crabs have a greater handling efficiency than small crabs. Although it takes a curlew longer to eat larger crabs, this is more than made up for by the increased amount of meat a large crab provides. Curlews never eat very small crabs, even when such crabs are very numerous; the curlews can see them, but refuse to eat them because it would not be profitable. Curlews have an average intake rate of 2 - 2.5 mg per second. A curlew's long bill can be rather clumsy when it comes to picking up small prey off the sediment surface so swallowing prey, even small prey, takes at least 2 seconds. This means that when curlews eat prey weighing less than 5 mg their handling efficiency is lower than the intake rate. So when a bird encounters such small prey, even if all it has to do is swallow them, they're not worth eating. The 2 seconds that it takes to swallow a small prey item would be better spent searching for larger prey that would provide more meat. Curlews are better off leaving common shore crabs smaller than 0.5 cm, and that's what they do.

The basic premise of optimal foraging theory is that birds try to maximise their intake rate. It follows then, that birds should only eat prey that have a handling efficiency higher than the intake rate. To investigate this, we need to determine the intake rate and the relationship between prey weight and handling time, so that we can calculate the handling efficiency of all the prey size-classes. Then we can check whether size-classes that have a handling efficiency lower than the average intake rate are refused. For most prey species, the handling efficiency is lower for small prey and increases with prey size, which makes it easier to predict which prey should or shouldn't be refused. This is the case when red knots feed on sand gapers, as the following example shows.

When sand gapers are a few months old, their shells are usually 5 - 10 mm long, although some are still just a few millimetres and others are already 15 mm. These young sand gapers make ideal knot food: they are small enough for the knots to swallow whole without problems and are easily gathered as around a quarter of them live in the upper 2 cm of sediment. Bigger sand gapers bury themselves deeper in the sediment, out of reach of the knot's 3.5 cm long bill. Small sand gapers can occur in densities of thousands of individuals per m². Knots swallow these little sand gapers shell and all, and leave small droppings full of shell fragments on the tidal flats. Along the edges of each shell is a hinge where the two valves were connected and, luckily for researchers, this usually passes through the knot undamaged. This allows researchers to measure the hinges found in droppings and estimate the size of the shellfish that were eaten. The size of shellfish eaten can then be compared to the size distribution of the shellfish living in and on the sediment (Figure 4.2). This reveals that knots don't eat sand gapers smaller than 4 mm, even though the birds would encounter them en masse.[77]

The data shown in Figure 4.2 were taken from a group of knots that foraged near an observation tower in the

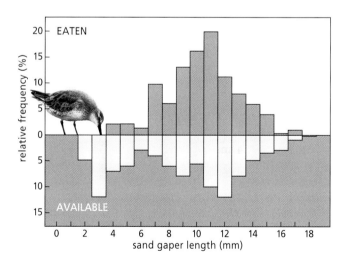

Figure 4.2 Size-selection by red knots feeding on sand gapers (*Mya arenaria*). The relative frequency of different size-classes of shellfish available in the sediment is compared to those eaten by knots at the same location on the tidal flat. Available size-classes were determined by core sampling and sieving the sediment, while the knots' diet was determined from shell remains in the knots' droppings. The size of the shells eaten could be determined by measuring recognisable parts of the shell fragments. Data were collected in the Mokbaai, on the Dutch island Texel, in October 1993.[77]

Mudflats where knots have been foraging or roosting (upper photo) are covered with droppings and, during moult, old feathers (lower photo). The droppings can be collected and analysed, allowing researchers to accurately reconstruct the knots' diet.

middle of their feeding grounds. The birds could be observed from this tower for a long time. They ate an average of five sand gapers per minute.[77] Faecal analysis showed that they ate prey that were on average 10 mm long, which would have consisted of 3.6 mg dry meat. This gives the knots an intake rate of 0.3 mg per second. The researchers measured the time that the knots spent handling sand gapers in the field, and found it depended on the prey size and their depth in the sediment. It took a knot 0.3 seconds to take small prey off the surface, but when these prey had to be extracted from the sediment the handling time increased to 0.5 seconds. Sand gapers that are 4 mm long contain 0.17 mg dry meat. This means that the handling efficiency of a 4 mm sand gaper was 0.33 mg per second when the prey was buried, and 0.51 mg per second when the prey was on the surface. In both cases, the handling efficiency was above the intake rate of 0.3 mg per second, so the prey were worth picking up. If the relationship between handling time and prey weight is extrapolated to include small prey that weren't eaten, the handling efficiency of a 3 mm sand gaper is half that of a 4 mm animal. This means that 3 mm prey should be refused, which is exactly what happens (as shown in Figure 4.2).

### 4.1.4 Why birds sometimes refuse large prey

It takes an oystercatcher 3 seconds to eat an 8 mm cockle. When a 35 mm cockle is eaten, the handling time increases twenty-fold. However, a 35 mm cockle contains 200 times as much meat as an 8 mm cockle, so the increase in meat obtained is much larger than the increase in the handling time. This relationship has been found in many different birds feeding on all sorts of prey, not only oystercatchers eating cockles or, as we saw above, curlews eating common shore crabs and knots feeding on sand gapers. But there is one important exception to the rule; the bird doesn't always succeed in finally being able to eat its prey. The oystercatcher that needs to hammer through a mussel shell before it can eat the flesh is a good example of this.[563] The larger the mussel, the thicker the shell, and the thicker the shell the more time it takes to hammer a hole in it, but also the greater the risk that the oystercatcher will not succeed. Oystercatchers give up more often when trying to get into large prey than small ones. Researchers calculating the handling efficiency per size-class must include this wasted time, known as 'negative handling time'. This means that for oystercatchers that hammer mussels the handling efficiency of the largest mussels is lower than that of somewhat smaller mussels.[561] So sometimes birds must refuse both the smallest and the largest prey.

There is another reason to refuse large prey: they can be stolen and the risk of this happening increases with increasing prey size. When larger prey are stolen more often than small prey, the average handling efficiency of the large prey can decrease so much that the birds choose to eat smaller prey when potential robbers are nearby. Food theft is discussed further in Section 4.6.4.

## 4.1.5 Why birds are more selective at higher intake rates

A bird decides whether or not to refuse prey based on its intake rate. In the example in Section 4.1.3, red knots in early autumn had an intake rate of 0.3 mg per second. However, by the end of winter, the sand gapers had almost disappeared and the knots would have had a much harder time in the intervening months. As the prey became depleted, the knots' intake rate would have slowly decreased. In autumn, when their intake rate was 0.3 mg per second, the knots only ate sand gapers larger than 4 mm. As the prey density decreased, so too would the intake rate. Once their intake rate had halved, the knots were better off lowering their acceptance limit by 1 mm, so that they could also eat the less profitable 3 mm prey. This sort of change in selection criteria is not incongruent with the previously stated rule, i.e. that a bird should only eat prey with a higher handling efficiency than the intake rate. In fact, when a changing intake rate causes birds to change their prey selection criteria, it shows that the birds are reassessing their options.

Almost all publications on prey selection are concerned with eating or not eating prey of different sizes. However prey of the same size aren't necessarily equal. Birds often use this to their advantage and selectively feed on prey that are easy to catch. For example, peregrine falcons that chase waders tend to catch more juveniles that are in poor condition and are probably less adept at escape.[113] Birds can also selectively feed on prey that have a short handling time. This is why oystercatchers that hammer mussels usually leave the mussels that are covered in barnacles, as these mussels would be more difficult to pierce.[127] Birds may also select types of prey that are easier to find and handle. Waders that feed on buried bivalves are a good example of this. They tend to select shallow prey, which are quick to find and handle. When waders reach a high intake rate, they become very selective and will eat only the shallowest prey, as is shown below.[909]

To investigate selection at high intake rates, wild oystercatchers were caught and held in an aviary, where they could be studied more intensely than if they were out on the tidal flats. In the wild, the oystercatchers had been feeding on the peppery furrow shell, so they were given the same food in captivity. The aviary was set up to resemble a tidal flat as closely as possible and had a large flat tray of sediment from the tidal flat on the floor. The researchers endeavoured to control any other factors that could have affected intake rates. For example, they sieved the sediment to remove any alternative prey such as the ragworm (*Hediste diversicolor*). The birds were only fed fresh shellfish, 35 - 36 mm long, and the oystercatchers opened the bivalves and removed their flesh, just as they do on the tidal flats. The researchers varied the prey density. At the lowest density, 10 prey were spread over the sediment tray. One animal was buried just under the sediment surface, the next at 1 cm depth, another at 2 cm depth etc., until the deepest was buried at 9 cm. For higher prey densities, the researchers added the same number of prey to each depth-class. The 'artificial tidal

Very small crabs are picked up and swallowed with ease (upper photo). Out on the mudflats spoonbills often catch flatfish that are simply too large to swallow. Some spoonbills, especially inexperienced juveniles, spend a lot of time trying in vain to swallow these tasty prey (lower photo).

flats' in the aviaries appeared very natural, but took a lot of preparation before the birds were allowed to forage on them. The researchers had to stop the bivalves from burying themselves deeper than was required for the experimental design, and managed this by placing pieces of wood of different heights beneath the shellfish.

The oystercatchers foraged just 1 m away from the observer. To ensure that any rapid movements weren't missed, the behaviour of one oystercatcher was filmed. By slowly watching this high speed film, it was found that it took the bird just 0.2 seconds to probe its bill less than 3 cm in the mud. The deeper it probed, the longer it took. It took 1.5 seconds for the oystercatcher to probe with its entire bill (7 cm long) in the sediment. The bird had to probe many times before it found prey, particularly at the lower prey densities, so whether it took 0.2 or 1.5 seconds to find each prey item made a big difference. The chance of an oystercatcher finding prey depends on the surface area of its bill tip and the prey's horizontal surface area.[417] The chance of an encounter is directly related to prey size and density, but not their burying depth. Since an oystercatcher has to probe as many times for deep prey as for shallow prey, the total search time is determined by how long each probe takes. If a bird forages on deep rather than shallow-living prey that occur in the same density and a deep probe takes seven times longer than a shallow probe, the average search time will also increase seven-fold.

As well as its searching time, the filmed oystercatcher's handling time increased with prey depth. When a shellfish was buried just below the surface the total handling time was 18 seconds, and this increased to an average of 43 seconds when prey were buried 5, 6 or 7 cm deep. Prey buried deeper than 7 cm were not eaten; the bird would go up to its eyes in the mud, but no further. Each shellfish contained an average of 274 mg dry meat. The handling efficiency for deeply buried prey was less than 7 mg per second, and was almost 2.5 times greater for shellfish buried just below the surface (Figure 4.3A). It was predicted that when the intake rate was less than 7 mg per second the bird would always eat all the available prey, but when the intake rate rose above 7 mg per second, the bird would ignore the deeper prey.

In the experiment, the density of the available prey (those in the upper 7 cm) varied between 6 and 437 shellfish per m². At a low prey density, the bird ate less than 1 shellfish per minute. This increased to at least 3 per minute at high prey densities, which means the intake rate increased from 5 to 15 mg per second. We can use Figure 4.3A to predict the depths at which the oystercatcher would still take prey depending on its intake rate. At the lowest intake rate, the bird would eat prey from all eight depth-classes to try and maximise its intake. But at the highest intake rate, the bird only needs to eat prey from just below the surface, which means that we predict the bird will refuse seven out of eight available prey items.

Was the bird really that selective at higher intake rates? Figure 4.3B shows the predicted limits between prey that should or shouldn't be eaten for four different prey densities. As predicted the oystercatcher would only eat prey up to 7 cm deep when its intake rate was low, and mostly ate prey from the upper few centimetres when it had a high intake. But the bird didn't do exactly what the researchers had predicted. At the highest prey density it was predicted that the bird should only eat prey within the upper 1 cm, but he also took prey that were 1 and 2 cm deep. This means the oystercatcher didn't reach the highest intake rate that was theoretically possible. However, this made very little difference. When you calculate the depth of the prey that can still be taken without causing the intake rate to decrease by more than 5%, you find that the bird stayed well within this margin. So this oystercatcher showed almost perfect depth selection.

From Figure 4.3 you might conclude that the filmed bird was somewhat less selective than predicted, but that this had little effect on its intake rate. Careful analysis of the footage showed that the bird was much smarter than the researchers had imagined. When the bird foraged on low densities of shellfish, it took all the prey it encountered, but with a higher prey density, an increasing number of prey encountered weren't eaten. The bird had become doubly selective. It didn't probe its bill as deeply, so it was ignoring all the deeper prey, but it also refused some of the shallow prey. The researchers found that at high intake rates the bird only ate prey that it could open quickly. It's almost certain that these were shellfish that were gaping, which allowed the bird to push its bill directly between the partially open valves. When a peppery furrow shell is closed, it takes so long to open the shell beneath the sediment, that it's worth going to the trouble of pulling it out of the mud and placing it on the surface, to open it and remove the flesh. It's likely that only a small number of the shellfish were gaping. But by becoming super-selective at high prey densities, the oystercatcher reached an intake rate 60% higher than would have been possible solely on the basis of optimal depth selection.

Another researcher was similarly surprised when he studied a captive oystercatcher foraging on cockles in the sediment. The researcher had previously erected a small tent over part of an exposed cockle bed, determined the cockle density and found that the cockles were all of the same size-class.[416] The oystercatcher was then observed feeding within the enclosure. It spent 30% less time handling cockles when they occurred in a high density than when they occurred at low densities. An oystercatcher foraging on cockles can either probe its bill randomly in the sediment or shallowly plough the sediment with its bill while walking along until it finds its prey. We can calculate how long the bird has to search before a cockle is found using both search methods. At low prey densities, the bird did indeed find the number of prey predicted on the basis of random search behaviour. But at higher densities the bird appeared to encounter only one tenth of the prey predicted. In fact the bird had also encountered the other prey but, unseen by the observer, was able to decide in a fraction of a second which prey were worth catching and which weren't. From the way

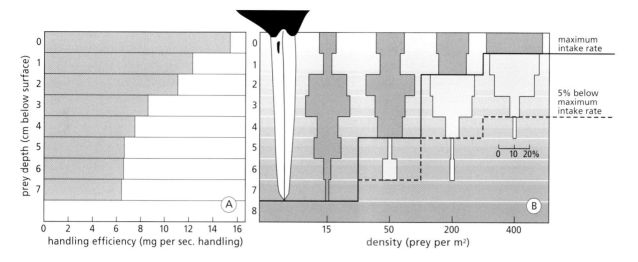

Figure 4.3 The effect of prey depth and density on oystercatcher foraging. A captive oystercatcher was offered peppery furrow shells (*Scrobicularia plana*) buried at different depths. Prey were all the same size and in each trial the same number of shellfish was offered for each depth. (A) The handling efficiency of the prey clearly decreased when prey were buried deeper, because it took longer to eat deeply buried prey. (B) The percentage of prey taken from different depths, at four prey densities. The two lines on the figure show the predicted depth that a bird should forage to in order to achieve the highest intake rate at each density (solid line), and the depth at which the intake rate is 5% below this maximum (dotted line). At low prey densities, prey were taken from the upper 7 cm (the full length of the bill). When prey densities were high, the bird restricted its feeding to the shallower prey, as predicted.

the oystercatcher ate the cockles, it was apparent that the bird began to select only gaping cockles at higher prey densities. So again at high prey densities the oystercatcher's intake rate increased and it became super-selective, choosing only to eat prey that were extremely efficient to handle.

### 4.1.6 Prey selection

Waders that forage on tidal flats lend themselves perfectly to studies on prey selection. They are out in the open where they can easily be followed with a telescope. Their prey choice is often easy to measure directly or can be determined later by analysing the prey remains. Furthermore, their foraging behaviour can be described precisely: the handling time of both the prey consumed and those not consumed are usually easy to determine. This sort of research is either much more difficult or completely impossible when you are interested in the food selection of forest birds or diving ducks.

To really understand the foraging decisions a bird makes, you also need to measure the prey's behaviour, especially if this directly affects searching time and handling efficiency. As we've seen, a bird takes longer to find and handle deeply buried prey. We also know that gaping shellfish have a higher handling efficiency than closed shellfish. The same applies to benthic animals that normally hide in the sediment but periodically appear on the surface. When these prey reveal their presence on the tidal flats, they are easily found and can be quickly eaten. So again the bird's handling efficiency depends on prey's behaviour. Section 4.5 has more examples in which both bird and prey behaviour have been measured.

We know that foraging waterbirds don't eat all the prey they encounter. They are selective, which makes it more difficult to estimate what food is really available for the birds. To find out which prey the birds are likely to refuse, you need to determine the intake rate, as well as the mass and handling time of all the prey. Handling time doesn't simply depend on the prey species and size-class, it can also differ depending on how the prey behave and position themselves. Once you have all this information, you can calculate the handling efficiency for each size-class of every prey species and, perhaps for benthic animals, for each depth-class. When you compare these handling efficiencies with the intake rate you can tell which prey provide enough energy to be selected. This doesn't mean that birds will select all the prey that you have calculated they would be 'allowed' to eat. Often, searching for one prey species is not compatible with searching for another. In this situation, if a bird continues to try to forage on several prey species at the same time, its intake rate might be lower than if it had restricted itself to one prey species. In these cases we'd expect the bird to focus on the prey species that provides the highest intake rate. This is discussed further in Section 4.6.

But before the relationship between birds and their food is discussed in Sections 4.4 and 4.5, we need to cover a few other topics. To understand prey choice, the intake rate must be known. Intake rate decreases when the food availability diminishes and eventually this can reach the point where the bird simply can't get enough food. This doesn't depend solely on the intake rate, but also on the time the bird has available to spend foraging each day. So we have to know how much food a bird needs per day (Section 4.2) and how many hours a day can be spent foraging (Section 4.3).

FOOD

## 4.2 Food requirements

How much food does a bird need to eat each day to survive without using its body reserves? It takes a lot of effort to measure the daily food requirements of birds in the field while simultaneously determining the variation in their body weight, so most of these measurements have been taken from captive birds. If the daily food requirements of a bird species have never been measured, they can be predicted reasonably accurately from a series of formulae that estimate the species' daily energy expenditure. The energy expenditure is then assumed to equal the energy intake in the form of food. This is only true of a bird that doesn't decrease or increase in weight (Figure 3.24), which means it is not using or storing body reserves. So food requirements can be determined via an estimate of the energy expenditure, but we still need to know how much of the food consumed is digested and how much energy it provides.

The energy expenditure of a resting bird (its basal metabolism) is closely related to its weight. This relationship is linear when weight and energy requirements are plotted on a double log scale. The basal metabolism of waders can be calculated using the following formula,[460] in which $E_b$ stands for the basal energy requirement (kilojoules or kJ per day) and M stands for the bird's mass (kg): $E_b = 437 \times M^{0.73}$

Figure 4.4 shows a bird's energy expenditure as watts (W; joules per second) rather than kJ per day. To express energy expenditure in watts the formula is rewritten as: $W_b = 5 \times M^{0.73}$

A bird that does nothing spends less energy, and therefore requires less energy, than a bird that flies for many hours a day. On average the daily energy expenditure of free-living birds ($E_v$, kJ per day) is two to three times higher than their basal metabolism.[103] For most waterbird species, there is little variation in the average amount of time spent on energetically expensive activities. Waders stand around sleeping half the day, do not usually fly for more than 0.5 hours per day, and obtain their food by walking and picking up prey. So daily energy expenditure for these free-living waders ($E_v$), is normally 2.5 times their basal metabolism: $E_v = 2.5 \times E_b = 1092 \times M^{0.73}$

Flying is the most expensive item in the budget. Birds in flight spend energy at 12 times their basal metabolic rate. So per day, a 30-minute flight equals 10% of total average living expenses, but a 2-hour flight equals up to 40% of the budget, making total living costs rise to 3.5 times the basal metabolism. Flying isn't the only expen-

When temperatures drop, life becomes much more expensive for birds as they need to spend extra energy to keep warm. A bird can save on thermoregulation costs by tucking its bill in its feathers, standing on one leg, or sitting down and pulling its legs up into its plumage. These oystercatchers (photos below and opposite above), curlew (photo opposite above) and shelducks (photo opposite below) are trying to reduce their heat loss behaviourally in icy conditions.

sive activity. When ambient temperatures drop, the bird must spend more energy keeping warm. It also takes extra energy to build up body reserves, as is discussed later.

Birds must always take in more energy than they appear to need. This is because the digestive tract is unable to process all the energy consumed. For birds that eat fish or meat, the digestive efficiency is usually around 80%, but can be somewhat higher if the food is very fatty. Digestive efficiency is lower when food contains organic material like hairs, or when the prey has a thick skin that the birds can extract little or no energy from.[125, 973] This isn't usually the case for most of the benthic animals that waterbirds feed on, so their gross energy intake is roughly the same as their net energy intake divided by 0.8.

The daily gross food intake also depends on the energy contents of the food. The energy content of benthic animals varies little and is usually 23 kJ per g meat (not including water or non-organic material like salt, sand and silt; see Section 4.4.2). This means that a bird that forages on lugworms, ragworms or Baltic tellins requires roughly the same amount of meat regardless of which species it forages on. If you want to calculate the required daily consumption ($V_d$, g dry flesh) of food with a digestive efficiency of 80% and an energy density of 23 kJ per g dry meat, you would use the equation: $V_d = E_v/0.8/23$, also expressed as $59 \times M^{0.73}$

Although dry flesh is a scientifically useful measure, it's much easier to picture an amount of fresh food. Like humans and other mammals, the flesh of live worms and

shellfish consists of 80% water. When benthic animals live in the sea, their body water contains the same salt concentration as the seawater, around 3%. Once they've been dried, prey that live in saline conditions consist of 80% meat and 20% non-organic material, mainly salt and sand. The proportion of non-organic material is much higher in crabs, shellfish or snails that are swallowed whole. On average, prey that live buried in the sediment have a calcium covering that weighs five times more than their dry flesh, but prey that live on the sediment surface are much more heavily armoured. Their armour weighs 10 times more than their dry flesh on average. If we assume that birds eat benthic animals without their armour, then the daily consumption in terms of wet meat ($V_n$, in grams): $V_n = V_d \times 100/20 \times 100/80$, also $371 \times M^{0.73}$

The daily consumption can now easily be expressed as a percentage relative to the bird's weight ($V_\%$): $V_\% = 241 \times M^{-0.27}$

Table 4.1 shows the daily food intake for five bird species of very different masses. A small bird weighing just 20 g, such as a little stint, must eat more than its own weight in food each day. Food intake as a percentage of the bird's own body mass decreases for larger bird species. Birds like the oystercatcher and the curlew that are heavier than 500 g need to consume less than half of their own body weight each day.

The formulae above, and the data in Table 4.1, provide an estimate of the expected average food intake per 24 hours. Variation in a bird's digestive efficiency or in the energy contents of the food can cause daily meat consumption to be 10 - 20% higher or lower. When birds eat commercially attractive prey, fishermen often greatly overestimate the amount of fish or shellfish eaten by the birds. In such cases, using our food intake formulae it's easy to show that the bird's daily consumption has been vastly overestimated.

*Increased food requirements*

There are three situations when birds must increase their food intake: when they must work hard, when they have to increase their weight, and when it's cold.

Birds are pushed to the limit when, for example, they raise their chicks.[187] At such times, their daily energy intake is equivalent to 4.5 to 5 times their basal metabolic rate. This level is believed to be the maximum achievable.[469, 518] In Figure 4.4 this energetic ceiling is plotted as a function of body mass. The bird spends its life with its energy intake in the area from two to just over four times the basal metabolic rate.

Birds that fly all day spend two to three times more energy than breeding birds that feed their young. These flying birds don't feed, so all the energy they spend on flight has to have been stored first. This is why birds put on fat to use as flight fuel and build extra muscle tissue before departing on a long flight (Chapter 3). Birds preparing for a long flight increase their daily food intake far above the level necessary to maintain constant body weight (Section 4.3.7).

Birds, with a body temperature of 41 °C, are almost always warmer than their surroundings. If birds didn't have feathers, it would take a lot of energy to keep their bodies at this temperature. But even though its plumage provides excellent insulation, as the ambient temperature decreases there comes a point where a bird will require extra energy to keep warm (called the thermoregulation costs). Above a certain temperature the bird doesn't need to spend any energy keeping warm, but below this temperature energy requirements increase steeply. The increase in thermoregulation costs is linear and can be accurately measured in aviary experiments (Figure 4.5). For a bird like a turnstone that weighs more than 100 g it doesn't matter if the ambient temperature is 20 or 30 °C, the cost of living remains the same. But once it gets colder than this, thermoregulation costs increase dramatically. For every 1 °C that the temperature drops below the critical level of around 20 °C the costs of living increase by more than 5% relative to the bird's basal metabolism. For a larger bird, like an oystercatcher that weighs around 500 g, the ambient temperature must drop below 10 °C, before the bird's energy expenditure starts to increase. It then increases around 3.5% of the basal metabolic rate for every 1 °C that the temperature drops. The rising costs of thermoregulation can quickly get very small bird species into trouble. Figure 4.4 shows the costs of thermoregulation for different-sized birds. It is almost impossible for a 20 g little stint to survive at temperatures around 0 °C.

In our estimates of daily consumption in Table 4.1 we didn't consider the effects of ambient temperature on food intake. Figure 4.5 shows how much the energy expenditure increases in cold weather, but what effect does this have on food intake? Birds are able to put on extra fat before the winter sets in. They can then use this extra

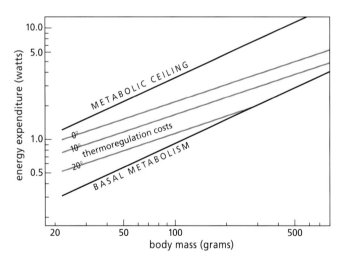

Figure 4.4 The costs of living increase logarithmically with a bird's mass. There is a minimum energy cost, the basal metabolism,[460] and a maximum level that is approximately five times the basal metabolism.[469] The 'normal' cost of living is around two to three times basal metabolism, but when temperatures drop below a certain critical level a bird must start burning energy to keep warm (Figure 4.5). This cost of thermoregulation is shown for temperatures of 0, 10 and 20 °C.[659] The smaller the bird, the higher the relative cost, and hence the food intake required if the bird is to survive in cold weather.

| Bird species | Body mass, g | Consumption, g dry meat per day | Consumption, g wet meat per day | 'Wet' consumption, % in relation to body mass |
| --- | --- | --- | --- | --- |
| Little stint | 20 | 3.4 | 21 | 107 |
| Dunlin | 50 | 6.7 | 42 | 84 |
| Redshank | 150 | 14.9 | 93 | 62 |
| Oystercatcher | 500 | 35.8 | 224 | 45 |
| Curlew | 750 | 48.2 | 301 | 40 |

Table 4.1 Daily food consumption by five bird species of differing body masses, predicted from different formulae (see text). Consumption is given as grams dry meat and as grams wet meat. The last column gives the consumption of wet meat as a percentage of the bird's body mass.

fat under harsh conditions, when they are unable to increase their food intake enough to keep up with the increased energy expenditure. Although this fat storage is a good buffer against extreme conditions, they can't survive on it for weeks let alone months. However when the temperature decreases, birds don't immediately start using stored fat. Instead, birds kept in aviaries eat even more than is required due to the increased thermoregulation costs. So when the temperature drops birds store more fat in case it gets even worse. This phenomenon was found in captive birds that could eat as much as they wanted,[305, 460] and in various waders in an estuary in Northern Scotland.[828] But other field studies found that waterbirds had to use their stored fat straight away during a cold snap.[152, 199, 803] These studies seemed to contradict each other. It's likely that waterbirds always try to maximise their food intake at the onset of harsh conditions. Sometimes they manage to keep up with the increased energetic costs and even put on extra fat. But often they fail to keep up, and die of starvation. The fact that waterbird mortality strongly increases as temperature decreases shows that the birds are often unable to keep up with their increased energy expenditure. The colder the winter, the higher the mortality amongst waterbirds.[124, 342, 556, 992]

## 4.3 Time budgets

We say 'as free as a bird', but what does this mean? Birds are free to decide which prey they will eat, and whether it's better to keep snoozing or go foraging at a particular time, but this freedom is very relative. Birds must meet their daily energy requirements to survive. Table 4.1 shows the amount of food required daily, but how much work does this involve and how long does it take each day?

The maximum foraging time available to waders on the tidal flats depends on how long the flats are exposed. At high tide the birds are forced to rest. When waterbirds have limited time in which to collect their food, their intake rate needs to stay above a certain level. The minimum intake rate required depends on whether the birds can also forage at high tide or during low tide at night.

What happens when a bird needs more food, perhaps because it's storing extra body reserves or needs more energy because of low temperatures? Does the bird increase its intake rate or does it maintain the same intake rate while increasing its total foraging time? How much the intake rate can increase depends on the food availability and the bird itself. The bird needs to process the food it consumes, and if the bird is already full to the brim, it has to wait before it can eat more. This means that it's important to know how quickly birds can process their food and how much can be stored in the digestive tract. When birds are unable to forage more quickly they must spend more time searching for food. But is this possible? In tidal areas, the feeding grounds are unavailable half of the time on average, although the exposure times vary from day to day. On stormy days, birds may be completely unable to forage, because the wind pushes water over the tidal flat. The birds must be aware of this and keep a strategic fat store. How big is this fat reserve and how long can they last without food?

### 4.3.1 Variation in exposure and foraging times

Waders that forage on the tundra or on inland grasslands are able to feed 24 hours per day, but when these birds forage on tidal flats the feeding grounds are only available for part of the day. When the incoming tide covers

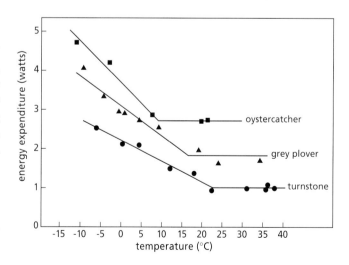

Figure 4.5 Minimal costs of living for three bird species: oystercatcher (500 g), grey plover (250 g) and turnstone (120 g) at different ambient temperatures. Measurements were made on resting captive birds.[460]

FOOD

Shorebirds that would roost on sandbanks and salt marshes at high tide in Europe, often have to roost among mangroves when they are in tropical Africa. Whimbrels search for single trees with bare branches (photo left). Bar-tailed godwits, knots and curlew sandpipers clearly prefer open areas within the mangrove forests, where they can stand on the sand in large flocks (photo below).

Quiet and safe high tide roosts are relatively scarce in tidal areas, and many species gather on the same roost site: on the top photo on page 159, dunlins and knots mingle, while gulls roost further back. As the tide comes in, waders like these bar-tailed godwits and dunlins (photo on page 159 below) often gather on the higher parts of the tidal flats initially. They only fly to the high tide roosts once the upper tidal flat is submerged.

158

the tidal flats, the birds will have to wait many hours until the water has receded enough for them to forage again. How long the birds will need to wait varies for the different species. Oystercatchers often fly to their high tide roosts 3 hours before high tide, while dunlins and redshanks often continue foraging along the edge of the incoming tide for another 1 - 2 hours. Then as the tide recedes, these latter species will often start to feed as soon as the first mud is exposed, while oystercatchers wait another few hours. Usually, smaller species forage longer than larger bird species. This relationship has been found in Europe,[408, 639] Africa[222, 984] and Australia.[148]

Why do oystercatchers and curlews tend to rest so much longer at high tide than dunlins and redshanks? Part of the explanation is that larger birds eat larger prey, and larger prey tend to occur low in the tidal zone. So larger birds are obliged to feed for shorter periods on the tidal flats. The smaller prey eaten by the smaller waders, occur in high densities nearly as far up as the high tide line. Oystercatchers and redshanks are a good example of this. Oystercatchers find little in the way of suitable food high in the intertidal zone[991] and when they do forage there, their success is low.[254] Mussels and cockles are the most important prey for oystercatchers, and mainly occur low in the tidal zone. In contrast, the most important prey for redshanks, the amphipod *Corophium volutator*

FOOD

and the mudsnail, often reach their highest densities high on the tidal flat.[66, 147, 283] There is also another reason why larger birds rest longer. Compared to smaller birds they can store more of their daily food intake internally, and then take it with them to their roosts and digest it there. This means they can manage with a shorter foraging time.

Birds don't spend all their time out on the tidal flats foraging. Every so often, the birds will preen, snooze, and argue amongst themselves. Large waders usually actively forage for 70 - 85% of the time and smaller waders for 80 - 95%. If the tidal flats are only exposed for a short time, the birds have no time to lose and hardly preen or rest at all during the foraging period. When the foraging area is available for more time, large groups of resting birds can form on the tidal flats.[984, 990]

### 4.3.2 Enforced rests at high tide

When the incoming tide drives waterbirds to the higher parts of the tidal flat, the birds are forced to stop feeding and will often search for conspecifics to roost alongside in dense flocks where they can preen and sleep. There are exceptions to the rule. Whimbrels that defend feeding territories on high tidal flats or the beach will often stay on the edge of their territory when it is submerged at high tide. Because of this individual whimbrels may be dispersed along the high tide line around 100 - 200 m apart. Whimbrels that aren't territorial or don't have a feeding territory bordering the high tide line do form high tide flocks.

All sorts of theories have been proposed to explain why non-foraging birds gather in flocks.[589] Maybe the birds would rather roost in quiet places with little disturbance and such places are scarce. This could be why birds go to the same high tide roosts and form flocks. Flocks of curlews and bar-tailed godwits are more often seen standing knee or belly deep in the water, than say, oystercatchers. Avocets can even be seen regularly in swimming flocks.

Probably the most important biological function of flocking is to minimise an individual's risk of being caught by a predator. In the past this seemed a rather unlikely proposition, particularly in areas where birds had been protected from human hunters for many years and other predators, such as weasels and rats, weren't present. There didn't appear to be any other predators that posed a threat to roosting waterbirds. However researchers then probably underestimated the predation pressure exerted by birds of prey, especially on smaller waders.[143] Twenty five years ago, the number of birds of prey had fallen to the lowest levels historically known because of human hunters and poisoned food. Back then it was possible to complete a high tide count of European waterbirds without a single disturbance. Nowadays peregrine falcons regularly cause the birds to panic (and the bird counters have to start counting all over again when the counted and uncounted birds mix).

There is another reason to roost in dense flocks in cold weather; being close to your neighbour helps save energy.[935] But the birds need to stand far enough apart to still have room to fly away in the event of a disturbance. The closer the birds stand, the greater the risk that they'll crash into each other on take off. Cannon netters know that this is more than just a theory. When a cannon net is shot over a few well-dispersed roosting birds, most of the birds can quickly fly away, escaping the net. When this happens to a dense group of birds, relatively few birds can escape. Generally, larger species like curlews and oystercatchers stand further apart than the smaller wader species. But knots, which are 2.5 times heavier than dunlins, form tighter groups than dunlins. Maybe this is due to differences in the amount of space the birds need to take flight without a hitch when danger threatens.

Different waterbird species differ in the extent to which they gather in large flocks. Turnstones often form many small groups spread along the high tide line, but all the spotted redshanks, greenshanks and red knots within an area will usually assemble in one flock. One disadvantage of all the birds gathering in a single large flock could be that it increases the distance between the feeding grounds and the high tide roost for many birds. Greenshanks have even been observed flying 13 km from the feeding grounds to the high tide roost and back, even though there were other roost sites available just a few kilometres away from the feeding grounds.[971]

There seems to be another reason for birds deciding to fly further than appears strictly necessary during the tidal migration. When birds have the choice of several different high tide roosts, they show subtle preferences. Greenshanks prefer to roost in marshes, particularly along the edges of the channels and pools and almost never roost on exposed sand flats at high tide. The greenshanks that flew 13 km up and down to a high tide roost foraged on a mussel bank 1.5 km from the coast. They continued to use this traditional high tide roost for many years until a nearby low dyke broke through and a

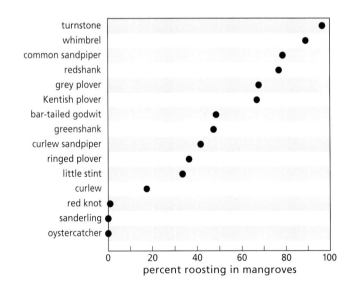

Figure 4.6 The proportion of wading birds that roost in mangroves during high tide in tropical tidal areas in Guinea-Bissau. Oystercatchers, sanderlings and knots always roost on muddy areas or sandy beaches, whereas turnstones and whimbrels usually spend high tide in the mangroves.[967]

small area of grassland became a marsh and provided a new, closer roost site. Unlike greenshanks, roosting knots are seldom found amongst dense marsh vegetation, it seems that even a small amount of glasswort (*Salicornia spp.*) is too much for them.[686] Knots are only seen inland of the dykes in exceptional circumstances. Oystercatchers, curlews, bar-tailed godwits and redshanks don't appear to have such a big problem with this, but also prefer to roost seaward of the dykes. Only when heavy storms cover the marshes and sand flats with water will they roost inland en masse. Inland roosts are only very rarely used at night. However when pasture has been disturbed or vegetation removed, more birds of more species can often be observed roosting inside the dykes. Grey plovers, for example, seldom roost in pasture at high tide, but regularly roost on bare agricultural ground.

Most waterbirds are world travellers, but once they have arrived in an intertidal area they often stay put for a long time. This has been convincingly shown by bird catches in the Wash in England,[721] where over the last 30 years a total of 200 000 waders were caught using cannon nets at different high tide roosts. Many of these birds were caught more than once. Ninety percent of the grey plovers recaptured were less than 2 km away from their first capture site. Dunlins were somewhat less site faithful and redshanks a little more.

Waterbirds also gather in large flocks at high tide in the tropics, and if the tide is not too high they can roost on beaches and on the higher parts of the tidal flat. In tidal areas that are surrounded by mangroves these roost sites aren't always available, as during spring tides all the areas free of vegetation are submerged. In West Africa some birds fly over the flooded mangrove forests to roost in the rice paddies created where the mangroves have been cleared. However most birds remain in the mangrove forests, where they balance on bare branches and aerial roots. For a bird watcher used to seeing waterbirds in temperate areas this looks very odd! But it's not so strange for people lucky enough to have seen the birds on their northern breeding grounds, where bar-tailed godwits sometimes also perch in trees. Each species has its own preferred perch. Turnstones, whimbrels, common sandpipers, redshanks and grey plovers often stand in rows on branches and aerial roots (Figure 4.6). Curlews and, to a lesser extent, whimbrels tend to perch high in the mangroves or on other big trees. The smaller species stand on aerial roots and bare branches close to the ground. Maybe they prefer bare branches as a way of avoiding predators. Snakes are very common in mangrove forests. Most roosts used at high tide during the day are not occupied at night. At night the birds fly many kilometres further to concentrate on areas such as uninhabited islands where they probably feel safer. Not all waterbirds roost in trees; oystercatchers, sanderlings and knots remain firmly grounded.

### 4.3.3 The minimum intake rate required

A bird's intake rate decreases when it has to feed on prey that occur in low densities and/or provide little food because of a low handling efficiency. When the intake rate becomes too low, birds die of starvation. How low can the intake rate go? This mainly depends on the amount of time that the birds can spend foraging each day. When the birds are unable to reach a high intake rate, they could compensate by extending their daily foraging time. In theory, there are other ways birds can avoid starvation.

First, they could try to cut down on their living expenses by, for example, flying as little as possible or using sheltered high tide roosts in bad weather. But when they try to avoid starvation this way, they expose themselves to other dangers. In harsh winters birds often only fly up at the last moment, making them easier for predators to catch.

Secondly, birds could start using their fat and protein reserves. This is only a temporary solution and is of no use if the food supply is exhausted in midwinter and will not improve for months. Body reserves are only a strategic energy store to get through short-term problems.

The amount of time that waterbirds spend foraging per 24 hours has been measured for 14 wader species that winter on the Banc d'Arguin, in Mauritania.[984] The three largest wader species (curlew, whimbrel and oystercatcher: 400 - 700 g) foraged for a relatively short time, only 6 hours of the 24. Ten wader species that were somewhat smaller (40 - 250 g) spent more time foraging, 7 - 10 hours. The smallest species, the little stint (22 g), foraged the longest, 13 hours per 24. To be able to estimate the daily foraging time, it was necessary to measure how long the birds were present on the tidal flats day and night, and what proportion of that time they were really looking for food. The amount of time that the birds spent on the tidal flats during the day had little effect on the observed variation in total foraging time, as that time was almost equal for all species. But there was a large difference in the amount of time that the birds were active on the tidal flats at night. The total time spent foraging (F, hours per 24 hours) could be described as a function of the bird's mass (M, in grams): $F = 20.5 \times M^{-0.22}$

Since larger species spend less time foraging than smaller species, the difference in intake rate between large and small birds is much greater than would be expected solely on the basis of their daily food requirements. In Section 4.2 we saw that when the daily food intake of birds is plotted against their body mass, the relationship is exponential, with an exponent of 0.73. The amount of time waders on the Banc d'Arguin spend foraging each day also has an exponential relationship with body mass, but is negative with an exponent of -0.22. So from that we can conclude that when the intake rate during foraging is plotted against the body mass, the exponent is around 1. The intake rate (I, mg dry meat per second) of waders on the Banc d'Arguin can be calculated as a function of their body mass (M, in grams) using the equation: $I = 0.04 \times M^{0.96}$

The intake rates predicted by this equation were compared with the intake rates measured for the three largest wader species and were found to be very similar. The intake rate wasn't measured for the smaller species, but would be around 0.07 mg per second foraging for a small sandpiper. This is worth knowing, as it shows that a small sandpiper can get enough food by eating prey so tiny that they completely escape the attention of researchers using the standard sampling techniques.[983]

This equation calculates the intake rate required when a bird wants to maintain the same body weight, given the observed time spent foraging. Perhaps the intake rate could be lower if the birds foraged more at the night, or preened and rested less allowing them to forage more during the day when the flats are exposed. But is this really possible?

## 4.3.4 Digestion limits food intake

When birds have only a limited time to find their food, they must increase their intake rate to meet their energy requirements. This can only happen when enough food is available, but food availability isn't the only factor involved. When birds eat quickly they reach a point where they are completely full and must digest the food already consumed before they can resume foraging. So how quickly can birds digest their food? When birds start to eat after a long rest, i.e. with an empty digestive tract, they usually start to defecate regularly after about 30 minutes. When eider ducks that have been dining on mussels choose cockles for their next course, it takes around 30 minutes for their droppings to change from black mussel grit to white cockle grit.[830] The amount of faeces excreted is a good measure of how quickly the food is being processed, particularly when the amount of calcium and the number of prey in every dropping can be measured. When a bird feeds on ragworms, the number of ragworm jaws in the droppings can be counted and divided by two to determine the number of prey consumed. The number of common shore crab pincers and plaice or common goby otoliths can be counted in the same way. When both the number of prey per dropping and how frequently the bird defecates are known, it's possible to determine how many prey were processed per time unit. If, in addition, you know the relationship between the size of the prey fragment and the total prey mass, it is possible to determine how much food has passed through the bird and at what speed. When non-foraging intervals are taken into account, the intake rate during foraging will never be higher than the speed at which the individual bird can process the food. At least, this is the case during long foraging bouts. Thus the period that the maximum intake rate can be maintained is partially determined by the amount of food that can be stored in the digestive tract.

The speed at which food can be consumed and processed can be accurately measured by letting a captive bird forage and monitoring (1) how much it eats, by

When oystercatchers eat large mussels, their intake rate is so high that they become 'full' after 2 hours and are forced to take a digestive pause.

On stormy days, small birds like these dunlins search for food along the edge of the incoming tide. Here they are feeding alongside brent geese that are eating seaweed that has been washed in.

measuring the decrease in the uneaten food, (2) the total amount of food the bird stores, by measuring the bird's weight change during foraging, and (3) how much undigested food it excretes, by measuring apparent weight loss due to defecation (weighing droppings). It would appear almost impossible to monitor all this at the same time, but it has been done with captive oystercatchers that were offered mussel flesh on a balance. In this way, the researchers could continually record both the food intake and the bird's changing mass.[462] They found that an oystercatcher can't store more than 80 g of meat in total or about 16% of its own body mass. Eighty grams of fresh mussel flesh equates to 12 g ash-free dry meat. Since the rate of food digestion seems to never rise above 4.4 mg fresh meat or 0.66 g ash-free dry meat per second, you can calculate that it would take a full oystercatcher about 5 hours to completely empty itself out. This is almost as long as the average rest period oystercatchers are forced to take on the mudflat at high tide.

When an oystercatcher begins foraging it can't consume more than 12 g dry meat in the first 30 minutes, as the bird has not yet started to defecate and it can only store 12 g of food in its gut. Once it has reached this limit, food cannot be taken in faster than it can be digested. This limit is represented by the shaded area in Figure 4.7, which shows that during each 6-hour low tide period a bird cannot consume more than around 25 g of food. This is only 70% of the amount an oystercatcher requires per 24 hours,[990] which means the bird needs to use more than one low tide period in every 24 hours to feed. This is why oystercatchers must also forage at night in winter. In Northwest European winters, the night is half as long again as the day, so usually there is only one low tide during daylight hours. So in winter, even with a very good food supply, oystercatchers would die of hunger if they didn't forage at night.

It's not possible to measure the rate of digestion in the field as accurately as you can with captive birds. But you can measure the change between foraging and resting periods, as well as the total food consumption in free-living birds. To illustrate this, the two black lines in Figure 4.7 shows how a male oystercatcher ate 25 g of food on two different days during a low tide. The bird was individually marked with colour rings and foraged near a high observation tower on the middle of the mudflats. This bird was continually watched by two observers. The oystercatcher ate Baltic tellins and peppery furrow shells alternately. The amount of shellfish consumed could be accurately determined. On both days the foraging area was exposed for 5.5 hours. On one day, the bird was completely full after 3 hours and took a break of 1.5 hours before filling itself again in the last hour that the

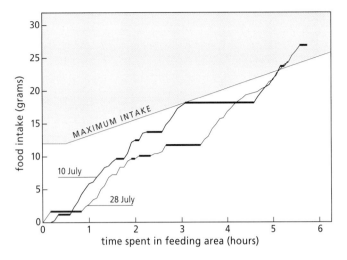

Figure 4.7 Accurate measurements of the food intake of a foraging oystercatcher.[462] Food intake is plotted cumulatively, and therefore shows how much food has been eaten over a given foraging period. The maximum possible food intake (derived from laboratory experiments) is shown by the junction of the grey and white panels. The two lines represent the cumulative consumption of a free-living male oystercatcher over 5.5 hours of possible foraging time on the mudflats for two low water periods.[990] In the field data the horizontal lines are caused by digestive pauses; for clarity these are marked in bold.

flat was exposed. On the other day, it also took a rest pause halfway through the foraging period even though it wasn't full. But it was full by the end the foraging period, so in both low water periods the bird ate the same amount of food. Such data are also available for other days and other individuals. From these data we can conclude that when birds reach a high intake rate, they are quickly forced to take a digestive pause.

Oystercatchers' intake rates have often been measured. By 1996, 240 measurements of their intake rate had been made,[989] and this number has since increased. When all these studies are combined, the intake rate appears to decrease when birds can spend longer on the feeding grounds. When the foraging period is less than 6 hours, the intake rate varies from 1 - 5 mg per second. But when the foraging period is more than 10 hours, this decreases to 1 - 2 mg per second. A similar decrease can be observed in their foraging activity. When the oystercatchers can only spend 2 hours on the feeding grounds, they forage for more than 80% of the time, sometimes even 100%. When the foraging period is longer than 10 hours, oystercatchers usually only forage for half of the time. So when oystercatchers have time to feed over a long period, they tend to take longer and more frequent digestive pauses and to forage more slowly.

An experiment using captive birds showed that oystercatchers really do slow down and relax when they have more time available. The oystercatchers were allowed to forage on a cockle bed set up inside an aviary where water levels could be manipulated with a pump.[841] When the birds were given a long period of low water, they foraged slowly. But when they were only given 2 hours of low water they fed quickly. The greatest difference was in handling time, when the birds had more time they were much slower to open and eat a cockle than when they were in a hurry.

The obvious conclusion is that a food's digestive speed determines much of the bird's feeding pattern. This is particularly so for larger bird species, which are often seen taking a long break after around 2 hours of foraging. In many cases these birds are probably forced to take a digestive pause. You could say that they have hit a limit determined by the rate of digestion (Figure 4.7) and that their total consumption depends on the speed at which the food can be processed and the length of the foraging period.

The total amount of food consumed is not always as much as is possible for any given foraging period and rate of digestion. Sometimes birds could eat more, but simply don't want to. Oystercatchers only need to eat 36 g of food per day. Once an oystercatcher has consumed 27 g in a 4.5-hour low water period (Figure 4.7), it only needs to eat 9 g in the following low tide. One of the days shown in Figure 4.7 had a low tide late in the evening and was in early July when the night is short. After a few hours sleep the observers returned to the observation tower at sunrise to see the bird follow the receding tide out again. In the following hours, the bird ate no more than 6 g of food, even though the feeding grounds were exposed long enough for it to have eaten 19 g. But over the last 24 hours the bird had eaten 33 g, nearly enough to maintain its weight for that day. Birds don't always need to reach their digestive limits.

Sometimes birds would like to eat more but can't. When there is little food available, the intake rate can become so low that even if the birds don't rest they remain beneath their digestive limits. But how often is the intake rate of wild birds limited by food availability and how often by the speed at which the birds can process the food? This has been studied in oystercatchers.[990, 995] In summer there is so much food available that the birds can always reach a high intake rate, through which consumption is limited by the bird's processing speed. But in winter, food is often so scarce that the intake rate is limited by food availability. The birds cannot eat faster than their limited food allows.

A similar situation is found in whimbrels in Africa. The whimbrel's intake rate can be limited by the speed at which it is able to digest its food. However this only happens in spring when their most important prey, the fiddler crab, is available en masse. From March onwards the fiddler crabs leave their burrows to graze and make an easy meal.[975] There's also more than enough food for the whimbrel's daily requirements in summer, but adult whimbrel don't make use of this as they begin to depart for their breeding grounds in early April. By the time the whimbrels return in August, the fiddler crabs have started to hide more and for most of the wintering period the crabs will remain in their burrows for days on end. Because of this, during winter months a whimbrel's intake rate often fails to meet the maximum possible rate set by its own digestive system. The whimbrels must often rely on their fat stores during this time and feed more on the few days when the fiddler crabs are present in large numbers.⁹⁶⁹

### 4.3.5 The significance of nocturnal feeding

Do waterbirds feed at night? In midsummer in Northwestern Europe it is only dark for 6 hours and is light for the remaining 18. In midwinter it is dark for 15 hours and light for 9. How much time can waterbirds spend foraging each day if they are not active at night? When birds only forage on tidal flats below sea level their feeding grounds are unavailable half of the time. This means that on average these birds can only forage for 4.5 hours a day in the winter compared to 9 hours in midsummer. The time that low tide occurs changes slightly each day, so during the moon's 28-day cycle, the exact amount of time available for foraging in daylight varies. Exposure times can also vary depending on the strength and direction of the wind. Tidal heights were continuously recorded in the Eastern Wadden Sea to calculate the exposure times of feeding areas over different timeframes.⁹⁹¹ In midwinter the feeding areas would be exposed for 2 hours or less per day on one of 50 days, assuming that the birds feed day and night. Birds that do not forage at night risk having 2 hours or less to forage on one in every five days. If the birds feed both during the day and at night, they would always be able to feed for 2 hours or more in any 48-hour period. But if they only forage during the day, there is a 20% chance that they will only be able to forage for 2 hours or less. So if waterbirds didn't feed at night, they would have little foraging time available in winter, and would require large fat reserves to carry them through extended lean spells.

There are a few ways to find out if waterbirds do forage at night. Birds that forage on mudflats leave behind telltale footprints. These footprints can be counted and compared with counts from daylight low water periods to estimate the intensity of night feeding. The number of fresh droppings on the mud, or the number of fresh regurgitates on a high tide roost, can be used in the same way. When droppings or regurgitates are monitored, researchers are able to compare the birds' food choice during the day and the night. Oystercatchers often leave prey remains in the form of empty shells on the mud, so examining the number of freshly emptied shells provides a combined measure for the number of birds and their foraging success.⁸²⁰ This has shown that at night oystercatchers eat smaller and fewer cockles than during the day. Fewer cockles being eaten may be due to birds being less successful or fewer birds being out on the mudflats. Another way to monitor nocturnal food intake is to monitor the weight loss of birds after they have been foraging. When birds arrive at the high tide roost with a full stomach, as long as they are still digesting, they lose weight faster than birds with an empty stomach.⁹⁹⁴ By catching the birds on the incoming tide both during the day and the night, any difference in the rate that body mass decreases can be detected. There appears to be no difference in the rate of weight loss in oystercatchers caught in winter, which suggests that their consumption during the day and night is similar. When a continually recording balance is placed beneath a nest it's possible to determine the weight of the individual breeding birds before and after each foraging bout during the day and the night. This technique has been used on oystercatchers that bred on the edge of a salt marsh and foraged on the tidal flats.⁴⁶³ The main finding of this research was that the birds could feed as quickly at night as during the day. Using radio transmitters that record a bird's activity is another useful way to gather foraging information.²⁷⁴

The ideal situation is when light-gathering or infrared binoculars are used and all the necessary information is recorded directly. Researchers can then count the number of birds on the tidal flat, and determine the proportion of time spent actively foraging or, in favourable conditions, even what the food intake is. Looking at the publications it appears that more data were collected in the (sub) tropics (for example ²⁷⁹, ⁷⁴¹, ⁸³⁴, ⁹⁶⁹, ⁹⁸⁴) than on cold nights in temperate egions (for example ¹⁴⁰, ⁹⁹¹).

On the Banc d'Arguin, it was found that typical tactile hunters like the red knot, bar-tailed godwit and dun-

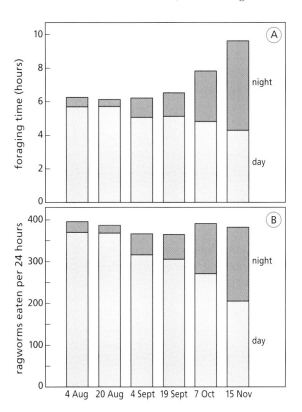

Figure 4.8 (A) As day length decreases in late summer and autumn, curlews increase both the amount of nocturnal foraging they do, and the total amount of time spent foraging in a 24-hour period. (B) The number of ragworms (*Hediste diversicolor*) eaten by curlews remains the same over this period. Because their foraging success is lower at night, birds compensate by feeding for longer periods during the night. These data were obtained from counts of curlews feeding around an observation tower on the tidal flats in the Eastern Dutch Wadden Sea, combined with long-term measurements of the food intake of these birds. During the day observations were made with telescopes; at night infrared binoculars and a light intensifier were used.¹⁴⁰, ⁹⁶⁰

lin were present in the same densities on the tidal flats day and night, while visual hunters like the Kentish plover, ringed plover, turnstone and curlew sandpiper were not out at night.[984] This difference may seem logical but wasn't entirely expected. First, some benthic animals are more active at night, probably because there is less predation risk at night than during the day.[196] Secondly, visual hunters have much better vision than the average tactile hunter and would still be able to see with even a little light.[740] Thirdly, the birds can use a different method to search for prey at night, as is found in oystercatchers feeding on cockles,[416, 1007] whimbrels that hunt fiddler crabs,[969] and dunlins that eat small amphipods.[577] It's mainly the long-billed waders that change from visual hunting to tactile foraging at night.[547]

So why feed at night? There are several explanations.[547] One is that it can be less dangerous at night because predators, such as humans, are not active then. This certainly explains why ducks in areas with strong hunting pressure feed exclusively at night.[848] Another reason is that prey may be more active at night and so easier to catch. As we saw earlier, birds also forage at night when they want to consume more food than can be taken solely during daylight hours. Ringed plovers kept in captivity in Guinea-Bissau and fed dried food pellets ate 6 - 12 g per 24 hours. During the 12 daylight hours the amount of food eaten remained quite constant at 5 - 6 g, but during the equally long night consumption varied from 0 - 6 g.[985]

Birds may also forage more at night if they have been unable to meet their daily energy requirements. This was found in curlews when their food intake rate and foraging activity from different months were compared. Tens of curlews that foraged around an observation tower on the mud were intensively followed for five months from July to November. By November, their foraging activity during the diurnal low water periods had increased from 65 - 80%. Even so, they were unable to compensate for the large reduction in the available diurnal foraging time. The amount of time spent foraging during the day had dropped from nearly 6 hours to over 4 hours. But over the same months the average number of hours spent foraging at night increased from 0.5 hours to almost 5.5 hours (Figure 4.8A). Throughout a 24-hour period the birds spent half as much time again foraging in November compared to August. They spent more time foraging because their intake rate had gradually decreased. All of the curlews around the observation tower fed almost exclusively on ragworms. In July and August they could find an average of 1.1 ragworms per minute, but from September onwards the rate gradually decreased until November when they could only find 0.8 worms per minute. They were less successful at night and their intake rate was a third lower than during the day. When the intake rates are combined with the number of hours spent foraging it shows that the birds managed to eat the same amount of worms per 24 hours each month (Figure 4.8B). Obviously, the birds tried to find as much food as possible at low tide during the day, but when that wasn't enough they also foraged at night, and because they were less successful at nocturnal foraging, it took them longer to get enough food. So birds prefer to forage at times that provide the maximum intake rate. When they are unable to find enough food during that preferred period, they will also start to forage at times when food is harder to come by. This allows them to consume as much food as possible in as little time as possible. An elegant example of optimal foraging.

The oystercatcher has been the subject of more studies on nocturnal feeding than any other waterbird species. These studies appear to contradict each other when it comes to nightly food intake. In summary, it appears that oystercatchers, like curlews, don't forage at night if they can consume enough food during the day. But when they are unable to get enough food diurnally, they start to feed at night. Breeding oystercatchers forage a lot at night.[463] Non-breeding birds have more time available and in summer will only forage during the day, unless their intake rate is very low. In one study the time budget and the intake rate were measured from summer through to early winter.[995] These oystercatchers fed on Baltic tellins and peppery furrow shells. Both prey species burrowed deeper as autumn progressed. This meant oystercatchers needed to work harder to find them, so their intake rate during foraging decreased. Like the curlews (Figure 4.8), when the oystercatchers were unable to get enough food during the day they began to forage more at night.

Inland pastures are important feeding grounds for curlews in autumn and winter. When stormy weather only allows the birds to feed on the mudflats for a short period, most of them feed intensively in nearby meadows.

In May large flocks of bar-tailed godwits sometimes leave the mudflats to forage on inland pastures. Recently mown fields are especially popular (upper photo). There they forage on crane fly larvae (lower photo). Bar-tailed godwits move to the mown meadows because they can reach a higher intake rate feeding on crane fly larvae than they can eating worms and shellfish on the mudflats. However the godwits must accept the cost of losing some of their prey to thieving gulls.

### 4.3.6 The importance of inland pastures

Waterbirds cannot forage on most intertidal areas at high tide. Of course there are exceptions to every rule. Whimbrels sometimes search for the leftovers of fiddler crabs caught a few hours previously high in the tidal zone, but usually little is found.[975] In some tidal areas birds can move to other feeding grounds such as neighbouring beaches, saltpans or pastures. If these alternative food sources were really attractive we could expect the birds to remain there rather than returning to the mudflats. The fact that the feeding areas inside the dykes are either visited sporadically or only at high tide suggests that this is not the case.

Without human intervention most of Northwestern Europe's intertidal areas would gradually continue above the high water line into extensive salt marshes and dunes. Over the last few centuries, dykes have been built seaward of many salt marshes, converting them into pasture. Salt marshes are not an important feeding area for most waterbirds, although ringed plovers forage on salt marsh insects in summer. However pastures are an important alternative and additional food source in spring for oystercatchers, curlews, bar-tailed godwits, redshanks and gulls. There the birds feed mainly on earthworms in winter and crane fly larvae in spring.

Oystercatchers begin to forage in pasture en masse when they have been unable to gather enough food dur-

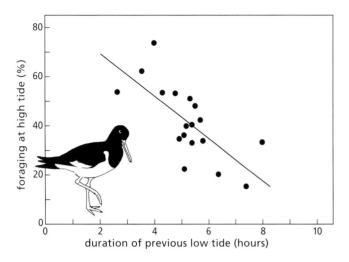

Figure 4.9 When tidal flats are exposed for only a short period at low tide, oystercatchers continue to feed during the following high tide. These data are from oystercatchers in the Wadden Sea that fed on a mussel (*Mytilus edulis*) bed at low tide and on inland pastures at high tide.[146]

ing the previous low tide.[146] Monitored oystercatchers foraged on a mussel bank at low tide and on a grassy polder (area of reclaimed land) at high tide. When the oystercatchers had foraged on the mussels at low tide for quite a while, they flew to the polder to rest on the incoming tide. But if the oystercatchers could only feed briefly on the mussels at low tide, during the following high tide they foraged in the pasture (Figure 4.9).

In months when inland pastures serve as additional feeding grounds, it's possible to predict that the birds will then be mainly found on the intertidal areas near pasture, as was the case in the Dutch Wadden Sea.[995] There, in the autumn the numbers of oystercatchers and curlews decreased in the intertidal areas without pasture nearby, and increased in the tidal areas that were close to pasture. Maybe the importance of pasture for waterbirds has increased in recent decades, as intensifying agriculture has caused earthworms and crane fly larvae to become more numerous.

### 4.3.7 Do time budgets change when food requirements increase?

Birds need more food each day when temperatures drop or when they want to gain weight. There are three ways to get more food. First, the birds can spend more time out foraging. They can do this by remaining on the tidal flat longer during the incoming and outgoing tides, or by foraging more at night or at high tide. Secondly, they can rest and preen less during their foraging periods. Finally they can try to increase their intake rate. The captive ringed plovers in Guinea-Bissau were a good example of birds prolonging their foraging period (Section 4.3.3). These birds never ate more than a certain amount of food during the day. The ones that were eating more did so at night. There are also observations that show how birds increase their food consumption in the field.

The body mass of waders on the Banc d'Arguin stayed constant over winter but began to increase in March by more than 1% daily.[241, 669, 985] To achieve this the birds had to increase their daily consumption by around 30%.[472] Dunlins did this by prolonging their foraging bouts by the same amount, mainly by foraging at night (Figure 4.10). Unfortunately, on the Banc d'Arguin dunlins eat such tiny prey[983] that their daily food intake could not be measured in the field. Because of this their actual food intake is unknown, but according to these data it could have remained constant and their recorded weight increase would still have possible.

The intake rate of whimbrels feeding on large crabs on the Banc d'Arguin could easily be measured. Whimbrels increase their daily food intake by 30% in the months preceding their departure for the northern breeding grounds. This is the same increase as was predicted for the dunlin. But unlike dunlins, the whimbrels got their extra food from a higher average intake rate both at day and at night, not by foraging for more hours per day.[975] They were able to increase their intake rate as their food availability markedly improved a few weeks prior to their departure.[969]

The foraging behaviour of red knots in the Wadden Sea in spring is another example. Knots that winter in the Wadden Sea begin to slowly fatten up in March and April. But birds that winter in Africa and don't arrive in the Wadden Sea until early May, fatten up much faster.[707] The total food intake of these African wintering knots is predicted to be twice as high as the Wadden Sea wintering knots. The African knots mainly achieve this through a higher intake rate, but also by spending more of their time foraging at low tide, and prolonging their total foraging period.[688]

The four examples above show three slightly different solutions. Ringed plovers and dunlins increased their daily consumption by foraging more at night and therefore prolonging their daily foraging period. Whimbrels achieved the same by increasing their intake rate, and knots did both, they foraged longer and faster. To understand these differences, we need to know the maximum rate at which each species can process its food. Perhaps a bird cannot eat faster because the rate at which it processes food is already limiting its food intake. The intake rate may also be limited by the food availability. In the period leading up to migration it may be, as was shown for the whimbrel, that a higher food intake is made possible by increased food availability.

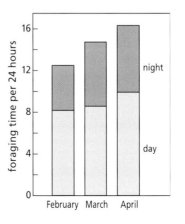

Figure 4.10 Increase in the daily foraging time of dunlins preparing to migrate from the Banc d'Arguin (Mauritania) back to their breeding grounds.[984]

## 4.3.8 Strategic stores

In winter, bullfinches put on 1 g of fat every day during daylight hours. This is exactly enough to get them through the night. As the nights get colder they put on more fat during the day.[598] It looks as the bullfinches assume that on the following day they will be able to find enough food to last them exactly one more night. Over the last decade, variation between diurnal and nocturnal fat stores has been found in a number of other songbird species. However a number of these more recent studies found that some species put on more fat than was needed for one night. Presumably these species have a less predictable food source than that of the bullfinches. Waterbirds cannot be too sure of the reliability of their food supplies either. The amount of time that they can spend foraging on the tidal flats per 24 hours varies daily in many places. However this variation is not that big. The biggest problem for birds that winter in cold areas is preparing themselves for cold snaps when the tidal flats are covered in ice.

Waders that winter in temperate areas begin putting on good fat stores in November and put on bigger stores in colder areas.[646, 991] In colder areas it's much more likely that during a cold spell the mudflats will freeze over. These birds need to have enough energy in reserve to be able to fly to ice-free areas when necessary. The birds may also decide to stay in the hope that the cold snap is short and not too severe. In this case, they will also need energy stores to survive the fast.[421] Birds that winter in the tropics don't need to take these precautions and have conspicuously low body masses.[177, 546, 991]

When they are not covered in ice, the regular tidal rhythm means tidal flats provide a very predictable food source for waterbirds. Daily variations in exposure time, like those caused by the effect of the wind on the tide, average out nicely when the exposure times are calculated over a period of a few days (Section 1.2.3). So we would expect that waterbirds don't need to plan for daily variations in the availability of their feeding grounds. But for oystercatchers, there are various data that suggest this is not the case.[991]

First, oystercatchers are heavier in late summer in intertidal areas with a smaller tidal amplitude than in areas with a large tidal difference. This is because the larger the tidal difference, the smaller the effect of wind on the exposure time, which means less daily variation in exposure time. Oystercatchers react to this by putting on 20 - 30 g less fat in these areas.

Secondly, breeding oystercatchers that forage low in the tidal zone are 20 g heavier than birds that forage higher in the tidal zone. The lower tidal flats are not only exposed for less time than the higher flats, but apparently also have a larger daily variation in exposure time (Figure 4.11).

Finally, breeding oystercatchers that forage on polders inside the dykes are 20 g lighter than birds that feed high on the tidal flats. Birds that forage inland have a very predictable food source and can in principle forage for 24 hours a day, while the coastal birds, even though they use the high tidal flats, still have a limited and less predictable foraging period (Figure 4.11).

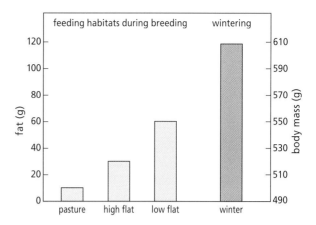

Figure 4.11 When daily food intake is constant, oystercatchers remain relatively light and deposit only small fat stores, but when food intake is less predictable, birds increase their fat storage. In this example, fat stores are shown for breeding birds from different feeding habitats (light grey bars). Inland-breeding birds can feed for 24 hours on pasture, but birds that feed on tidal flats can only forage for around half of this period, and tidal flat exposure can vary greatly. This especially affects birds that feed on the lower part of the tidal flats (which consequently have larger stores). In winter (dark grey bar), the wind causes large variation in tidal flat exposure times, and tidal flats can become unavailable because of cold spells and ice formation. Masses are from birds that were caught in the Eastern Dutch Wadden Sea.[991]

## 4.4 Food availability

### 4.4.1 Density

Although cockles and lugworms are rather different beasts and difficult to compare directly, that's no reason not to count them, weigh them, and figure out their densities. In any case, that's what everyone does when they want to determine the food availability on the tidal flats for birds. Sometimes prey density is simple to determine. In summer lugworms feed on substrate and leave behind squiggly droppings shaped like piles of squirted toothpaste on the sediment surface. Counting these droppings within a square of say 1 m x 1 m provides an easy estimate of the lugworm density. Assessing the density of most other benthic animals requires much more work, as you need to take core samples. Then you must take a corer with a diameter of 10 cm or more often 15 cm and push it 30 - 40 cm into the ground, before digging it out or raking it up. The core sample is then sieved in water over a 1 mm sieve, either in the lab or in the field. If the sample is not going to be sieved or sorted until a later date, it can be stored in diluted formalin to preserve the benthic animals present.

A corer is ideal for finding the density of deeply buried benthic animals, but isn't good at detecting prey that live on the surface especially if the animals are mobile or rare. Shrimps and common shore crabs are much easier to sample using a big open quadrat of say 50 cm x 50 cm. The quadrat is placed on the sediment, and the upper centimetres of sediment are removed and sieved. A similar technique can be used to measure the density of swimming prey in shallow water. For this a big cage of say 1 m x 1 m x 1 m made of fine mesh with the top and bottom open is used. Two people use two long sticks to hold the cage above the water, before dropping it and using a scoop net to catch all the fish inside. Although this method isn't often used it has a big advantage over the usual drag and push nets, as the exact fish density in the shallow water can be determined at every moment of the tide cycle. The method is ideal for shrimps and small fishes that bury themselves just beneath the sediment surface. It doesn't work for fishes that swim away quickly when disturbed.

*Determining density: how many samples?*
In the laboratory, benthic animals are identified, counted, measured and weighed. Counting them allows us to calculate their density, which we can then combine with their size and mass to determine the 'flesh' weight or total biomass per surface area. All this begins with the corer out on the mud. Here we need to decide exactly how many samples we need to take, so that later we can accurately estimate the density of the different benthic animals. The number of samples depends on how accurate you want your density estimate to be and the variation you find between the different samples. It's usually accepted that density estimates contain around a 20% error. You can use a pilot study to determine how much variation there is between samples. This variation is related to the spatial distribution of benthic animals over the intertidal flat. Suppose that rather than having a random distribution, the prey have a completely homogenous distribution. If you know this before starting your main sampling, you'll realise that you only need to take a few samples, as each sample contains about the same amount of prey. If, however, the animals are obviously clustered, you will have to take considerably more samples than if they showed a random distribution. Variance is a statistical way to describe the variation in the number of animals in the samples. When the distribution is completely random, the variation will equal the average. For most benthic animals, the variation is larger than the average, which means that they have a clustered distribution. The less homogenous the distribution, the bigger

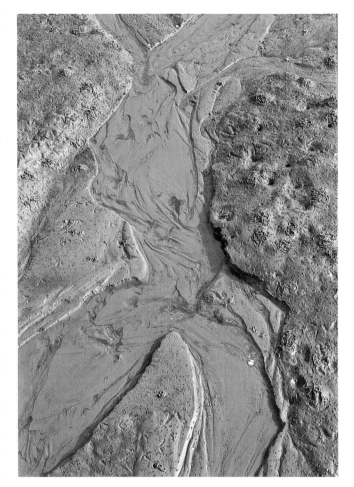

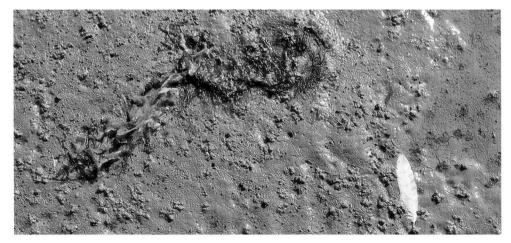

A gully just tens of centimetres wide (upper photo) looks like an aerial photo of an eroding river some kilometres wide. Only the tracks of the benthic animals betray its true size. The only surface evidence of shellfish like these sand gapers and Baltic tellins is their empty shells (middle photo). Tracks of different worm species are visible on the sediment surface, particularly on sandier parts (lower photo).

Researchers use a hollow tubular core sampler to collect sediment samples, then sieve the mud over a fine mesh to find out which prey species occur on the mudflats, and at what densities (upper photo). The only way a dunlin can find prey and measure prey density is to repeatedly probe its bill into the sediment surface (middle photo). Black-headed gulls stir up muddy sediment with their feet to reveal hidden prey (photo lower right), while a bar-tailed godwit inserts its whole bill into the sediment (photo lower left).

the variation in the numbers found between samples, and the more samples you'll have to take.

At first glance the toothpaste heaps of the lugworm appear to have a homogenous distribution, but when you begin to measure them you find they are randomly spread. To determine the micro-distribution of animals that live hidden in the sediment within a small area of tidal flat of almost 1 m x 5 m, seven lots of 37 13.3 cm x 13.3 cm samples were taken.[968] The sampling cores were square so that the samples could be collected side by side. The 13.3 cm square corer covered the same area as the standard round corer with a 15 cm diameter. Figure 4.12 shows the distribution of the catworm *Nephtys hombergii* in this seemingly homogenous piece of mud. On average, each sample contained 0.48 worms. However the variance at 0.54 was somewhat bigger than the average, suggesting that the catworms had a slightly clustered distribution.

Having 0.48 worms per sample equals a density of 27 worms per m². Figure 4.12 shows that if within this 1 m x 5 m area only one sample had been taken, our estimate of the catworm density could have varied from 0 - 170 worms per m². How many samples were needed to get a reliable estimate? If the variance equals the average meaning that the distribution is completely random, you have to keep sampling until 25 prey animals have been collected. At least this number is required if beforehand you've accepted that there should be a 95% chance that your estimate is within 20% of the true average. If you decide to accept twice as big an error (i.e. 40%) you will only need to collect a quarter as many animals (i.e. six). However, if you'd like to halve the error (10%) then you need four times as many animals (i.e. 100). In the catworm example, you can calculate that to stay within the 20% error a total of 28 worms must be collected. This means you will need to take 58 samples to be able to say that the actual catworm density is between 21.6 and 32.4 animals per m² with 95% confidence.

Most benthic animals are similar to the catworm and have a random or lightly clustered distribution. Mudsnails are the exception to this, and are very heterogeneously distributed even on a very small scale. Fortunately, for the people who collect the samples, most benthic animals are much more numerous than the catworms. When cockles exist in a density of 500 per m², you will find an average of 8.6 cockles using a standard 15 cm diameter corer, so you will only need three or four samples to stay within the 20% margin of error.

Juvenile benthic animals often have a clustered distribution, which disappears as they age. This may be due to predators' reaction to the clustering, if as soon as a predator finds an individual prey item it immediately searches for others nearby. This is also why most foraging birds check out the areas with the highest food availability. As a result of this selective predation, areas of previously high density are depleted and the prey's distribution becomes more random. This was nicely illustrated by some research on the annual density of sand gapers in a permanent sampling site in the Wadden Sea. In 1979, first-year sand gapers were present at a density of 378 per m². A year later only 29 per m² were left and the following year only 0.3 per m². In the first year the variance was 6.7 times greater than the average, the second year only 2.5 times, and in the third year only 0.9 of the average. Even though third-year sand gapers were 1000 times less abundant than first-year animals, it wasn't necessary to take 1000 times more samples. If we accept an error of 20% for the sand gapers, eight samples must be taken of first-years, 41 of second-years, and 33 of third-years.

### 4.4.2 Determining prey size and mass

To assess prey size, it is usual to measure the length of bivalves, the height of snails and the carapace width of crabs. There are different methods for assessing the lengths of shrimps and worms. Shrimps and amphipods are normally measured from head to tail i.e. not including their antennae. Worm lengths vary greatly; a ragworm crawling along a ruler can extend or contract its length by 40%. The maximum 'crawling length' is 27% more than the length of the same worm dead in alcohol.[968] Naturally, it's much easier to measure a dead worm, but dead ragworms often break. It is still possible to estimate their length however, because there is a strong relationship between the length of ragworm and its width just behind its head. A lugworm's length can be assessed either with or without its tail. It's better not to include the tail, because tail length is very variable, independent of the length of the rest of the body.

When you assume that there is a strong relationship between the volume of a prey item and its mass, the mass can be easily estimated in the field by dropping the prey into a measuring cylinder filled with water. Live prey can also be weighed themselves, but the problem with using live or wet weights is that the water content can vary between 75% and 82%. This doesn't seem much, but if in addition the dry flesh or meat content varies from 18 - 25%, that's a difference of almost 40%. Much of this variation comes from the way in which the animal was weighed. A worm that is taken straight from the sea is usually several percent heavier than one that has first been placed on filter paper. Both methods are seldom used nowadays, but have the advantage of not requiring the animals to be killed. The methods discussed in the following paragraphs do require sacrificing the animals being studied.

Problems with variable water content can be solved by drying the prey in a 60 or 70 °C oven for a few days, until all the water is lost. However this 'dry weight' is still not very accurate, as the prey animal may also contain sediment and other non-organic material (particularly salt) that can account for up to 10 - 25% of the dry weight. This variation can be reduced to 10 - 15% by storing the live prey animals in clean and preferably running seawater immediately after they reach the laboratory.[970] These percentages are found in samples consisting entirely of meat, such as worms and bivalves without their shells. In

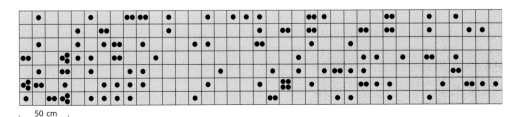

Figure 4.12 The distribution of a polychaete worm, *Nephtys hombergii*, in a seemingly homogenous piece of tidal flat. In an area of 93 cm x 492 cm, 259 adjacent 13.3 cm x 13.3 cm samples were taken. Each dot represents one worm.[968]

common shore crabs and other crustaceans, sediment and salt are not the only source of inorganic material. When the common shore crab's calcified outer skeleton is included, inorganic matter accounts for 35% of the dry weight. These percentages are much higher in shellfish if the meat and shell aren't separated: 70 - 80% in Baltic tellins and sand gapers, and 90% for cockles.[970]

There are two ways to solve the problem of dry masses including varying amounts of inorganic material. First, a subsample can be incinerated at 550 ºC. At this temperature all the organic material is burnt and the inorganic matter is left behind as ash. By subtracting the ash weight from the original dry weight, we get the ash-free dry weight. This is the weight of the meat without any water or inorganic material. For the rest of this book we use the term 'dry flesh' to mean 'ash-free dry weight'. Secondly, the energy content can be assessed with the help of a bomb calorimeter, which converts the prey weight into joules, the units in which energy is expressed. For either method, it's best to remove the calciferous parts first if possible.

With shellfish we are particularly interested in their flesh weight. Even though shell weight is closely related to the shell size, it's always best to weigh the shellfish after you have removed the shell. The main reason for this is that as well as the flesh, a lot of muddy water may be shut inside the shell. The flesh can be easily separated from the shell by putting the animal in boiling water for a few seconds and then cutting out the meat.

A lot of things can go wrong when you try to assess flesh weight and, depending on the method used, different outcomes may be produced. It's important to store the animals in seawater as soon as possible after collection and to keep them at a temperature a few degrees above freezing, at least for temperate benthic animals. This stops the animals from losing any weight. The alternative is to freeze them, but that can lead to weight loss when they are defrosted because the nutrients from the damaged body cells leak out with the associated seawater. Similarly, storing in alcohol or formaldehyde damages body cells leading to loss of nutrients and a potential weight loss of more than 40%.

The energy content varies little and is therefore not usually determined.[64] However, there are small but systematic differences between the species.[982] On average, worms yield a bit more energy per unit weight than shellfish. Energy also varies between shellfish species. For example, mussels have a larger energy content than cockles. The lowest values are found for the amphipod *Corophium volutator*, shrimps and common shore crabs (20 - 21.8 kJ). This is partly due to the low energy content of the shell of these species. Also, although burning most materials in the bomb calorimeter gives off energy, it actually takes energy to burn calcium. The presence of calcium in a flesh sample is not usually taken into account, causing a systematic error in the results obtained. However differences in the energy content of crustaceans, worms and shellfish are to be expected. Compared to worms and shellfish, crustaceans don't have much fat. Fat has a higher energy content (39 kJ per g) than fat-free dry meat, which mostly consists of protein (23 kJ per g) and to a lesser extent carbohydrates (17 kJ per g). The energy content of all species of benthic animals from the Wadden Sea is lower than that of benthic animals from the Baltic or in the far north of Canada. These types of regional differences in energy content are, like the differences between species, caused by differences in fat content.

### 4.4.3 Variation in prey weight

Once it has caught its prey, the amount of flesh a bird consumes depends on the prey's size and condition. These vary through the season, as when benthic animals grow and increase in mass, they increase in length, width and height. For this reason it is expected that mass is a function of length cubed. This is not completely true, because the shape of the animals can change slightly as they grow. Sand gapers and cockles will become rounder in shape, so when weight is plotted against length, the exponent is not 3 but 3.1 or nearly 3.2. This means that mass rapidly increases with increasing length. A 2 mm cockle contains only 0.06 mg dry meat, but once it is 10 or 20 times longer, its meat weight is 1500 or even 13 000 times greater respectively.[970]

The body composition of benthic animals also changes during their lives. During a growth period they build up body reserves that are used in the non-growing period. Also, benthic animals make gonads that are broken-down through the summer (after the reproductive season). As benthic animals age, the gonads make up a larger proportion of their mass. When ageing animals invest more in reproduction this should reduce the resources available to invest in growth. Sure enough, old animals only grow a little annually.[970] You might also expect that the weight variation in juvenile benthic animals would be smaller than that of the older year-classes that can invest more or less in gonads. However the opposite seems to be the case. Baltic tellins, mussels, cockles, sand gapers and peppery furrow shells all show the largest variation in body weight in smaller individuals. Given the relatively low gonad production of juveniles, the annual variation in the size of their body reserves must be considerably greater than that of older animals.[970]

A condition index is often used to describe this variation in mass, independent of prey size. The index is calculated as mass divided by length$^3$. Despite the fact that this index is often used, it is not correct, because the exponent systematically deviates from three and also varies through the year. It is better to calculate the relationship between length and weight for each set of samples. The best way to do this is to calculate the average weights per size-class and convert both to logarithmic values. After this, the average body mass predicted by a regression can be used as a measure of the condition. By an average prey animal we mean an average-sized prey such as a 15 mm Baltic tellin or a 20 mm cockle.

The condition of almost all species of benthic animals in the Wadden Sea peaks in May and June, apart from mussels which peak in July. After this peak, all species show a gradual decrease in mass until the end of winter, after which mass begins to increase quickly again. The maximum weight of most benthic animals lies around 70% above their minimum, but is only around 50% in Baltic tellins. Even if birds did not have to eat more in winter than in the summer because of the cold, they would still need to find more prey in winter to compensate for the poorer prey condition.[970]

The seasonal variation in the body condition differs from year to year. The maximum summer mass of a 15 mm Baltic tellin, for example, varies annually from 37 - 60 mg dry meat, and the minimum winter mass from 25 - 32 mg. Similar differences are found in other benthic animals. A 20 mm sand gaper's winter mass can vary from 80 - 130 mg annually, and its summer mass from 160 - 230 mg. Annual differences in winter masses are due to variation in the temperature. Shellfish must live off their reserves during winter, as there is little or no food available. Curiously, the colder the winter is, the less weight is lost. When the winter sea temperature is around 9 - 10 °C, sand gapers, peppery furrow shells and cockles lose 20% of their body mass, but when the sea is 2 - 3 °C their mass doesn't change. At lower temperatures a shellfish requires almost no energy and doesn't need to use its body reserves, but in mild winters it must. So in cold winters, warm-blooded birds must spend a lot more energy trying to keep themselves warm, but their cold-blooded prey are in better condition, at least they are if they don't freeze to death.[970]

How much does the mass of the same benthic animals vary in British or French tidal areas? In winter when there is no food available for the benthic animals, the milder temperatures in these areas should lead to poorer winter condition. A comparison of all available data found that the winter condition did not decrease in areas where the average seawater temperature is higher.[970] The problem is that such an effect is not easy to detect. In the Wadden Sea, the size of the annual variation is known from annual measurements, but for other areas in Northwestern Europe there are usually just a few individual records available and data have seldom been collected over more than a year. At least we can say the available data don't show that the winter weights in France or Great Britain are lower than in the Wadden Sea. In areas with higher average seawater temperatures the benthic animals probably find enough food to compensate for the higher energy expenditure. On average, the seasonal trends are similar, but there are subtle differences. The Baltic tellin reaches its peak mass twice as early in the Gironde (Southwestern France) as in the Baltic Sea or in Nova Scotia (Canada). The mussel reaches its maximum mass in summer in the Wadden Sea, but not until autumn in the Dutch Delta area and five English estuaries. With this in mind, it seems that oystercatchers feeding on mussels should move from the Wadden Sea to England in September.

As well as condition varying seasonally, annually or by region, it can also be very different locally. For example, the condition of cockles that live high in the intertidal zone is lower than that of their conspecifics that live nearer the low tide line. This is because cockles must filter their food from the surrounding seawater, so when

Three important prey species for waterbirds in Northwestern Europe: the common shore crab (upper photo), ragworm (middle photo) and cockle (lower photo). Large crabs migrate to the tidal flats with the incoming tide and return to the channels again with the outgoing tide. The small crabs stay on the tidal flats over low tide just under the sediment surface, and are an important food source for waterbirds. Ragworms are even more important prey. These worms live in deep burrows, but regularly come up to the sediment surface or even crawl out of their burrows. Once on the surface they put themselves at great risk, as few birds would let such an opportunity pass. Cockles are the main prey of oystercatchers. Some individuals are visible at the surface but most are buried in the sediment so that their siphons just reach the surface.

they are exposed they are unable to feed, and they are exposed longer the higher they are on the tidal flats. But even benthic animals collected on the same day from the same spot can show relatively high variation in mass between individuals. Some of this variation is due to the effect of the following five factors.[970] (1) Older Baltic tellins (aged by annual growth rings) contain more meat during the winter than younger animals of the same length. (2) Parasite-infected Baltic tellins are heavier than parasite-free individuals, especially in winter. This is not because infected shellfish are in better condition; it's because the parasites are included in their weight. (3) Peppery furrow shells that have developed large gonads are heavier in the summer than conspecifics that didn't. (4) Ragworms, Baltic tellins, peppery furrow shells and sand gapers that live near the sediment surface are in worse condition than deeper-living conspecifics. (5) Shellfish with a relatively small inhalant siphon are in a worse condition than the ones with a larger siphon. The last two points are discussed fully later in this chapter.

The factors affecting prey condition as listed above aren't of equal importance to the birds feeding on the prey. Oystercatchers can not possibly feel if a 30 mm Baltic tellin is two, three or four years old, so are unable to select the older, meatier year-classes. But oystercatchers can identify infected Baltic tellins.[417] Although deeper-buried prey may be in better condition, birds can only reach benthic animals that haven't buried themselves too deeply. Often researchers endeavour to determine the average prey weight, but birds feeding on this prey can only get the shallow-living thin ones within bill reach. For this reason, researchers often greatly overestimate birds' food intake.[982]

### 4.4.4 The life cycle of the benthic animals: mortality and growth

In Northwestern Europe massive numbers of juvenile benthic animals appear on the intertidal flats in summer. In early summer they are still so small that they slip through a 1 mm sieve, but by July the little shellfish are already a few millimetres long. They continue growing until late summer, by which time most have already been eaten by shrimps and crabs. After late summer, they stop growing until early the following spring. Young benthic animals don't appear on the mudflat every summer. Years may pass without a single juvenile being found. The 1983 sand gaper class was the most numerous seen in the Western Wadden Sea for 30 years (the year-class of 1963 had been bigger). The 1983 class has dominated the biomass of benthic animals in the Wadden Sea for more than 10 years. Similarly, peppery furrow shells from a massive spatfall in 1976 were still common six years later in coastal parts of the Dutch Wadden Sea.[982] The cockle, in contrast, is a short-lived species. In the first winter 85% are lost on average, and every following year a further 75% disappear. In harsh winters, up to 100% may be lost.[53] But if the year-class is strong enough, the winters are not too harsh, and a large enough area is sampled, it's still possible to find six-year

On the Banc d'Arguin (Mauritania) oystercatchers eat giant bloody cockles that are 5 - 10 cm long. The prey lie on the surface at densities of tens per m² (photo below) and are easily found by the birds. The problem for the oystercatcher is to find one that hasn't closed its valves firmly, so that the oystercatcher can push its bill between the valves and sever the adductor muscle.

old cockles.[59] Baltic tellins live longer. The annual mortality is conspicuously constant; on average 44% disappear each year.[55] This means that after 10 years, 1.5% of the original population still survives.

It appears that prey that grow slower live longer. In the Baltic Sea all sorts of shellfish live many years longer than the same species found in the Wadden Sea. But shellfish can also reach a ripe old age in the tropics. At least this is the case for the tropical giant bloody cockle (*Anadara senilis*) which must have had a massive recruitment on the Banc d'Arguin in 1968.[949] Twelve years later, in 1980, these animals were 65 mm in size and in good spots could be found in densities of 56 per m². The population has been sampled several times since, most recently in 1997. By then, these cockles were 29 years old, 85 mm long and their density had decreased to 7 per m².[996] Extrapolating from this, the last of this generation of bloody cockles should reach more than 40 years of age. Oystercatchers feed on these giant cockles which, including the shell, are heavier than the birds themselves. An oystercatcher can easily reach 20 years of age (Chapter 5), so the prey are sometimes older than the birds.

### 4.4.5 Seasonal variation in total food availability

On the tidal flats of Northwestern Europe, birds encounter considerably more food in summer than in winter. In summer, the biomass of all benthic animals is twice as big as in the winter. Biomass peaks in June and reaches a minimum in March (Figure 4.13). This seasonal pattern differs somewhat depending on the species. The growing season of Baltic tellins starts earlier than that of other shellfish. Cockles and ragworms reach their highest biomass in July and August rather than June.[982] The variation in biomass also differs depending on the species. A cockle's maximum biomass is 3.3 times higher than its minimum, but a peppery furrow shell's is only 1.6 times higher. For most species, around 60% of the seasonal variation in biomass is due to changes in body condition. The rest can be attributed to the arrival of new generations and the growth of the populations.[982] For almost all species, the seasonal pattern of biomass parallels the changes in body condition. However the condition of cockles and mussels decreases in late summer (like that of other species), while their total biomass continues to increase. This is because in these two species, the shell continues to grow while the flesh either doesn't or doesn't grow as fast.[982]

Shrimps and crabs are usually very abundant on the tidal flats of Northwestern Europe, but they are absent in the winter months. Common gobies also disappear from the tidal flats in autumn and only return in the course of summer. Birds like the greenshank and spotted redshank that mainly forage on these seasonal prey are then left without a food source in the Northwest European tidal areas in winter.

### 4.4.6 Annual variation in total food availability

Can birds count on finding a good spread on arrival in the Northwest European tidal flats, or must they wait and see what's available? We know from long-term data collected in the Dutch and German Wadden Sea that food availability can vary greatly from year to year. Most prey species show a high annual variation that causes significant changes in the species' contribution to the total food available. In March in the early 1970s the total biomass of all bivalves on the Balgzand (in the far west of the Wadden Sea), was 14 g dry meat per m² (Figure 4.14).[68] The 1963 year-class of sand gapers made up 10 g

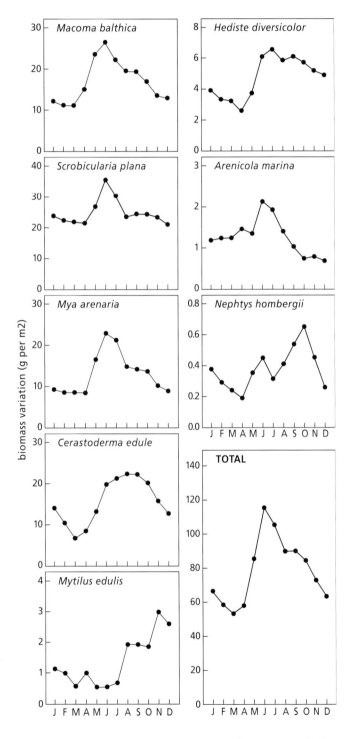

Figure 4.13 Seasonal variation in the biomass (total flesh mass per m²) of a variety of benthic animals in the Eastern Dutch Wadden Sea. Average values were calculated from monthly samples taken over seven years (1979 - 1986) and are shown for each species separately, and for their total biomass combined.[982]

of this. Over the following years these animals slowly died out. During the last three decades, the total biomass of sand gapers has varied from 1 - 13 g per m². Cockles (0 - 16 g per m²) and mussels (0 - 8 g per m²) have also shown a large annual variation. But when the biomass contributed by the different species is combined, the annual variation is much less, at 8 - 28 g per m² (if we exclude one year when intensive shellfisheries lowered the bivalve biomass to under 5 g per m²).

A different, shorter data set from the Dutch Wadden Sea also found a large annual variation.[982] In 1981, 70% of the total biomass of all benthic animals (85 g per m²) was due to five year old peppery furrow shells that had also dominated the biomass in previous years. The following year, the peppery furrow shells had disappeared. But two years later the biomass was back up to 80 g per m², half of which consisted of cockles.

Such large variation in biomass is probably characteristic of the Wadden Sea. In the Eastern Schelde, an estuary in the Southern Netherlands, the variation in the cockle stock appeared much smaller,[76] and mussel biomass was remarkably constant in the Exe estuary in Southwest England.[549] This might be due to the milder climate and more sheltered locations of these areas. The effects of harsh winters on the abundance of many benthic animals are clear. When the mudflats freeze over, there is a massive mortality of cold-sensitive species like the polychaete *Pygospio elegans*, and cockle and mussel beds can be wiped out by drifting ice.[52] Maybe even more important than the high mortality of benthic animals during the cold snap, is the massive recruitment of all sorts of benthic animals some months later.[56] Plus after a cold winter shrimps and crabs don't appear on the tidal flat until late spring and are less abundant, giving young benthic animals a better chance of survival.[60, 61] Cold winters in the Wadden Sea are responsible for much of the variation in the biomass of the benthic animals and the dominance of different species. This effect appears to be smaller in areas where the climate is somewhat milder. We can therefore assume that the annual variation in biomass will be smaller in areas with higher average winter temperatures. If this really is the case has yet to be determined.

## 4.5 Variation in food accessibility

### 4.5.1 Why don't benthic animals stay safely buried deep in the sediment?

The main issue for birds is whether the food on the tidal flats is actually within their reach. When a sand gaper is buried 20 cm deep, no bird can reach it, not even the curlew with the longest bill. This means researchers must do extra work to find out what food really is available to the birds. Additional questions must be answered, such as at what times do amphipods crawl over the sediment, how deeply are the shellfish buried, how thin must a shell be before an oystercatcher can make a hole in it? It's useful to try and take the prey's perspective when searching for the answers. Almost all of the prey in the tidal

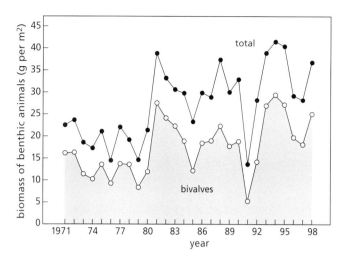

Figure 4.14 Annual variation in the biomass of benthic invertebrates in March at Balgzand, in the Western Wadden Sea. Values are averages from 15 sampling stations.[49, 62, 68]

zone live buried in the sediment. Even though they are safe there, they cannot remain hidden all of the time. They must get oxygen from the water above the sediment and, unless they eat other benthic animals, their food must come from the water or the sediment surface. Substrate feeders must also excrete non-digested material at the sediment surface. Benthic animals must decide how deeply to bury themselves and how often they will come up to the surface. They must choose between starving but being safe, or eating and being able to grow but facing the associated risks. This trade-off is different for each species, as we will see for those discussed below.

### 4.5.2 Lugworms

The lugworm (*Arenicola marina*) lives in a 30 cm deep u-shaped tube.[982] Down in its burrow the lugworm is safe from predators. It creates a small water current to bring food-rich sediment down through the funnel-shaped entrance of its tunnel, where it can then feed safely. However, every 1 - 2 hours the lugworm has to excrete the ingested substrate, and must crawl up its burrow to deposit the droppings on the surface. This takes several seconds, and many birds wait for just this moment.[750] If a curlew is standing 1 m away from its burrow, the lugworm won't survive its toilet stop. If the curlew is 2 m away, the lugworm will escape more than half of the time, or will only lose part of its tail to the curlew. A curlew that tries to make a 3 m dash won't catch a defecating lugworm. Although lugworms are out of reach 99.9% of the time, they run a very high risk for the remaining 0.01% but they have no choice. It is like knowing a large shark could be waiting in the toilet bowl, but sooner or later you'll still have to go.

Lugworms could avoid being preyed on by birds if they fed and excreted their food during high tide. However this isn't much of an alternative, as the birds are simply replaced by fish.[899, 900] The numerous juvenile flatfish that swim around the tidal flats in early summer will quickly grab any lugworms they see defecating nearby. Even if the fish only grabs the end of the lugworm's tail, it will eat whatever bit it can get. Although losing the end of its tail is not fatal, lugworms can lose considerable amounts of body tissue in this way, as their tail lengths show. Lugworms' tails get visibly shorter through the summer.

### 4.5.3 Ragworms

The ragworm (*Hediste diversicolor*) also lives in a u-shaped tube, but uses different techniques to get its food.[259] One way is to crawl out of its burrow and eat substrate. If a bird wanders past, whether the ragworm will make it back into its burrow in time becomes a matter of life and death (for the ragworm). The ragworm also uses a less risky foraging technique. It makes a small net out of slime threads just inside the burrow entrance, then pumps water through it. The worm then eats the net and all the tiny particles of food caught in it, before making a new net and starting all over again. Although the worm stays in its burrow, this feeding technique is not completely risk-free. Pumping the water down causes air bubbles to develop and this changes the water level in the tunnel opening. Birds can use these subtle clues to search for ragworms. The ragworm must make a compromise between the differing risks and benefits provided by feeding up on the surface or down in its burrow. The food yield differs depending on the method used, the substrate, the stage of the tidal cycle, and the season.

There are two reasons why ragworms in a sandy sediment tend to pump more and feed on the surface less often than those in silt.[259] At low tide on a silt flat ragworms have nothing to pump when there is no water left on the surface. It can continue pumping on a sand flat, as sand flats are rippled and have small puddles (Figure 1.5D) In addition, sandy sediments contain less food than silty sediments, which have a high organic content, so it's less attractive for a ragworm to eat sandy substrates. On silt ragworms only pump around low tide. After this they have no choice; if they want to feed they must leave their burrows.

Ragworms face similar decisions depending on the season. The amount of food available varies more in the water

Fiddler crabs are conspicuous inhabitants of (sub) tropical intertidal areas. They live in deep burrows but must come up to the surface to feed. They scrape up the food-rich sand, chew on it with their mouthparts, and make small balls of the leftover sand that they didn't swallow (upper photo). They forage close to their burrows so that they can beat a hasty retreat when danger approaches. Fiddler crabs live high in the tidal zone, but on spring tides thousands of large crabs leave the upper flats and march down to the lower flats. After foraging for a few hours, they walk back up the beach. While in transit they have no means of escape from hungry whimbrels, gull-billed terns, and turnstones (lower photo).

than on the sediment. In winter there are almost no algae in seawater, but food still remains on the sediment surface.[428] So in winter there is nothing to be gained from pumping water, and again if the ragworm wants to feed it must leave its burrow and graze on the sediment surface.

The ragworm is the staple food of many waterbirds in Northwestern Europe.[259, 978] These birds alter their behaviour to match the foraging behaviour of their prey. Curlews that encounter many ragworms out grazing around their burrows on the sediment surface walk quickly with an upright stance.[978] When the ragworms stay in their burrows the curlews walk slowly, looking down. Curlews can't see ragworms in their burrows, but watch for telltale signs such as rising air bubbles or the slight changes in water level that signify a burrow entrance. The curlews must watch very intently to detect these subtle clues, which is why they walk so slowly. If they kept the same pace they use for finding surface-grazing ragworms, they would miss many of the signs and catch less prey. Curlews hunting surface-feeding ragworms can probably see them from some metres away and will encounter more prey if they walk quickly.[290] There is a reason to keep a good pace: ragworms rapidly retreat to their burrows when they see approaching danger and if the curlews aren't quick enough they'll miss out.

### 4.5.4 Fiddler crabs

It takes a lot of time and energy to determine the variations in lugworm and ragworm availability. Their availability can be recorded using photographs or video footage, but this simply delays the problem and before you know it you've spent days analysing images collected in just a few hours. Fiddler crabs (*Uca spp.*) are large enough to be easily seen from a distance when they surface. The big advantage of this is that you, the observer, can record the behaviour of the prey (the crab) and the predator (the bird) at the same time. You can watch birds feeding on fiddler crabs and immediately see how the other crabs react to this.

Fiddler crabs are common in tropical tidal areas. They live in deep burrows, where they are reasonably safe. Birds that try to extract large fiddler crabs from their burrows need long bills, and in West Africa only curlews,

A whimbrel waits, concentrating on a fiddler crab that may emerge from its burrow at any moment (upper photo). Strike! The whimbrel has succeeded in catching a large fiddler crab (middle photo). The large crab must be dismantled, shaken until its legs and pincers have fallen off and can be eaten separately (lower photo). While this is happening, turnstones and grey plovers often try to steal a loose leg or pincer. Gull-billed terns, which swoop down on fiddler crabs, also have to be wary of thieves (photo right).

FOOD

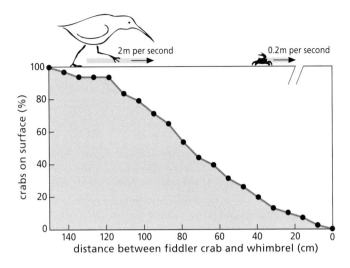

Figure 4.15 The game of risk: foraging fiddler crabs (*Uca tangeri*) versus foraging whimbrels. Fiddler crabs forage a short distance from their burrows, retreating to them when danger threatens. Whimbrels forage on fiddler crabs, and can move 10 times as fast as a 2 cm long crab can. Crabs stay within 10 cm or so of their burrows, and start to retreat when whimbrels come within 1 m of them. Data collected on the Banc d'Arguin.[960]

bar-tailed godwits and sacred ibises can do this. Small crabs don't burrow as deeply as the larger crabs, so are often eaten by smaller-billed birds such as redshanks, turnstones and common sandpipers.[965, 967]

Fiddler crabs leave their burrows to forage. They eat substrate near the burrow entrance and will race back to the safety of their burrow at the slightest disturbance, only venturing back out some time later. Birds like whimbrels try to catch the crabs by sprinting short distances. The outcome of this race between the whimbrel and the fiddler crab depends on a few simple factors: (1) how far the crab dares to venture from its burrow, (2) how close the sprinting whimbrel gets before the crab sees it, (3) how quickly the crab runs, (4) how quickly the whimbrel moves.[960]

Small crabs forage within a few centimetres of their burrows, but the area that the crabs graze daily increases exponentially with their body size.[245] That's why on average, large crabs venture further away from their burrow than smaller crabs. The larger crabs can walk faster than smaller ones, but not fast enough to compensate for the greater distance they usually need to cover. So it's dangerous for a fiddler crab to be big. The crab reduces its risk of being caught if it can make it back to its burrow in time during a disturbance. Exactly how far does a crab venture from its burrow and at what distance can it see danger? To find out 30 fiddler crabs were followed continuously using a video camera set up on a 3 m high platform. Unfortunately, not a single bird wanted to pass beneath the camera. So a ball the size of a whimbrel was rolled across the beach past the camera to allow the crabs' reaction time to be accurately measured. The first fiddler crabs reacted when the ball was 1.5 m from them, and the last at 0.2 m. However it is even more important to know how many crabs had disappeared down their burrows by the time the ball arrived at the burrow. The quickest crabs were gone when the ball was still 1.4 m away, half of the crabs were back in their burrows at the safe distance of 0.8 m and the last crab disappeared when the ball was just 7 cm away (Figure 4.15). So these fiddler crabs all made it back to their burrows in time. However it could still take a crab another half a second to disappear completely down its burrow, so the late crabs could be making the last dash of their lives.

A whimbrel must walk much faster than a crab to catch it. A whimbrel walking at a good pace takes three steps per second, each covering an average of 18 cm, so it travels at a speed of 0.5 m per second. At this speed, whimbrels can't surprise any fiddler crabs. But every now and then whimbrels quickly sprint, taking eight steps per second, each of 30 - 35 cm. They then have a speed of almost 2.5 m per second. Crabs running for their lives reach 0.2 m per second. When whimbrels can sprint 10 times as fast as the crabs can run, the distance at which the crabs can see the whimbrels should determine how far they graze from their burrows. The video footage showed that half of the crabs reacted when the danger was 0.8 m away; this means that they shouldn't graze more than 8 cm from their burrows. Sure enough, this is the maximum distance that average-sized fiddler crabs keep between their burrows and grazing sites.

When danger approaches, fiddler crabs react not only to the threat itself, but also to the escape behaviour of their neighbours. When a whimbrel runs into a group of grazing fiddler crabs, the fiddler crabs that see the whimbrel immediately dive into their burrows, as do their neighbours who have noticed that something is wrong. Everywhere else on the mudflat crabs are out and about, but a no-man's-land several metres in diameter extends around the running whimbrel. If a whimbrel continues running around its chance of encountering crabs just gets smaller. To solve this problem, every now and then the bird stands still for a while and waits until peace has

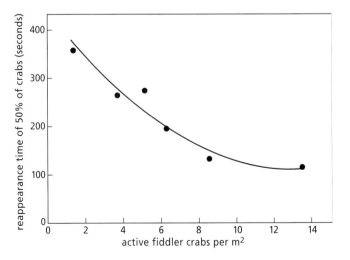

Figure 4.16 The waiting game: who can last longest? A fiddler crab (*Uca tangeri*) that has escaped a whimbrel sits in its burrow. It will return to the surface more quickly if there are already other crabs grazing there. The whimbrel's intake rate is low if there are few crabs around. It will then wait longer for a crab to re-emerge from its burrow. The crab, in turn, tries to wait even longer. Data collected on the Banc d'Arguin.[960]

returned. Eventually, the crab-free circle around the bird becomes smaller. A whimbrel standing in an area of foraging crabs needs to get up to speed from standstill and before this can happen the whimbrel has caused so much panic that any chance of catching a crab is lost. Measurements show that a whimbrel must take at least 50 steps before reaching its top speed of more than 2 m per second. Whimbrels have a clever way around this: they stand outside areas of high fiddler crab density and run into them. The experiments with the ball confirmed what had been seen in the field. On the edge of an area with fiddler crabs, the average distance between the ball and the burrow at the time that the crab dived back into its burrow was 55 cm. At 1 m from the edge, this distance had already increased to an average of 80 cm. Fiddler crabs that lived 2 m from the edge had already reacted when the ball was still 1 m away. So crabs living near the edge of a cluster run the highest risk of being caught.

When the whimbrel's sprints don't work, a new competition can develop: who can wait longest. The crabs that have dived back in their burrows want to come up and start foraging again. However the whimbrel stands stock-still with its bill 1 cm above a burrow, waiting. A whimbrel can sometimes stand like this for 30 minutes, but often the crab has already lost or the whimbrel has given up waiting before then.

The trade-offs that must be made by the birds that eat fiddler crabs, and the fiddler crabs themselves, can be seen through patient observation in the field. The behaviour of the bird and its prey is almost always measurable and through this hopefully can be understood. Here are a few examples of this. When it is cold by subtropical standards, i.e. below 15 °C, the crabs stay in their burrows. The cold makes them so slow that they would immediately lose any race against a whimbrel. In winter they stay in their burrows for days, coming up to the surface en masse at spring tide. However, even then they only come out for a few hours at low tide, never around high tide. In this way each crab decreases its individual risk of being caught.[245]

Whimbrels that continually sprint are more successful when there are more crabs on the surface. This shouldn't surprise anyone. At first, it seems a bit odd that when more crabs are active, whimbrels are also more successful than when they stand poised and waiting for a single crab to surface. The reason for this is not immediately obvious, but the more crabs are active, the stronger their urge to quickly return to the surface (Figure 4.16). It seems that fiddler crabs estimate how long a whimbrel will wait, and base this on an estimate of the density in which their conspecifics are grazing on the surface. Whimbrels will only have to wait a long time when few crabs are active and their own foraging success is very low. This time spent waiting is part of the whimbrel's handling time. Prey that require a long wait to catch have a lower handling efficiency. Whimbrel will only eat these prey if their intake rate is already low (as was explained at the beginning of Chapter 4).

If the birds fed solely on fiddler crabs they wouldn't survive, as fiddler crabs can go without food and remain in their burrows for weeks on end. When the crabs do come up, they all come up at the same time and the larger crabs venture far from their burrows. These burrows are often on the higher part of the mudflat. Fiddler crabs are not present on the lower tidal zone, as there they would be at greater risk of being dug out at high tide by fish and large swimming crabs. However the fiddler crabs can find more food in lower areas. This is why fiddler crabs make their burrows high on the tidal flat, then march in columns down to the food-rich areas at the low tide line. Before the tide turns, they must return to their burrows. They run a huge risk while out foraging, as predators can just grab them. They probably make the trade-off that each individual's risk is smaller when they migrate together in their thousands. Mostly large crabs take part in this tidal migration. The crabs' energy requirements increase with increasing body size. This is why larger crabs must graze a larger area around their burrows, putting them at greater risk than their smaller conspecifics.

Fiddler crabs show a large variation in their foraging behaviour and through this, their availability as prey. This variation is probably large in fiddler crabs that live in the subtropics and temperate regions, and small in their conspecifics in the tropics. The only species of fiddler crab found in West Africa, *Uca tangeri*, reaches the northern limit of its distribution along the south coast of Portugal

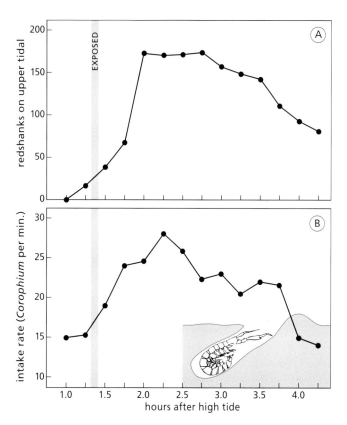

Figure 4.17 (A) As the tide recedes, many redshanks feed on amphipods (*Corophium volutator*) high on the tidal flat near the edge of the salt marsh. An hour and a half later most move to lower parts of the mudflat. (B) The intake rate when feeding on amphipods is low when the flats are still submerged, highest when the flats are exposed but wet, and decreases thereafter as the sediment surface dries. Data were collected in April and May at Schiermonnikoog in the Dutch Wadden Sea.[962]

The redshank's favourite food, the amphipod (*Corophium volutator*), and the curlew sandpipers's favourite, the ragworm, reach their highest densities on the upper mudflats. This is why redshanks and curlew sandpipers start to feed along the water line immediately after high tide.

and Spain. Here, in the winter, these crabs rarely leave their burrows outside the low tide periods during spring tides. Even though they are often eaten by waders during the spring and autumn migrations, they are not the staple winter food. In winter, well-known fiddler crab eaters such as whimbrels head further south. The furthest north that whimbrels winter is the Banc d'Arguin. In winter, they are not present in the most northerly 2000 km of the fiddler crabs' distribution. Even on the Banc d'Arguin, whimbrels would starve in winter if they were totally dependent on fiddler crabs. In winter, they complement their menu with other crab species that they find under the seagrass.[975]

It's obvious that without detailed knowledge of fiddler crab behaviour, we could never understand the whimbrel's winter distribution or the variation in its food intake. An observer can easily see whimbrels feeding on crabs when the fiddler crabs are out and about, or see the same birds meet with little success when the crabs are not active. The behaviour of other benthic animals also varies, but requires much more effort to observe.

### 4.5.5 Amphipods

The amphipod (*Corophium volutator*), like the fiddler crab, lives in burrows and feeds on the mud surface. Sometimes the amphipods only partly emerge from their burrows and graze in the immediate vicinity of the burrow entrance. Often, however, they crawl around on the sediment surface. When the amphipods have eaten enough they creep back to their burrows, which they frequently close with a lid of mud. In areas where the amphipods are abundant, you can actually hear a funny rustle across the mudflat as millions of their little burrows are closed.

Birds feeding on the amphipods deal with three types of prey, those that (1) crawl around on the surface, (2) sit in burrows with an open and visible entrance, or (3) are hidden in the substrate, with their burrows shut and invisible. Amphipods out on the surface are easy prey and redshanks can consume one amphipod per second. Redshanks can't feed anywhere near this fast on buried amphipods, which is why they seldom eat them. It's not very profitable for a redshank to spend so much time handling such tiny prey, apart from the very largest individuals. Plus the buried amphipods are much harder to find, so the searching time is also longer.

The amphipods are so small that redshanks will only ever try to get the very largest individuals out of their burrows; otherwise it takes too long for too little gain. This explains much of the observed variation in intake rate. Amphipods remain in their burrows when sediment temperatures are below 6 °C. At these temperatures, the redshank's intake rate is very low, so the birds switch from amphipods to ragworms and Baltic tellins.[310] Water level also affects amphipod behaviour. When the receding tide has just exposed the flats, amphipods are very active. Their activity then steadily decreases during the remaining time that the flat is exposed. Redshanks play on this by walking just behind the retreating tide (Figure 4.17A). Birds that forage in shallow water (and therefore ahead of the receding tide) are less successful than those walking behind the tide line, as are those that forage in areas that have already been exposed for some time (Figure 4.17B).

Figure 4.17 shows that as the tide retreats most redshanks forage in the area that provides the highest intake rate. A high intake rate attracts other redshanks, but this doesn't mean that when many redshanks are foraging their intake rate is high. In fact we'd expect the opposite. A classic experiment showed that when a redshank walks by, all the local amphipods retreat to their burrows and wait 2 - 3 minutes before resurfacing.[312] This means it's better for a redshank to keep walking rather than return to the same spot too quickly, and finding a 'recently-unused' spot becomes much more difficult in areas where many redshanks are foraging. Observations suggested that redshanks that foraged near other redshanks had lower foraging success.[313] Subsequently, this was shown even more clearly when a quadrat was marked out on the mudflat and the food intake of each visiting redshank was measured. As expected, redshanks were less successful if a conspecific had recently walked through the area (Figure 4.18).

If a high density of redshanks leads to a lower intake rate (Figure 4.18), why is the maximum intake rate reached when the highest numbers of redshanks are present (Figure 4.17)? Although the required data on amphipod behaviour are lacking, the most likely reason is that the tidal variation in the amphipod's grazing activity is big enough to cancel out the effects of the presence of all the redshanks. Not only are many amphipods out feeding on newly exposed sediment, but it's likely that they will remain in their burrows for a shorter period after a disturbance. We are really just speculating about amphipod behaviour here, but a similar effect is known in fiddler crabs. If a single fiddler crab is active on the surface, the others will stay down for a long time after a disturbance. But when many crabs are out foraging, the others quickly come out again (Figure 4.16).

### 4.5.6 Burrowing bivalves

*If you're not deeply buried you need to be thick-skinned*
When a bivalve is disturbed, it retracts its foot and closes its valves. This renders it immobile, making it relatively easy to take detailed measurements of the relationships between these prey and birds. To confirm that Baltic tellins, peppery furrow shells and sand gapers did not move even a millimetre after disturbance, they were taken to a laboratory and released into an aquarium with tiny threads glued to their shells.[960] The threads were attached to an automatic recorder above the aquarium, and the shellfish were allowed to bury themselves. The aquarium was made

Peppery furrow shells have two tubes, called siphons, which they use to stay in contact with the outside world. The longest tube is used to gather food and can be extended to about 25 cm. This allows the shellfish to bury themselves deeply. Most large individuals live 5 - 15 cm below the sediment surface. Out on the tidal flat, all you can see is the part of the siphon that sticks out of the mud.

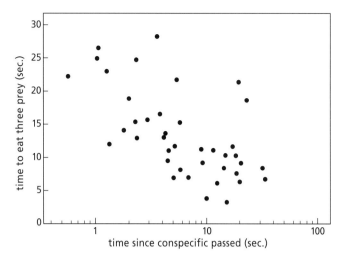

Figure 4.18 It takes redshanks longer to find amphipods (*Corophium volutator*) if another redshank has walked there recently.[781] Food intake is expressed here as the time taken to eat three prey (almost exclusively amphipods). Along the horizontal axis, the time since another bird has walked by is shown on a logarithmic scale.

as natural as possible, with simulated high and low tides (these 'tides' didn't move the shellfish). The shellfish were kept for some months and seemed happy enough, or at least stayed in good condition. Some of the animals buried themselves more deeply over time, but others did the opposite and came up to the surface. The data gathered in this experiment allow us to calculate what percentage of these prey would have been within reach of a bird's bill, if you know how deeply the birds can probe.

How deeply shellfish bury themselves can also be determined by taking cores, as you would to measure the density of benthic animals. But instead of the mud core being sieved, it is placed down very gently on its side. Then one of two techniques can be used.[966] The core can be cut into small (for example, 1 cm thick) slices with a sharp knife, and then each slice sieved separately. If the benthic animals are large it can be simpler and more accurate to break open the core, find the prey one by one, and measure the distance from the surface to the upper limit of each animal. Whether this will work depends on the substrate. If the substrate is very loose or soft, it's very difficult to determine where the sediment surface was as it usually collapses when the core is being laid down. This problem can be solved by pushing a flat piece of metal with a long pin into the sediment before the core sample is taken. This piece of metal can then be used as a reference point when the sample is being sorted.

A rather different method involves attaching small threads to benthic animals, similar to what was done in the laboratory experiment. Shellfish are dug up and a thin thread of known length is glued to each animal. At the other end of the thread is a piece of numbered plastic. The animals are released back into the sediment and the site marked with tall stakes. Some time later, the numbered threads are found again and the amount remaining above the sediment surface measured. Frequent measurements of the threads over some weeks or months allow the variation in each individual's depth to be determined accurately. However this method only works in relatively silty sediments in sheltered areas.

Burying themselves in the sediment seems to provide benthic animals with a good defence against avian predators, but is there any evidence for this? To find out, a number of peppery furrow shells were dug up and placed on the sediment surface. A small piece of mesh was glued across each animal's foot, preventing them from burying themselves. Oystercatchers were quick to find this ready meal and within a few hours all the benthic animals were eaten.[78] Peppery furrow shells that had escaped the oystercatchers for six years, were gone in just 3 hours! The obvious conclusion is that these benthic animals protect themselves from predators by hiding in the sediment. In this case the prey had been laid on top of the sediment so they were both visible and in easy reach. In a second experiment the shellfish were buried just beneath the sediment surface. The prey were not visible, but were within reach of every bird's bill. This time it took some hours before an oystercatcher found the shellfish, but once discovered they were all quickly eaten.[979]

Mussels always live on the sediment surface and cockles are only shallowly buried. How do these animals escape the continual threat posed by hungry birds, when peppery furrow shells and sand gapers can't? The answer is simple: mussels and cockles can shut their valves more tightly than sand gapers, and have much stronger shells. For every 1 g of flesh, shellfish that bury themselves deep in sediment invest 4 - 6 g of calcium in their shells. This ratio is 1:7 or even 1:10 for shellfish that live on the surface. Compared to their flesh weight, the protective calcium layer of mussels and cockles is twice as heavy as that of more deeply buried shellfish.[990] If you don't live deep in the sediment, you must be thick-skinned to survive.

Shellfish living in the sediment have different depth distributions. Usually, the deepest you'll find cockles is 2 cm, Baltic tellins 5 cm, and peppery furrow shells 10 cm. These data were measured on a fairly silty mudflat.[980] Shellfish can be buried deeper in sandy tidal flats, as can ragworms, whose burrows are often a few centimetres deeper in sand than in silt.[259]

*How siphon nibbling helps waterbirds*
Bivalves use their shells to protect themselves from hunting predators, but as we've seen this protection is relative. Bivalves smaller than a few millimetres can be swallowed shell and all by shrimps, and slightly larger bivalves are crushed by common shore crabs. The next size up can be swallowed whole by red knots and herring gulls, or dropped onto a hard surface and broken open by corvids (members of the crow family). However without their hard shells they would be totally defenceless. When cockles die from extreme cold or lack of oxygen and lie gaping on the surface, many birds including grey plovers, turnstones and dunlins gather to take advantage of these now defenceless animals.[995]

Shellfish also need to feed and they can't do this when their valves are tightly shut. To forage they must open their valves a little. Cockles and mussels keep their valves securely closed at low tide, but as the incoming tide inundates the flats with water, driving the birds away, it's the bivalves' turn to feed. If the water comes in slowly, oystercatchers have a chance to quickly feed on some of the now gaping and therefore relatively easy prey.[976]

Bivalves can be divided into two groups depending on how they feed, filter feeders or sediment feeders. Surface-dwelling shellfish, like mussels and cockles, filter their food from the water. Buried shellfish have two long siphons: the inhalant siphon is a long tube that they can stick out from their shell to take in food; the exhalant siphon gets rid of undigested material. Sand gapers' two siphons are fused together to form one. Sand gapers are typical filter feeders. They keep pumping even when the tide is on its way out and little water remains on the flat, and pump so vigorously that whirls are visible in the shallow water. This puts the mud around the siphon opening into suspension, which is probably their aim. However there is one disadvantage to this technique: the sand gapers have obvious siphon holes that betray their presence. This makes it easy for birds to find them. These holes are particularly visible in calm weather when there is not much wave action, and curlews walk from one siphon hole to the next, probing down to find what's below. Usually, the curlews are unsuccessful as most sand gapers live deeper than a curlew's bill can reach.

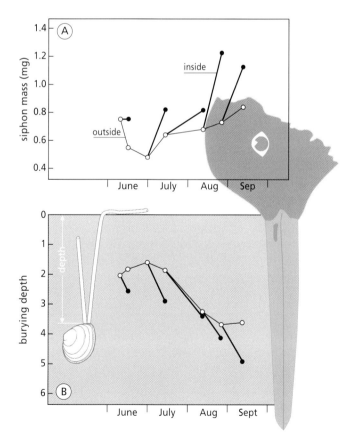

Figure 4.19 (A) Average siphon mass and (B) average depth of 15 mm Baltic tellins (*Macoma balthica*) from June to mid-September. The thin lines show the 'natural' seasonal changes in the wild; the thick lines show what happens when a fine-mesh cage that excludes siphon nibblers such as shrimps (*Crangon crangon*) and common shore crabs (*Carcinus maenas*) is set out for 2 - 4 weeks. Data were collected in the Eastern Dutch Wadden Sea.[972]

Baltic tellins and peppery furrow shells can also use their siphons to filter food from the water, but their siphons are flexible and can stretch over the sediment surface to vacuum up any food sitting there. How far these animals can reach across the surface with their siphons has been measured. There seems to be quite a lot of variation between individual shellfish. The same applies to the depth at which they have buried themselves. Very thin threads were glued to one valve of some Baltic tellins and peppery furrow shells that were then allowed to bury themselves into the sediment.[987] The burying depth could be measured by the length of thread remaining above the surface. The total siphon length equals the burying depth plus the distance that the siphon can reach over the surface when the animal is grazing. The siphon looks rather like it's made of rubber, but its elasticity is limited. When the end of a Baltic tellin's siphon is cut off the animal's initial reaction is to not extend its siphon as far across the mudflat; later it will decrease its burying depth. This is a rather clear example of the trade-off the shellfish must make between increasing its foraging opportunities and decreasing its risk of being eaten. A shallow animal can use more of its siphon to cover a larger distance across the mudflat, so it can consume more food. But if the shellfish has to reduce its burying depth to do this, it reduces its chances of survival, as it will put itself within reach of more bird bills. On the other hand, if it uses its whole siphon length to keep itself safely buried, it can't scrape food off the surface and can only consume what it is able to filter from the water. How good the shellfish's body stores are determine the decision it will make. Compared with conspecifics in good condition, very thin animals take more risks and bury shallowly. They have little choice, because they need food. In contrast, shellfish with a short siphon but good body stores avoid this risk and bury themselves deeply.[981]

As shellfish grow, their siphons get longer and they can bury themselves deeper. Sand gapers a few millimetres long are usually found in the upper centimetres of the substrate. When they are larger than 1 cm, for every additional centimetre they grow they bury themselves an average of 0.3 cm deeper, so that when they reach 5 cm long, they can bury themselves 14 cm deep. From then on, there is little increase in burying depth. Sand gapers invest a lot of tissue in their siphons, around 45% of their body mass when they are 30 - 50 mm long. When they are very young this percentage is much less, and it slowly decreases again once they are larger than 50 mm. So it seems that sand gapers begin to invest heavily in their siphons at the same time as they start to bury themselves increasingly deeply. The same applies to other shellfish that have been investigated. In these species the relationship between siphon size and burying depth can be described with an 's' curve; their siphon tissue increases the most when the increase in burying depth is largest.[980]

Sand gapers and cockles live at the same depth throughout the year, but Baltic tellins and peppery furrow shells bury themselves twice as deep in winter as in summer. This is because sand gapers and cockles must be able to reach the mud surface with their siphons year-round, even in winter when little food is available or required. However Baltic tellins and peppery furrow shells use their siphons to graze during the growing season and come nearer the surface to be able to graze as far afield as possible. When their food requirements decrease with decreasing temperatures, they don't need their siphons to reach further than the surface. So the part of the siphon used in summer to extend their grazing areas is used in winter to increase their burying depths.

Benthic animals have a particularly short growing season in the tidal areas of Northwestern Europe. Almost all growth occurs between mid April and mid June. When you find out how much of the siphon is used to graze and how much to increase the burying depth on a monthly basis, two months stand out, May and June. In May and June much of the siphon is used for grazing. June is also when peppery furrow shells and Baltic tellins live closest to the sediment surface. In May and June many young flatfish just 5–15 mm long are out feeding on submerged tidal flats. One of their main food sources is the siphons of grazing shellfish. It's not fatal for the shellfish to lose the end of their siphon. But when it happens repeatedly, they must spend a lot of energy regrowing a new siphon. The small flatfishes aren't the only predators on the tidal flats that view bivalve siphons as tasty bite-sized portions: the number of shrimps on the tidal flats peaks between June and September, juvenile common shore crabs are at their most abundant in July and August and common goby numbers peak in August and September. The shrimps and young shore crabs are far more abundant than both fish species, so a bivalve is at the greatest risk of losing part of its siphon between June and August. In April and May, Baltic tellins have a 1.4 mg siphon, which has already halved in June. Then from July onwards, their siphon mass slowly begins to increase again, even though there are still lots of predators around. In July and August Baltic tellins use their siphons to graze on the sediment less frequently than in the preceding months, allowing their siphon mass to increase even though siphon nibblers are abundant.

Seasonal fluctuations in the depth of peppery furrow shells and Baltic tellins vary annually. The reason for this is, within a season, burying depth increases with better body condition. So in years that these shellfish have good body condition, they live deeper. Animals in good condition also have relatively heavy siphons. Siphon mass can be very variable annually. Peppery furrow shells were monitored through nine winters, and the annual average siphon mass of 35 mm long individuals varied from 8 - 20 mg. In winters with a low average siphon mass the animals lived at a depth of 7 cm. But in years when the siphons were heavy, the winter depth was 14 cm. The annual variation in the siphon mass could be attributed to the large annual variation in the abundance of siphon nibbling predators.

How large an effect the siphon nibblers have on benthic animals has been determined experimentally. Cages 1 m x 1 m wide and 0.3 m high were made from thin

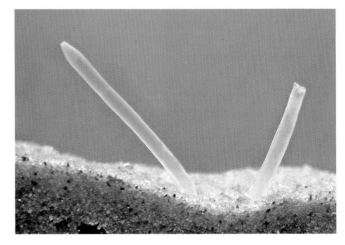

It takes a knot less than 1 second to swallow a small shellfish whole (upper photos, taken from a video). So its calcium skeleton gives the shellfish little protection from knots. The only way it can protect itself is to bury itself out of reach of the knot's bill. From a safe depth the bivalves can extend their siphons up to scrape food from the sediment surface or to filter food from the water (lower photo). The disadvantage of this is that they often lose a part of their siphon, as siphon tips make tasty morsels for flatfishes, shrimps and common shore crabs. Birds can also nibble on extended shellfish siphons.

Both the avocet (upper photo) and the ringed plover (lower photo) depend on prey found in the upper layer of the sediment. The short bill of the ringed plover is unable to reach deeply buried prey. The avocet has a long bill, but can't use this to probe deeply. Instead, it is swept sideways through the upper layer of soft silt.

pieces of wire. The top and sides were covered with fine mesh. The cages were put out on the mudflats to exclude young common shore crabs, shrimps and fish. After 14 days, and sometimes longer, the burying depth and siphon mass of Baltic tellins inside and outside the cages were compared. Even in this short time, the siphon masses of the caged Baltic tellins had increased (Figure 4.19A), which enabled the caged animals to bury deeper (Figure 4.19B). How far they could reach across the surface with their siphons wasn't measured directly, but could be estimated from the relationship between siphon mass and siphon length. The siphon length above the surface was determined by subtracting the burying depth from the total siphon length. This indicated that Baltic tellins in the cage were able to stretch their siphons further across the tidal flats than their conspecifics whose siphons had been grazed by predators. The costs involved in making new siphon tissue are clearly shown by the fact that the animals inside the cages were much heavier than those outside after just two weeks.

Without siphon nibblers, Baltic tellins would lead very different lives. They wouldn't have to spend so much energy endlessly rebuilding siphon tissue, their siphons would be heavier just when their own feeding opportunities on the sediment surface are at their best, and they could use their longer siphons to bury themselves deeper in the summer. The combined efforts of the fishes, crabs and shrimps ensure that peppery furrow shells and Baltic tellins live nearer the sediment surface in summer, increasing their risk of being taken by predators. In winter oystercatchers and knots hardly take any peppery furrow shells or Baltic tellins, but in summer they consume many. But in the summer there are far fewer oystercatchers and knots on the tidal flats, so each individual bivalve has a lower risk of being eaten in summer than in the winter. Baltic tellins and peppery furrow shells can take the risk of living near the surface in the summer, as their most important predators are seldom present.

## 4.6 Exploiting the available food

### 4.6.1 Which prey can be harvested?

Are tidal flats really a feast for waterbirds? When we look at the enormous densities of benthic animals present the answer appears to be 'yes'. But if you think about how much of the prey can't be eaten for one reason or another, the answer could well be 'no'. Total food availability for the waterbirds can be measured by counting and weighing all the benthic animals found per unit surface area. But to really determine which benthic animals are actually available to the different bird species we need to know: (1) which prey the birds can access or reach, (2) which prey can be found, (3) which can be swallowed, (4) which can be digested, and (5) which are profitable to eat. Only when a benthic animal meets these five requirements can it be considered potential prey. Such prey are said to be 'harvestable'. So let's take a look at these five criteria in detail.

*Prey must be accessible*

Not all prey are suitable food for waterbirds; some are inaccessible because they are protected by a hard calcium skeleton or live out of reach of the bird's bill. Most benthic animals bury themselves deeper as they grow larger. That's why many small prey remain within bill reach, but the fraction of accessible animals decreases for larger individuals (Figure 4.20).

Curlews don't go to the trouble of trying to eat sand gapers larger than 5 cm. These larger individuals are almost always buried 15 - 20 cm deep, where they are out of the reach of even the curlew's long bill. This is why curlews only consume large sand gapers that are shallowly buried. But almost all sand gapers of 3 cm or less live in the upper 12 cm of sediment, where they are within the curlew's reach. A curlew feeding on sand gapers selects gaping individuals. First the curlew pulls the siphons off, then eats the flesh from inside the shells. Curlews feeding on large sand gapers use so much force that a little crater is left on the mudflat. These craters can then be surveyed. The curlew leaves the empty shells behind, buried in the sediment, so with a corer you can determine how deeply these consumed sand gapers were living. In one study, the sand gapers consumed by a single female curlew were mapped as accurately as possible. The tidal flat could be mapped since thousands of numbered stakes had been temporarily placed in the sediment 2.5 m apart and the female could be identified by her individual colour rings. She was observed on different days for hours each day, making it possible to find many sand gapers that she had eaten recently, and determine the depth at which they had been living.[979] Most of the prey that she had eaten were 4 - 5 cm long. Sand gapers of this size live 10 - 20 cm below the surface, but the female curlew only ate sand gapers that had been living in the upper 14 cm. Curlews don't usually leave any flesh behind in the shell, but some flesh did remain in the shells of the single animal eaten at 14 cm depth. Since this female had a 14.3 cm long bill, the obvious conclusion is that bill length determines which prey are within a bird's reach and which are not.

Oystercatchers feeding on peppery furrow shells sometimes probe so deeply that the mud reaches their eyes. Prey living deeper than 7 cm below the surface are beyond an oystercatcher's reach. The depth that peppery furrow shells bury themselves varies seasonally (Figure 4.21). After the burying depths of around 20 individual peppery furrow shells had been measured every 14 days for 3.5 years, a clear pattern seemed to emerge. On average, during the three summers the shellfish lived 6 - 8 cm deep

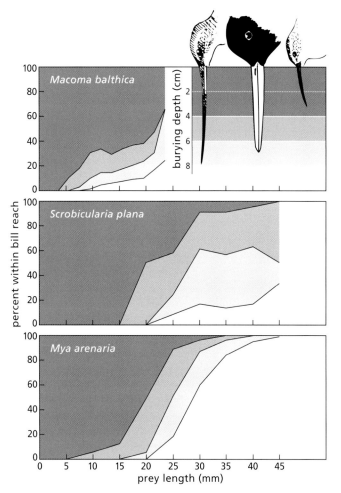

Figure 4.20 Shellfish such as Baltic tellins (*Macoma balthica*), peppery furrow shells (*Scrobicularia plana*) and sand gapers (*Mya arenaria*) live deeper in the sediment as they grow older. From depth distributions it is possible to determine the fraction that is accessible to birds with different bill lengths (shown in different shades of grey). Data were collected in summer in the Eastern Dutch Wadden Sea.[988]

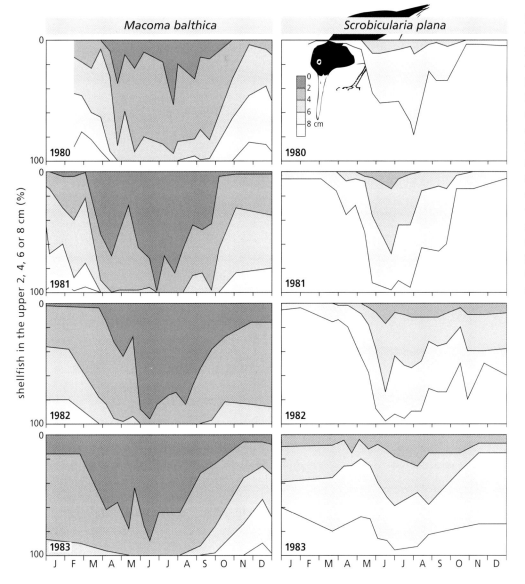

Figure 4.21 Burying depth of shellfish varies both within and between years. The figures on the left show the proportions of 15 mm long Baltic tellins (*Macoma balthica*) living 2, 4, 6 and 8 cm deep in the sediment. The figures on the right show equivalent data for 35 mm long peppery furrow shells (*Scrobicularia plana*). Each year, both species bury shallowly in summer and deeper in winter. During the first two winters most individuals of both species were out of reach even of oystercatchers, but in later years the shellfish lived much nearer the surface. Data were collected in the Eastern Dutch Wadden Sea.[982]

and in the four winters 12 - 14 cm. But, at the start of the fifth winter the animals did not return to their normal winter depth. Most of them stayed within reach of an oystercatcher's bill and the birds fed on them in huge numbers. In less than a year, these six year old peppery furrow shells almost completely disappeared.[995] This really shows the importance of monitoring for an adequate time period.

Baltic tellins don't live as deep as peppery furrow shells. In summer the burying depth of most adult Baltic tellins is just 2 cm. In winter they live somewhat deeper, but even then most specimens aren't more than 4 cm below the sediment surface. As Figure 4.21 shows, their depth distribution can vary annually. Like the peppery furrow shells, Baltic tellins lived deep in the sediment in the first two winters that they were sampled, but even then half of them would have been within reach of an oystercatcher bill.

It seems that with the right measurements it should be easy to determine the fraction of prey that are accessible, but this is often not the case. Oystercatchers searching for Baltic tellins only probe half of their bill (4 cm) into the sediment, even when the prey are buried deeper.[995] So should we then just include Baltic tellins

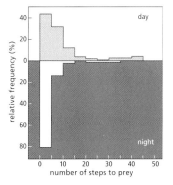

Figure 4.22 The grey plover's foraging strategy is to stand almost motionless, waiting for prey to appear on the surface of the tidal flat. When that happens, the bird runs quickly to catch it. The bird runs further in the daytime than at night because visibility is better during the day and the bird can see further. Data are from South Africa. At night a large light intensifier was used.[882]

from the upper 4 cm of sediment in our calculations? But if oystercatchers switch to feeding on larger peppery furrow shells or sand gapers at the same site, they will probe up to their eyes in the mud. This means that Baltic tellins in the upper 7 cm are within the oystercatcher's reach, but that the birds refuse to feed those living at a depth of 4 - 7 cm for some reason. We discussed the reason for this back in Figure 4.3, when we examined depth selection by oystercatchers: the handling efficiency of these deep-living prey is too low so the birds refuse to eat them. However it's not always clear whether birds actually decide

FOOD

Three birds with a common shore crab: a juvenile Arctic tern (upper photo), a herring gull (middle photo) and an oystercatcher (lower photo). Herring gulls can swallow large crabs whole, but oystercatchers break them into pieces to eat the meat. Terns look for crabs that have just moulted, as when the calcium skeleton is still soft the prey can be swallowed whole.

not to try to eat the prey or if they simply can't deal with them. Oystercatchers that smash open mussels refuse many of the large mussels (see Section 4.1.4). Maybe they can't access these prey because their shells are too thick and hard, but it is more likely that it would take the oystercatchers too long to hammer a hole through the shell.[561] If so, the meat of these large mussels would be accessible but not profitable.

*Prey must be detectable*
The grey plover is a good example of a 'sit-and-wait' predator. It stands stock-still, waiting intently for a suitable prey item to surface, and then dashes after it. Grey plovers usually only make short runs. Why is this? Can't the grey plover see any further or is it no use running after distant prey, as the bird will always arrive too late? The distance the bird can see certainly plays a limiting role, especially at night when these runs seldom exceed five steps. In daylight hours, plovers will run two to three times as far (Figure 4.22).

A lugworm that comes up to defecate within 1 m of where a curlew is standing usually leaves the last toothpaste pile of its life. But the lugworm will survive if the curlew happens to be looking the other way. Birds have their eyes on the sides of their heads, but they cannot see what's behind them. Plus a curlew probably needs to see the growing sand pile with both eyes to be able to locate it in a fraction of a second, which decreases their usable field of view considerably. When you watch curlews feeding on lugworms, you notice that each curlew only runs after the lugworms that come to the surface more or less right in front of its bill. Their field of view is possibly only around 120°. They can slightly adjust their behaviour to try and increase the number of prey they can see. When curlews hunt lugworms they stand as tall as possible. They probably do this to increase their field of view. As we saw earlier, curlews that feed on sand gapers have an easy time of it when the weather is calm and the siphon holes are clearly visible in the sediment surface, allowing the birds to hunt visually. But when the siphon holes are not visible the curlew has to walk across the tidal flat with its bill slightly open, searching for the sand gapers by touch, which takes much more time.

Oystercatchers also search for buried prey by touch. It

is logical that the bird doesn't need to probe as often when the prey density is higher, as has been shown in controlled experiments.[417] Like other touch-foraging waterbirds, oystercatchers hold their bills slightly open to increase their chances of contacting a prey item. They can find large prey more quickly than small ones, as large animals have a bigger surface area.[417] It takes an oystercatcher one tenth the number of probes to find a 30 mm long cockle compared to a 5 mm cockle.[986] The good thing about tactile-hunting waterbirds is that with these data available you can calculate precisely how often a bird needs to probe to find prey at various densities of the different size-classes. This makes it possible to accurately predict the search times. However, you do need to know how long a single probe takes. Oystercatchers probe so quickly that it can be almost impossible to measure directly. But by filming a foraging oystercatcher and then reviewing the images frame by frame, it was found that a shallow probe takes around 0.1 second. The deeper the bird probes its bill in the sediment, the longer each probe takes. Probing with the entire bill in the sediment takes 1.5 seconds.[909] So it takes birds longer to find deeply buried prey. If we really want to know how long it takes a tactile-foraging waterbird to find buried prey, we need to know exactly how deep the prey are living.[995]

*Prey must be able to be swallowed*
Curlews swallow small common shore crabs whole, and shake the legs off larger individuals (Figure 4.1). They ignore crabs larger than 4 cm, as even without their legs and pincers they are just too big to swallow. Smaller waders such as greenshanks and redshanks have a much lower upper size limit: they usually eat crabs smaller than 1 cm, but can eat 1.5 cm crabs. At the other extreme, herring gulls can swallow even 5 cm common shore crabs, which are now the largest specimens found on the tidal flats of the Wadden Sea.[964]

Which prey can be swallowed depends more on their circumference than overall size, as is clearly seen in fish-eating birds. A cormorant can still swallow a large eel, but can't swallow a fat bream of the same mass. Birds must be careful handling large prey that they can only just swallow. A bird can damage its oesophagus, or may even die if the prey gets stuck, as is sometimes seen in cormorants.

Red knots can swallow 16 mm long Baltic tellins but spit out any larger ones. The maximum circumference of 16 mm Baltic tellins is 30 mm. To find out if 30 mm is the red knot's size limit for prey it can swallow, we can compare the tellins to other shellfish eaten. Cockles are rounder than Baltic tellins, so 12 mm long cockles have already reached the predicted 30 mm limit. Mussels are more slender, so according to the 30 mm rule, could still be swallowed when they are 20 mm long.[974] Field observations concur with these predictions.[330, 686] It seems that gullet size determines which prey the red knot can and cannot swallow.

Sometimes birds want to eat prey larger than their gape width allows. Oystercatchers solve this problem by opening their prey and eating the flesh in pieces when

Eiders eat bivalves and crabs whole. They don't take the largest cockles or mussels because they would be unable to swallow them; they might also be unable to crush such large individuals in their stomachs. Eiders regularly produce large droppings full of crushed shells from their prey.

necessary, although opening the shells is also not without risk.[419] Whimbrels go to great lengths to open large fiddler crabs ventrally. Crows feeding on tidal flats must drop shellfish onto hard surfaces to access the flesh inside. In all three cases, it's obvious that the extra time required to open the prey makes the prey less profitable.

Arctic terns feed on very large common shore crabs, so large they are rarely eaten by curlews. The terns even feed these large prey to their young.[571] However the terns don't have a problem breaking through the crab's defensive armour, as they choose crabs that have recently moulted and whose carapace is still soft. Crabs have an external skeleton, so to grow they must moult into new armour a few millimetres larger than their old set. They do this six times in their first months of life, but only moult annually once they are larger. Until the new skeleton has hardened, the common shore crabs are defenceless against predators and usually try to stay hidden. Out on the tidal flats, newly moulted crabs are often found under rocks, coarse shell debris, or seaweed. Arctic terns forage in the sea rather than on exposed mud, so they probably just select crabs that are peeling and floating on the water surface. This means the terns restrict themselves to a very small part of the total prey population.

*Prey must be digestible*
Unlike a bird that swallows armoured prey whole, the bird that opens its prey to extract the flesh doesn't need to spend energy crushing its prey in its stomach. It takes a lot of effort to crush even a small mudsnail just a few millimetres long, and the force required increases for larger animals.[687] This sort of work requires strong stomach muscles. This is why a 'crusher' like the red knot needs a muscular stomach that is almost twice as heavy as that of other similar-sized waders that feed on soft prey. Purple sandpipers also swallow their prey shell and all, but tend to select smaller individual prey animals than red knots.[816] Compared to the red knot, the purple sandpiper's stomach is lighter and not as strong.

Stomach mass seems to vary a lot between different individuals of the same species. This is partly due to differences in diet. Bar-tailed godwits that feed on ragworms have smaller stomachs than those that swallow shellfish whole.[687] It's possible that only a bird with a strong stomach will choose prey that need to be swallowed shell and all, but it's more likely to be the other way around, i.e. birds' stomachs adapt when the birds feed on different food. Of course, this takes time. When captive knots have been fed soft food for a long time it takes four to five days before they will accept hard food that needs to be crushed.[687]

Birds that only eat soft meat are not necessarily any better off than birds that have to crush their prey internally. It seems that the digestive process requires some hard material to run smoothly. Birds that eat only meat also consume grit.[512] Individual bar-tailed godwits followed continuously for 10 hours or more, feeding solely on crane fly larvae, sometimes picked up more than 100 small stones and shell fragments.[77] A study on curlews with a mixed diet of common shore crabs and sand gapers showed that the need for 'mill stones' increases when the menu is based solely on soft meat. The fewer common shore crabs a curlew had eaten per hour, the more bits of loose shell it picked up from the mud.[73] This habit can have the unfortunate consequence of leading researchers completely up the garden path, when they try to determine prey choice from faecal or regurgitate analysis. Curlews that feed solely on sand gapers can produce regurgitates that appear to show that the bird fed exclusively on mussels and cockles or other shellfish that were swallowed whole, shell and all.

*Prey must be profitable*
Section 4.1 contains some examples of prey that were refused because their handling efficiency was too small (i.e. it takes too long to handle them). Handling efficiency often increases with increasing prey size. This relationship between prey size and handling efficiency holds for all prey species eaten by oystercatchers. However, the increase in handling efficiency dependent on prey size differs between the prey species.[989] Figure 4.23 shows all the measurements of handling efficiency that have been taken for 13 oystercatcher prey species combined. The smallest prey that observers ever saw oystercatchers eating were 8 mm cockles that contained 4 mg of dry flesh. The largest prey were big sand gapers and giant bloody cockles that contained 1000 times as much meat. Although it's clear that heavy prey animals yield more than the lighter prey, there is a lot of variation. Depending on handling time, a 100 mg prey item can provide 1 - 40 mg dry flesh per second. Much of this variation is due to how the prey protect themselves against predators. The only protection earthworms, crane fly larvae, ragworms and lugworms have is to stay

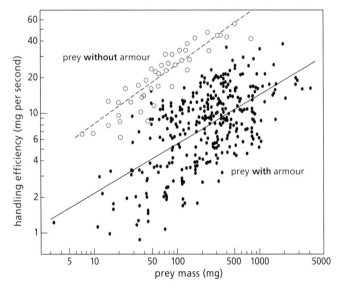

Figure 4.23 Oystercatcher handling efficiency in relation to prey mass. Note that both axes are on a logarithmic scale. This figure summarises work undertaken by tens of researchers, who measured the handling time of oystercatchers when feeding on 345 prey of different sizes, of 13 different species. Data are presented separately for prey that have a protective calcium layer (bivalves, snails and crabs) and those do not have such armour (such as worms and crane fly larvae).[988]

underground as much as possible, but once an oystercatcher reaches them, they are totally vulnerable and quickly swallowed. On average, the handling efficiency of these unarmoured prey is 4.5 times higher than that of shellfish, snails and crabs, which must be opened by the oystercatcher. This makes these armoured prey much less profitable. Amongst armoured prey, much of the variation in handling efficiency depends on how heavy their protective layer is. Buried shellfish with relatively thin shells yield a handling efficiency two to three times higher than prey that live at the surface and invest rela-

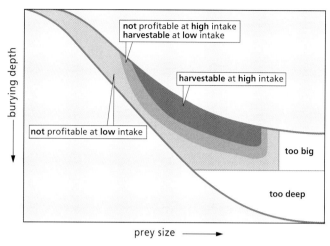

Figure 4.25 Which buried prey are harvestable? The two thick lines show how deeply prey of different sizes are buried. The deepest prey are out of reach of the bill, and large prey may be too big to be swallowed. On top of this, the small, the deeply buried, and sometimes also the largest prey are not profitable. As intake rates increase, birds become more selective and reject the less profitable prey.

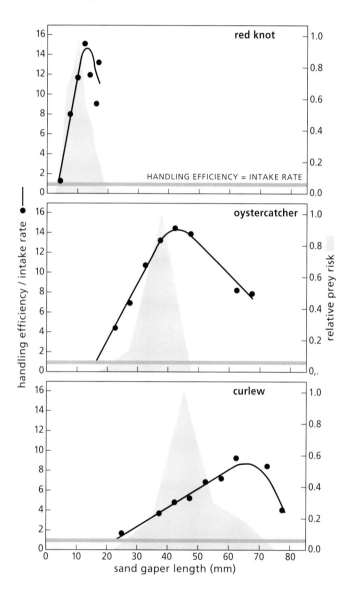

Figure 4.24 Handling efficiency for three bird species – knot,[77] oystercatcher[979] (intake rate corrected for size overestimation[989]) and curlew[979] – feeding on sand gapers (*Mya arenaria*) of different sizes. Handling efficiency is given as a multiple of intake rate, which was 0.3, 1.6 and 2.5 mg dry meat per second for the knot, oystercatcher and curlew respectively. As prey are not profitable when the handling efficiency is smaller than the intake rate during foraging, this cut-off is shown with a horizontal grey line. The size-selection of the three bird species is given in terms of relative prey risk (right-hand axis), shown by the grey areas. Prey smaller than the lower limit are not eaten. Prey that are about twice as large are eaten most, even though prey that are three times as large have the highest handling efficiency. While these large prey are profitable, they are also out of bill reach more often.

tively more in their protective calcium skeleton. Of the buried bivalves, the individuals that live near the surface are the most profitable. How profitable a shellfish is also depends on the technique used to open it. It takes a lot of time to hammer a hole through a mussel or cockle, which is why these prey can be handled more quickly if they are gaping slightly, allowing the oystercatcher to quickly push its bill between the two valves.

Oystercatchers usually reach an intake rate of 2 mg per second, but the rate can vary from 0.5 - 5.0 mg per second. Prey are not profitable when their handling efficiency is lower than this intake rate. We can see from Figure 4.23 that it can still be worthwhile for an oystercatcher to eat worms weighing just a few milligrams, but not shellfish of this size. The shellfish must contain at least tens of milligrams of meat before it is worth the oystercatcher's while to open and eat them.

Red knots ignore sand gapers that are smaller than 4 mm (Figure 4.2). When oystercatchers eat the same species they refuse individuals smaller than 17 mm. Curlews are even more selective and only eat the meat from sand gapers larger than 25 mm.[979] In all three cases the lower limit can be explained by the rule that prey whose handling efficiency is lower than the intake rate are refused (Section 4.1.2). Many studies have found that small prey are under-represented in the diet of many waders and that larger bird species are more selective than the smaller species. It takes smaller bird species longer to handle the same-sized prey than larger birds, but when their smaller intake rate is taken into account, the smaller prey may still be profitable for the smaller birds. For example, grubs are 1.5 times less profitable for little stints than for dunlins, because it takes the little stints 1.5 times longer to handle the same prey.[511] But because of the difference in their body masses and therefore also in energy requirements, a little stint only needs to eat half as much as a

FOOD

Fish-eating birds such as sandwich terns (upper photo) catch their prey in the upper few centimetres of the water column. Since almost all fishes swim deeper than this, only a very small fraction of the fish stock is within reach of the terns. This available fraction varies hourly and daily depending on the wave action caused by wind and tidal movement. Additionally, feeding grounds are often situated many kilometres from the breeding colonies, so only large prey are profitable food for the chicks. In the Arctic there is sometimes little choice: this Arctic tern feeds its chick crane flies that were caught close to the nest (photo below).

dunlin per day. If you also take into account the fact that on average little stints spend more hours foraging per day than dunlins, (Section 4.2.3) their intake rate is 2.5 times higher. This allows their handling efficiency to differ by the same factor. This allows the little stint to profitably eat smaller prey than its larger relative, the dunlin.

When the same prey is eaten by different bird species the best way to compare each species' handling efficiency is as a proportion of the average intake rate, as has been done in Figure 4.24. Here we can compare the handling efficiency of red knots, oystercatchers and curlews feeding on sand gapers, divided by each species' intake rate. For all three species, when the most profitable prey are being handled, the handling efficiency is 10 - 15 times higher than the average intake rate during foraging. For knots, the most profitable prey are 1 - 1.5 cm sand gapers, for oystercatchers, 4 cm animals, and for curlews, 6 cm. For each bird species the smallest prey animals being eaten have a handling efficiency that is barely larger than the intake rate during foraging.

The larger the bird, the bigger its prey. This rule is often true, but not always. When larger birds can consume many prey items quickly, there is no reason for them to refuse small prey. Herring gulls are larger than oystercatchers, but will eat smaller mussels. Herring gulls swallow the mussels whole, which usually restricts them to mussels of 2.5 - 3.5 cm. However they can eat these little mussels quickly and can swallow a clump of young mussels in one gulp. In contrast, oystercatchers eat mussels one by

one, and must use a time-consuming method to separate the flesh from the shell. Consequently, it takes an oystercatcher around 20 seconds to eat a 2 cm mussel. These small prey only yield 20 - 30 mg flesh, so their handling efficiency is 1 - 1.5 mg per second. This is below the oystercatcher's average intake rate. This is why oystercatchers often ignore mussels that are smaller than 2.5 cm.

Waders' tendency to peck at one prey item at a time makes them very selective. Although red knots can swallow two mudsnails per second,[293] it is hardly worth the effort of picking up mudsnails of 2 mm or less. The shelduck is 10 times heavier than the red knot and, on average, has an intake rate six times higher over an equally long daily foraging period. Nevertheless shelducks will rapidly gobble up snails that were considered too small by waders. The shelduck's bill works rather like a sieve allowing them to consume more snails per second than any wader.

*The remaining prey: the harvestable fraction*
Birds refuse prey that are too small to be profitable. They cannot eat large prey that are too big to be swallowed or too deeply buried to be accessible. What remains for them to harvest? It was calculated that knots foraging in August on the food-rich flats of the Wadden Sea[974] had six prey animals to choose from, with a total biomass of 139 g flesh per m². For the knots, 110 g of prey included in this total biomass were too big to be swallowed and could not be included. Of the remaining 26 g of biomass, 6 g were too deeply buried and 4 g too small to be profitable. So for these red knots the harvestable biomass was 16 g, 10 g of which was in the form of small cockles. The knots weren't eating these small cockles, even though they sometimes select such prey in other situations. The knots confined themselves to 1 - 1.5 cm Baltic tellins. So 11% of the total biomass could have been harvested by these knots, but they chose to eat just 3% of the biomass.

The harvestable biomass indicates the proportion of the prey that could be eaten. The red knot example shows that birds may choose to select only part of the harvestable biomass. This is in accordance with the optimal foraging theory. The birds make choices that will maximise their intake rates. The knots could reach a high intake rate on Baltic tellins alone and may have reduced their intake rate if they had added cockles to the menu. The need for a high intake rate makes birds selective.

Figure 4.25 shows the different variables that determine which prey are eaten. The bird can choose from buried prey that live at different depths. Smaller or deeper prey will yield less, as will very large prey. When its intake rate decreases the bird will begin to eat more prey that are smaller or more deeply buried, if its bill length allows. Figure 4.25 shows that birds that feed on buried prey can become more selective when their intake rate increases. This principle is also true in other situations. When many lugworms are out producing toothpaste piles, a curlew will only feed on the lugworms that come up right under its nose. But when few lugworms are defecating, the birds will also try to catch lugworms that surface further afield, even though they have only a slim chance of success.[750]

Large prey may provide more food than small prey, but they also take longer to eat. This is because the bird has to swallow repeatedly to move the prey from its bill tip to its throat. Plus large prey struggle more and it often takes the bird more than one attempt to get the prey correctly aligned in its throat. The photos show a redshank eating a common shore crab.

FOOD

### 4.6.2 Which prey can be exploited?

When a prey animal is not buried too deeply for a bird and has a high enough handling efficiency, it is said to be harvestable. But the same animal is not worth eating if it occurs in such a low density that the bird spends ages searching and cannot attain the required intake rate (Section 4.2.3). In this situation, the 'harvestable' prey is said to be 'not exploitable'. In what density must prey occur to be worth eating? Low prey density doesn't necessarily exclude exploitation, especially not if the bird can search for other prey in the meantime. For example, imagine a foraging wader searching for worms that live in the top 3 cm of substrate. If this bird encounters a rare worm species that happens to live within the top centimetres of substrate, there is no reason to assume that the

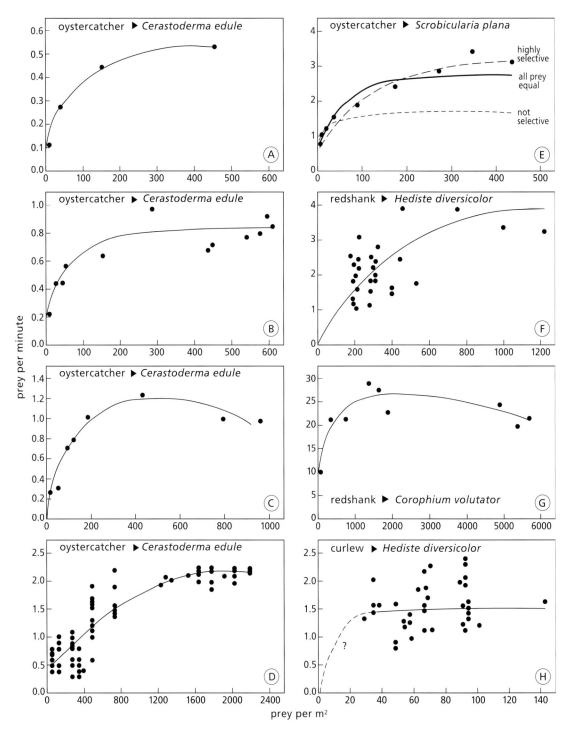

Figure 4.26 Intake rate as a function of prey density for a range of waders and prey: oystercatchers eating cockles (*Cerastoderma edule*; A,[416] B,[819] C,[314] D[871]) and peppery furrow shells (*Scrobicularia plana*; E[909]), redshanks eating ragworms (*Hediste diversicolor*; F[315]) and amphipods (*Corophium volutator*; G[317]), and curlews feeding on ragworms (H[257]). Three lines are drawn for oystercatchers eating peppery furrow shells: the predicted average intake rate when we assume all prey take the same time to eat (solid line), and the predicted functional responses for search and handling times in situations when birds select only the most profitable prey (long dashed line) and when they eat all prey regardless of the profitability (short dashed line).

bird wouldn't eat it. If this rare prey always lives 6 cm deep, the bird would need to always probe its bill twice as deep in the sediment to avoid missing this single prey item. This would probably lower its intake rate, as the bird would encounter less of the more common prey. There are other examples that show it's not so easy to feed on more than one type of prey at the same time. Oystercatchers walk faster when they are feeding on ragworms than when they eat Baltic tellins.[244] Curlews feeding on crabs walk at a different pace depending on whether they are eating large crabs or small ones. They walk twice as slowly when feeding on juvenile crabs as they would for larger crabs.[237] To maximise their intake rate the birds must choose which prey type they will eat. This also applies to prey that, as is often the case, have a variable distribution over the mudflats. Birds choose to forage in an area where one species is very numerous and so mainly feed on this one species. There are still plenty of examples in which birds eat more of one prey because another is present nearby. In late summer many crabs and shrimp are present on the mudflats. Even when redshanks or curlews are out looking for other prey, if they happen to encounter a nice bite-sized crustacean they'll seize the opportunity to gobble it up.

*Functional response*

When waterbirds forage on one prey species, three factors determine their intake rate: prey mass, handling time, and average search time. Search time mainly depends on the prey density. A bird doesn't need to search as long when there are many prey items, so its intake rate will increase. When the prey density is very high it reaches the point where the bird no longer has to search for prey. As soon as the bird has eaten one prey item, it can immediately start on the next. In this situation, the intake rate equals the handling efficiency. Even if the prey density increases further, the bird can't eat any faster. This is why when intake rate is plotted against prey density the intake rate initially increases, but will level off at a certain prey density. If you know the handling time and prey mass and you also know the relationship between search time and density, you can calculate the relationship between intake rate and prey density precisely.[406] The curved relationship between intake rate and prey density is known as the functional response.

The functional response of waterbirds has been measured several times (Figure 4.26). The curves often level off as predicted, but on closer analysis this is not due to a decrease in the required search time. This is easy to check. You expect search time to approach 0 seconds at the highest density. This makes the predicted handling time simple to calculate. The intake rate of redshanks feeding on amphipods didn't exceed 27 prey per minute (Figure 4.26G); this makes the predicted handling time 2.2 seconds. However, the actual handling time was 0.2 - 0.6 seconds.[314] The predicted handling time of redshanks feeding on ragworms (Figure 4.26F) was 12 seconds, but the observed time was just 0.4 - 6 seconds. The difference between the predicted and observed handling times is just as big for oystercatchers feeding on cockles (Figure 4.26A), 140 seconds and 21 - 29 seconds respectively. And there are plenty of other examples.[909] On average the predicted and observed plateau values differed little in the laboratory but greatly in the field.

So why does the intake rate level off and why was the theory behind it wrong? The main reason is that birds are not stupid. The model assumes that birds search for prey in exactly the same way independent of prey density, and that they eat all of the prey encountered and do so at the same speed. If all prey were exactly the same, as can sometimes occur in laboratory experiments, this might be true. But in the wild prey are not identical even if they belong to the same species and are the same size. Oystercatchers offered cockles in a controlled situation could eat them faster than expected at high prey densities.[416] This was because at a high prey density, and therefore a high intake rate, the birds ignored the cockles that they couldn't immediately open (Section 4.1.5). The researcher thought that by keeping all the factors constant and only varying the density of identical prey, everything was under control, but the bird could detect the difference between gaping and closed prey and based its behaviour on this.

When birds forage on prey that vary in their handling efficiency, we should always expect the birds to make the best of the situation. Holling's functional response[406] is not false, but it is too simple. The functional response usually assumes that only identical prey are involved but when it is extended to include different prey species, each with its own search and handling time required, the theory more closely resembles reality.[563, 690, 909, 995]

The search and handling times for prey in each

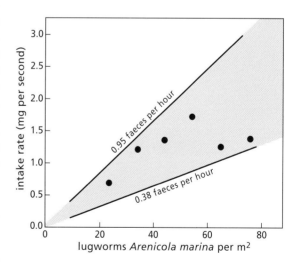

Figure 4.27 The relationship between intake rate and prey density for curlews feeding on lugworms (*Arenicola marina*; measured values shown with dots). The grey panel shows the predicted range of values based on variation in defecation rates of the lugworms; it is assumed that curlews catch all the lugworms that surface within 1.2 m of a bird, except those worms that surface behind the bird (a 240° arc), giving a total detection area of 1.5 m². The lines cover the maximum range based on recorded daily variation in defecation frequency of lugworms from June to September.[750]

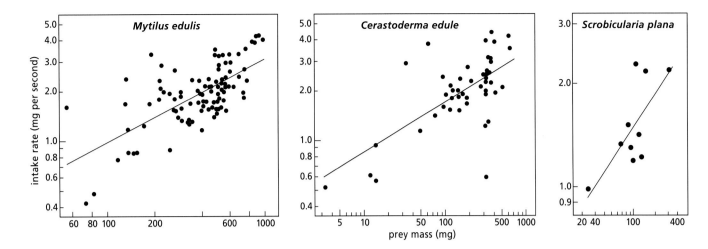

Figure 4.28 The intake rate of oystercatchers increases when they eat large prey, a relationship demonstrated here in birds that ate mussels (*Mytilis edulis*), cockles (*Cerastoderma edule*) and peppery furrow shells (*Scrobicularia plana*).[989] Both intake rate and prey mass are plotted logarithmically.

Grey plovers wait motionless until buried prey such as ragworms come to the surface (photo below). Cockles form an important food source for oystercatchers. Each cockle must be opened before the oystercatcher can extract the meat (photo right).

depth-class were known for the oystercatcher discussed in Section 4.1.5. Figure 4.26E shows three possible functional responses. The solid line is based on the assumption that all of the prey buried at different depths were considered equal by the bird and required the same handling and search time at equal densities. The dashed lines give the predicted functional responses for birds selecting prey in two ways. The upper line assumes that birds select only the most profitable shellfish (those closest to the sediment surface; see Figure 4.3), whereas the lower line assumes that birds take all shellfish, regardless of their depth. If the oystercatcher was not selective at high densities and ate all of the prey, it would only achieve half the intake rate it could reach by being selective. So the bird is better off when it changes from one functional response to another when prey density increases (and the bird apparently knows this!). How the optimal functional response curve finally appears can only be predicted after many calculations.

The really interesting part of the data summarised in Figure 4.26 relates to intake rates at very low prey densities. When there is little food it takes the birds a long time to find it, which may cause their intake rate to become too low. Oystercatchers need 36 g of dry food per day.[990] If they want to achieve this in 10 hours of foraging; they are in trouble when their intake rate drops below 1 mg per second. This is exactly what happens when oystercatchers feed on 300 mg cockles and there are less than 30 cockles per m$^2$ (Figure 4.26A). Cockles of this size have a handling efficiency of 40 mg per second, so they are almost always profitable. They live just below the sediment surface, within the oystercatcher's reach year-round. In short, if present they can always be included in the harvestable fraction of the prey stock. But when their density is below 30 per m$^2$ the search time required is too lengthy and they can no longer be exploited.

It would be useful to know the lowest densities in which other prey species can occur and still be suitable for birds to exploit. However this will only make sense if we specify the prey's accessibility and the consequences of this on handling efficiency and search time. So this is what must be done. The curlew foraging on lugworms is a good example. The defecating frequency of the lugworms is known, so we also know how many lugworms

FOOD

will surface around each bird. We also know that the curlew can catch all the lugworms that surface within 2 m of it. When we take into account the curlew's 120° field of view, and don't include all the lugworms that surface in the remaining 240°, we can calculate how many lugworms the curlew can eat at different lugworm densities. The predicted increase in intake rate at higher densities nicely matches what has been observed. The curlews do take all lugworms that surface within their reach (Figure 4.27).

*Intake rate and prey size*
Intake rate is closely related to prey density because of the relationship between search time and prey density. Intake rate also depends on the mass of the prey consumed. Or, more specifically, we expect the intake rate to increase when the prey that are eaten per second handling time yield more. If prey take a long time to handle, even if they are very numerous, the intake rate will remain low. Since handling efficiency increases with increasing prey mass we would expect the intake rate to increase when birds can eat larger prey. The relationship between prey size, handling efficiency and the intake rate is known for the oystercatcher. In fact, a total of 253 studies on this were available, some as published articles on the topic, others needing to be fished out of unpublished reports or even researchers' field notebooks.[989] Figure 4.28 is a summary of all this information and shows how the intake rate increases (as expected) when the oyster-catchers eat larger prey of three different species.

So we already know that oystercatchers need to maintain an intake rate higher than 1 mg per second. When they feed on cockles and peppery furrow shells lighter than 20 - 30 mg they start to have problems maintaining a high enough intake rate (Figure 4.28). As Figure 4.28 shows this critical weight limit is even higher for mussels. It is difficult for oystercatchers to cover their daily food requirements when they feed on mussels containing less than 100 mg of flesh. There is a lot of variation around the predicted values, which is expected given the large numbers of studies and the variation in prey densities. In addition, when oystercatchers feed on buried benthic animals the different burying depths encountered can affect the intake rate. This doesn't apply to cockles as they always live just under the sediment surface.

The intake rate of oystercatchers feeding on cockles has been measured 38 times and every time both the mass of the prey consumed and their density has been calculated. In every case the intake rate depended more on the prey mass than prey density.[989] When oystercatchers feed on cockles containing less than 20 mg flesh their intake rate remains well below 1 mg per second, so it doesn't really matter whether these little cockles have a density of 1000 or 10 000 per m². If old cockles weighing more than 400 mg are available, oystercatchers can reach an intake rate of 2 mg per second even when there are only 5 or 10 old cockles per m², and their intake rate doubles at around 50 per m². We can use these data to calculate the cockle densities required for oystercatchers to reach an intake rate of 1 or 2 mg per second for different-sized cockles (Figure 4.29). The relationships shown in Figure 4.29 are calculated from all the available data. However there weren't any data available for some combinations of density and prey size since, for example, oystercatchers have never been observed foraging on very small prey that occur in low densities. So the relationships shown in Figure 4.29 are partly based on extrapolations of what is known. Plus, the few observations of oystercatchers foraging on large prey that were available contradicted each other. Field studies suggest that oystercatchers still take in enough food when they can eat large cockles that occur in a density of less than 10 per m². In contrast, in the only study where an oystercatcher's intake rate was measured after the researcher had manipulated the cockle density himself, the bird was already running short of food when the density of large cockles was below 50 per m². So which measurements should we believe? It's difficult to say. On one hand, field measurements are less accurate than those collected under experimental conditions. On the other, the experimental bird was forced to forage on the same few m² for hours, something birds in the wild would never do. That may have decreased the captive bird's intake rate dramatically. In normal conditions oystercatchers feed along the water's edge, which allows them to continually forage on wet tidal flats. New, precise field measurements, preferably on individually marked birds, are needed to determine the minimum prey densities at which cockles can still be exploited by oystercatchers.

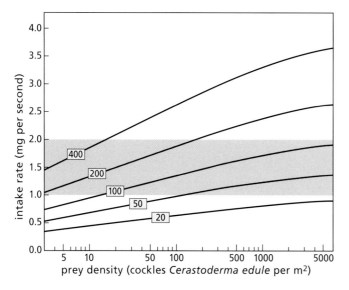

Figure 4.29 The intake rate of oystercatchers feeding on cockles (*Cerastoderma edule*) relative to cockle density (note that prey density is plotted logarithmically). Prey mass also has a large effect on intake rate. An intake rate of 1 - 2 mg per second (grey panel) can be achieved at quite low densities when prey are in good condition (numbers in the boxes give the mg dry meat per cockle), but cockles with little meat must occur in high densities to be worthwhile for oystercatchers.[989] An oystercatcher requires an intake rate of at least 1 mg per second on average to stay alive, but at that rate it would have to forage for the entire period that the tidal flat is exposed. On average, oystercatchers achieve 2 mg per second.[989] The figure shows the cockle densities required by oystercatchers (for different qualities of cockles) for it to be worth exploiting them.

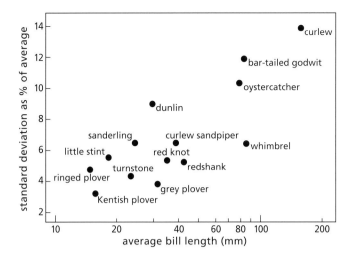

Figure 4.30 Individual variation in bill length tends to be large in long-billed species and small in short-billed species. This variation can be expressed as the relative standard deviation, the standard deviation as a percentage of the mean bill length. The standard deviation is a statistical measure that describes the spread in which two thirds of the values lie (i.e. two thirds of the values lie one standard deviation either side of the average value). Amongst the short-billed species it is obvious that sandpipers show more variation in bill length than plovers. Bill length data are from waders caught on the Banc d'Arguin and in Guinea-Bissau. [954]

### 4.6.3 Each bird is unique

We've discussed the proportion of prey oystercatchers, redshanks and other waterbirds can harvest. But how far these generalisations hold depends on how much the birds within each species differ. It's important for wading birds to have long legs and long bills. Long bills are also required by birds that feed on buried prey. Leg and bill length can vary between individuals. We also need to remember that not all individuals are equally experienced, which can affect both the handling efficiency and the search time for different prey for each bird. This in turn, can affect the proportion of the prey that is profitable for different individuals.

*Learning*
In early July, when curlews from the breeding grounds arrive on the tidal flats, with a little effort it is possible to identify the young birds by their shorter bills and slightly different plumage. But once you've seen them foraging, the juveniles are easily distinguished from the older birds. The old birds know how to catch one unsuspecting benthic animal after another and make it seem easy to disembody a crab or swallow an uncooperative shrimp. But young birds repeatedly drop difficult prey, breaking them into little pieces, or get completely on the wrong track and fish up a small twig or branch or some other inedible item. However, a few months later they seem just as experienced as the old birds. The learning process takes longer in oystercatchers (Chapter 5), which is the main reason that oystercatchers, how flexible they might be, show a striking degree of food specialisation. Every bird becomes expert in handling a limited number of prey types in certain ways. Adult birds continue this learning process. At least this could be concluded from experiments in which a few oystercatchers were offered peppery furrow shells and sand gapers. Some birds were handier with one species and some with the other. Over a few weeks these differences disappeared, as the birds became better at handling the prey with which they had previously had less experience.[910]

*Size differences*
For all birds that forage on buried prey the length of their bill determines how far down they can reach, which immediately affects the proportion of prey that are accessible (Figure 4.20 and 4.21). The same food can differ in accessibility for different individuals. The size of this difference can be determined by measuring the bills of as many birds as possible and seeing how greatly bill size varies. Bill length varied to greatly differing extents in 14 wader species that winter on the Banc d'Arguin or Guinea-Bissau (Figure 4.30). This variation is expressed here as a standard deviation. The average bill length of the bar-tailed godwit is 84 mm, but there is a 10 mm standard deviation. That means that the bill length of two out of three bar-tailed godwits is within 10 mm of the average length of 84 mm. So the standard deviation of the bar-tailed godwit's bill is 12% of the average length. This is quite large compared to standard deviation of the smaller wader species' bills, which are around 3 - 6% of the average. You can see in Figure 4.30 that birds with the biggest bills have the greatest variation in bill length, and that the curlew holds the record. Perhaps the variation in bill length increases in larger bird species, because the individual variation in body size also increases with species size. This doesn't seem to be the case. Wing length is a useful way of describing a bird's size. To determine

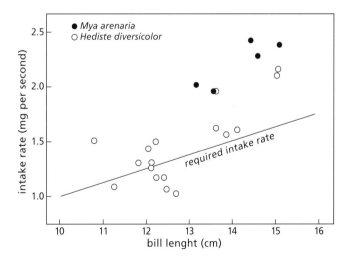

Figure 4.31 The longer a curlew's bill, the higher its intake rate. Curlews with longer bills are also larger and heavier and therefore need more food; how much extra they need is shown by the line on the graph. Prey choice also varies with bill length, and long-billed birds eat fewer ragworms (*Hediste diversicolor*) and more sand gapers (*Mya arenaria*). These data are from individually marked birds feeding at one site in the Eastern Dutch Wadden Sea from August to November over two years. The relationship was the same in both years.[237, 257]

whether species with a large variation in bill length also differ more in their body size, researchers have checked whether species with large individual differences in bill length also differ greatly in wing length. No relationship has been found between the two. However within a species, wing length and bill length are correlated. Amongst curlews, bar-tailed godwits and oystercatchers the larger and heavier birds have a longer bill.[954, 993]

Why do birds have different bill lengths and why does this vary more within some species than others? If individuals with different bill lengths choose to feed on different prey, you might expect variable bill lengths to reduce competition for food between conspecifics. But does prey choice really differ between birds with long and short bills? When we ask this, we're really asking whether food choice differs between the sexes, as in the three species just mentioned females have the longest bills and males the shortest.

The male oystercatcher has a shorter and thicker, and because of this stronger, bill than the female. Females eat more worms and buried shellfish than males.[78, 200, 422] This is why in winter it's mainly females that forage on earthworms in fields at high tide.[960] When oystercatchers eat peppery furrow shells the birds with longer bills have a higher intake rate, which is why so few short-billed birds eat these shellfish when they are buried deeply.[960] Baltic tellins live closer to the surface in summer and can then be eaten by both long and short-billed oystercatchers. The birds with short bills use more brute force to break open the shell and often pull the shellfish out of the sediment and open them on the surface. In contrast, the birds with long bills usually try to open Baltic tellins while they are still in the sediment. This requires a longer search time as they try to find the few shellfish that gape slightly. But once they have found one, they are quick to open it. The short-billed oystercatchers are less selective than the long-billed birds and have a lower handling efficiency when feeding on Baltic tellins. In the end, the short and long-billed birds achieve the same intake rate.[425]

Oystercatchers with a long, thin bill often have a pointy bill tip and birds with a short, strong bill often have a blunt bill tip.[200, 423] Bill shape can indicate the type of prey that different individuals eat and how they attack their prey. A bird with a short, stout bill uses brute force to hack holes through bivalve shell. A bird with a pointed, chisel-shaped bill will mainly eat soft prey, and when feeding on shellfish will search for individuals that are gaping slightly so that it can push its bill between the valves. The specialisation doesn't end here. Amongst the oystercatchers that feed on mussels, some individuals mostly hammer on the left valve and others on the right valve, some on the top and others on the underside of the mussel. The advantage of hammering on the underside it that the mussel is attacked at its weakest spot, but this means the bird must first pull the mussel loose, turn it around and put it down again, all of which take extra time.[612] Birds that hack into the top of the mussel don't need to pull it

It's easy for common gulls to rob oystercatchers. They just wait until the oystercatcher has opened the mussel, then quickly pounce. But they must be quick, as oystercatchers need little time to extract the meat from a gaping mussel. The oystercatcher tries to gobble the meat as fast as possible, especially when a robber waits nearby.

loose, so they can immediately start chipping away at the shell, but on average this takes them longer because the shell is thicker here than it is on the underside. Presumably each individual makes a trade-off, choosing the foraging method most profitable for him or her.

The prey choice of curlews with long and short bills also differs. The large sand gapers that are profitable for curlews are rarely eaten by the short-billed birds.[238] They simply can't reach them. In winter these large sand gapers are often the only prey eaten by long-billed curlews. Both male and female curlews feed on ragworms. The birds with the longer bills tend to catch more worms that are still down in their burrows, while the birds with the shorter bills concentrate more on the ragworms found at the surface. This is logical since longer-billed birds can reach further and access more ragworms that are still down in their burrows.[259, 978] Curlews with longer bills are at an advantage when feeding on ragworms. When individual curlews foraging in the same area at the same time were intensively followed, the longer-billed birds had a higher intake rate. However the birds with the very longest bills were not feeding on ragworms, as for them sand gapers made a more attractive meal (Figure 4.31).

Why don't all curlews have long bills, if long bills can help them reach a higher intake rate out on the tidal flats? A long bill isn't always an advantage. It's of no use when the prey can be pecked off the sediment surface. Plus, it has the disadvantage of requiring prey to be transported further before reaching the throat, which could be a problem when birds feed on small prey that must be gathered quickly. In this situation handling time could increase with increasing bill length. It is known that bird species with short bills can eat small prey faster than species with long bills,[982] but whether the same applies when conspecifics are compared is still unknown. A longer bill is heavier and must be thicker to remain strong, which is why it's usually large individuals that have a long bill. In curlews there is a very strong relationship between an individual's size and its bill length, which is why in Figure 4.31 a single line could be drawn to describe the relationship between a curlew's bill length and its predicted intake rate. The predicted intake rate is the rate required to maintain the same body mass, assuming that the birds with long and short bills spend the same amount of time foraging. The strong relationship between body mass and bill length allows the predicted intake rate to be shown as a function of bill length. From this we can conclude that large curlews need longer bills to find enough food.

Curlews that forage on Western European tidal flats have bill lengths of 10 -16 cm. Their conspecifics that winter in West Africa have bills that are more than 2 cm longer, and are also more variable in length (12 - 19 cm).[954] When you're used to watching curlews on European tidal flats, the curlews on tropical tidal flats seem to be carrying a rather top-heavy bill. Most curlews that breed in Western Europe also winter on the European coast, but birds that breed further east use tropical wintering grounds. Across the curlew's breeding grounds there is a systematic increase in bill length: for every 10

Black-headed gulls are no big deal for curlews. But if a gull attacks a curlew unexpectedly from the air, the startled curlew may drop its newly caught ragworm (upper photo). Just like the common gull, the black-headed gull must time its act of piracy precisely.
Waterbirds are not only robbed by gulls. They also try to steal from each other. This oystercatcher (lower photo) has unsuccessfully tried to steal a crane fly larva from another.

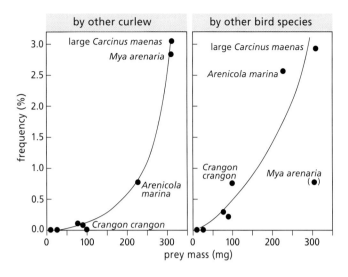

Figure 4.32 Curlews lose more large prey to theft than small prey. This is true for prey stolen both by other curlews and by other species, particularly herring gulls. In the graphs, each point represents a different prey species. The two smallest prey are the siphons of peppery furrow shells (*Scrobicularia plana*) and first-year common shore crabs (*Carcinus maenas*). Sand gapers (*Mya arenaria*) are stolen by conspecifics but not by other bird species.[242]

longitudinal degrees east, the males' bills become 2 mm longer and the females' 3 mm.[138] It would be pure speculation to try and explain why this is. But curlews that winter in West Africa make good use of their super-long bills. These birds feed on fiddler crabs and are able to pull up the smaller individuals from down in their burrows.[965] This food source would be out of reach for most of the short-billed European-wintering curlews.

### 4.6.4 Robbing and being robbed

Not all birds on the tidal flats search for their own food themselves. A number of birds leave it to other individuals and steal their prey just when it has been pulled from the sediment or broken open. This sort of food parasitism occurs both within and between bird species. Waders are most often robbed by gulls. Once an oystercatcher has broken open a mussel it takes little effort for a black-backed or herring gull to chase the oystercatcher away. It takes more effort for a common gull to do this, and every now and then the common gull is rewarded with a big peck from an angry oystercatcher. The black-headed gull is too small to chase an oystercatcher, but will sometimes dive-bomb one, giving it such a fright that it drops its prey, which the gull can then grab. Black-headed gulls also use this technique to steal ragworms off curlews.

The number of prey that birds lose through theft is very variable. The highest proportion occurs in oystercatchers feeding on giant bloody cockles on the Banc d'Arguin. These oystercatchers lose half of their prey to lesser black-backed gulls once the prey have been opened.[834] The lesser black-backed gulls are unable to open these giant shellfish. So they just sit patiently waiting for an oystercatcher to put its bill between the two valves and cut through the adductor muscle that keeps the bloody cockle shut. This doesn't take long, just enough time for a lesser black-backed gull to wander over and chase away the oystercatcher that has just finished preparing its meal. The gull can then simply tear out large chunks of flesh from the opened shellfish. Herring and common gulls sometimes get all their food by robbing oystercatchers. This can often be seen when oystercatchers are feeding on large mussels. Each common gull keeps an eye on 25 to 50 oystercatchers on the mussel bed and chases away any conspecifics that try to steal from one of its oystercatchers.[960] In total, the oystercatchers lose about 10% of the mussels they have opened to gulls.[478, 976] They would lose even more if the gulls didn't chase each other away. In the Exe estuary, oystercatchers also lose mussels to thieving hooded crows, but this is only a tiny percentage of their total prey caught.[234] The crows also find their own mussels which they can break open by dropping onto hard surfaces.

Birds of the same species also steal from each other, usually when they are feeding on large prey that require a long handling time. This is why oystercatchers steal mussels more often than cockles from each other, and curlews steal large crabs and sand gapers from each other more often than ragworms or peppery furrow shell siphons (Figure 4.32).[242] Deciding whether to steal or not is a trade-off that depends on the expected result. This choice, like other feeding decisions, can be investigated as an optimisation problem. A black-headed gull doesn't steal amphipods from a redshank because the yield is too low and it could find them faster by itself. But when a redshank encounters a 1.5 cm common shore crab that takes it tens of seconds to handle, it's very likely that a nearby black-headed gull will make the most of this opportunity. The yield of a robbery depends on the prey mass, how much food the victim has already eaten, and the time it takes to steal the prey and then eat it. The robber must also make a trade-off on its chance of succeeding. The likelihood of robbing successfully depends on the victim's handling time, the distance to the victim and the thief's dominance over the victim. When a curlew eats a large common shore crab there is a reasonable chance that it will have to give this prey to another curlew or another bird such as a herring gull. But when the same curlew is trying to eat a sand gaper, only other curlews will try to steal its food (Figure 4.32). It wouldn't make sense for other birds to chase a curlew feeding on sand gapers. The curlew eats sand gapers that are still deep in the sediment so the robber would need a bill as long as the feeding curlew's.

When talking about food theft we shouldn't just think of robberies involving a struggle. These robberies are often very subtle and hardly discernible to a human observer. Sometimes a bird suddenly walks off just when it has found a prey item. It's very likely that a passing neighbour will quickly try to find the prey in the same spot. You can see this happening when waders feed on earthworms and crane fly larvae in meadows or when birds eat buried prey on tidal flats, such as curlews feeding on sand gapers.

Robberies are carried out by birds that are dominant

over other birds.[234] Birds that are high in the pecking order can afford to be thieves. Birds lower in the pecking order are always the victims.[235, 327, 332] Trying to escape their assailants should help the potential victims, but how can they do it? First, they should try to stay away from their attackers. When birds forage in a flock, it's always worthwhile for less dominant birds to stay away from their dominant flockmates. Amongst bar-tailed godwits that feed on crane fly larvae in meadows, the females are nearly always dominant over the males. It's advantageous for the males not to forage too close to the females, so the males often end up on the edge of a flock or aren't really part of the group. When bar-tailed godwits are individually colour-ringed we can see that the same applies to certain individuals. Some individuals never steal, but regularly lose their prey to thieves. Others are never robbed, but regularly steal themselves.[77] Birds that often steal usually forage in the middle of the flock, while the potential victims are mostly found around the edge.

Oystercatchers also have a clear pecking order, which explains their age distribution in tidal areas. On mussel beds, young birds have no chance of opening and eating a mussel undisturbed when older birds are around. So in winter the young oystercatchers are usually seen out on the mud where they feed on peppery furrow shells and ragworms. The young birds only get the chance to feed on the mussel beds once the older birds have left for the breeding grounds in summer.[324, 325, 326, 327]

The likelihood of successfully robbing another bird also depends on the number of birds nearby that can be robbed. There is more theft when birds forage in a tight flock. Dominant oystercatchers feeding on mussels acquire on average 15% of their food by theft. When birds forage in a tight flock, the dominant birds have more opportunities to steal, and act on this. Consequently, the less dominant birds lose more of their prey. Plus, the less dominant birds spend more time looking around and trying to escape conspecifics at higher densities. So when oystercatchers forage on mussels in a high-density flock, the intake rate of dominant birds increases, and that of the less dominant birds decreases (Figure 4.33).[235, 327]

When it's not possible to avoid thieves, there are other ways to reduce the risk of being robbed. Potential victims can try to swallow their prey faster or, if possible, eat them while they are still buried in the sediment. Any prey found could be taken away to suitable dining area a safe distance from thieves, and it may be worthwhile to look around more often while eating.[126, 234] The risk of being robbed can also be reduced by switching to prey that are less profitable for potential robbers. Lapwings and golden plovers start to eat smaller worms when they are being harassed by thieving black-headed gulls. These small prey are less profitable for the lapwings and golden plovers as well as the black-headed gulls, but when the risk of losing prey is taken into account, it is worthwhile to eat smaller prey.[32, 851]

## 4.7 The distribution of waders across the tidal flats

### 4.7.1 Bird numbers and average foraging density

*Counting birds*

If you want to know how many blades of grass there are in a field or how many cockles in a mudflat you don't need to count them one by one. Instead, you can simply take a few random samples. You can even calculate the maximum number of samples required to get a reliable estimate of the total stock (Section 4.4.1). The same techniques can be used to estimate waterbird numbers. Instead of taking samples with a corer, you count the birds out on the flats at low tide over a known surface area. However this method is seldom used.[967, 983] The mudflats are heterogeneous, bird density varies between sites, and the birds move around while foraging, all of which make it difficult to take representative random samples. Plus the birds often tend to take flight just when they are being counted. It's quite easy to count waterbirds from some places, such as a high dune, a cliff or a dyke. Another option is to build a viewing tower out on the feeding grounds and ensure that you are in the hide before the birds appear on the outgoing tide. With a good telescope in a 5 m hide even the smallest shorebirds can be identified from up to 1 km away, and can easily be counted. If you walk across the tidal flat you can't see as far and the birds often take flight when you are still hundreds of metres away, making it difficult to determine the bird density on foot. But walking has the advantage of allowing you to cover more ground, so you can count the birds over a much larger area in one low tide period.

Almost everywhere in the world, waterbird counts are based on estimates of the total number present rather than counts of smaller random samples. At high tide, waterbirds are almost always concentrated in large flocks, which makes it easier to count all birds present in an intertidal area in one go. The disadvantage of this method is that the birds are almost never really counted, even though it's said that they are, instead only rough estimates are made. A flock of 50 000 dunlins cannot re-

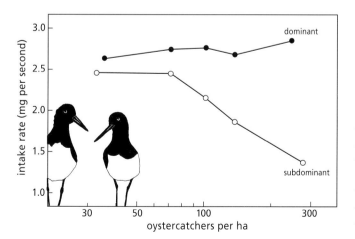

Figure 4.33 The intake rate of oystercatchers that eat mussels (*Mytilis edulis*) changes when the birds forage in high densities. Birds low in the pecking order (subdominants) suffer a large decrease in intake rate, while dominant birds manage to increase theirs slightly.[235]

ally be counted. Usually, the person counting doesn't want to disturb the flock and will remain at least 100 m away. Individual birds may be counted one by one at the front and side of the flock, but it's impossible to see individual birds at the back, and even the middle is just a thick streak of bird backs and heads. The only solution is try and count groups of, for example, 500 birds. You have to try and determine which part of your telescope's field of view corresponds to 500 dunlins at a certain time, and then extrapolate from there. If you repeatedly end up estimating that there are around 100 groups of 500 dunlins you then assume that 50 000 birds were present. You can check the accuracy of your estimate by recounting the birds from a different perspective, by having different observers count the same flock independently, or by comparing the number of birds that fly up to the roost at high tide with those that fly down to the mudflats at low tide, or the other way round.

Time is almost always a limiting factor when counting birds. Ideally all the birds should be counted within the same high water period. Experienced bird watchers can sometimes count thousands of birds per minute, which means they can count many tens of thousands of birds during one high tide period. When the flocks aren't this large, it's possible to check the accuracy of the estimate by counting all the birds in the flock one by one, or by taking a photo and individually counting the birds at a later stage. The errors involved in large-scale waterbird counts on tidal flats have been systematically investigated.[718] The average counting error was 37% for birds that roost in flocks and 17% for flying birds. These are large margins of error. Fortunately, on average the observers didn't do too badly as the birds were not systematically over or underestimated. So we can comfort ourselves with the thought that when the final count is compiled from many sub-counts these 'non-systematic' errors should balance each other out.

But errors don't just come from the actual counts. Birds can also be missed in the field, because flocks were not seen or moved during the count. If a flock moves it may also be mistaken for another flock and counted twice. There can also be problems identifying the different bird species. When large flocks comprise different bird species it can often be quite hard to determine the internal ratio in numbers, particularly when their winter plumage makes it difficult to distinguish the different species. In this case it's a good idea to recount the birds when they fly between the mudflats and the high tide roost, or to look at the species breakdown at low tide when the birds are out foraging on the mudflats.

In large tidal areas it's not always possible to visit all of the high tide roosts in one day, even when many people are out counting. It can be useful to census large tidal areas from a plane. However small waders can't be identified from a plane and you have to be satisfied with counts of groups of species, such as 'small sandpipers'.

Using high tide counts to estimate the number of birds that forage on the tidal flat can be problematic when waterbirds continue to forage at high tide. The birds counted at high tide may not all go down to the mudflats at low tide. This can be a real problem at high tide when intertidal birds mix with birds that forage on saltpans or in fields. For example, in autumn many thou-

At high tide, waterbirds concentrate in dense flocks. These can be hard to count accurately, and the error in estimating numbers is twice as large when birds are on the ground (upper photo: dunlins and knots in the foreground with black-headed and common gulls in the background) as when birds are flying (lower photo: bar-tailed godwits). Counting flocks of just one species is easier than counting mixed flocks. Birds can be accurately counted while they forage (photo opposite upper: shelduck, curlew, oystercatcher and redshank) or when they roost in loose flocks on the mudflat (photo opposite lower: bar-tailed godwits and oystercatchers).

FOOD

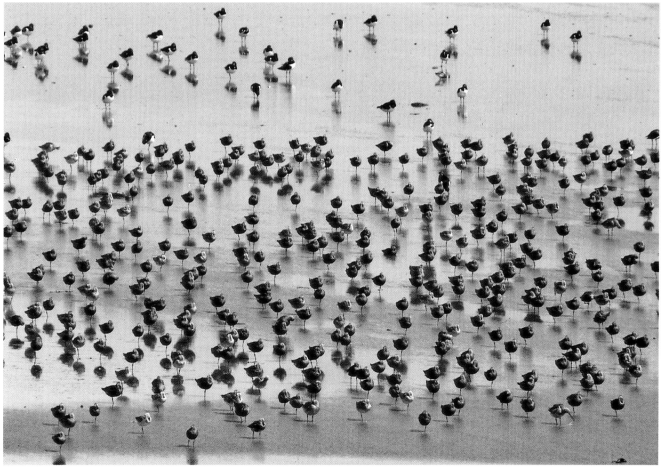

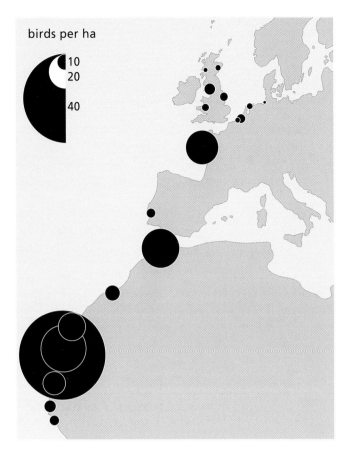

Figure 4.34 Foraging densities of waders in midwinter (January), expressed as the total number of birds per ha of tidal flat. The Wadden Sea is divided into Dutch, German and Danish parts (densities approach zero in Denmark). British areas are summarised for East and West Scotland, Northwest England, East England and Wales.[16, 704, 777, 785, 946, 967] The highest densities are reached on the Banc d'Arguin and the Baie d'Arguin in Mauritania (46 and 24 birds per ha respectively).

sands of curlews from the Dutch tidal flats gather with tens of thousands of curlews that forage inland in the north of The Netherlands.[971]

In many tropical tidal areas it's impossible to count waterbirds at high tide as the birds disappear into the mangrove forests. The only way to estimate the number of waterbirds in tropical West Africa is to count the birds at low tide. Fortunately, detailed topographical maps and marine charts of the tidal areas in Guinea-Bissau are available and the presence of rock formations, gullies and other landmarks make it much easier to count the birds in clearly demarcated sections. On average it proved possible for a team of two people to count the birds in 300 ha of mud flats during one low tide period; on sand flats with few birds twice as big an area could be counted, but on bird-rich silty areas only half as much.[967] Six years later, using a similar technique, it was possible to count birds over an area that was, on average, three to four times as large per day.[762] However this time, only the birds near the observers were counted and the distance over which birds had been counted had to be estimated. This made it possible to quickly estimate the density of foraging waterbirds. However, the disadvantage was that it introduced a different and probably much larger error, namely the estimate of the area where birds had been counted.[983] This method isn't really worth repeating.

*Average bird densities*

Along the European and African coasts there is a total of around 900 000 ha of tidal flats used by around seven million waders. Hundreds of thousands of cormorants, terns, gulls, geese, ducks and flamingos also use these areas. Some of these birds forage in channels below the low tide line or in the marshes and nearby cultivated land above the high tide line. This makes it difficult to determine the density of birds foraging in the intertidal area. Overall, there is estimated to be an average of 10 waterbirds per ha tidal flat, eight of which are waders. In the following discussion we limit ourselves to the waders.

Waders avoid cold tidal areas in the winter. In Northwestern Europe the average winter density is just 0.4 - 8 waders per ha, compared to 5 - 40 waders per ha in Southwestern Europe and Northwest Africa (Figure 4.34). The millions of waders that winter in Africa use the tidal areas of Northwestern Europe during their spring and autumn migrations. This is why the bird densities from April to May and July to September in Northwestern Europe are twice as high as in winter. Few birds winter in Northwestern Europe because there is little food available during winter. There is nothing suitable for waders that feed on crabs, fishes and shrimps. Plus, as we saw in Chapter 3, it's much more expensive energetically to live at low temperatures and there is even a risk that birds could die of starvation if the mudflats freeze over, making it impossible to feed. It's usually only large waders such as curlews and oystercatchers that take the risk of staying up north for the winter. A decrease in the ambient temperature has less of an effect on the energy budget of a larger bird (Figure 4.5).

Waders can often reach high densities in small intertidal areas.[162] The likelihood of encountering extremely high or low densities increases when densities are calculated for smaller tidal flats, but also on average bird densities are lower in large tidal areas. A comparison of all British intertidal areas showed that the wader density is 10 times greater in areas of less than 100 ha than in areas of more than 10 000 ha of tidal flat. A possible explanation is that the larger the intertidal area, the larger the average distance between the high tide roosts and the feeding grounds. If intertidal areas that lie far offshore are systematically visited by fewer birds, this could cause bird densities to decrease in areas far from high tide roosts even if these areas could provide relatively large feeding grounds. We do know that when there are few birds in an intertidal area they will forage close to the high tide roost, and that the feeding grounds further away will only be visited when numbers increase.[318, 961, 963, 964] This suggests that the closest feeding areas are preferred. Assuming a flight speed of 1 km per minute[604] and flight costs that are equivalent to 12 times the basal metabolic rate, it is possible to calculate how much extra it costs to fly further. Waders fly up and down between the high tide roost and the feeding grounds twice a day.

This means that for every extra kilometre between the sites, they have to spend 1.3% more energy. So from an energetic view point it makes little difference whether a bird has to fly an extra few kilometres or not, so this doesn't explain why foraging densities are so much lower in larger intertidal areas.

There is however another explanation for why more birds forage in small intertidal areas. Large tidal areas are often much less sheltered than small areas, so their sediment usually consists of more sand than silt. Wader density is clearly related to sediment composition.[94, 955, 967] This is no surprise since silty mudflats contain more food than sandy flats (Figure 1.6). The relationship between the density of waders and their food is well-known, and is discussed further in Section 4.7.2. But even if the food availability in sandy and silty areas was exactly the same, you would still expect a relationship between the presence of waders and the sediment composition. If sanderlings tried to walk on soft silt, they would quickly sink up to their bellies in the mud, yet avocets could easily walk across these areas (Chapter 2). Avocets couldn't scythe their bills through the upper layer of a sand flat because there would be too much resistance, but this technique works well on silt flats.[861] We know that avocets are birds of the silt flats and sanderlings are beach birds.

In general, we can conclude that wader density decreases in winter when the average sea temperature is below 4 °C. In areas where the average temperature is higher, the chance of a tidal flat freezing over is almost nil and the average density of waders is 13 birds per ha. There is a lot of variation around this average. This variation can be explained by differences in food availability, sediment composition and the size of the area. These three variables are interconnected, which makes it worthwhile reanalysing the available data.

### 4.7.2 Bird distribution across the tidal flat

It sometimes seems as if the huge numbers of waders that were present on the edge of the tidal flats at high tide completely disappear at low tide. The birds have distributed themselves across the tidal flat. The average density is around 10 birds per ha or 1000 m² of tidal flat per bird. If the birds spread themselves equally across the tidal flat you would expect to find one bird every 30 m. However, this is not the case. Different species more or less follow the waterline with the incoming and outgoing tide, so they only use part of the total feeding grounds at any particular time during low tide. This is why it's sometimes not possible to find a single bird across many hundreds of hectares of upper tidal flat at low tide. Later, hundreds of birds per hectare will forage on these upper tidal flats during the incoming and receding tide, when the low tide feeding grounds are covered in water.

Would it be possible for more birds to forage on the tidal flats than there are now? Would the birds still have enough food from an average area of say 500 m² rather than 1000 m² that would provide only half as much food? This is discussed further in Chapter 6 in relation to the carrying capacity of the tidal flats for waterbirds. The question of how many birds can live in an area can only be answered when you know how much food is available for the birds and how much they will consume. You also need to know if the food supply is consistent and how the birds will react to food shortages. Much of the information that we've covered in this chapter can help determine the carrying capacity of an area: we already know how much food the waterbirds need per day (Section 4.2) and how much time they can spend foraging (Section 4.3). The total food availability on the tidal flat varies seasonally and annually (Section 4.4 and 4.5) and a varying proportion of this can be harvested by the birds (Section 4.6). Section 4.6 also describes the relationship between intake rate and food density. All we need to know now is how the bird density relates to the variation in food sup-

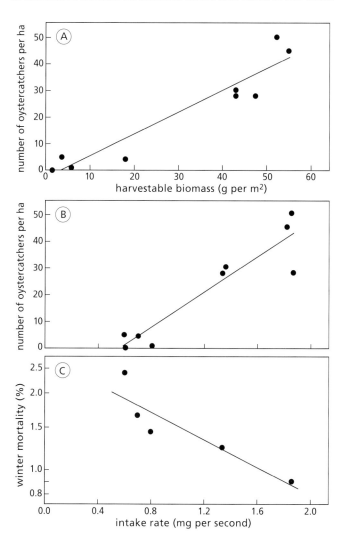

Figure 4.35 In winter, oystercatchers leave a feeding area and go elsewhere when (A) little food is available and (B) the intake rate achieved is very low. Birds probably respond immediately to a low intake rate, and the strong relationship between bird density and prey biomass should be seen as an indirect consequence of the high correlation between prey biomass and intake rate, and intake rate and bird density. (C) Mortality of oystercatchers is higher in winter when foraging conditions are bad. Most birds try to escape to another site, but it is not necessarily any better there. These winter data represent five mild winters; in particularly cold winters the relationship can be much more extreme. Data were collected along the Frisian coast of the Dutch Wadden Sea.[995]

ply and how much food the birds consume in total. The relationship between the density of birds or other predators and their food is known as the numerical response.

*How birds react to annual variation in food supply*
Most waterbird species leave the tidal flats in spring and return a few months later. They never know from year to year how much food there will be, or which prey species they can forage on. No waterbird species can afford to limit its food choice to one or two prey species. If they did, they would quickly become extinct. Redshanks prefer to feed on the amphipod *Corophium volutator*, but when the amphipods aren't around the redshanks have to survive on ragworms or mudsnails.[310, 311, 315] The same applies to red knots that prefer to feed on Baltic tellins. In some years there are none of these shellfish, but more often the tellins are too large or are too deeply buried and the knots are forced to feed on small cockles and small mussels or mudsnails.[686, 986] Different birds must deal with differing degrees of annual variation in their food supply. For decades, mussels have provided a regular food source for oystercatchers in the Exe estuary in Southern England, just as amphipods have for redshanks in the Ythan estuary in Scotland. But redshanks and curlews in the Wadden Sea must deal with high annual variation in prey availability (Section 4.4.6).

Switching from one prey species to another allows waterbirds to return to the same tidal areas annually. But do they remain site faithful when little food is available? There are large annual variations in mussel stocks in the Wadden Sea.[62, 68] From 1968 to 1973 almost all oystercatchers on tidal flats south of Schiermonnikoog (an island in the Dutch Wadden Sea) fed on mussels. But after 1973, when almost no live mussels remained, the oystercatchers disappeared from the mussel beds. Throughout this time, the total number of oystercatchers in the area remained constant at about 10 000 birds. The oystercatchers didn't leave the area, because they could still feed on cockles on the same tidal flat south of Schiermonnikoog.[976] This is not always the case. From 1975 to 1988 the number of oystercatchers on tidal flats south of Ameland (another Dutch Wadden Sea island) grew from 20 000 to 60 000 birds. In the following years, the number of oystercatchers dropped back to their previous levels. The change in oystercatcher abundance was due to changes in the surface area of mussel beds south of Ameland. But at this site the oystercatchers were unable to switch to alternative food sources, and when no mussels remained they left the area.[465, 717]

On the Frisian coast of the Dutch Wadden Sea, annual low tide counts of waterbirds on a 400 ha area of tidal flat were made and food availability was monitored for 11 years, starting in 1977.[982] The oystercatchers at this site ate five species of shellfish, whose total winter bio-

Researchers use colour rings - sometimes with numbers or letters that are visible from afar - to identify individual curlews.

Most oystercatchers winter in Northwestern Europe, but some thousands of birds winter in tropical West Africa. In front of these oystercatchers in West Africa are bar-tailed godwits, curlew sandpipers, royal terns, sandwich terns, little terns and a Caspian tern.

mass varied from 40 - 90 g per m². The variation in biomass was even larger when only harvestable prey were included. In some years almost no harvestable prey were present, especially since prey living deeper that 6 cm could not be included. Over these 10 winters, the total harvestable biomass fluctuated between just over 0 g per m² and 55 g per m². When there was no food, there were also no oystercatchers, but in the years with plenty of food there were 50 oystercatchers per ha (Figure 4.35A). Oystercatcher winter abundance depended on the abundance of surface-dwelling prey. Buried prey like Baltic tellins and peppery furrow shells often live out of bill reach in winter and when they are accessible their handling efficiency is too low.[995] Similarly, a study in the Eastern Schelde (an estuary in the Southwestern Netherlands) found that cockle biomass determined oystercatcher numbers.[560] Cockle biomass in winter ranged from 0 - 200 g in a 1 ha study plot, while oystercatcher numbers varied accordingly from 0 - 65 birds per ha.

Unlike the oystercatchers, little variation was found in the number of wintering curlews at the same research area on the Frisian coast. During winter an average of 400 - 600 curlews were present on the 400 ha of tidal flat. Each year, most of the curlews fed on ragworms and sand gapers. In late summer and autumn they also ate common shore crabs and lugworms. Their annual food availability varied, but this did not cause a variation in curlew abundance. The birds were site faithful. Several hundred curlews could be recognised by individual colour rings. These marked birds foraged at exactly the same sites for years in a row and some even defended the same 2000 - 10 000 m² of tidal flat that formed her or his feeding territory for eight months of the year, year after year. If in a particular year there was not enough food within its own territory, the bird often kept foraging within 1 km of that site.[140, 238, 239]

Redshanks and red knots were at the other end of the continuum. These birds only returned to the Frisian research area if there was plenty of food available. Many thousands of redshanks foraged at the research site in July, most of which were seen leaving the Wadden Sea for their African wintering grounds in early August.[681] But sometimes there were only a few hundred redshanks in the research area in late summer. Redshank regurgitates were collected annually and showed that the birds fed on five different prey in summer: shrimps, amphipods, small common shore crabs, ragworms and mudsnails. However there was no relationship between redshank abundance and the total food availability of all five prey species. Redshanks were only abundant when amphipods were abundant or, to a lesser extent, first-year common shore crabs were present. When amphipods and common shore crabs weren't present, the redshanks fed on ragworms and mudsnails. But although this alternative food may have been abundant, it didn't appear to be an attractive option.[960] Whether these transitory birds could find a better food source elsewhere in the Wadden Sea is not known.

We know a little more about the red knots. Like the redshanks, the number of knots passing through the research area in July was 10 times higher in some years than in others. They were only abundant on the 400 ha feeding site when there were plenty of Baltic tellins. The Baltic tellins had to be 10 - 16 mm long and not buried deeper than 3 cm below the sediment surface. A nearby high tide roost was used by knots that foraged over an area around 10 times larger than the research area. The high tide numbers at this roost remained fairly constant annually. This meant that if the knots were unable to find enough food at a particular site, they could find an alternative food source nearby and did not need to move to a completely different tidal area.[986] In this way, these knots were similar to the oystercatchers on Schiermonnikoog. But what happens when food availability varies on a large scale? In this case the birds disappear, as was found with red knots in the Western Dutch Wadden Sea.[686]

We can already draw a few conclusions from these examples. The first is that if there is a food shortage and the birds move to another site it's necessary to start research at this other site too, to determine how well the birds are managing over there. The main problem with this is that we often don't know where the birds have gone. The Wadden Sea is vast. The obvious advantage of this for the birds is that they if they need to move, there are likely to be plenty of other areas where their preferred food is still abundant. But the disadvantage for the birds is also a disadvantage for researchers. It already takes a huge effort to determine the food availability, which prey the birds are eating and what the birds' intake rate is over a few hundred hectares, and it would be hopeless to attempt to also gather this data for other possible alternative feeding grounds. Researchers working in smaller tidal areas also face this problem. Birds can leave these smaller sites and move to completely different tidal areas that are sometimes very far away. The most extreme example of this is the oystercatchers in West Africa. Usually, most of these birds were concentrated on the Banc d'Arguin where they feed almost entirely on one prey species, the giant bloody cockle. When this prey began to decrease over the years, some of the oystercatchers disappeared. At the same time the number of oystercatchers on the tidal flats of Guinea-Bissau, around 900 km to the south, increased[996] and more oystercatchers were also seen in Senegal, 700 km away.[777]

Secondly, it's not that easy to relate bird density to food density. Birds eat many prey species and the total food availability can't be determined by simply adding the total flesh mass of all of these species. Even when the harvestable biomass of each prey species is known, the sum of this doesn't always provide a good prediction of bird abundance. Individual birds decide where they will go to forage based on their expected food intake rather than food availability.[336, 344, 995] In other words, bird abundance is predicted to be related to their intake rate and since intake rate is often related to food availability, the abundance of both the birds and their food are linked (Figure 4.35A and B).

The third conclusion is that bird abundance is not only linked to the food availability at that site, or rather the intake rate that can be reached there. It also depends on the situation elsewhere. When the birds can find a better place to forage at another site, it's expected that they will go there.

In general, examining bird abundance is only of partial assistance if we want to determine whether food availability is limiting bird numbers. This is not the case if we know more about migration or mortality related to varying food availability. For example, ringing research near the Irish Sea showed that it was mainly young oystercatchers that move from site to site depending on food availability.[821] Juvenile oystercatchers move around more than older birds, sometimes giving the juveniles an advantage. However roaming birds will encounter food shortages more often than site faithful birds. Young grey plovers that are able to support themselves in one particular area for their first winter will become site faithful and return to the same area every year. If the plovers are unable to settle at one site in their first winter they will search for other wintering grounds, which may be hundreds of kilometres away, and from then on only use the estuary as a stopover during migration.[867]

Bird ringing also makes it possible to determine whether annual mortality increases when food is scarce. If you only counted the birds in years with poor food availability you wouldn't know whether the low number of birds was due to the birds deciding to go elsewhere or increased mortality. The only problem with analysing ringing data is that annual mortality does not depend solely on food. The mortality of waders wintering in Northwestern Europe also depends on winter temperatures (Section 4.2). In the research on oystercatchers along the Frisian coast that was discussed earlier, oyster-

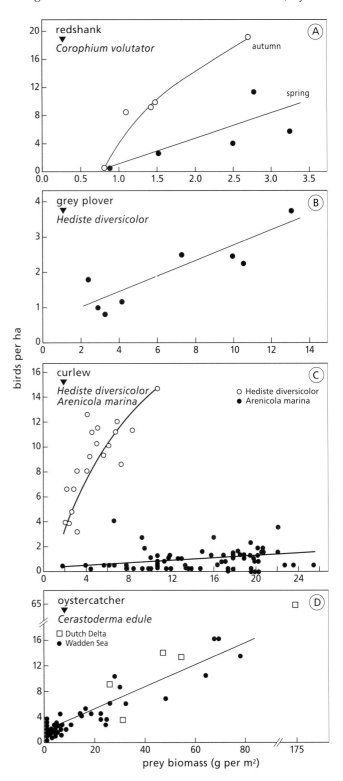

Figure 4.36 The relationship between bird density and food availability as measured in six studies: (A) redshanks foraging on amphipods (*Corophium volutator*) in autumn and spring at the Ythan Estuary in Scotland,[311] (B) grey plovers foraging on ragworms (*Hediste diversicolor*) at the Berg Estuary in South Africa,[452] (C) curlews foraging on lugworms (*Arenicola marina*) or ragworms in the Dutch Wadden Sea[257, 750] and (D) oystercatchers foraging on cockles (*Cerastoderma edule*) in the Dutch Delta[560] or the Wadden Sea.[964]

catcher mortality more than doubled when little food was available (Figure 4.35C). The birds fanned out to other places (Figures 4.35A and B) but apparently it was not much better at these other sites. The levels of winter mortality shown in Figure 4.35C are only for mild winters. In harsh winters many more birds die, 5 - 20% of the older birds and even more of the young birds. The large variation in annual mortality appears to be related to both food and winter temperatures.[76, 124, 995] Ringing data combined with year-round measurements of bird density and food availability provide the information needed to determine the carrying capacity of intertidal areas for birds (Chapter 6).

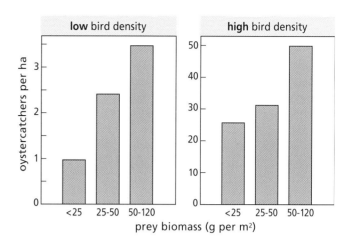

Figure 4.37 How do oystercatchers distribute themselves across a mudflat where the peppery furrow shell (*Scrobicularia plana*) density varies? Most birds are concentrated in the richest part of the feeding area. At low bird densities almost no birds forage in the poor area, but at high densities the marginal part of the feeding area is visited relatively more. Based on low tide counts along the Frisian coast in August.[964]

*How birds react to spatial variation in food availability*

The relationship between bird numbers and the availability of their food can be determined by comparing many years of food availability data with bird counts from the same site. The relationship between the density of the birds and their prey can also be investigated by trying to relate the bird distribution across the mudflats with the spatial distribution of their food at a certain time of the year. You then find exactly what you would expect: birds are most abundant in the areas where they can find the most food (Figure 4.36).[100, 254, 257, 311, 329, 331, 339, 452, 561, 686, 750, 964, 967, 986] All you have to do to find this relationship is mark out the mudflats with stakes, count the birds inside the marked area and measure the density of prey being eaten. The data you end up with lead to many questions, the most important of which are why aren't the birds only present in the richest feeding areas, and why do some birds keep foraging on parts of the tidal flat with little food where their intake rate is very low? Before we try to answer these questions we need to remember that data for all of these studies were collected in the field, which causes a number of problems with methodology.

On tidal flats the areas that provide the most food are often situated low in the tidal zone. When these areas are under water during the incoming and outgoing tides, the birds will forage higher on the tidal flat where there is less food out of sheer necessity. When counts taken some time either side of low tide are used to calculate the average bird density this causes some bias. This problem can be solved by calculating the bird density from low tide counts made when all the areas being counted are exposed. But estimating total predation pressure from low tide counts can underestimate the importance of the areas lying higher in the tidal zone. Plus different bird species follow the tide line and leave the tidal flats when they become too dry at low tide. In such a dynamic system, how you determine the numerical response can become a bit arbitrary. The problem can be solved by relating the bird counts to the density of the active or visible prey rather than the total prey density.

There's also another problem encountered in the field. Free-living birds often don't limit themselves to just one prey species. This makes it more difficult to compare bird and prey densities. In fact, for all of your monitoring sites you need to know not only the bird and prey densities, but also how much each type of prey contributes to the birds' diet. You'd expect that if the density of one prey type is low, the diet will comprise a higher proportion of alternative prey species. This was found in red knots that foraged on Baltic tellins. When there were only low densities of Baltic tellins, the knots switched to cockles.[686]

Figure 4.36C shows the numerical response of curlews to ragworm and lugworm abundance. This study covered an area of 10 ha divided into counting and sampling sites of 1000 m². Much of the area was a sand flat where only lugworms were present, but in a lower more silty area the lugworms were replaced by ragworms. If it weren't known that in the plots with low lugworm densities the birds were foraging on other prey, the curlew densities in these plots would have seemed unusually high. Ideally, for the other graphs in Figure 4.36 you should know to what extent other prey species are eaten, especially at low densities of the graphed prey. Currently, the graphs give the impression that birds continue foraging on certain prey even when that prey is at zero density. In reality, a prey species must occur in a certain density before the birds will forage on it. A solution to this problem is to plot bird density against the biomass of different prey species and to determine the extent to which the distribution of the bird density can also be explained with variation in the densities of the alternative prey species.

But we still don't know why all the birds aren't foraging on the best feeding sites where they would have the highest intake rates. Maybe the birds don't know exactly where the good and bad areas are, so a numerical response only develops because the birds stay around longer on good spots than on the bad ones. If food availability varied on the scale of tens of centimetres one would excuse the birds for not being able to learn the exact position of good and bad spots. However at the

When waders and gulls disperse across the tidal flat at low tide to forage, they have an average density of about 10 birds per ha. The density varies greatly from place to place, depending on the food resources present. On and around mussel beds the density can be up to 70 birds per ha. On mussel beds, only oystercatchers actually harvest the mussels. Common and herring gulls live off the mussels they steal from the oystercatchers. Curlews and redshanks search for common shore crabs amongst the mussels. Black-headed gulls and greenshank feed on shrimps in the shallow pools and channels. Bar-tailed godwits and dunlins search for ragworms in the soft silt around the mussel beds.

scale of hundreds of metres, the birds should be able to learn where the good and bad areas are.

We expect all birds to search for a foraging site that will give them the highest intake rate. But if all the birds went to the same site, they would get in each other's way. They may need to avoid neighbouring birds to reduce their chances of being robbed, as is the case for oystercatchers feeding on mussels (Figure 4.33; Section 4.7.1). Even when birds don't try to steal from each other, they can still have other negative impacts on their neighbour's intake rates. Prey may hide in their burrows because another bird has just wandered past. This can be a problem for redshanks feeding on amphipods (Figure 4.18) and for whimbrels pursuing fiddler crabs (Figure 4.16) and there are probably many other examples. You would expect higher densities of foraging birds to cause a lower foraging success rate for any birds that forage on prey that withdraw when disturbed, for example black-headed gulls that eat the siphons of peppery furrow shells. These shellfish stretch their siphons over the sediment to feed, but quickly retract them when they are disturbed, perhaps by feeling a slight vibration on the tidal flat. So black-headed gulls that feed on the siphon tips are less successful if their conspecifics have just walked past.[580]

On one hand you expect that the birds will go to wherever their intake rate will be highest. On the other hand you expect that a high bird density leads to a lower intake rate, which will cause the birds to spread out to areas where there are fewer conspecifics. These two effects work against each other. When all birds are equal and they all make the right decision about where they should forage, the intake rate will be independent of the food density and most, but not all birds, will forage on the richest part of the feeding area.[285, 286] When this is the case, you would predict that the birds will distribute themselves differently depending on whether there are only a few birds or many birds. When there are few birds, they will mainly forage in the richest parts of the feeding grounds and when there are more, their numbers will mainly increase in the marginal areas. This is exactly what has been found for curlews[257] and oystercatchers.[343, 564, 964] Figure 4.37 shows how oystercatchers distributed themselves over an area of 7 ha when there were either only a few birds foraging there (less than 100) or many (200 - 300). The birds were counted in 1000 m² plots and there were large differences in the density of their two main prey, peppery furrow shells and cockles, between the plots. At high bird densities the oystercatchers apparently began to select for plots where cockles were scarce. When there were few birds, the foraging densities in the rich plots were more than three times higher than those in the poor plots, but when there were a lot of birds this difference was more than halved. In other words, in rich plots at a high bird densities there were 13 times more birds than at low bird densities, but in poor plots there were 24 times more birds at high than at low bird densities. To really understand why the numerical response differs depending on whether there are many or a few birds we need to determine how the numerical response arises, which is what we'll do in the following section.

### 4.7.3 Are good spots visited more often and/or for a longer period?

Average bird density has been plotted against prey density in Figures 4.36 and 4.37. The average bird density is caused by several processes: birds come to an area in certain numbers, they stay a certain time and they return at a certain frequency. One way of analysing this is return to the rough count data and look at the frequency distribution of the number of birds counted. There are two very large data sets available for this. In the first, birds were counted in 13 different 1 ha plots on and near mussel beds every 15 minutes during the whole low tide period. This was done 7316 times over three years (1971 -

As the tide recedes, many waders temporarily forage on the upper tidal flats. There, many birds feed in close proximity: bar-tailed godwits (upper photo); spotted redshank, bar-tailed godwit, curlew sandpiper, dunlin and grey plover (lower photo).

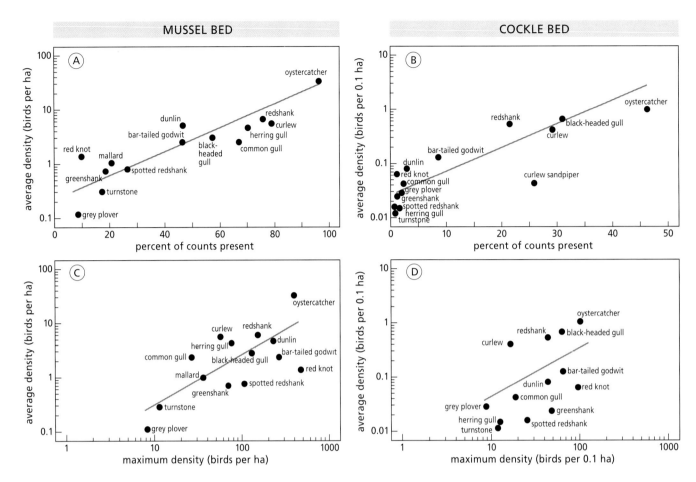

Figure 4.38 The relationship between the average density and frequency of occurrence for a variety of bird species feeding on mussel (*Mytilus edulis*) beds (upper, left) and cockle (*Cerastoderma edule*) beds in The Netherlands (upper, right). The lower panels show the relationship between the average density and the maximum density recorded for those species. Data were collected from mussel banks south of Schiermonnikoog from 1971 - 1973 and on cockle beds along the Frisian coast in 1980 and 1981. Birds were counted in 1 ha (mussel) or 0.1 ha (cockle) plots from observation towers during late summer over many entire low tide periods, every 15 minutes on mussel beds and every 30 minutes on cockle beds.[960]

1973). The second data set is based on bird counts made every 30 minutes in 521 plots of 1000 m² spread out over six years. A total of 65 800 counts made over two years (1980 and 1981) between July and September were selected from this second data set. Most of these counting plots were identified as cockle beds.

On purely mathematical grounds you would expect that the bird species that are often present will also have a high average density. This was the case for oystercatchers, the most dominant species on both the mussel and the cockle beds (Figure 4.38A and B). They were the most numerous species and were also present most often. The relationship between frequency of presence and density also applied to redshanks, black-headed gulls, and curlews, to a somewhat lesser extent. But there were striking differences between the species, for example, both knots and grey plovers were observed in less than 10% of all mussel bed counts, but the average density of knots was 12 times that of the grey plovers. Figures 4.38C and D, which show the average density of each species compared to its highest observed density, also show striking differences between species. Here you would also expect a positive relationship on purely statistical grounds, and indeed one is present. But while the average density of red knots foraging on the shellfish beds was just one twentieth of the average density of oystercatchers, the observed maximum densities of both species were very similar.

You can tell from Figure 4.38 the extent to which each bird species foraged in flocks. The red knot was clearly the most social bird, followed by the bar-tailed godwit, dunlin, and the three 'shanks': greenshank, spotted redshank and redshank. The curlew is the least inclined to forage with its conspecifics followed by the herring and common gulls, grey plover, and turnstone. This might not surprise any bird watchers, but perhaps the high densities that the different species can reach would. In late summer an average of 66 birds per ha foraged on the mussel beds. The densities on the cockles beds were less than half this, at 30 birds per ha. These are extremely high densities when you consider that across the whole intertidal area an average of around 10 birds per ha is expected in late summer. When you consider that the mussel beds plus the surrounding area of mussel silt make up just 5% of the tidal flat, you can calculate that the remaining birds forage on the rest of the tidal flat at a density of 7 birds per ha or less. This is just one tenth of the average bird density on and around the mussel beds.

If you keep counting the birds for long enough, you

will eventually find 500 - 600 birds foraging on 1 ha or 100 birds on 0.1 ha of a rich cockle or mussel bed. Red knots reach their highest densities just when the flock has landed and begins to forage. However, within 5 - 10 minutes these birds have already fanned out and their density is less than 100 - 200 birds per ha. Black-headed gulls, spotted redshanks and greenshanks also forage in very high densities and often remain in close flocks for a long time. They seek each other out because hunting together increases their foraging success. This behaviour, called social foraging, is described in Section 4.7.6.

*Birds that forage in flocks*
Different bird species differ in how often they return to the same site, and whether they do so as part of a flock (Figure 4.38). Do the birds reach a higher average density on the richer parts of the feeding grounds by returning to these sites more often, or by more individuals being present. Few relevant data are available. For red knots there was no relationship between the average number of birds present in 0.1 ha plots and food availability. However food availability had an obvious effect on the frequency in which the knots were seen. In plots where densities of their preferred prey were low (less than 5 g per m$^2$), knots only stayed for short periods before moving on, and did not return. In plots with more food (twice that of the low-quality plots), birds were present in half of the low water periods.[986] So the numerical response of these knots was due to the birds staying on the good spots longer and returning to these sites in subsequent days. This is what you would expect to find when birds can't immediately tell how much food is available. Knots must probe the sediment with their bills to find out how much food is around. They leave areas with little food, but continue foraging when they find a good spot. The fact that they are observed on the same good spots for days on end suggests that they can remember where on the vast mudflats these good spots are.

Even if the birds weren't fully aware of where the good and bad foraging areas were, you would still expect more birds to forage on the richer parts of the feeding ground. When a reasonable part of the foraging time is spent handling prey, each bird will spend longer in places where many prey are found. The relationship between prey and bird densities is further strengthened if birds consider that finding a prey animal signals that this is probably a good site and search for more prey near where the first animal was found. It has been known for 30 years that crows and thrushes can adapt their search path in this way.[791, 792, 860] But whether waterbirds do this has hardly been investigated. That is a pity since the search paths followed by individual birds along the silty mud can often be clearly seen. It's rather a muddy job, but you can even make an exact image of this by putting a large piece of clear plastic over one such search path and tracing over all the marks that were left (Figure 4.39).

In some ways, the world is quite simple for knots that eat Baltic tellins. When a knot walks across a tidal flat, the most a Baltic tellin will do is retract its siphon and firmly shut its valves. But this is no protection against the knot. When a Baltic tellin isn't buried too deeply to be pulled out of the sediment and isn't too big to be swallowed, the knot can eat it shell and all. So it doesn't matter to a knot if one of its conspecifics has just walked over the same spot and, since they don't bother each other, knots can forage in high densities. The situation is very different for birds that hunt prey which retreat to their burrows when disturbed, putting the prey temporarily out of the birds' reach. These birds must figure out how to avoid their conspecifics while still being able to forage on the richest part of the feeding grounds where, naturally enough, their conspecifics want to be too. Some individual birds solve this problem by occupying a specific part of the tidal flat and chasing the other birds away. The next section deals with these territorial birds. However most birds on the tidal flats forage in loose groups and simply try to make the best of it, as curlews do when feeding on ragworms.

To investigate whether curlews that ate ragworms foraged longer on rich spots, a large number of 5 m x 5 m adjacent plots were set out on a tidal flat close to a high observation tower. For each curlew that walked into a plot, researchers recorded how long the bird stayed and how many prey it ate while there. After the researchers had spent some weeks collecting these data from the observation tower, they accurately determined the prey density in each plot. Prey densities varied from 40 - 140 worms per m$^2$, but to the surprise of the researcher, the lengths of the curlews' visits to the plots were independ-

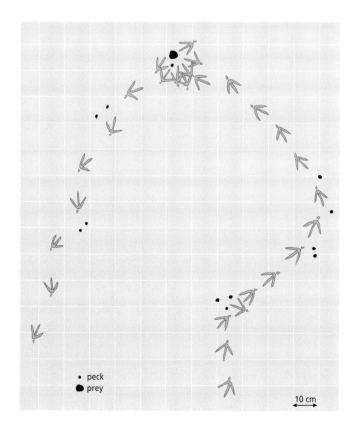

Figure 4.39 The search path of a curlew looking for ragworms (*Hediste diversicolor*) on a silty tidal flat. As well as footprints, the mud also retained imprints of the bird's bill tip where it had pecked or probed. Only one ragworm was caught, as could be seen from the deep crater that was left.[73]

Knots are the most social shorebirds. They are almost always found in large dense flocks, even while foraging. This may help them find areas where food is abundant more quickly. Large flocks also offer better protection against predators.

ent of the prey density. A curlew foraged for 50 seconds on average in a 25 m² plot.[257] When the researchers monitored the curlews' visits to 500 m² plots, there was still no relationship between length of stay and ragworm density. The birds only foraged for 30 minutes on average in these larger plots. However the lack of a relationship between the prey density and the length of stay could be explained retrospectively. The intake rate for these birds was not related to prey density (Figure 4.26H). So for the birds all of the plots were equally good and there was no reason to stay longer in a rich plot. But the curlews did show a clear numerical response (Figure 4.36C). This response was not caused by spending more time on good spots during any particular visit. It was caused by returning to the good spots more often. On average, rich and poor areas had eight and four visits respectively from individual curlew per low water period.

At first glance it seems strange that the curlews returned to plots with high prey densities more often, even though these plots didn't give them a higher intake rate. Maybe the curlews' intake rates would have been related to the prey density, if it wasn't for their nearby conspecifics disturbing the prey. But did the curlews get in each other's way more in the rich plots than in the poor less-visited plots? To determine this, the researchers regularly measured the distance to the nearest conspecific. They found that the curlews did have a lower intake rate on average, when the distance to their nearest neighbouring curlew was small. When conspecifics have a negative effect on intake rate, this effect is greater in food-rich plots where many birds are foraging. This may explain why the intake rate of these curlews was not related to the ragworm density. This is also what you would expect to find theoretically.[285, 286] Ragworms retreat to their burrows at even the slightest vibration on the sediment, and only resurface some time later. So you would certainly expect a negative relationship between bird density and foraging success. Maybe this also explains why individual curlews that forage on ragworms don't have longer visits in places with a high prey density. Their presence causes the local foraging conditions to become poor.

Even if there are many prey, when these prey are all hiding deep in their burrows, it's better for the curlew to keep walking and come back later when there are less birds. You can see the same thing happening right in front of you when you watch whimbrels hunt for fiddler crabs. The crabs run back to their burrows when a bird approaches and only return to the mud surface somewhat later.

*Territorial birds*
Whimbrels that feed on fiddler crabs don't have problems with their conspecifics. They ensure that while they're foraging they will not be bothered by others. All conspecifics are systematically chased away and any whimbrel that dares fly past will be greeted with an angry "bibibi" call that lets it know that another whimbrel already has a feeding territory there. Some individuals of other wader species also defend feeding territories on the tidal flats. When sanderlings, whimbrels, curlews, and grey plovers were individually marked, it was found that the same birds defended exactly the same areas for months, and sometimes even returned to the same areas for years.[239, 589, 867, 969] Redshanks may also defend feeding territories.[312] The fact that bossy territory holders demand part of the tidal flat for themselves has many consequences for the way that bird densities are related to food availability. We will return to this later, as first we will discuss observations about how territorial birds use feeding areas. Again, we want to know whether the birds stay longer on better foraging sites and whether they come back to these sites more often. This is easier to investigate in territorial birds rather than birds that forage in groups, because there are no conspecifics nearby to negatively influence the intake rate. What is most interesting about territorial waterbirds is how they react to short-term variation in prey availability.

Pied wagtails that defended feeding territories along river banks had low foraging success when they returned to a spot where they had foraged the day before. They were feeding on washed-up insects. As a territory holder foraged, it ate all the prey it could find along the water's edge in that area. It only makes sense to return to a site once the river has brought in new food. In a situation like this, where the food source renews itself continuously, a territory holder can increase its intake rate by not returning to places that have only recently been picked clean.[164, 167] Avoiding these recently visited areas is called systematic foraging. Do some waterbirds also deal with a self-renewing food source, and if so, are these birds particularly territorial?

Fiddler crabs, amphipods and ragworms all live in the sediment, come to the surface to feed, and retreat to their burrows when danger threatens. When a bird forages nearby, all of these prey temporarily return to their burrows, and the bird must wait for at least a few minutes before it can try foraging again on the same spot. So this food source renews itself very quickly.

Many benthic animals live hidden in the sediment but betray their presence when they forage. Worms and amphipods crawl up to the surface and shellfish stick out their siphons. These activities create little holes on the surface that the birds can see. How long these signs remain visible has been investigated for sand gapers.[237] Most siphon holes can be seen just as the tidal flat is exposed by the receding tide. After that, they become less and less obvious, and 4 hours later almost all of the siphon holes have collapsed. So here we are dealing with a food source that renews itself every low tide. Curlews walk from one siphon hole to another and probe their bills down the holes. They can probably see how large the buried sand gapers are, since large sand gapers have thick siphons and so also make bigger siphon holes.[980] It's likely that the siphon holes of large sand gapers remain visible longer than those of the smaller individuals, as the smaller siphon holes would fill up with sediment quicker than large ones. The prey selection of the visually-hunting curlews supports this theory. Over the course of the low tide period, curlews ate fewer small sand gapers.[73]

How do curlews react to this changing sand gaper availability? You'd expect them to forage systematically and not visit an area more than once within the same low tide period. Curlews do stay longer in areas where they have found a sand gaper,[73] and they don't return to

Curlews that winter in Northwestern Europe find plenty to eat on tidal flats in late summer, such as common shore crabs (upper photo). In the winter the crabs have left the tidal flats, and the curlews run the risk that (part of) the tidal flat may freeze over (middle photo). Yet all of the curlews that breed in Western Europe winter on the European mudflats. Curlews that winter in tropical Africa can always find enough crabs around the mangroves. Those birds come from a breeding area that stretches from Eastern Europe deep into Siberia. Note the extra-long bill that is characteristic of these more easterly curlews (lower photo).

# FOOD

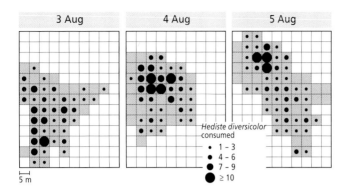

Figure 4.40 The feeding territory of a curlew, male 'G', was marked out with numerous stakes that allowed researchers to divide his territory into hundreds of 5 m x 5 m plots. He was observed continuously from a nearby observation tower for an entire low tide feeding period. During this period, the time he spent in each plot and the number of prey he ate there could be recorded. This figure shows how many prey he ate over low tide on three successive days. During the study he fed exclusively on ragworms (*Hediste diversicolor*). This research was conducted on the Frisian coast of the Dutch Wadden Sea.[140]

these sites more often. However they don't have less success in an area that has already been visited.[73, 237, 239] This is because the curlews don't give up easily. When they can no longer see the siphon holes, they start to search for their prey by touch. This means we aren't dealing with a self-renewing food source, and all of our grand theories don't apply.

We run into the same problem when curlews feed on ragworms. The curlews change their search strategy to suit the ragworms' foraging behaviour.[259, 978] When ragworms are out on the surface grazing around their burrows, curlews walk by with a measured pace. The curlews are dealing with prey that retreat to their burrows when the bird approaches, and resurface a little while later. How long the ragworms stay away isn't known. But when they remain in their burrows to feed, the curlews slowly walk along searching for burrow entrances and other signs on the surface. Whether these signs disappear during the low tide period, like the sand gaper's siphon holes, is unknown but seems unlikely. Ragworms actively forage throughout the low tide period, so they are continually rebuilding their entrance holes. How curlews exploit this food source has been researched extensively.[140, 237, 257] Since ragworms that forage on the sediment surface retreat to their burrows when disturbed, we'd expect the curlews to be less successful if they return to an area that they had recently walked through. Sure enough, it was found that a second visit within 5 minutes of the first yielded less. However this relationship could only be shown once and was also often not found. Perhaps this isn't so strange. You would expect a lower intake rate if the curlews quickly returned to same spot when they were hunting surface-grazing ragworms, but not when they are searching for clues on the mud.

The research on curlews foraging on ragworms and sand gapers yielded some unexpected results. Each bird foraged in a different part of its territory every day. This can also be considered as systematic foraging, as it reduces the bird's chance of foraging in areas that have already been depleted. Figure 4.40 shows the areas where one curlew, called Male G, ate ragworms on three consecutive days within his territory. It's obvious that he foraged in a different area each day. Male G defended a very small territory of just 1100 m². Most curlew territories are larger, some cover more than 1 ha (10 000 m²).[140, 239]

Birds don't only use different parts of their territories from day to day; they can also use different areas from month to month. This was nicely shown by many weeks of data collected over three years on a female curlew called 20Y. This female almost exclusively fed on sand gapers within her territory. When observations collected through a whole season were compiled, it became apparent that 20Y slowly foraged less in the northwest and more in the southeast parts of her territory over time. There was no relationship between the amount of time spent in the different parts of the foraging area and the prey availability on each day. This isn't that unexpected since 20Y only used a limited part of her available foraging area each day. But there is also no relationship when all 31 observation days from July to December are combined. During this time, 20Y foraged from 3 - 9 hours in areas of 1000 m², but the variation in time spent wasn't related to an area's food availability. Few other individual birds have been studied as intensively as 20Y was; in one year alone she was watched for nearly 200 hours. Maybe a relationship would have been found if the feeding areas used by 20Y could have been recorded continuously. The researchers' enormous time investment did provide them with a beautiful functional response. 20Y's intake rate within her territory varied from 1 - 2.5 mg per second depending on where she foraged. This variation could be

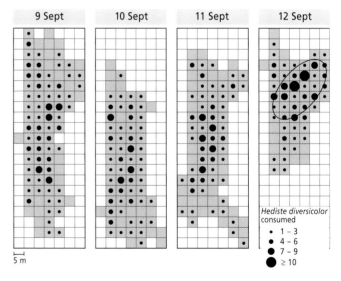

Figure 4.41 Similar data to Figure 4.40, but now for G's neighbour, male 'F', who also ate only ragworms (*Hediste diversicolor*). This male's territory was twice as large as male G's, and he searched twice the area his neighbour did each day. Bird 'F' was followed over four consecutive days. On the fourth day the bird's foraging shifted to the northern part of his territory, where he caught many prey. This was because the researchers had spread a delicious shellfish soup on the sediment that lured the ragworms out of their burrows.[140, 978]

explained by sand gaper availability, which was between 5 and 40 g per m² at the beginning of the season.

We still don't know exactly why birds forage at one site one day and another site the next. The way that 20Y slowly moved southeast through the month suggests that her movements were rather arbitrary. Maybe they were, but this doesn't mean that territorial birds aren't skilled at using subtle variations in local food availability to their advantage. There are two clear indications that this is the case. A male curlew called male F had been foraging for weeks in his 2000 m² territory under the observation tower. He was then followed continuously for three days. On the fourth day the researchers boiled up a soup of peppery furrow shells. Then, on the outgoing tide just before the curlews arrived, a 30 m long garden hose was laid across the tidal flat. When the tidal flat was exposed, 40 litres of shellfish soup were pumped from the observation tower through the hose, and spread across the tidal flats.[978] The researchers knew from laboratory experiments that ragworms would then come out of their burrows en masse to feed.[259] Male F quickly noticed that something unusual was happening in part of his territory, and caught all of his prey for that tide from the soupy area (Figure 4.41).

Female 20Y's prey choice showed that she too was aware of how to maximise her intake rate. A small gully ran through her territory. She rarely foraged there but thought it made an ideal place to preen. However on some days she did feed on shrimps in the gully. When 20Y fed on sand gapers on the tidal flat her intake rate varied daily between 2 and 2.5 mg per second. When she ate shrimps the daily variation was much higher, from almost zero to 4 mg per second. Figure 4.42 shows that she only left the tidal flat and foraged in the gully on the days when the shrimps yielded more than sand gapers.

### 4.7.4 Landlords and free birds

Claiming part of a tidal feeding ground for themselves is a widespread strategy amongst waders. Territorial behaviour on the feeding grounds has been observed in two of every three wader species found in America,[591] and has also been observed in other European and African species. No species are consistently territorial. Territorial behaviour tends to be shown by only some individuals. Redshanks that foraged on the open tidal flats of the Ythan estuary in Scotland didn't appear to have feeding territories, but individuals that foraged in small bays, along channels, narrow silty tidal margins, and near the edge of a salt marsh did.[312] The same was found in grey plovers in the Tees estuary in Northeast England.[868] The thousands of avocets that winter in the Tagus estuary in Portugal have also been well studied.[75] These avocets forage in groups on large mud flats, but divide the upper tidal flat, comprising a small-scale mosaic of small channels and salt marshes, into a number of adjacent territories. Such observations suggest that waterbirds only defend a territory when the local terrain provides suitable landmarks to demarcate the area. But we shouldn't let this fool us. Some curlews defend territories day after day

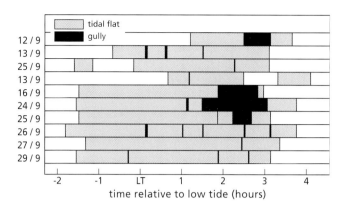

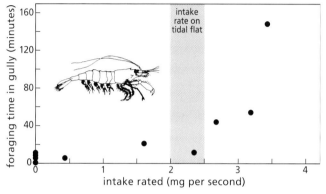

Figure 4.42 The amount of time that a territorial female curlew, 20Y, spent in a tidal flat gully changed daily. When she ate sand gapers (*Mya arenaria*) on the tidal flat her intake rate was rather constant at 2 - 2.5 mg per second (grey panel). In the gully she ate shrimps (*Crangon crangon*) and had a highly variable intake rate. 20Y only continued foraging in the gully when it provided a higher intake rate than the tidal flat.[237]

within vast tidal flats where the only structures are millions of toothpaste squiggles made by lugworms, yet the borders of their territories do not change.[750]

Are there any species that never establish feeding territories on tidal flats? Of all the waders, you would expect the red knot to be the most likely contender. The red knot is the most social wader species and almost always forages in tight flocks. But even in this species, territorial behaviour has been observed on a very large tidal flat.[77] After birds foraging around a viewing tower had been observed for weeks in a row, a large team of researchers measured the food availability during a few low tide periods. The researchers repeatedly walked between the various sampling sites and a small channel where the samples were sieved. The sediment was very soft and the researchers often sank up to their ankles, and sometimes even over their calves. A few days later, when the sampling was completed and observations from the viewing tower resumed, it was still possible to see where the people had been walking. To everyone's surprise almost all of the knots began to feed on Baltic tellins in the trampled areas. This was in August and most of their prey were already buried out of the knots' reach.[986] The footprints gave the knots a unique opportunity to find Baltic tellins. The depths of the Baltic tellins inside and outside the trampled areas had not been measured, but the only place on the mudflat that Baltic tellins could be found at the surface was in the trampled area. So we can assume that there were many tellins available in the trampled area. The researchers were surprised that the knots had discovered this opportunity, but an even bigger surprise was in store: the knots defended the footprints against their conspecifics. A few days later the footprints had gone and the knots foraged in flocks again like the old days.

The choice of whether to defend a territory or not is related to the prey availability. There is little point in acquiring an area where food is scarce, so there are no territories in poor sites. Conversely, there is usually no point in trying to make the richest part of the feeding grounds into a territory, as such a site would have too many intruders. So territories are not found in the poorest or the richest parts of the feeding grounds. This is the case both for curlews in the Wadden Sea[239] and sanderlings in California.[591] For the sanderlings, however, there were also some years when not a single bird maintained a territory.[589] This was because a pair of merlins were periodically in the area, forcing the birds to feed in a flock. It's likely that territorial behaviour would have put individual birds at greater risk of being caught by hunting merlins.

There are costs and benefits associated with defending, or not defending, a territory. One obvious cost to being territorial is the time spent chasing away intruders. Territorial curlews usually spend around 5% of their total foraging period chasing other curlews away. Many territories border each other. Birds in these territories have relatively few problems with intruders, because their neighbours have often already chased the intruders away. Instead, they must spend a lot of time defending their territories against their neighbours. These border incidents can be quite time-consuming and sometimes last for 15 minutes or more. There is a lot at stake for the curlews at this time, as if they don't put up enough resistance they slowly lose parts of their territory. Every day, on the apparently endless tidal flats, the curlews must go to war for a few square metres of ground.

So far, we've only discussed the short-term advantage of having a territory: a higher intake rate can be achieved in the absence of conspecifics. There is also a long-term advantage. Territorial birds can reserve more food for themselves. This only happens when the foraging density inside a territory is lower than outside of it (in the 'free area'), which usually seems to be the case. When both territorial and 'free' curlews fed on ragworms, curlew densities were higher in areas with higher prey densities. But the foraging densities within the territories were clearly lower than in the free area (Figure 4.43). Ragworm availability within the territories was always between 4 - 10 g per m². So within the territories food was neither too scarce, nor so abundant that the territory couldn't be defended against intruders. In territorial areas there were an average of two curlews foraging per ha. In 'free' areas with 4 -10 g ragworms per m², there were twice as many curlews.[237] There may be another long-term advantage to having a territory. A territorial bird can save the best

Aggression between foraging grey plovers (photo page 222), oyster-catchers (upper photo) and turnstones (lower photo). The grey plovers are two territorial birds having a border dispute. The left-hand oystercatcher is only pretending to be sleeping. The raised neck feathers tell the observer (and the other oystercatcher) that it will not take much for the oystercatcher to explode into action. The two turnstones are fighting over a large item of food.

feeding sites in its territory for bad days. Observations of a territorial grey plover suggested that in normal conditions the bird avoided the parts of its territory which would provide the highest intake rate, so that it could spend more time in these good spots on very cold days when it was more difficult to feed.[197]

These observations reinforce the fact that when some birds are territorial and others forage in flocks we shouldn't expect to find clear relationships between the foraging density of the birds and prey density. Viewed in this light, territorial behaviour complicates the work of researchers trying to relate bird densities to food availability. On the other hand, if we are trying to ascertain how many birds can forage on the mudflat, it is easier to determine this for territorial birds than for birds that forage in flocks. The real question is what happens when the number of waterbirds increases. Will the territories become smaller or will the extra birds end up in the undefended areas?

### 4.7.5 Competition between species

Some birds find it a nuisance to have conspecifics feeding nearby. Having other birds around can cause problems if prey retreat to burrows when disturbed, as is the case for black-headed gulls plucking peppery furrow shell siphons, redshanks feeding on crawling amphipods, curlews hunting surface-grazing ragworms, whimbrels chasing fiddler crabs, and motionless grey plovers

waiting for a ragworm to surface. It doesn't really matter if the disturbance was caused by a conspecific or a member of another species. So if a redshank is feeding on amphipods, a passing black-headed gull is just as a much of a nuisance as another redshank. In which case, we'd expect some birds to try and avoid both their conspecifics and members of other waterbird species. Perhaps this is why grey plovers often stand all alone on tidal flats. As the tide recedes, many birds follow it out or fly to feeding grounds that will soon be exposed, leaving the upper tidal areas practically empty. Grey plovers often remain behind in these upper tidal areas; finally they have their kingdom to themselves.

There is another reason for different bird species to

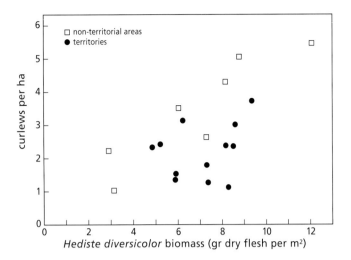

Figure 4.43 When curlews feed on ragworms (*Hediste diversicolor*), they reach their highest densities in areas where their prey are numerous. When a distinction is made between areas defended by territorial curlews and 'free areas' that are not claimed by birds, the bird density in territories is systematically lower than in the free areas. The curlews were counted regularly during late summer and autumn in a feeding area of 15 ha that was divided in counting plots of 0.1 ha. The food availability was also measured in all 150 plots. Since the feeding territories were known exactly, it was possible to determine the density of the curlews and their prey for all territories and free areas. All observations were from the Frisian coast of the Dutch Wadden Sea.[237]

try to keep out of each other's way on the tidal flats. Once prey are caught, birds often have to hand over their hard-won meals to other birds, both conspecifics and birds of other species (Section 4.6.4). Gulls often steal from waders. It's therefore conceivable that when birds have a choice between two more or less equal feeding areas, one with more gulls than the other, they'll choose the area with less thieves.

It's difficult to show that the bird distribution over the mudflats is related to the presence of other bird species on the basis of field measurements. To do so requires many additional measurements to make sure that any differences in distribution are not actually being caused by some other factor. For example, the distribution of two bird species may be closely related to the height of the water level that varies daily. If one species forages in shallow water and the other doesn't, systematic counts on the feeding grounds could suggest that they are trying to avoid each other. In this case, the prudent step would be to search for a relationship between water level and foraging success for each species. In another example, when bar-tailed godwits feed exclusively on lugworms and redshanks exclusively on amphipods the chance of an encounter between the two is almost nil. This is nothing to do with the birds themselves. It's because the amphipods can't settle in areas where lugworms live, so the two prey species never co-occur. These are obvious examples, and hopefully no-one would think there was active competition between these bird species. The situation changes when different parts of the feeding grounds have similar food availability, and we find that at low tide at the same water level one species is increasing uncommon when another bird species is very abundant. This is what low tide birds counts on mudflats in the Haringvliet, in the southwest of The Netherlands showed (back when the Southwestern Netherlands still had extensive brackish tidal areas). Here, the avocets, curlews and black-headed gulls all fed solely on ragworms. The researchers divided the area into quadrats and regularly counted the birds within the different quadrats. So each quadrat contained a certain percentage of all the birds counted of each species. That percentage was found to be lower if many birds of other species were also found in the quadrat. The clearest relationship was between the curlews and the black-headed gulls. The counts suggested that the curlews tried to avoid the gulls.[964] This is exactly what you would expect given the black-headed gulls' tendency to steal food from curlews.

Gathered prey can only be stolen or available prey disturbed if the birds are in the same place at the same time. You can calculate the likelihood of this by analysing the tens of thousands of plot counts that formed the basis for Figure 4.38. The chance of two non-conspecifics encountering each other is extremely small. One species may be more abundant in July (the black-headed gull), the other in August (the greenshank). One species usually forages along the water's edge as the tide retreats (the greenshank), the other mainly on the incoming tide (the black-headed gull). One species likes to forage on mussel beds (the oystercatcher), the other species prefers the silt between the ridges of mussels and outside the mussel bed (the dunlin). To calculate the odds that two birds belonging to different species will meet in the same 10 000 m² plot, the probabilities of these different events must be multiplied by each other. The end result is a minuscule number, with many zeros after the decimal point. So we can conclude that the different bird species on the tidal flats seldom get in each others' way.

Potential competition between species is often determined by comparing their choice of food. It's then obvious that there is little overlap for many species. However this is not always the case for waterbirds. Ragworms are important prey for many different waders, both large birds like the curlew, oystercatcher and bar-tailed godwit, and small birds like the dunlin, redshank, and grey plover.[104, 110, 258, 266, 316, 317, 868, 953, 978] On tropical tidal flats most waders also eat the same prey species, the fiddler crab.[965] Every year in the Bay of Fundy, an important stopover area for millions of Canadian waders on autumn migration, amphipods are the main prey for various species.[88, 89, 350, 399]

So there certainly appears to be some overlap in prey choice between the different waterbird species. However the picture changes when we consider the size of prey selected by the different species. The Baltic tellins eaten by red knots are too small for the oystercatchers.[977] The second-year common shore crabs that curlews feed on would be much too big for redshanks, and there are many other similar examples. But of course the fact that the different species eat different-sized prey doesn't mean that there is no competition. A bird that eats small individuals deprives a bird that eats the larger ones of the option of eating those individuals at a later stage. This is what herring gulls do when they consume all the mussels of one year-class before the oystercatchers can start on them.[977] Red knots can eat all sand gapers that would be large enough for the oystercatchers in a year's time, and when the knots do leave enough for oystercatchers, they in their turn eat the sand gapers that would have been large enough for curlews to eat after another growing season (Figure 4.44). So it's a case of first come, first served. The species that can take the first slice of the available prey have an advantage over those that must wait. Larger birds tend to eat the larger prey, which puts them at a disadvantage compared to smaller bird species.

One consequence of this may be that when the predation pressure exerted by the birds on the tidal flat increases, for example if the surface area of the tidal flat declines, a species like the curlew that eats large and therefore older prey animals could be harder hit than a species like the red knot that eats smaller shellfish.

### 4.7.6 Social foraging

The previous section may have given the impression that foraging waders are only interested in avoiding each other on the tidal flats. Of course, this is not true. Sometimes they forage in groups as this makes it easier to find the good spots. There are also other advantages to foraging with friends. Avocets are widely dispersed across the tidal flat when they search for ragworms, but gather in very dense groups to chase shrimps. Why do they do this? Shrimps and gobies usually lie half-buried in the sediment, but rapidly swim away when touched. This is their only means of escape from birds, but it makes them very visible, which is why shrimps try to bury themselves again as soon as possible. A solitary greenshank hunting shrimps in a pool has to race to catch any of the escaping shrimps. Usually the greenshank fails, but if it continues to walk around, more and more shrimps are disturbed, and there can be several hundred shrimps per m² in these pools. Even though most shrimps are too quick for the greenshank, there will still be some it can catch, particularly if the greenshank keeps foraging in the same pool until the shrimps are exhausted. Escaping the bird is vitally important to a shrimp, so it swims off with greatest possible speed. Racing off causes the shrimp to suffer a short-term oxygen shortage, and it takes a while before it can repeat such a performance. As the greenshank continues walking, it becomes more likely that the shrimp will be chased again before it can repay its oxygen debt, which causes the shrimp to react very slowly. If the greenshank can repeatedly chase the same shrimps within a short time, it can increase its food intake. So when several greenshanks forage in the same pool, they help each other. This explains why different wader species that hunt shrimps or gobies do so in groups. There is one disadvantage with foraging in a group; your neighbour might catch the prey that was right under your nose. However this seems to be more than compensated for by more prey becoming exhausted and easier to catch.

Social foraging is a good way to catch fleeing prey, which explains why this behaviour is widespread in fish-eating birds like pelicans, herons, egrets, spoonbills, grebes, mergansers and cormorants. The majority of waders feed on prey that at most try to make for their burrows, which explains why few waders forage socially. Those that do in Northwestern Europe include the greenshank, spotted redshank and avocet. Greenshanks often form loose groups of single birds, seldom more than 20 or 30, that run around more or less randomly. In contrast, spotted redshanks form a tight group, which usually includes all of the birds present locally. Spotted redshanks cooperate far more than greenshanks, often forming lines with almost military precision, something

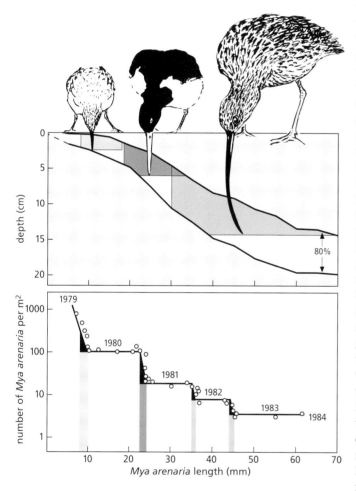

Figure 4.44 (A) The prey choice of knots, oystercatchers and curlews doesn't overlap when they eat sand gapers (*Mya arenaria*). This figure shows how deep sand gapers of different sizes bury themselves, and which portion of the potential prey are eaten by the three bird species. The bill length limits the fraction of the prey that is accessible. Since the bill lengths of the three species differ, this accessible fraction also varies. Of the accessible sand gapers, the smallest ones are not profitable (Figure 4.24) and for a knot the large ones are not ingestible. This explains the limited overlap in prey choice of the birds. (B) Over several years, the sand gapers grow, but also decrease in number. Shellfish that have survived their first months of life are suitable prey for knots. A year later they are large enough to be eaten by oystercatchers. After yet another growing season they are suitable for curlews. Over time, these three bird species successively deplete the same year-class of sand gapers.[979]

never observed in greenshanks. Avocets also forage in tight groups, before fanning out in long lines, and then regrouping in a concentrated mass - a spectacular sight! Around 6000 avocets could be seen doing just this every day in late summer on the Ventjagersplaten, before the Haringvliet in the Southwestern Netherlands was dyked. The avocets would be spread out across the tidal flat foraging on ragworms at low tide, until a few birds began to hunt shrimps in a shallow channel, when suddenly all the avocets would leave the exposed sediment to join the shrimpers. The excited calls of gathered avocets could be heard from hundreds of metres away, which must have signalled to the other avocets that it was time to leave the tidal flats and fly to the channel. Something similar can still seen in the Oostvaardersplassen, a shallow freshwater lake area in the middle of The Netherlands, where thousands of avocets gather in late summer. These birds feed on chironomid larvae in the shallow water but often switch to chameleon shrimps. Just like on the tidal flats, a few birds begin to chase the chameleon shrimps in a tight group, calling loudly, at which point all the nearby avocets abandon the chironomids to join the shrimpers. These flocks can sometimes contain thousands of birds. Other bird species sometimes join the avocets on the Oostvaardersplassen as they once did on the Ventjagersplaten. Any spotted redshanks in the vicinity always join in, and shovelers and sometimes mallards will also take part.

Socially fishing waterbirds help each other out by foraging in groups: spotted redshanks mix with little egrets and a reef heron (inland Niger Delta, Mali; upper photo) and greenshanks search with a redshank in a tidal pool between mussel beds (lower photo).

Turnstones, sanderlings and purple sandpipers find a rich food source in the debris at the boundary between the dry and the rippled sand. Foraging in groups makes it easier for them to find these rich food sources.

Social foraging is very common amongst spotted redshanks and greenshanks, but why is it never seen in redshanks? Analyses of regurgitates from redshanks have shown that shrimps are an important local prey. If you ever did see a redshank with a group of socially foraging greenshanks or spotted redshanks, you'd see how much slower the redshank is than the other two 'shank' species. It doesn't seem to pay for the redshank to try and hunt jumping shrimps. Redshanks must take advantage of prey that can be caught before they try to escape. A redshank mows back and forth with its bill or pecks at the sediment to find shrimps that are still buried when danger approaches.

Curlews sometimes also eat shrimps, and like the redshanks, they don't do this in groups. They're just not fast enough. Compared to socially foraging waders, curlews hunt shrimps very quietly and cautiously. They hunt visually, sometimes going after escaping prey, but usually only taking the individuals pecked off the sediment on their first attempt.

Finally, black-headed gulls also feed on shrimps. When they do so on foot, the gulls are spread out, just like redshanks or curlews. But the gulls also gather in tight groups to plunge into the water after shrimps. Black-headed gulls are very skilled in flight, which enables them to flush their prey out of hiding by lurking above the shrimps in a tight cluster, and repeatedly plummeting into the water.

## 4.8 Predation pressure of waterbirds on benthic animals

### 4.8.1 The average predation pressure

One curlew eats as much food as 14 small sandpipers do. To measure the predation pressure the birds exert on the tidal flats we need to relate the densities of the different species to their daily food intake. We can use the formulae from Section 4.2 to do this. When this is calculated for all of the tidal flats along the west coasts of Europe and Africa, every day the waders eat around 140 g dry flesh per ha in winter. There is a lot of variation in predation pressure from the waders, just as there is in the birds' densities (Figure 4.34). The amount consumed varies from 10 - 100 g per ha in Northwestern Europe and 50 - 400 g per ha in Southwestern Europe and Northwest Africa.[967]

An average predation pressure of 140 g per ha per day means that less than 0.014 g per $m^2$ is eaten each day (Figure 4.45A). If, for convenience, we assume that bird predation from 15 August to 15 April is on average the same as in January, then during those eight months an average of 3.4 g dry meat is consumed per $m^2$. But is this a lot?

Whether average winter predation of 3.4 g flesh per $m^2$ is high or not depends on the annual prey production. The total combined mass of all benthic animals varies greatly. On sand flats there are only a few grams dry flesh per $m^2$, on siltier tidal flats this increases to 10 - 50 g per $m^2$, and in the richer areas, which usually lie midway between the high and low tide line, there may be 100 - 300 g per $m^2$ (Chapter 1). Mussel beds may even contain 1000 g dry flesh per $m^2$. When a tidal area's biomass is calculated across all of the surface exposed at low tide much of this variation disappears. However we'd also lose many of the sampling points from Figure 4.45A, as there are many sites where we know the predation pressure from the birds but not the local food availability. Both predation and food availability were measured in eight tidal areas. This enables us to calculate how much of the available food the birds removed from these areas in winter. Predation pressure by wintering birds was around three times higher in tropical tidal areas than in temperate ones (Figure 4.45B).

There are a number of reasons to assume that waders that winter in the tropics eat a relatively higher proportion of the available food than waders in the temperate areas. Unfortunately, predation pressure is usually expressed relative to the average biomass of benthic animals. It would be better if we could compare predation pressure with the production of benthic animals, but too few data are available. If a prey species has three generations per year, its annual meat production can be several times higher than its average biomass. But when a species only produces a new generation once a decade, its annual production cannot exceed 10 - 20% of its average biomass, especially if the animals are already quite old. Large benthic animals live longer on average than smaller species. Because of this, the annual production of large benthic animals is smaller than their average biomass, while that of smaller benthic animals is larger than their average biomass. The ratio of production to biomass for all of the benthic animals on Northwest European tidal flats is about 1:1. On average, waterbirds eat around 10% of this biomass during the eight winter months, which is 10% of the annual production. In general, tropical tidal flats are dominated by small, short-lived species. The ratio of production to biomass for all of the benthic animals on a tropical mudflat hasn't been measured, but it is likely to be larger than 1:1. On tropical tidal flats, waterbirds eat 40 - 50% of the biomass during the eight winter months (Figure 4.45B), but this may be less than 20 - 30% of the annual production.

By late April, most waders have left the African tidal flats and few birds remain.[16, 195] Most of the migrants won't return until August or September. This is why the total annual predation on African tidal areas is little more than the winter predation pressure just calculated. The situation is very different on Northwest European tidal areas, where wader numbers peak in spring and late summer when the African migrants mingle with the European wintering birds. When predation pressure is calculated annually rather than just during the eight winter months, there is less of a difference between the African and European tidal areas.

There is another reason why predation pressure from waders in the tropics isn't as high as it might appear. The biomass of the benthic animals is usually determined by sieving sediment samples through a 1 mm mesh. This misses shellfish as big as 2 mm long and worms up to 4 cm (Chapter 1). The biomass of temperate areas is dominated by shellfish, but in the tropics it's dominated by worms. Using standard sampling techniques means that more benthic animals are missed in the tropics than in temperate areas. This problem gets worse, because birds wintering in Africa eat relatively more of the missed prey than those in Europe. There are more small waders in Africa during the winter that eat more small prey. For example, the little stint is one of the most common species on the African mudflats, but isn't found on European tidal flats. Plus waders in Africa feed on very small prey compared to their conspecifics in Europe. Nearly all prey of the most abundant wader on the Banc d'Arguin, the dunlin, would slip through the standard sampling sieves.[983]

### 4.8.2 Predation pressure and food availability

A graph like Figure 4.45 is easily made, but a lot of research still remains to be done before it's clear whether birds take a higher proportion of the available prey in the tropics than in Europe. There are many traps and pitfalls associated with such research. In general, taking a broad approach means that some of the most interesting information is lost. For example, it doesn't account for the waterbirds' food availability varying seasonally and annually and from place to place (Section 4.4). We know that when there is more food the birds' foraging success increases (Figure 4.26), as does the density in which they forage (Figures 4.36, 4.37, 4.43). This allows us to calculate how much food the waterbirds consume at each site and whether this predation pressure is related to variation in the food availability.

We'd expect predation pressure to be higher at sites with high prey densities. So if predation pressure is expressed as a percentage of prey density, what do we find? Redshanks eat just 4% of amphipods when they occur in a density of 1000 prey per m$^2$, but this increases to 10% when the prey density is 2500 per m$^2$. When the prey density increases further, relative predation pressure drops to around 7%.[316] The same was found in curlews feeding on lugworms.[750] In late summer and autumn, the curlews eat 5% of the prey when there are 40 - 60 lugworms per m$^2$, but only 2% of prey at 30 lugworms per m$^2$ and 3% of prey at 70 - 80 lugworms per m$^2$.

It's not surprising that relatively few prey are eaten when prey densities are low. At such sites, the birds have little or no success, and will only forage in these poor areas in low numbers. It's not immediately obvious why predation pressure should decrease at extremely high prey densities, but on closer inspection this doesn't seem so strange. A bird's intake rate doesn't increase in direct proportion to the increasing prey density, and often

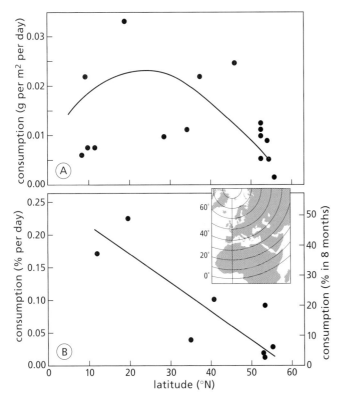

Figure 4.45 The predation pressure of waders on the tidal flats expressed as (A) the biomass per m$^2$ consumed by all the waders in the winter, and (B) their consumption as a percentage of the total biomass of all benthic animals, expressed on a daily basis as well as summed for the entire winter (15 August to 15 April).[967] Predation pressure is plotted against latitude and is based on the same data used in Figure 4.34.

doesn't rise beyond a certain point, no matter how much prey density increases (Figure 4.26). So as prey density increases, the relative predation pressure of each individual bird decreases. However an increasing number of birds visit these rich feeding sites, which is what finally determines whether the relative predation pressure continues to increase at higher prey densities or not. Unlike the redshanks feeding on amphipods and the lugworm-eating curlews, the percentage of prey consumed by curlews feeding on ragworms was constant for all prey densities.[257]

The examples we've discussed so far are of birds that use good and bad feeding grounds that lie in close proximity. It's a different story when we compare the predation pressure in good and bad years. When conditions are bad and birds cannot escape to alternative feeding areas, the total amount of food consumed will be independent of any variation in food availability, with one exception. When the food shortage is so extreme that individual birds cannot meet their daily food requirements and die of starvation, the predation pressure will decrease. When this is not the case, but food is still short, the birds will eat much of whatever food is available. This is especially the case when the food availability is considered on a large scale, for example, all tidal areas in Northwestern Europe. There are few data on this, but one dramatic example shows that waterbirds may have limited options. In 1991 there was almost no food for oystercatchers in the Western Wadden Sea, as there had been no recruitment of cockles and mussels for a few years and, to make the disaster complete, fishermen had cleaned out the last of the natural mussel beds. Shellfish biomass in 1991 reached a previously unknown low (Figure 4.14). That winter, the birds tried desperately to get enough food and began to eat prey that they don't normally eat in the winter.[62] In spite of this, there was conspicuously high oystercatcher mortality.[124] This is discussed further in Chapter 6.

When the relationship between annual predation pressure and food availability is measured on a smaller scale, we usually find that birds move into an area in good years and go elsewhere in bad years.[341, 560, 995] This results in very low relative predation pressure when very little food is available, and increasing predation pressure as the food supply increases. When there is a lot of food, relative predation pressure decreases again. There are two explanations for this. First, only mobile birds will discover that more food is available at another site. Most site faithful birds won't move en masse to another site, even if it has plenty of food. The second explanation is more theoretical. If we assume that bird numbers on the tidal flats are limited by fluctuations in food supply, then more food must be leftover in good years (Chapter 6).

### 4.8.3 Predation pressure on the exploitable biomass

The predation pressure of waterbirds on benthic animals has been measured many times. Sixteen studies were discussed in a 1984 review[320] and as more recent reviews show,[451, 559, 843] many more have been completed since.

Most studies found that the birds consumed 25 - 45% of the available food during winter. However numbers like these aren't very meaningful in isolation. The real question is to what extent the amount of food consumed leads to a reduction in the birds' intake rates, reducing their chances of surviving. When the relationship between intake rate and prey density is known (Figure 4.26), it's possible to show how much the intake rate will drop depending on what proportion of the prey has been consumed. In the Scottish Ythan estuary, the intake rate of redshanks decreased by 23% through the winter as the birds consumed 45% of the amphipods present.[319]

Each study on predation must define exactly what is meant by predation pressure. Was predation pressure calculated for all prey, or just the prey that the birds select, and if so, how were the selection criteria determined? Foraging birds are always selective. They refuse prey that are not profitable, which means they will almost always eat the larger individuals of a prey species and take the most shallowly buried of any subsurface prey (Sections 4.1 and 4.6). If you determined the predation pressure of red knots, oystercatchers and curlews on sand gapers without taking size and depth selection into account (Figure 4.44), the outcome would not be very informative. As birds repeatedly select the most profitable prey their intake rates will drop more rapidly than expected based on non-selective predation. The oystercatchers from Figure 4.44 consumed 80% of prey during the winter.[979] They ate almost 100% of the shallow prey, but only half of the most deeply buried prey.[995]

The fact that waterbirds systematically remove their pick of the prey makes it difficult to discuss the 'food availability' and 'predation pressure on the food supply'. Prey refused because of their low handling efficiency when food is plentiful may be consumed when the intake rate drops too low. Research models dreamed up behind a desk may seem accurate in a laboratory experiment, but seldom describe the real situation. Again and again it has been shown that birds can adapt to make the best of a situation. If they weren't so flexible they would have become extinct long ago. But this doesn't mean that there are no limits. Birds can only use the morphology and experience that they have at any given time: a red knot can never catch a deep-living Baltic tellin, even though the shellfish is within an oystercatcher's reach. In addition, birds always require a certain amount of food and have a limited time to forage, so they must reach a specific intake rate. When the mortality rate increases with a decreasing intake rate (Figure 4.35), it's a sure sign that food supplies can limit waterbird populations (Chapter 6).

# 5. Reproduction

## 5.1 No reproduction, no life

### 5.1.1 The moment of truth

The breeding season really is the moment of truth. Completing impressive migratory journeys, capturing numerous prey, escaping the clutches of birds of prey and all the other feats a waterbird must accomplish are in vain if the bird doesn't succeed in producing offspring. Birds that complete every task well throughout the year but fail in the breeding season have blown their chances of passing their genes on to offspring. Every bird alive in the world today is, without exception, the descendant of individuals that succeeded in reproducing. In fact, these birds are the descendants of animals that were exceptionally successful. In the fight for existence, good isn't good enough: only the best survive. 'Survival of the fittest' and the 'fight for life' are well-known phrases, but are not universally accepted. Not everyone believes that natural selection is the most important designer of the anatomy, physiology and behaviour of organisms. However it's not the purpose of this book to present evidence and arguments for the absence of a sound alternative explanation for the complex design of organisms. Other authors have already done so.[170, 173, 541, 938] What we would like to do here is show how it's possible to gain a better understanding of waterbird behaviour by putting on evolutionary-tinted glasses. Accepting natural selection as the designer-in-chief isn't automatically accompanied by a good understanding of the subject. Research is required to shed light on this. So this chapter considers the design of the reproductive behaviour of waterbird species in this light.

A bird cannot reproduce on its own. There are animal species that can reproduce asexually, such as stick insects, but none of them are birds. Birds must interact with each other for long enough for the male's reproductive cells to be successfully transferred. Shellfish can simply pump their reproductive cells into the water, but birds must actually make physical contact. To successfully copulate, a bird must push its cloaca against another bird's. This is the only personal contact that a reeve and a ruff have in their lives. Oystercatchers are a different story; the same pairs will defend the same breeding territory together for the entire breeding season, for years in a row.

### 5.1.2 What is fitness?

To start with, it is important to realise that a failed breeding attempt is not the same as being an absolute failure. Even if a bird has made all the right decisions its brood can fail through sheer bad luck, perhaps due to a sudden change in the weather or predation by an Arctic fox that happened to pass by. Fortunately for the bird there are second chances. Maybe the bird can attempt to breed again in the same season and even if it can't, if the bird survives it can always try again next year. This has two important consequences. When reproduction incurs a cost to the parent(s), as everyone assumes it does, extra effort can only be put into the current breeding attempt at the expense of the future possible breeding attempts. This is why parent birds don't always do their utmost to defend and feed their young. A clear example of this is the oystercatcher, which can live to be 40 years old. A small increase in an oystercatcher's chance of dying (or of losing its territory) leads to a large decrease in the number of offspring it could potentially produce. So when unfavourable conditions mean that it would require a huge effort for the oystercatchers to feed their young, they often don't make that effort.[458] Although the oystercatcher is rather an extreme example, all of the species discussed in this book can reach a reasonable age.

On the marshy tundra, a breeding knot is perfectly camouflaged (photo left). Only the knot's reddish breast is conspicuous (photo below), and knots hide this by pressing themselves to the ground when danger approaches.

The trade-off between investing in a current breeding attempt and future breeding attempts affects all of these species and is discussed further at the end of this chapter.

So each individual breeding attempt is not simply a matter of raising as many young as possible. Unfortunately it's not that easy to explain what it really is about in just a few words.[594] It is often said to be the maximisation of fitness, but what does fitness mean in this context? A bird's reproductive fitness is made up of different components, such as the number of eggs produced, the number of nests, the chance that an egg will become a fledgling and the adult's chance of survival. To discuss fitness, all of these different components must be measured and combined into one single fitness measure. Not surprisingly, there are different ways to do this. One important consideration is whether all offspring are considered equal, or if the offspring that a bird raises at the start of its life should be considered differently from offspring that the bird brings up near the end of its life. Just recently, it has become clear that this depends on how the population is regulated.[594] One easy method of assessment is to simply sum up all the offspring that a bird raises during its life, known as the lifetime reproductive success, or LRS for short. This method is conveniently simple and, in many cases, fits with theory. Slowly, actual measured values are starting to be included in measures of LRS. For this, individual birds must be followed from the cradle to the grave, which requires individually marked animals and much tenacity on the part of the researchers.

Curiously enough, a difference in two individuals' LRS doesn't necessarily mean a difference in their fitness. One bird may well have had a lot of bad luck, while the other was far more fortunate. It's only considered a difference in fitness when the variation in the LRS can be attributed to a systematic difference in the animal's behaviour. Fitness is a characteristic of a group of animals with the same strategy, rather than of an individual animal. So to consider fitness you have to calculate the average LRS of a group of individuals. The group with the highest total number of offspring on average (or in other words the highest LRS) has the highest fitness. In short, fitness is the expected LRS for a certain strategy, not the actual measured LRS of any particular individual.[347, 938]

5.1.3 No individuals, no results

There are three requirements for determining how many young an individual animal raises throughout its life. First, the animal must be individually recognisable. Secondly, it must be possible to repeatedly locate that animal. Finally, the researchers must be able to follow the animal continuously.

In an ideal situation, the researcher learns to recognise individuals and follow them without needing to lay a finger on them. This is possible for many mammals. Every visitor to the zoo knows how easy it is to distinguish chimpanzees from each other. It's a bit more difficult to distinguish individual buffaloes, but it is possible to eventually be able to recognise every member of a herd of hundreds of animals by using a sketchbook. You simply make a sketch of each animal's distinguishing features: its horns, snout, and frayed ears.[705] It's believed that many, if not all, bird species can recognise other members of their species individually, although this has been proven experimentally for only a few species.[304, 937] Unfortunately, the birds differentiate on the basis of characteristics that aren't readily perceived by the human researcher. Usually, humans are only able to differentiate a single bird from its conspecifics on the basis of very aberrant plumage or calls. Hence the stories of completely or partially albinistic oystercatchers that are seen at the same sites for many years. The sketchbook method can work for a small number of bird species. Individual Bewick's swans are recognisable on the basis of the yellow and black pattern on the bill.[778] Turnstones show striking variation in their plumage both in summer and winter.[929, 930] Grey plovers in breeding plumage show a large individual variation in the black markings, however the differences in their winter plumage are much less obvious. In winter, adult oystercatchers have a white neck ring that varies in size. That is handy for differentiating 10 oystercatchers foraging on a small mussel bed, but not so useful when tens of thousands of oystercatchers are standing on a high tide roost. Also their plumage is moulted annually and you can never be certain that the new plumage will be sufficiently similar to the old plumage for you to recognise the animal.

If you want to study birds you need to mark them individually. Birds' legs are very well suited to ringing. Waders are particularly good ringing candidates as they have very long legs, and walk around in open habitats where they are easily visible. Birds that breed in burrows or in colonies are often just banded with an individually numbered metal ring. Some birds are so tame that reading their rings simply requires picking them up off the nest and gently placing them back down. Some extremely keen birders make a sport of reading the metal rings of free-living birds with a telescope. This is not an easy task and requires a very high tolerance to frustration! Particularly when the ringer has accidentally put the ring on upside-down. Metal rings are not made to be read in the field or to measure an individual's LRS. Originally their main use was to help investigate the migration patterns of birds that were found dead. Nowadays, recoveries of dead individuals and resightings of recaptured animals are mainly used to learn about mortality and site faithfulness.[607]

A much better way to monitor an individual throughout its life is by using coloured rings. Colour rings can be put on in all sorts of combinations, although it's a good idea not to use too many different colours. It can also be difficult to differentiate some colours when the light isn't good, which is why it's not a good idea to use both blue and green rings. If you use six different colours and put four rings onto each bird then you can ring $6^4 = 1296$ different birds with unique combinations. This may seem like plenty but it isn't really. If the study is on a long-lived species that is being studied by a large group of researchers, the combinations would be quickly used up. The oystercatcher is one such species, which is why it is sometimes called 'the great tit of the mudflat' (great tits have also been the subject of numerous ringing studies). Oystercatchers are very site faithful, which might lead you to think that there wouldn't be any problems in allowing each research area to use the same 1296 combinations. That wouldn't be a good idea! For years an oystercatcher that bred on the roof of the IBN (the Dutch Institute for Forest and Nature Research) on Texel in summer spent the winter on the mudflats south of Schiermonnikoog, another Dutch Wadden Sea island 50 km away. There are many other similar examples. One way around this problem is to use more complicated rings. Code rings are larger and may be marked with letters, numbers, or coloured bands of varying height. When each code ring has space for three bands, a tall band, a short band, or no band can be used, creating $3^3 = 27$ different possibilities. If we have five colours that can be used as the background colour of the ring, and four colours for the foreground colour, we have 20 possibilities. This gives us a total of $20 \times 27 = 540$ different code rings. If the code ring is put on the tarsus and an extra colour ring is used that can be put on above the tarsus, we have six different possibilities. If we have colour rings in six different colours, we get a total of $6 \times 6 \times 540 = 19\,440$ different combinations. And we haven't even considered putting colour rings below the code ring. Or two code rings, in combination with two colour rings etc. In short maybe we couldn't individually mark one million oystercatchers, but we could come pretty close if we wanted to. Of course we don't want to, as there would be no point!

An overview of the different colour ring programmes currently running in Europe is available on the internet (www.cr-birding.be). Table 5.1 shows the number of wader studies using colour-marking per country, per species in Europe in 1998. Every one of the 21 wader species central to discussions in this book was the subject of a colour-marking programme somewhere in Europe. Most of this research took place in the United Kingdom, Germany and The Netherlands. This is not surprising because these are the countries that have the largest intertidal areas. The International Wader Study Group administers the various wader colour-ringing programmes to try to ensure that the schemes used by different groups do not overlap. Information about this is available from the Wader Study Group Colour-marking Register, The National Centre for Ornithology, Thetford, Norfolk, IP24 2PU, United Kingdom, E-mail wsg@bto.org. In spite of all this, in every edition of the Wader Study Group Bulletin there is a long list of resightings of colour rings that cannot be ascribed to any known program. Part of the problem is due to incorrect ring readings. It has now become very

| | Belgium | Denmark | Germany | Finland | France | Hungary | Italy | Netherlands | Norway | Ukraine | Poland | Portugal | Russia | Spain | United Kingdom | Iceland | Sweden | Total |
|---|---|---|---|---|---|---|---|---|---|---|---|---|---|---|---|---|---|---|
| *oystercatcher* | | | 1/1 | | | | 1/1 | 4/- | | | | | 1/- | | 3/3 | -/1 | | 10/6 |
| *avocet* | | 1/- | 1/- | | | 1/- | 1/1 | 1/- | | | | | | 1/- | -/2 | | 1/1 | 7/4 |
| *ringed plover* | | | 1/- | | | | | 1/- | | 1/- | | | | | 1/6 | | 2/1 | 6/7 |
| *Kentish plover* | 1/1 | | 1/- | -/2 | 1/- | | -/1 | 1/- | | | | -/2 | | 3/5 | | | -/1 | 7/12 |
| *grey plover* | | | 1/2 | | | | | | | | | | | | 2/3 | | | 3/5 |
| *golden plover* | | | -/1 | | | | | | | | | | | | -/4 | | | -/5 |
| *red knot* | | | 1/1 | | | | | 1/1 | | | | | | | 1/- | | | 3/2 |
| *sanderling* | | | 1/1 | | | | | | | | | | | | 1/1 | | 1/- | 3/2 |
| *curlew sandpiper* | | | 1/2 | | | | | | | | | | | | -/- | | | 1/2 |
| *dunlin* | | 1/1 | 1/1 | | | | | -/1 | | | | | | 1/- | 1/4 | | 1/4 | 5/11 |
| *little stint* | | | -/1 | | | | | -/2 | | | | | | | -/- | | | -/3 |
| *purple sandpiper* | | | -/1 | | | | | 1/- | 1/2 | | | -/1 | | -/1 | -/2 | | 1/1 | 3/8 |
| *bar-tailed godwit* | | 1/- | 2/1 | | | | | | | | | | | | 1/2 | | | 4/3 |
| *black-tailed godwit* | | | 1/- | | | | | 2/- | | | | | | | 1/5 | | -/1 | 4/6 |
| *curlew* | 1/- | | 2/1 | -/1 | | | | 1/2 | | | | | | | 1/4 | | -/1 | 5/9 |
| *whimbrel* | | | -/1 | | 1/- | | | | | | | | | | | 1/1 | | 2/2 |
| *spotted redshank* | | | | | | | | -/1 | | | | | | | | | | -/1 |
| *redshank* | | | | | | | | | 1/- | | | | | | 1/6 | | | 2/6 |
| *greenshank* | | | | 1/1 | | | | | | | | | | | 1/1 | | | 2/2 |
| *common sandpiper* | | | | -/1 | | | | | | | | | | | 1/3 | | | 1/4 |
| *turnstone* | 2/- | | -/1 | | | | | -/1 | -/1 | | | | | | 1/2 | | | 4/5 |
| Total | 4/1 | 3/1 | 14/15 | 1/5 | 2/- | 1/- | 2/3 | 12/4 | 1/7 | 1/- | 1/- | -/3 | 1/- | 5/6 | 16/48 | 1/2 | 6/10 | 72/105 |

Table 5.1 The number of European colour-marking programmes on waders per country per species in 1998. The first number is based on information from the internet site www.cr-birding.be. The second refers to the numbers of programmes registered with the Wader Study Group in 1998.

difficult for an inexperienced observer to read bird rings correctly.[224] But that's not the only reason. Table 5.1 shows the discrepancies between the programmes listed on the internet site and those officially registered with the Wader Study Group. Contrary to what might be expected, the electronic highway has made it more difficult, not easier, to integrate the various colour-ringing programmes.[97] Because of the internet site, many observations are sent directly to the ringer and not to the Wader Study Group. This creates problems when the ringers concerned have not registered their programmes with their national central banding office, or if the ringers cannot place the observation within their scheme and fail to forward it to the Wader Study Group. Another problem is that observations sent via email don't always include all of the necessary information. This is seldom the case when observations are submitted using a standard form. But there is no way off the electronic highway now! The Wader Study Group will simply have to create an attractive website too, including electronic forms that can be used to report colour ring observations.

The limited life spans of the rings themselves can be a real problem for ringing programmes. When colour rings give up before the bird does, you gather interesting data on the life expectancy of the rings but not much about the age of the birds. Colour rings made of substandard plastic can change colour. White can become creamy yellow, and yellow can fade to white. Or, even worse the rings may wear down and fall off. Some metal rings may become illegible when they get worn down by abrasion. For this reason, it's advisable to use stainless steel rather than aluminium rings for long-lived waterbirds like the oystercatcher. The drawback with stainless steel rings is that it takes much more force to pinch them shut. Bird legs are very fragile and it takes nerves of steel to push hard enough on the ring pliers to get a stainless steel ring properly closed. It's also advisable to put the rings above the tarsus, where they will be subject to less wear. That means that if a bird has lost its plastic colour rings, it can still be correctly identified from its metal rings, and have its colour rings replaced.

Researchers also have a finite life span. Clutton-Brock's[133] overview of reproductive success contains two photos of Professor Dunnett. One shows the professor as a young researcher holding a fulmar in 1951. The adjacent photo shows Professor Dunnet as an old man, holding the same fulmar in the same research area in 1984. Professor Dunnett has recently died, but the photographed fulmar may well be still alive. One of the very first long-term studies on individually marked animals was started by Huib Kluyver in 1933. Huib Kluyver was first student in The Netherlands to produce a thesis

As well as a numbered metal ring, this knot carries colour rings that allow observers to recognise him in the field. He is also equipped with a small 1.5 g radio-transmitter. The transmitter's antenna is visible as a thin black line protruding from under his tail.

Around the Wadden Sea breeding waders such as redshanks and oystercatchers really hit the panic button when researchers want to ring their chicks. This purple sandpiper on the Canadian tundra, however, remains very quiet. There are still three beautifully camouflaged chicks in its nest.

based on a field study. His thesis was on great tits that used nest boxes on the Oranje Nassau's Oord estate in The Netherlands. Many of the long-term bird studies summarised by Newton[599] began much later, in the 1970s, and many of these studies have stopped in the meantime (the Oranje Nassau's Oord study was also interrupted). Long-term studies are often driven by one key person. As soon as that person drops out of the project, perhaps because of retirement or a new job, the study is over. Sometimes a research area changes hands and the new owner is less than enthusiastic about the research. This is how a study on oystercatchers on the Welsh island Skokholm when began in 1963 came to an abrupt end in 1977. It's no small feat to keep a long-term study going, particularly if funding is withdrawn by the government and other sources of funds have to be found.

What about the birds? Don't all those rings cause them problems? This question takes us back to the reason why the birds were ringed in the first place. We want to be able to recognise individual birds, but the rings should not influence their natural behaviour, otherwise these birds aren't really representative of the wider population. But it is a difficult question to answer, as there has been little research on this area to date. Birds that are released after being ringed often let their legs drop a bit as they fly off, which isn't normal behaviour. They can also be seen pecking at the rings for the first day or so. But within a short time the birds seem to behave just like the unringed birds. Wearing rings probably does carry extra risks for the birds, just as it does for humans. A soccer goalkeeper once lost a finger when he was inadvertently hung up by his ring finger in the goal. However these sorts of weird accidents happen so rarely that there is no point in worrying about them. And this also seems to apply to waterbird rings. For its first 12 days of life, a juvenile dunlin is continually accompanied by its father. An extensive study found no difference in the survival of ringed and unringed juveniles over this period.[853] Ringed adult oystercatchers have such a high survival rate that it is unlikely that the unringed birds survive any better. Sometimes people report seeing a bird that has deformities on its legs, which seem to result from ringing. Although such reports can cause a lot of fuss, there is often no evidence that the bird's rings caused the problem. There are also birds that have never been ringed walking around with all sorts of leg deformities.

To die because you've been ringed would be a very drastic outcome. Ringing may have far more subtle effects. There has been little research on the extent to which colour rings influence a bird's attractiveness as a partner amongst waterbird species. However there has been extensive research on the effects of colour rings on captive breeding populations of zebra finches.[107, 108, 109] Males with red rings and females with black rings raised significantly more young than birds with rings of other colours. Males with green rings and females with blue rings did particularly badly. This is probably because birds with an attractive ring could get partners that put more effort into their young. These attractive partners didn't have to work so hard, which led to better survival and therefore the possibility of more breeding attempts. So over the years, their reproductive success was considerably higher.

### 5.1.4 How, where, when and with whom?

For a bird to maximise its fitness, it must always pick the best options for the many choices that arise during each breeding attempt. To start with there is the question of parental care. What sort of care should be given and how much? And where, when and with whom should it breed? Let's take a look at these main questions one by one, before considering what they amount to overall.

(1) How much care? There is a striking amount of variation in parental care among waders. Wader chicks are precocious, but without their parents to brood them and warn them when danger threatens they wouldn't last a week. Oystercatcher chicks are not just warned, they are fiercely defended and fed by the parents until well after they have fledged. There are also extreme differences between species in how parental tasks are distributed between the sexes. Ruffs (the male birds) just stand around looking beautiful on their leks and reeves (the females) only visit to get some sperm. This is the ruff's entire contribution to his offspring! In phalaropes these roles are reversed. Of course the females have to lay the eggs, but the males do all the rest. Oystercatchers are equal-opportunities parents; the males and females work equally hard for their eggs and chicks.

(2) Where? The tidal flat is not a suitable place to make a nest and raise chicks. A small number of species (oystercatchers, avocets and redshanks) nest in salt marshes bordering the tidal flats. However the majority of species migrates to the taiga and tundra areas or other areas where the vegetation is very low, such as dunes and paddocks. Paradoxically, these birds must conduct their most important life's work in places where they don't really seem at home. Maybe their choice of a nest site is simply a compromise between the best site to raise a chick and the best site for parents who are better adapted to life on the tidal flats.

(3) When? Many wader chicks are capable of foraging on insects in the vegetation by themselves. So a simple answer could be: when there are the most insects for the young. This is almost right.

(4) With whom? It's impossible for an individual to breed successfully without the contribution of one or more partners, even if the partner's contribution is just supplying semen, as it is for the reeve and the ruff. Without even considering the genetic contribution, the importance of choosing the right partner becomes blatantly obvious when the partner is expected to make a significant contribution to the parental care. Also for a pair bond to arise, the two animals must choose each other.

These four questions cannot be answered separately. When a bird chooses to go to a certain breeding area, it automatically limits its choice of potential partners to the other birds present at that site. Similarly, the best time to breed depends on the location. The more northerly the area, the later it will thaw and the sooner winter will return. And perhaps not all of the birds of the opposite sex will feel ready to breed at the same time. The species-specific breeding strategy must be a well-integrated unit.

Even though each species follows its own strategy, we can sometimes see large differences between individuals within one species. This is partly because the advantages and disadvantages of a certain behavioural choice cannot be seen separately from the choices made by conspecifics. First of all, there is competition. It seems logical that all animals should want to settle on the best breeding spot. However when increasing densities of breeding animals at this top location begin to interfere with their breeding success, some animals will find it profitable to go to an area that isn't quite as good. By moving to a less densely populated site, that perhaps wouldn't seem so good if the other 'better' site wasn't so overcrowded, a bird can increase the number of offspring it produces (see Section 5.3.3 for more discussion on distribution and competition). There can also be competition for partners, as quality differences between individual birds are practically inevitable. This means that what would be the best choice for one individual cannot be viewed separately from the choices being made by its conspecifics. To make it more complicated, the preferred partner also has choices to make. Just because Female B is the preferred choice of Male A doesn't automatically mean that the reverse applies.

This brings us to the second situation where the best choice for one individual depends on choices made by another animal: the two parents must cooperate when they raise their young. But raising chicks is an exhausting and risky business. It seems logical that if an animal takes it easy this season, it has more chance of being able to breed again next season. Each parent would then feel that it's in their best interests to let the other one do as much of the work as possible, leading to inevitable conflicts over task distribution.[509, 872] However this doesn't have to mean that every breeding attempt is a running battle. Task division may have already been settled over the course of evolution, as it has been for the ruff or the phalarope. It would probably be impossible to get ruffs to incubate, unless they were given intensive hormone treatment. In species where both parents care for the young, evolutionarily stable behavioural rules may have arisen that balance the optimal level of effort for one partner with the effort of the other.[509] In which case we would expect that tasks would be distributed differently from pair to pair. So as well as competition, cooperation and conflict may also cause individuals of the same species to behave differently.

Finally the prospect of future offspring doesn't only depend on determining the optimal effort and task division in one particular year, but also on when, where and

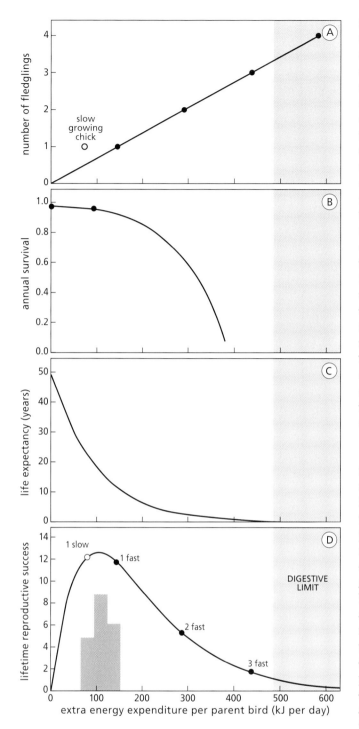

Figure 5.1 The reproductive costs of oystercatchers in poor territories, where it is very energy demanding to transport food to the chicks.[458] Such oystercatchers (called leapfrogs, see Section 5.3.3) raise just one chick at best. A leapfrog without chicks has an estimated energy consumption of 590 kJ per day. If a leapfrog increased its workload so that it had a consumption of 665 kJ per day, it could raise one chick, but that chick would only grow slowly. The adult would require 733 kJ per day to raise a chick that grows well. Knowing the extra daily energy requirement for raising a well-growing chick, we can calculate how much additional effort is required to raise a particular number of young. It would be impossible for a leapfrog to raise more than three young: the bird would have to spend more energy per day than it can take in through its digestive system (A). There are also likely to be hidden costs in putting extra effort into breeding, resulting in lower survival over the subsequent winter. These survival costs could take the form shown in (B), which relates the annual survival rate to the extra effort put in during breeding (kJ per day). From the survival rate, life expectancy can be calculated (C). By multiplying the life expectancy (C) with the number of chicks that fledge (A) we get the total number of chicks that the bird raises in its life (D). There is no point in a leapfrog running itself ragged to raise two or even three chicks in one breeding season, as it will pay later in reduced possibilities for successful breeding in the future. This conclusion results largely from the steep drop-off in survival with high energy expenditures. While it would be good to have direct measurements of this, in the wild oystercatchers never live at such high energy levels: the frequency distribution in (D) shows that leapfrogs never exceed 150 kJ extra effort per day.[458]

this is oystercatchers that breed in poorly located territories where they have to transport food for the chicks over large distances. These 'leapfrog' birds are discussed further in Section 5.3.3. Until the chicks can fly and accompany the parents to the tidal flat, the leapfrog parents must fly food in to their chicks. Flying is energetically expensive. So expensive in fact, that differences in energy expenditure between different oystercatchers in the breeding season are mainly due to differences in the amount of time spent flying.[457] Leapfrogs seldom raise more than one chick, and even this chick doesn't receive the best treatment. If the parents bring less food, the chicks don't grow as fast, but the parents don't have to put so much effort into collecting and transporting food each day. It's rather sad to see a hungry chick jump up to ask its parent for food, while the parent nonchalantly puts its bill between its feathers and goes to sleep. Hungry chicks don't tend to stay under cover as well, so they probably run a higher risk of being caught by a predator. Poorly nourished juveniles are probably also more vulnerable to disease.[756, 757] So it's not surprising that a chick's chance of fledging mirrors its growth rate: most of the poor little weaklings don't make it.[459] It is difficult to understand why the leapfrog parents don't do more to help their chicks. If they put in more effort, they would be able to raise two or maybe even three chicks. Oystercatchers could take in over 1000 kJ per day if they chose to, so they would be able to spend this much energy on breeding.[462] If they put in that maximum amount of energy it would be possible for leapfrogs to fledge three chicks. And not only that, the chicks would grow quickly (Figure 5.1A).[458]

The only reasonable explanation for the leapfrogs'

with whom the bird breeds. Is it better to choose a new partner each year, or are there advantages to making another breeding attempt with the previous partner? Is it better to return to the same breeding area as last year, and if so, why? The answers to these questions must be viewed in terms of the individual's lifetime reproductive success.

5.1.5 Costs of reproduction

Raising offspring is often hard work. It seems logical that the harder the parent(s) work, the more offspring will be produced. To raise the maximum number of offspring would require the maximum effort. So do the parents do their utmost for their offspring? Strangely enough, the answer is no. A good (if rather harrowing) example of

# REPRODUCTION

apparent laziness is that an extra effort brings extra costs. Working harder costs more energy, but if that were the only expense there wouldn't be a problem. If the extra effort required puts the parent's own survival at jeopardy, it's a completely different story. Putting extra effort into the present brood could reduce future possibilities to raise offspring. As oystercatchers are very long-lived, a small increase in the mortality rate quickly equates to a considerable decrease in future reproductive output. It is very difficult to determine these costs of reproduction well, and trustworthy measurements are lacking. It is however possible to play around a bit with the data from a long-term study on the island Schiermonnikoog.[458] In 1990 a storm tide washed away all the eggs and all but one of the chicks from a salt marsh on Schiermonnikoog. The following winter, the parent birds had an extremely high survival rate, 98%. In normal years when they raised chicks, the parent's survival was 95% on average. This may seem a tiny difference, but it isn't. Annual survival of 98% corresponds to a life expectancy of 50 years, while annual survival of 95% means a life expectancy of 20 years (Figure 5.1B and C)! On the basis of these two points, we can make a rough estimate of the relationship between effort and survival (Figure 5.1B). Rough, because survival doesn't only depend on parental effort, but also on winter temperatures and food availability.[124, 597, 992] Plus, two points is nowhere near enough to estimate the shape of the curve. However, it seems plausible that survival decreases with increasing effort. If the estimated

Oystercatchers are unique amongst waders; they feed their chicks, and males and females play in equal role in chick rearing. When the chicks are still small, their food is taken to them (below) and prepared; here the adults remove the flesh from a bivalve for the chicks (top photo page 239). Large chicks follow their parents, trying to get each prey item as quickly as possible (lower photo page 239). Oystercatchers that breed on the edge of a salt marsh (above) have to put less effort into feeding their chicks than oystercatchers in breeding territories far from the tidal flats that have to transport each prey item over a large distance. This oystercatcher is brooding chicks and is wearing conspicuous colour rings. These rings are used in a long-term study on Schiermonnikoog, in the Dutch Wadden Sea.

life expectancy (Figure 5.1C) is multiplied by the annual chick production (Figure 5.1A), we obtain the total number of offspring that could be raised over a parent's entire life for a varying amount of effort (Figure 5.1D). The curve that this produces peaks at the effort required to raise one chick. Going that extra mile just doesn't pay.

So can we conclude from Figure 5.1D that leapfrogs that are economical with their parental energy can raise at least 12 young during their lives? No we cannot! The problem is that so many eggs and small chicks are lost to predation that many leapfrogs have already lost their entire brood before parental effort becomes a problem. Our calculation in Figure 5.1D only really applies to a small hypothetical group of lucky leapfrogs. Good luck is difficult to attract. This might explain why so many leapfrogs lay three or even four eggs when one chick would be the ideal brood size. Having two large young in the nest means the chicks must compete with each other for food. This delays the growth of both of them, even though a hierarchy quickly develops where the dominant chick gets the most food.[244, 458, 756] But if the leapfrogs had laid fewer eggs, it's very likely they would have no chicks left at all by day 10.[458]

The fortunes of an individually marked oystercatcher population were also followed on Skokholm, an island off the Welsh coast, for many years. This study would probably still be running if the island's new owner hadn't decided that he didn't want to have researchers on his land. This brought research that had started in

1963 to an abrupt end in 1977. In spite of the untimely end to their study, the researchers still had a sufficiently extensive database to estimate the costs of reproduction. They came up with a surprising result: animals that had not reproduced had the lowest survival.[760] Still we can't conclude from this that reproduction doesn't bring survival costs. The problem is that all the birds were grouped together in the analysis, and no distinction was made between the quality of the individuals or their territories. Birds that are well-off can afford to spend up large. So the positive relationship between effort and survival may result from the natural variation between individuals.[608] The only way to solve this problem is experimentally.[145, 508, 856] For example, you can increase the parental effort of one group experimentally and decrease the parental effort of another. Although there will be variation between the individuals in both groups, this variation will not be important when the two groups are compared as long as group membership was determined randomly.

Experimentally manipulating parental effort is easier said than done. The usual method is to take away or add a number of eggs or young. After that, it's up to the parent bird to decide whether or not to increase or decrease its efforts. This means that it is important to actually measure the parental effort and not to simply assume that animals with experimentally enlarged broods work harder. While there is much literature available on the underlying theory, results and limitations of these sorts of experiments,[508, 739] few such experiments have been carried out on waders.[908] It is often thought that species

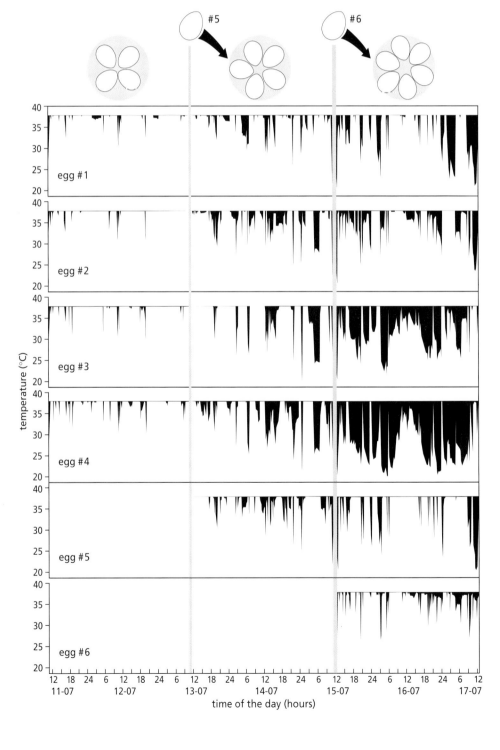

Figure 5.2 Can a monogamous pair of oystercatchers incubate more than four eggs in a clutch? This was investigated by temporarily swapping the four real eggs of a pair with replica copper eggs that could record their temperatures. When an egg is being well incubated, its temperature is around 38 °C. When the adults are off the nest (perhaps as a result of disturbance) the temperature of the eggs drops. When four eggs were in the nest (left hand panels), cooling periods were infrequent and brief. After two days an extra egg was added to the clutch, and cooling periods became longer and more frequent. Cooling was not synchronous between the eggs, indicating that rather than being due to the adults being absent, the cooling resulted from difficulties in keeping all five eggs warm at the same time. When a sixth egg was added the effect became even more extreme.[388]

that do not feed their young have an easy life, but the costs of incubating, accompanying and guarding the young should not be sneezed at.[908]

Despite the lack of knowledge about the extent and the nature of reproductive costs in waders, in the following pages we will continually return to the idea that such costs exist, for two reasons. First, to assume that reproduction does not carry fitness costs would go against the foundation of current evolutionary theory. If reproduction isn't costly why hasn't natural selection led to animals that reproduce infinitely fast? Of course it's not impossible that our present theories are wrong. No matter how elegant the lines of reasoning, in science it's the empiricist rather than the theoretician who has the last word. However, and this brings us to the second reason, there is hard evidence for the existence of reproductive costs in species that have been more thoroughly researched. You can, for example, select for fruit flies (*Drosophila*) that live longer, but this irrevocably leads to animals that have a lower reproductive rate.[959] It appears that female fruit flies with a hectic sex life may pay a price in the form of lower survival because of chemicals transferred by the male.[129] Even in humans there are indications of a programmed trade-off between the costs of reproduction and becoming older. One study examined the well-recorded family tree of the English nobility, including 5499 of the Prince of Wales' ancestors going back to the year 740 AD. Only noble women that had no or few children lived to a ripe old age.[924]

## 5.2 Reproduction: a broad overview

There is a lot of variation in the amount of effort waterbird parents make for their offspring, both within and between the different species. Before we discuss this variation, it's helpful to step back and take an overview of reproduction. So let's consider three questions that every breeding wader faces: (1) How many eggs should be laid? (2) How do you ensure that the egg hatches? (3) How do you raise a downy chick so it becomes a fledgling capable of completing the journey south under its own steam?

### 5.2.1 Clutch size

When clutch size is discussed, proximate and ultimate explanations can easily be mixed up.[908] A proximate explanation for the clutch size considers factors that have a direct influence on the number of eggs laid. Do females in better condition lay more eggs? The ultimate explanation for the clutch size concerns the selective forces that act on clutch size, the predictable consequences of clutch size for the fitness of the individual. Do females in better condition lay more eggs because good condition at the time of egg laying indicates that they will be good at raising young? The ultimate explanation is about the fitness costs that accompany reproduction. But these fitness costs are always made via some sort of proximate mechanism.

Here's an example. The parents of altricial chicks (which include species like blackbirds, starlings and great tits) have to transport a lot of food to their helpless young day after day. It's easy to imagine that there is a limit to how hard the parents can work and that this limit determines how many young they can raise.[489, 490] But would this be a proximate or an ultimate explanation? At first sight it seems like a proximate one. But think about an athlete that has to deliver a top performance and just cannot work any harder. This sort of inability is often psychological. Perhaps the pain limit could be pushed even further or the physiological reserves could be even further exhausted. To understand why a bird stops working after a certain effort level we can't avoid an ultimate explanation. Working harder probably leads to a higher risk of mortality. In the most extreme case the bird works so hard that it drops dead at the end of the season. The bird has sacrificed all its future options for raising offspring for the chance to raise its present clutch. We've already seen that this would be a poor choice for oystercatchers in bad territories to make (Figure 5.1). According to the ultimate explanation, the chosen effort level is the best compromise between the number of young that can be raised this season and the parent's own survival. But of course, there has to be a proximate mechanism to ensure that the parent bird doesn't exceed the optimal effort level, and clutch size must also be adapted to the optimal effort level.

The precocial chicks of many waders can forage independently soon after hatching. You might think that this allows the parents to raise more young. It's true that waders that feed their young, like the oystercatcher, lay less eggs on average than the species that don't feed their young, like the dunlin.[908] However, even the waders that don't feed their young never lay more than four eggs. The shelduck and eider duck don't feed their young either, but these ducks lay considerably more eggs than the waders. Most eider clutches have four to six eggs. Shelducks have even larger clutches, many containing 10 or more eggs. Even blackbirds, starlings and great tits that have to work so hard for their young have clutches far larger than four eggs.

Why do waders only lay four eggs? One suggestion is that they are unable to incubate more eggs properly.[19, 450, 847, 908] Nests with more than four eggs are sometimes found, but without exception more than one female has laid her eggs in these nests. On average a pair of avocets incubates a four-egg nest for around 23 days, but two pairs that incubated six-egg nests took 30 and 34 days.[409] This could result from some of the eggs having been laid much later than the others, or from the development of the eggs being slowed through poor incubation. In oystercatchers, there is evidence that extra-large clutches are not incubated very well; once a clutch exceeds four eggs, the likelihood that the eggs will hatch decreases strongly. Heg and van Treuren[391] placed copper eggs containing automatic temperature recorders in the nest and found that when there were more than four eggs in the nest, the eggs were exposed to long periods of cooling. Figure 5.2 shows the results of an experiment that recorded the temperature of all of the 'eggs'. In a four-egg clutch all of the eggs could be brooded adequately, but once the clutch size was larger there was always at least one egg that cooled down.

This incubation problem doesn't really give us an ultimate explanation for wader clutch size. If ducks can brood more than four eggs, why can't waders? The waders could, for example, have evolved smaller eggs or larger brood patches. A better ultimate explanation is that even waders that do not feed their young are faced with costs that increase with the number of chicks.[908] These costs certainly exist, even though they are not yet measurable in terms of survival and their relationship with clutch size is still unclear. A parent bird cannot forage while it's keeping a lookout to protect its young and warn them if danger approaches. And feeding is completely out of the question for a parent that would have to go to outside the breeding territory to be able to forage properly. This is the case for golden plovers that sometimes have to travel large distances from the breeding territory to forage. Golden plover parents, and especially the fathers, spend a lot of their time brooding and guarding their chicks, particularly when the chicks are newly hatched (Figure 5.3). So the complete explanation for the waders' clutch size of four is that the optimal clutch size lies around four, and secondary adaptations have arisen that make it disadvantageous for waders to lay more than four eggs.[908] These secondary adaptations include extra-large eggs allowing higher survival of newly hatched chicks, and a pointed shape to the eggs so that they fit together well in the nest.

Snowy owls have a large influence on the breeding success of waterbirds that breed in the far north. Here, a female snowy owl is about to feed a lemming to her young. In years when lemmings are scarce, snowy owls and other lemming-eating predators such as Arctic foxes and skuas are a real threat to growing waterbirds. This photo clearly shows why a female snowy owl can incubate a much larger clutch than a wader, which restricts itself to four eggs. The owl has plucked the feathers from much of her breast, creating an extra-large brood patch. An incubating female owl can do this because her mate catches all the prey she and her young need and brings it to the nest. Waders are not fed by their partner and must leave their nests periodically to forage. If they had a large bald brood patch like the owl they would lose too much heat while searching for food.

# REPRODUCTION

This turnstone's nest on the Canadian tundra (above) is simply a shallow depression in the ground. It's obvious that the incubating parent only just covers the four large eggs with its body.
Avocets usually decorate their nests with dried grass and prefer to breed on (relatively) safe islands (middle photo). Oystercatchers sometimes have rather eccentric taste when it comes to nest sites, and may be seen nesting on flat roofs or even on top of a mooring pile (adjacent photo).

## 5.2.2 The risk of egg predation

Rather than considering why birds use what initially appears to be a rather impractical way to reproduce, let's delve deeper into how waterbirds solve the problems they face with their eggs.

Eggs must be laid in a nest, if you can call it that. Most wader species simply use their feet to scrape a bit of soil or vegetation away, creating a small hollow that they decorate with a few shells or bits of grass. Leaving their eggs uncovered out in the open on the ground seems to be asking for trouble. And sure enough, even with extensive anti-predator adaptations, ground-breeding birds often suffer high egg losses because of predation.[38, 270, 605]

## REPRODUCTION

In a study on the east coast of Britain, almost no ringed plover nests hatched due to high levels of predation.[642] In Hungary, more than half of all Kentish plover nests were lost to predators.[847] The same applied to a population of dunlins in Southern Sweden.[443]

Although usually only the clutch is lost, parent birds also run the risk of being eaten. There is no night during the Arctic summer. This means that near the Arctic Circle predators such as the Arctic fox, which like to eat eggs or even adult birds are active during the day. In more temperate latitudes many of the dangerous predators such as foxes, stoats and owls are only active at night. This makes the night shift a rather dangerous time to be incubating, especially since the parent bird's visibility is reduced in the dark. Observations through a night vision scope of an incubating oystercatcher showed that birds find it frightening to incubate at night. The oystercatcher was barely able to escape from a cat, but didn't return to its nest that night. Similar observations have been made of curlews.[774] This isn't necessary fatal for the embryo as cooling may simply delay its development. However when the embryo is more developed its risk of being seriously damaged by cooling increases, and the egg must be incubated at night. It's interesting to see which parent ends up with the night shift in species where both sexes incubate. In curlews, godwits and broad-billed sandpipers the male is invariably on duty,[774] in golden plovers and dunlins it's the female's job,[116, 443] and in oystercatchers and knots there is no regular task division. The reasons for these differences are as unclear as they are interesting. But we can assume that incubating during the day also has disadvantages. In North Holland (a region in The Netherlands) female curlews that had spent the day incubating could only feed at sunrise and sunset. At night they roosted in the safety of the North Sea beaches. The females ended up losing much more weight over the incubation period than the males.[77]

Redshanks nest in relatively tall vegetation that they bend over the nest with their bills. This makes the incubating bird practically invisible. If danger approaches, an incubating redshank will only fly up at the very last moment. Redshanks, very cunningly, often nest near other species that actively chase or try to lure potential robbers away rather than trusting in camouflage. Larger waders like the grey plover, avocet, lapwing, godwit and oystercatcher are particularly good at making a huge uproar and chasing off potential predators such as herring gulls.[303] It's understandable that herring gulls beat a hasty retreat when an oystercatcher is in hot pursuit, as oystercatchers use their bills to hack open shells that are considerably thicker than bird skulls. Why a herring gull would let an avocet bully it isn't quite as clear. Maybe the avocets make use of the gulls' in-built flight reflexes that prevent collisions with obstacles. Or maybe it's the way they beat the gulls with their wings. Avocets often breed in colonies so a large group of avocets raising the alarm can make a gull's life miserable. It probably makes it very difficult or even impossible for the gull to concentrate on searching for eggs. All the attention that the predator gives to the alarmed parent birds flapping around its

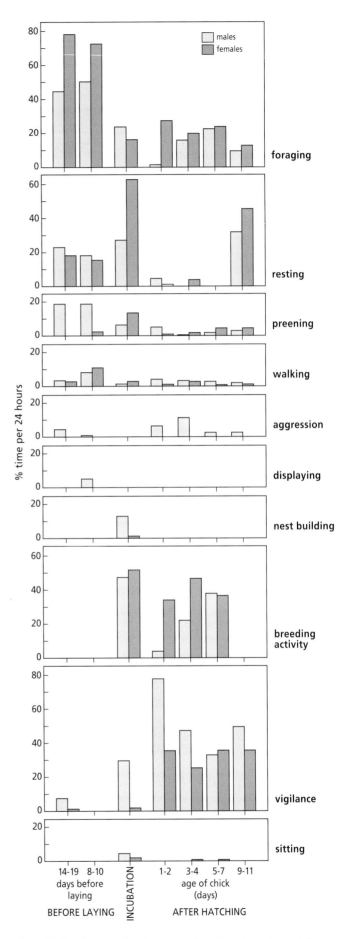

Figure 5.3 Time budgets of male and female golden plovers through the breeding cycle. Data were collected from 1980 - 1981 in the Hardangervidda, Norway. The breeding activity category comprises the nest building and sitting categories plus part of the resting and vigilance categories.[116]

head cannot be spent searching for eggs or chicks.

There is also one small wader species that boldly pesters predators in an attempt to prevent them from stealing its eggs. Rumour has it that turnstones have specialised in pecking into the cloacae of gulls and skuas. This would be such a painful experience that any predator that had fallen victim to an attack would be terrified whenever it saw another turnstone.

Attacking is not the only way to deflect predators, and other birds try to distract or mislead would be nest robbers. This requires abandoning the nest, which makes for a rather critical moment. A sudden movement from the bird could reveal the location of the nest to a predator. So it's important that the parent leaves the nest as early and with as much stealth as possible, while the potential predator is still some distance away. So a good view is important. The alarm calls of other birds nearby can also be of great assistance. A multitude of predators eye up the eggs of the Temminck's stints that breed along the Botnische Gulf: hooded crows, common gulls, turnstones, mustelids and foxes.[479] The stints are quick to leave their nests when danger approaches, as long as they have good visibility or other birds have raised the alarm. They then try to make a run for it as inconspicuously as possible. Incubating oystercatchers take a very different approach. An oystercatcher on a nest is easily identified by the way it alertly looks around and holds its tail up at a slight angle. Oystercatchers that abandon their nest when danger approaches will often sit down again some distance from the real nest and pretend to incubate using the same posture. The oystercatcher may also try another tactic, staggering off as if it has a broken wing, with its half-spread wings flapping on the ground. It seems that oystercatchers mainly try this tactic when the nest has already been found by a potential predator (in this case the researcher). The oystercatchers' broken wing displays could be their last attempt to save their eggs; they try to delude the predator into thinking that if it chases them it would end up with an even larger meal (a big fat parent bird). Golden plovers treat humans as one of the many ground predators. Usually a plover will sneak off its nest when the humans are still far away (as a Temminck's stint would). Sometimes though, the plover remains on the nest until the very last moment when the danger is less than 10 m away, before it suddenly explodes off the nest. The plover then tries to lure the threat from its nest using a broken wing display.[116]

So there appears to be a pattern here. If you have time it's best to sneak off; if not, stay where you are and only fly from the nest making a spectacle of yourself at the very last moment. So the best choice depends on the situation. If you don't notice the danger until it's too late for the first strategy, there is always still the second strategy. But there are differences between species. Plovers tend to try sneaking off, while many sandpipers will stay sitting for as long as possible.[774] When tundra-breeding waders finally abandon their nests because of danger they have another trick up their sleeves (or wingpits) – the rodent run! The bird runs off zigzagging away from its nest, making peeping noises. They fluff up their feathers making them look rather like a rodent's fur coat. The dark strip running up from the tail of many waders even manages to look just like the dark stripes down a fleeing lemming's back.[303] The bird's jerky way of walking makes it look even more like a lemming. The purple sandpiper holds the honours in the rodent run.[691]

Various factors make it difficult to accurately measure predation risks. Gulls steal eggs in a flash, and a researcher's chance of seeing this is very small. Sometimes predators leave clear signs, but often they don't. All the researcher can do is determine that the nest is empty. Only seldom can the remains of predated eggs be found in or around a plundered nest. But an empty nest doesn't always leave the researcher without clues. Perhaps the nest is empty because the eggs have hatched. If so, there would usually be tiny fragments of eggshell remaining in the nest. However if the nest is empty long before the eggs were due to hatch, it's probably due to predation.

Another problem facing researchers who would like to measure predation risks is that their visits to the nest could actually be increasing the chance of nest predation. Researchers searching for nests on the tundra sometimes get the feeling that they are being followed by gulls and skuas that fancy an egg or two.[808] Predators might also find the nests by watching the parent birds fly off when disturbed or returning afterwards, and take an egg.[927] Perhaps the predators can learn to recognise the marks that researchers leave near the nest site. Human visitors leave a scent trail, something that predators with a well-developed sense of smell like the fox and the Arctic fox could home in on. Götmark[346] reviewed studies on the effects of human disturbance on nesting birds and noted that in around half of the studies there had been an obvious effect. Colonial birds seemed to be especially sensitive to disturbance, while crows and gulls were the predators that made the most use of the disturbance. However there is little evidence that mammals such as foxes use human scent trails to find nests. In a study in Massachusetts, foxes devoured an estimated half of all nests of the piping plover (*Charadrius melodus*). The researchers then decided to put radio transmitters on the foxes to determine how they had found the nests. Not even one of the foxes followed a human scent trail.[528] It's not hard to imagine why. A fox that follows the scent trail of a botanist or a hiker won't find much to eat.

Let's assume that we know how to visit nests and mark them without increasing their risk of predation. Even then trying to determine nest success is no simple matter. Something Albert Beintema, a Dutch biologist, has emphasised. In his words: "Many people think it's simple to determine nesting success: you simply find lots of nests and monitor them until they either hatch or disappear. So if you find 90 nests, of which 74 hatched and 16 were eaten, you have a nest success of 88%. But this is not correct. It may seem almost unimaginable that something so simple can be wrong, but it is. There are many pitfalls in the determination of nest success. This is such a common problem that it really deserves to be highlighted."[41]

So what's all the fuss about? The key point is that rather than using the percentage of eggs that hatch, biolo-

REPRODUCTION

Some waders put on a real show when a predator approaches their nest or chicks, in the hope of distracting the predator. This grey plover (above) acts as if it has a broken wing. This knot (left) scuttles off like a lemming. However incubating birds will stay on their nests as long as possible, relying on their beautifully camouflaged plumage. As long as the purple sandpiper remains sitting on its nest in a clump of mountain avens (*Dryas sp.*) it is well hidden (lower photo page 247). A dunlin is well concealed amongst taller vegetation (upper photo page 247).

gists should focus on the probability of survival,[37, 440, 543, 544] which is commonly called the Mayfield method. The main problem is that few nests are found at the absolute beginning and nests often disappear before they can be found. Hatching percentages don't take these complications into account and almost always end up overestimating the hatching probability. Survival probabilities take these complications into account, since the calculation only includes the actual number of days that each of the nests was monitored. So using this method, a nest that is found the day before hatching and then hatches carries much less weight than a nest that was found during egg laying and is monitored until it eventually hatches. Here's another example. Suppose that you want to estimate the daily survival probability based on observations from just one nest (this would, of course, be a complete waste of time but it's easy to calculate and so makes a good example). If that nest disappears the following day the estimated daily survival probability is $1/(1+1) = 0.5$. But if the nest disappears after 19 days then the estimated daily survival probability is $19/(19+1) = 0.95$. The two estimates of the daily survival probability are very different, but in both cases the classic nest success is 0%.

Another good thing about the Mayfield method is that an estimate of one risk factor (like predation) is not influenced by the size of the other risk factors (like being trampled). For example, think of two sites with comparable numbers of predators, so that the risk of nest predation is roughly the same. When a herd of reindeer grazes in one of the sites, many of the nests are trampled. Once a nest has been trampled the eggs can no longer be eaten, so there will be fewer nests lost to predators in the reindeer site than in the site without reindeer. This could lead you to conclude that there is less risk of predation in the reindeer site. Using the Mayfield method would prevent you from jumping to the wrong conclusion.

The survival probability is usually expressed on a daily basis, but can also be expressed per hour. An advantage of the latter is that you can then determine whether the predation risk differs by day and night. There has been little research on this, but you might expect a difference since some predators are more active during the day, and others are more active at night. We've already

# REPRODUCTION

## Box 5.1 Nest predation and nest guarding

Far more nests are lost while eggs are still being laid than when the clutch is complete.[38, 226] There are several reasons why this could be so. Perhaps the risk of predation changes over the course of egg laying and incubation. Or maybe there are simply some nests that were always at a higher risk of predation, being poorly located or camouflaged, or not defended well. These nests might quickly disappear, leaving only the nests that were better camouflaged, or better defended or whatever. So as nesting progresses a random sample of nests will contain fewer of the more vulnerable nests, and the apparent risk of predation will become smaller. According to this explanation, the predation risk of an individual nest doesn't change, but the composition of the random sample of nests that we would analyse does.

Although not difficult to understand, this explanation isn't easy to prove experimentally and has never really been tested. But it is almost certain that some such effect occurs. For years researchers on the Dutch island Schiermonnikoog watched in dismay as a common gull colony in their study area increased. Common gulls may look much friendlier than herring gulls, but they too are merciless egg thieves. The oystercatchers that bred among the common gulls hardly succeeded in raising any young. The empty eggshells scattered everywhere at the end of the season bore silent witness to the oystercatchers' failed attempts.

There are also indications that bird behaviour, and therefore perhaps predation risk, changes during egg laying. Waders do not begin to incubate as soon as the first egg is laid; they wait until the whole clutch is more or less complete. What effect this has on the nest's camouflage and predation risk is unclear. Maybe an incubating bird is more easily found than an unbrooded clutch. Maybe the predators react to the movements of birds to and from the nest. In any case it's obvious that an undefended nest runs more risk than a defended nest.

Oystercatchers that breed on the salt marsh will go off together to forage, leaving their nest unguarded for hours on end during the egg-laying period. As more eggs are laid, the level of nest guarding improves, so later eggs have a much lower risk of being eaten than the first egg.[226] Once the clutch is complete, the parents' carefree behaviour ceases. From that time on there is almost always one parent bird incubating. Meanwhile the other bird will keep guard nearby, or go to the tidal flats to forage. If their first egg is eaten, the parents often move to another nest scrape and lay the rest of their eggs in this new nest. Maybe this helps to mislead any predators that return to the old site for another tasty titbit. It also explains why instead of making just one nest scrape, oystercatchers often create a whole series of practice nests before they start to lay eggs.

But why don't the oystercatchers do a proper job of defending their nest as soon as the first egg has been laid? It could be that the males are not very observant and don't realise that there is an egg in their nest. If so, this is rather short-sighted of the females. It's in the females' best interests to make sure that the males know that there is an egg in the nest, as soon as that egg is laid. It certainly seems unlikely that males in experienced pairs would not know when the first egg is laid. The few times that I (B. Ens) have seen an experienced female lay an egg it seemed a big event for the birds, and the male stayed nearby making soft calls. The only inexperienced female that I have ever seen laying an egg did this while her male was involved in a territorial conflict some distance away. This egg did not last long. Another explanation is that rather than the males being oblivious to what is going on, they know but have made a choice between leaving their nest alone and leaving their mate alone. If they leave the nest they run the risk that the egg will be eaten. If they leave their mate they run the risk that she will copulate with another male who will fertilise her next egg. Not having offspring to raise (because the egg has been eaten) is not as bad as raising someone else's offspring, when raising offspring reduces your chance of survival.

It is possible to come up with other explanations for why nest defence intensifies over the course of the breeding cycle, rather than being maximal immediately. The big challenge is to find a way of determining how much of this 'egg laying effect' is due to risk differences between nest sites and how much is due to changes in the parent birds' behaviour.

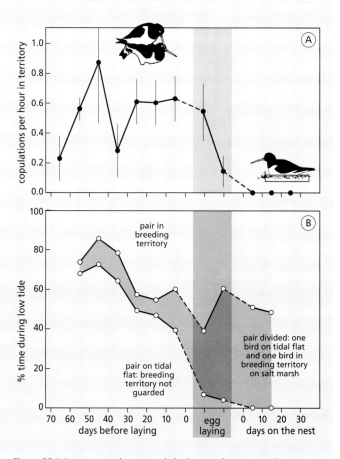

Figure B5.1 An oystercatcher can only be in one place at one time. Before the eggs are laid the male and female stay together and copulate frequently (A: number of copulations per hour that the birds are in the breeding territory together). They regularly leave their breeding territory totally deserted during this phase (B: territory attendance, represented as the percent of time that birds are out on the tidal flat, away from the nest). Once the clutch is complete, there is almost always at least one parent near or on the nest. While the eggs are being laid, the territory is often left unguarded and a considerable number of eggs are lost to predation.[226]

discussed how such differences could make it more dangerous for birds to incubate at night, but it may be that for eggs the opposite applies. On the Dutch island Schiermonnikoog, gulls are the main predators on oystercatcher eggs, and gulls are usually active during the day. Sure enough, the predation risk turned out to be much higher during the day than during the night.[223] It's possible that there was no predation at night, but that couldn't be confirmed. Nests could only be monitored while it was light, but the period between the evening inspection and the next inspection the following morning did not consist solely of night.

### 5.2.3 Egg trampling and other threats

Most waders breed in areas with low vegetation. These are often the areas where herds of large grazing mammals spend their time. On the tundra most of these mammals are reindeer. In pasture they're mostly cows, horses and sheep. The problem with large grazing mammals is that while they trudge along they sometimes stand on a nest, with devastating consequences.[41] Lapwings or oystercatchers might be able to make a lone cow change direction in time by pecking it on the nose or carrying out a dive-bombing mission. But less aggressive species can do little more than watch from a safe distance as their nest is stomped flat. And if a whole herd of cows rushes past, even oystercatchers and lapwings have to just stand by helplessly. Unsurprisingly, the more intensive the grazing, the higher the risk of trampling. But the type of stock also plays a role. Yearlings are more dangerous than dairy cows, because yearlings are more likely to frolic. All cattle are more dangerous than sheep, because cattle have larger hooves. All in all, a nest in an intensively grazed area has only a minuscule chance of surviving the many stomping hooves. This is why nests are often protected with cattle excluders.

The density of large grazing animals is much lower on the less productive tundra than on the intensively fertilised pastures of Western Europe. So overall, waders that breed on the tundra probably don't lose that many clutches to trampling. Problems may arise locally when waders breed in places where reindeer like to graze. Sometimes these are the very rich feeding grounds or places where mosquitoes are less abundant. On the Alaskan tundra, in the area studied by Declan Troy,[873] caribou rather than reindeer roam. The caribou prefer to forage inland, rather than on the coast where more waders breed. However every year there is a temporary mosquito plague inland caused by the relatively warm weather. During these warm periods the temperature can rise to 20 °C in calm weather. As soon as the mosquito plague sets in, the caribou migrate en masse to the coastal areas, where there are far fewer mosquitoes. Nest losses are high when the caribou migration occurs while the waders are still nesting.

Wader nests are made out in the open on the ground, so in principle they could be washed away after heavy rain when hollows fill up with water or when rivers flood. The first seldom happens, which may mean that the birds can assess the risk of an area being inundated by looking at the local vegetation. You might think that flooding would be a problem on the tundra, where the water drainage is poor because of the frozen ground causing melting snow and ice to create large marshy areas. However there seems to only be a single documented case of a nest being flooded (a grey plover).[116] In fact, birds that breed along the coast in salt marshes, on stony beaches or sand spits are at greater risk.[270] The birds choose nest sites that normally remain dry at high tide, but strong winds from the wrong direction can cause problems. In the Wadden Sea area strong northwesterlies push extra water into the Wadden Sea and prevent this water from flowing back out, causing much bigger high tides than normal. Spring is usually a peaceful time, but every so often a northwesterly storm wrecks havoc. Chicks drown and the eggs that are near hatching float away on the rising water. Newer eggs (which do not float) have a better chance; they are only rolled away if the current is very strong. If the eggs remain in the nest they often still hatch, even though they spent several hours in the cold seawater. Storm floods are second only to predation as a cause of egg loss for avocets that breed in salt marshes.[409] Floods are equally damaging to the eggs of Kentish plovers that breed on the Wadden Sea islands.[876] Sandwich terns prefer to breed on the highest and driest parts of the island Griend in the Dutch Wadden Sea and these are the sites that fill up first in spring. Later breeders have to make do with lower-lying areas on the edge of the colony, where nests are at more risk of being washed away during a storm flood.[895]

Washing away, flooding and predation are risks that breeding waders have had to face for many, many years. This has given natural selection ample opportunity to do its work and, as we've seen, birds have a number of adaptations that help to minimise their losses. Whether this also applies to cattle trampling is less clear. Like predators and floods, large grazers have been around for a long time. But the extremely high densities of grazing mammals now found in artificially fertilised fields are completely unnatural. The dispersal of these grazers is also unnatural: the farmer can simply decide to move the herd from one field to another at any time. The modern day Dutch field looks more like the top of a billiard table than vegetation revealing anything like a natural grazing pattern and the accompanying risks to nesting birds.

From a nesting bird's point of view, the worst aspect of modern agriculture must surely be farm machinery. Large tractors, mowers and muck spreaders have come into use so recently that there hasn't been time for the birds to develop behaviours to cope with these threats. But what behaviours could they possibly develop? As soon as a mower approaches it's already too late to do anything, so presumably the birds should only nest in places that aren't going to be mown. To do this the bird must be able to figure out, before building its nest, the likelihood that a particular area will be mown within a certain period. Is this possible? If the birds tend to return each year to their old breeding site (as most species do), then they could learn from experience and not return to fields that they have previously been mown out of. This

A spotted redshank sits on its nest amongst scrubby vegetation and rocks, near a small lake close to the tree line in Lapland. Here it is reasonably well protected against the three main threats facing wader nests: predation, flooding and trampling.

sort of behaviour is also useful in natural areas if the birds are unable to properly assess all the potential risks in advance. However where farm machinery is concerned, avoidance behaviour only helps if the fields receive the same treatment each year. Unfortunately for the birds, this is not usually the case.

Modern crops also cause problems for nesting birds. Oystercatchers and avocets prefer to nest in low vegetation, or on barren sandy or stony ground. Many crop fields are barren in spring and look as if they would be very suitable breeding sites for these species. If a bird nests in a crop field and survives the initial crop management, it faces a new threat, the unexpectedly fast growth of the crop. Suddenly the nesting bird is surrounded by high vegetation obscuring its view. Oystercatchers and avocets don't like this and often abandon such nests. In their natural breeding habitat short vegetation remains short.

### 5.2.4 Hatching

It's important that the eggs aren't eaten, trampled or washed away, but the real aim is to ensure they hatch successfully. The Dutch biologist Drent[186] describes the incubation phase as a rather monotonous time for many field biologists, wedged between two far more exciting periods: first, pair bond formation and territory acquisition, and later, the chick raising. However, he notes that this dullness and predictability also makes incubation a very profitable phase for research. A hide can be built near the nest, a thermometer can be placed between the eggs, not to mention all the other latest monitoring techniques. But there's always room for innovation, so Drent constructed tiny temperature recorders[185] that could be built into artificial eggs. He tunnelled beneath a nest, and placed a piece of clear Perspex on the nest floor, so that he could lie underneath and see exactly what the incubating herring gulls did with their eggs (An earlier attempt with a sheet of glass had failed when the parents kept covering the glass with extra nesting material). While Drent lay in his nest tunnel watching the eggs, his supervisor Professor Baerends watched the gulls from a nearby hide (filled with dread at the thought that his PhD student might be suffocating in the stuffy tunnel).

Drent found that by the end of the incubation period the eggs lay horizontally with a permanent upper and underside. Early in incubation, the parent rolls the eggs to ensure that the embryo doesn't become stuck to the shell. Later in incubation, rolling the egg makes sure that the heaviest side is always at the bottom. This helps the hatching chick, as it can peck at the upper side and doesn't get into breathing difficulties.

The parent bird usually incubates the egg to keep it

# REPRODUCTION

Oystercatchers regularly swap shifts during incubation. This series of photos shows the male bending down as he approaches the female. In the third photo he settles himself on the eggs, while the female tosses a piece of grass as she walks off, which is also a typical nest building behaviour. Oystercatchers lay their eggs one to two days apart, but as the parents only begin to incubate once the clutch is complete the eggs hatch at around the same time (lower photo). A hatching chick has a small sharp bump on its bill, the egg tooth, which it uses to crack the shell at the round end of the egg.

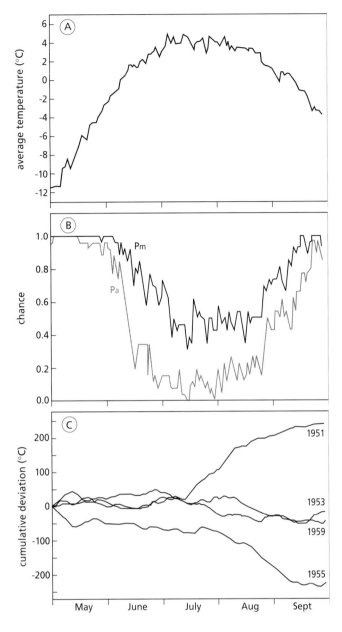

Figure 5.4 The summer weather on the tundra in Barrow, Alaska.[598] (A) Average daily temperature, calculated for the period 1950 - 1975. (B) Probability that the average daily temperature ($P_a$ - grey line) and the minimum daily temperature ($P_m$ - black line) will drop below freezing. (C) Cumulative deviation from the average temperature in 1951, 1953, 1955 and 1959.

that are found at an unknown stage of incubation.[622]

The incubating bird develops a bald spot on its belly where its feathers have fallen out, called a brood patch. This brood patch enables it to transfer heat to the developing eggs. The brood patch is very well supplied with blood, allowing further heat transfer. To incubate, the bird must press its naked belly against the cold eggs. This sounds like an energy-consuming pastime, but is it?

On the High Arctic tundra the subsurface soil is permanently frozen year-round (hence the name permafrost). When the upper layer is not thawed, the birds lay their eggs on top of the ice. Since the parents don't put down a generous insulation layer in the floor of their nests, you would think that it would take a lot of heat to keep the eggs warm enough. And in that part of the world, the air temperature is also often on the cold side. Analysis of weather data collected at Barrow in Alaska[590] showed that the average air temperature doesn't rise above freezing until June and that winter returns in early September (Figure 5.4). And that's not all. Throughout the season, there is a good chance that the minimum temperature will fall below 0 °C, even at the height of summer, when it can periodically freeze and snow (Figure 5.4). So it would seem that it must be very energy-demanding to incubate eggs on the tundra. To find out how demanding it is, Piersma and Morrison[674] caught turnstones on the nest and injected them with doubly-labelled water. Doubly-labelled water sounds very radioactive, but the isotopes it is made from are completely inert and harmless. Using doubly-labelled water enables researchers to estimate an animal's energy consumption; the speed with which these isotopes disappear from the body is determined by taking two blood samples and provides a measure of the metabolic rate, and thus the energy consumption. For the method to work well, the researcher must wait until the doubly-labelled water has spread through the animal's body before taking the first blood sample. After that the released animal should behave normally, as measurements from a completely stressed animal would be useless. Luckily, turnstones are tough little birds and all of the injected birds returned to their nests very quickly. Like many waders that breed in the Arctic, turnstones are surprisingly tame and could be recaptured for the second blood sample without difficulty. The energy expenditure of the incubating turnstones was calculated to be around 4 watts. This might not sound like much, but the energy expenditure of a resting turnstone at a comfortable temperature (so it's not spending extra energy keeping warm or moving) is only 1 watt. Spending four times more energy than its basal metabolic rate takes the turnstone close to the maximum that is physiologically possible (Chapter 3). It wouldn't be possible to spend much more energy, so perhaps there are years when the weather is so bad that the birds cannot incubate their eggs successfully.

Sitting out in the permafrost with a naked tummy is probably not the biggest problem facing the incubating birds. After a while, the eggs would be warmed up and the air between the eggs and the bird's body would also become warmed. Once the eggs were warmed, as long as the bird sat still it wouldn't need to spend too much energy.

warm, but sometimes it must actively protect the eggs from overheating. The egg's water content must also be well regulated for the embryo to develop properly. The embryo must neither drown in its own liquid nor dry out. The eggshell is porous, so oxygen can be taken in and carbon dioxide and hydrogen given off. The developing embryo uses some of the egg contents as building materials, and the rest as fuel for development. During incubation the egg gets lighter and lighter. Since the contents have remained the same, the specific gravity must have decreased. A recently-laid egg is heavier than water and sinks. An egg that has already been incubated for some time is lighter than water and floats. This test can be used to make a rough estimate of the date that eggs were laid for eggs

## Box 5.2 The weight method

It sometimes seems that field biologists spend all their waking hours, and some when they should have been asleep, chasing innocent animals to gather apparently (to the general public anyway) useless data. It's true that some techniques such as attaching transmitters to animals' heads can look rather ridiculous. It's also true that field biologists have developed some rather ingenious methods to gather data on free-living animals. But it's not true that these measurements are useless, nor do the biologists want to chase animals as frequently as possible. In fact, the whole idea is to study the animals in their natural environment with as little disturbance as possible. Ideally the data would be gathered without the researcher needing to lay a finger on the study animals. And occasionally, this can be done.

It's possible to bury a balance under a nest during the incubation period. This does make the nest look a bit different, but that doesn't seem to disturb a ground-nesting bird like the oystercatcher. Burying a balance does cause a sudden change to the environment, but the oystercatchers quickly get used to stepping over the small gap beside the nest. The gap is necessary since the nest sits on a tray that needs to be able to move freely for the balance to work.

However the tray isn't the only problem. The electronics also have to work, and there is plenty of potential for things to go wrong. Instead of being used in a dust-free dry space at room temperature, the equipment is now expected to work on a damp windy salt marsh full of animals. Hares are fond of nibbling cables, so it's a good idea to bury the cable between the balance and the recording unit. Strong winds can also cause problems, as the set-up shakes so much that accurate readings are impossible. But the most annoying problem would have to be an unexpected storm flood. It is very difficult to make the set-up completely waterproof and being submerged in salt water does little to improve electronic instruments.

Only a determined researcher with a high tolerance for frustration can obtain data of this kind. What do these data look like? The simplified graph in Figure B5.2 reconstructs the mass changes of an oystercatcher taking a break from incubation to forage. The plot starts with a completely empty oystercatcher (i.e. nothing in its gut), whose mass (M1) can be worked out from the combined mass of the nest plus the bird. Once the bird is away feeding, we can get no more weight measurements until it returns. But we know how much food the gut can hold (80 g)[462] and how long it is before the bird does its first dropping (28 minutes - their digestive efficiency is much higher than that of humans). The bird will lose weight not only through defecation, but also through water loss from its salt glands. The salt glands are essential for a marine bird like the oystercatcher, as their prey are as saline as seawater, and they also drink seawater. Salt droplets are excreted through the nostrils (they can sometimes be seen hanging from the bill tip), and shed with a quick flick of the head.

The bird foraged for around an hour (during which time it put on weight), then snoozed for around an hour (during which time it will have lost weight) before returning to the nest (bold lines in the plot). It lost mass at a constant rate once it was back on the nest, at 0.23 g per minute, 0.16 g of which is faeces. So we can say that as soon as the bird stops eating it will lose 0.23 g per minute. This value enables us to 'back-calculate' how much weight the bird has put on (and used up) between when it did its first dropping and when it returned to the nest. The calculation would be most accurate if the bird came straight back to incubate after it stopped feeding, but this seldom happens. So from the second mass measurement (M2) we can start to figure out how much food the bird has actually eaten during its lunch break. A problem can arise if the bird completely empties its gut before being returning to the nest. In that case it will only be a little bit heavier, which will reflect the amount of food that has actually been absorbed into its body. The amount absorbed is estimated to be around 11% of the total food intake[462] so in theory you could simply multiply the increase in mass by nine to get the total food intake. In practice, the measurement errors are too large for this method to give an accurate result (as the measurement errors should also be multiplied by nine). Measurement errors also make it difficult to accurately determine how mass slowly decreases through burning of energy (rather than the excretion losses shown here).

In conclusion, even when the equipment finally works well, the birds can spoil everything by swapping shifts at inopportune moments, meaning the food intake may not be able to be estimated well. Only a persistent researcher comes up with results, but these results are well worth having. The food intake at night seems to be as high as it is during the day.[463] Something that is difficult to determine on free-living animals using other methods, but see Sitters.[1007]

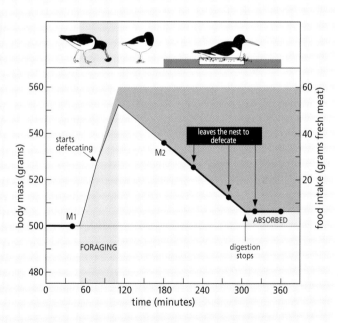

Figure B5.2 Stylised diagram of the mass changes of an incubating oystercatcher as it forages and then digests its meal, based on data collected with a balance buried beneath the nest. The bold line shows the changes in body mass of the bird on the nest (measured continuously). Dots mark when the bird starts sitting, or leaves the nest. The thin line is the reconstructed mass of the bird while the bird was off the nest (see text). The left-hand y-axis gives the body mass of the bird, the right-hand axis the total food consumption.

REPRODUCTION

Chicks have to be regularly brooded by their parents to keep warm, particularly in the first days after hatching. This is true both for red knot chicks being brooded in the far north (photo left below) and for redshank chicks being brooded in more boreal areas (photo left above). When danger threatens, the chicks press themselves to the ground and stay as quiet as a mouse (photo right above). They must rely on their beautifully camouflaged plumage, which is a perfect match for the colours of their surroundings, to keep them safe. The bar-tailed godwit chick's grey-brown fluff blends in well with the grey stones and lichens of the far northerly tundra (photo right below).

The problems would come when the birds leave the nest to forage in the bitter cold. The few doubly-labelled water data available on incubating knots showed that the birds that were incubating throughout the measuring period had significantly lower energy expenditure than knots that stopped incubating to forage.[879] As a rough estimate (based solely on the tiny sample of one foraging bird and two incubating birds), foraging on the High Arctic tundra may require twice as much energy as incubating.

The situation is very different for oystercatchers that nest on the salt marshes of Schiermonnikoog in spring. The energy expenditure of the incubating oystercatchers has been estimated in several ways:[458, 463] (1) using doubly-labelled water, (2) by converting observed time budgets to energy consumption, (3) by converting observed prey catches to energy intake and (4) by estimating energy intake from changes in the birds' mass (see Box 5.2). All of these studies came to the same conclusion: the oystercatchers' energy expenditure during incubation was minimal. Even oystercatchers in captivity had a higher energy expenditure than oystercatchers incubating on the salt marshes at the same time. Just like in the Arctic, it's cheaper to sit quietly incubating than it is to walk around. The explanation for the much higher energy expenditure of Arctic breeding waders has two components. First, it's much colder on the tundra than in temperate zone. Secondly, the red knot and turnstone have lower critical temperatures than larger birds like the oystercatcher. This means that red knots and turnstones

must start using extra energy to keep warm at a higher ambient temperature than oystercatchers (Figure 4.4).

How long must the parents sit on the eggs before they hatch? The egg's size is one factor. The larger the egg, the longer it takes before it hatches.[186] Hummingbirds incubate their eggs for less than two weeks, while albatrosses must wait more than two months. The incubation times of the waders discussed in this book vary from around three weeks for sandpipers to around four weeks for the curlew. Larger animals take longer to reach adulthood and this size effect begins with growth in the egg. But as usual, it's not quite that simple. In some species, the chicks are more developed when they hatch than in others, which requires a longer incubation period. Precocious wader chicks have to be highly developed when they hatch, as they immediately need to move and feed independently. There are also differences between species in the amount of yolk left at the time of hatching. A large yolk at hatching allows the incubation period to be reduced. The chick's yolk sac provides it with an energy buffer should it not do so well on its own straightaway, rather like the fat stores that adult birds put on to survive periods when food is unavailable.

Temperature also plays a role. Chemical reactions occur more quickly at higher temperatures and growth is essentially a very complex chemical reaction. The growth of the embryo therefore speeds up when the temperature is higher, unless the reaction rates get out of control due to the temperature becoming too high. In general, most species incubate their eggs at a similar temperature, but there are large differences in when the parents start to incubate the eggs properly. There are also large differences in how often the eggs can be left alone and are exposed to periods of cooling.

These differences in incubating behaviour are also found within species. The parents that incubate irregularly have to incubate longer in total than those that sit continuously on the eggs. How quickly the embryos develop also depends on the weather. If it's very warm, the eggs cool down more slowly when they are not being incubated and there will be little delay in their development. It usually gets warmer through the breeding season. So you would expect the incubation period to become progressively shorter through the season. Incubation times of 19 to 34 days have been observed for avocets.[409] As predicted, the incubation time decreased over the course of the season. And this can be partly explained by differences in the ambient temperature. However the ambient temperature doesn't seem to be the only factor involved. Even when differences in temperature are taken into account, later clutches still hatch faster. So what's happening? Although every avocet chick has to leave the egg sooner or later, his or her survival is higher as an egg than as a chick. Hötker[409] concluded from his data that there were advantages to the chick in staying in the egg for as long as possible early in the season and for as short as possible at the end of the season. This is connected in a rather complicated way to the embryo's, and later the chick's, rate of growth and the way that predation risks change over the course of the season. Early eggs survive better than later eggs, and early young survive better than late young.

Whether or not we believe this complicated story about the optimal time to hatch, it's clear that there is a lot of variation between eggs in the hatching time. This means it's a good idea for the incubating parent bird to keep sitting even when the eggs seem slightly overdue for hatching. And that's what they do. Unfortunately, this sometimes causes the parent birds to spend days and days incubating eggs that are never going to hatch because the embryos have died. The record for continuous incubation without interruption is held by a pair of oystercatchers who sat for 46 days,[193] although the normal incubation time is around 28 days. Maybe they would have sat even longer if the observers hadn't felt sorry from them and replaced the dead eggs with a live egg from another nest. However the pair probably wouldn't have carried on with their sad attempt endlessly. Oystercatchers have never been seen incubating dead or infertile eggs into autumn. Somehow they must have an idea of the maximum incubation period.

Remarkably enough, oystercatchers don't seem to have any idea of the shortest possible incubation period. When eggs that have just been laid are swapped for eggs that are due to hatch, the parent birds will start to feed their adopted young as soon as they hatch.[389] Maybe the birds are a bit surprised but they don't seem to treat their 'instant young' any differently. And maybe this isn't so surprising. These sorts of exchanges don't occur naturally, as when a different female has laid eggs in a pair's nest (which only happens rarely), her eggs will hatch later than the pair's, not earlier. Young that hatch slightly on the fast side of the normal incubation period don't show any growth delays. They are probably very healthy and well worth the parent's hard work.

### 5.2.5 Raising young

Waders are precocial birds and their chicks must stand on their own two feet right from the start. When the chicks hatch, they still have some yolk left, which they use as an

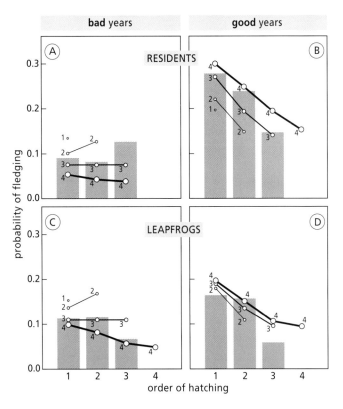

Figure 5.5 What effect do siblings have on an oystercatcher chick's chance of fledging, and is the hatching order important? The plots show the probability of fledging in relation to hatching order, averaged across all clutch sizes (grey bars), and shown for clutches of two, three and four eggs (based on a statistical model: dots joined by lines). Results are shown for birds in resident and leapfrog territories, in both poor (A and C) and good (B and D) years.[392] It is much less energy demanding to raise chicks in resident territories than in leapfrog territories. Types of oystercatcher territories are discussed further in Section 5.3.3 and Figure 5.11.

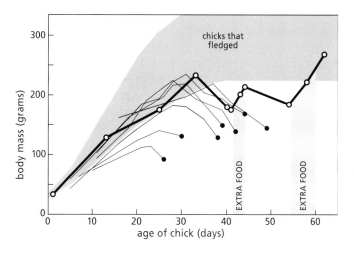

Figure 5.6 Growth curves (body mass in relation to age) of oystercatcher chicks from the salt marsh on the Dutch Wadden Sea island Schiermonnikoog. The shaded area represents the body masses of chicks that successfully fledged. Growth curves of chicks that eventually starved to death are shown with thin lines. The thick line shows the body mass trajectory of a starving chick that was fed experimentally twice, and immediately increased in mass as a consequence.[459]

energy supply. They leave the nest soon after they hatch, but this doesn't mean that their parents have abandoned them. One or both of the parents continues to care for the chicks until they fledge (or are very near fledging). The parents will (1) brood the chicks, especially little chicks, to prevent them from getting chilled, (2) protect the chicks from danger, which varies from keeping a lookout and raising the alarm to aggressively chasing off any potential predators, and (3) guide the chicks to rich and safe feeding grounds where they can find food for themselves. Oystercatchers take their parental duties a step further and actively feed their young till well after they have fledged. All waders eventually abandon their offspring on the breeding grounds. The young birds then have to make their own way to the wintering grounds.

Many chicks never need to find the wintering grounds. Newly hatched wader chicks may look like cute little balls of fluff, but Mother Nature isn't known for being kind to little fluff-balls. Although many eggs are predated, relatively more lives are lost at the chick stage.[39, 270] Even in good years, only a small minority of the very young chicks will actually fledge. Nature lovers who go out once in a while and gaze around through their binoculars won't really notice this. Only the poor old field worker who closely monitors the eggs and the chicks will be confronted daily with new disappearances. By the end of the field season, when more than half of all the research plans have failed and the accumulated sleep deficit begins to reach unbearable levels, this loss of young life can sometimes lead to dark thoughts. But once this person has had a good sleep and is back in the office a more cheerful state of mind returns, which is why scientific papers nearly always discuss breeding success rather than breeding failure. So what are the factors that determine breeding success?

The first critical point is hatching. The chick first starts to peep while it is still in the egg. This helps it make contact with its parents. Peeping may also be a way for the young to communicate amongst themselves and synchronise their hatching. The fast individuals could hatch a bit slower and the slow individuals could speed things up a bit. It's not a good idea to hatch first and have to wait for your parents to finish incubating before they can give you any attention. It's also not a good idea to hatch much later than your siblings; your parents might think you're not going to hatch so they and your siblings might leave the nest without you. Every so often, field workers find a nest where all but one of the eggs have hatched and sometimes a cold little bill sticks helplessly out of that last egg. You'd expect hatching to be well synchronised in precocial birds, and it is. It takes grey plovers a week to lay their eggs, but there is less than a day between when the first and the last eggs hatch.[116] Eggs only begin to develop when they are incubated so by waiting until the clutch is complete before incubating, the parents help to ensure their chicks hatch at the same time.

Synchronised hatching seems to be less important for altricial birds and there can be large differences in the times at which the young hatch. Lack[489] suggested that these differences are due to the parent birds beginning to incubate before the clutch is complete, and that this may well be intentional on their part. The chick that hatches last would always be less developed, and would lose further ground if there were food shortages. In raptors, such as the barn owl,[139] the difference between the first and the last chick can be especially large, and when there are food shortages the smallest chick is sometime torn to pieces and eaten by its elder siblings.

Oystercatcher chicks are precocial but the way they are fed by their parents is more typical of an altricial bird.[759] Chick feeding makes the oystercatcher unique amongst West European waders and has far-reaching consequences for the oystercatcher chicks. When the parents don't provide enough food, the cute little fluff-balls must engage in a life or death battle with their siblings. The breeding territory is then less like a happy family home and more like a prison where the inmates must fight for food to survive. Providing enough food for the chicks is always a problem in territories where the birds have to transport the food over a large distance (see Section 5.3.3). But even in good territories the parents may be unable to bring enough food in bad years.[392] The eggs that were laid first usually hatch first. These chicks are then dominant over the others that hatch somewhat later; they get more food, grow better and are more likely to fledge.[359, 392, 756, 757] Or at least, this is what happens in good years (Figure 5.5). In bad years the probability of fledging is much lower as you'd expect. But in these difficult years, it matters little whether a chick has hatched first or last. And when you determine which chicks are dying at what time, the results are difficult to reconcile with the idea that asynchronous hatching helps the chicks establish a hierarchy, allowing the weaker chicks to quickly drop out during food shortages. The dying chicks live longer in the bad territories than in good territories. In bad ter-

Most wader parents warn their chicks when danger approaches. Only some species, like the avocet, actively attack the potential threat, in this case a shelduck (photo left). This turnstone is also on the attack (photo below) and is diving at the photographer, who has approached the chicks too close for the turnstone's liking.

ritories, only one chick at most will fledge (see Section 5.1.5 about the costs of reproduction). But in bad territories, there are often still two chicks alive 10 days after they have hatched. If there is still more than one of these chicks alive after 25 days, none of them will survive.[458] Oystercatcher chicks are real survival artists, and the starving chicks can hang on to life for a very long time. Chicks that have grown well fledge after four weeks. Chicks that don't grow particularly well will take around five weeks. But sometimes a starving leapfrog chick will last seven weeks before finally dying. (Figure 5.6).[459]

The oystercatcher's 'drop off' system is far from perfect, so why don't they just lay fewer eggs? Oystercatchers don't seem like the type of birds that should be trying to use a drop off system anyway. These systems are meant to be prevalent in species that are unable to predict during egg laying how much food will be around for their chicks. It would be a waste for such birds to only lay a few eggs and then find that there would have been enough food to raise a larger brood. However if they find they have laid too many eggs compared to the amount of food available, they require a system to quickly get rid of the young that they won't be able to raise. Perhaps owls and falcons aren't very good at predicting food availability, but oystercatchers must be able to predict what the situation will be in a few weeks time, based on what they see in early spring. The large changes in the abundance of benthic animals due to spatfall and mortality happen in autumn and winter; during spring the animals simply grow. Similarly, pairs in bad territories wouldn't suddenly discover that their territory is far from the feeding grounds once their chicks hatch. It's obvious which territories are good and which are bad from the instant that they start to defend them.

The oystercatcher's main problem is that it can't accurately predict how many eggs and chicks will be lost through predation. Oystercatchers' fierce attacks may frighten and distract gulls, but they can't keep them away indefinitely. Gulls are numerous and very skilful fliers. Oystercatchers on Schiermonnikoog lost an estimated 50% of their eggs and 90% of their chicks to predators.[392] Oystercatchers lose so many eggs and chicks that they must start with plenty in order to have a few left at the end. The parents cannot predict how many chicks will escape the predators. So, sometimes too many survive and there is not enough food for them all.[392, 458]

Getting your own meal and not becoming someone else's are two important challenges for a chick. A third problem is keeping warm. The embryo already begins to produce its own heat in the egg. By the end of the incubation, a developing herring gull embryo produces more than half of the warmth it needs to be able to hatch.[185] But the eggs still need to be brooded to remain at the right temperature, as do small chicks. As they develop, the young become increasingly resistant to cooling off, and sooner or later, the time comes when they don't need their parents' help to keep warm. How long it takes to reach that point varies greatly between different spe-

# REPRODUCTION

Wader chicks must be able to stand on their own two feet as soon as they have hatched. Hence the relatively well developed legs and feet of this whimbrel (photo top right), which is just a few hours old. The chicks must also be able to find their own food. Like many wader chicks, this knot (photo above) feeds on insects that it finds amongst vegetation. Young brent geese feed on small plant shoots, rather than insects that could fly away. They don't have to worry about their food being accidentally chased away by their siblings, so they can stay close to each other and their vigilant parents (photo below).

cies. Plover chicks grow slowly and need their parents to be available as heaters for a longer period. Sandpipers grow much faster and are independent of their parents as far as heating is concerned much earlier. Why?

A growing chick is actually changing from a reptile that has no internal thermostat to keep the heaters running at the required level, into a bird that can maintain its own body temperature. The advantage of being a reptile is that your energy consumption is low. The disadvantage is that you can only be active on days when you can warm up in the sun adequately. The advantage of being a bird is that you can always be active, but the disadvantage is that it requires a higher energy consumption. A bird that cannot find enough food quickly runs out of stored energy. If it's difficult to find food, the chick is better off staying like a reptile for as long as possible. But if there is plenty of food, there is no need for economical growth. The suggestion is that the slow-growing plovers, like the lapwings, have evolved in poor dry areas, with regular food shortages. The quick-growing sandpipers, like the godwit, would have evolved in areas with better food availability. People don't often think of the tundra as providing a plentiful food supply, but seen through the eyes of insectivorous chicks, it is a rich feeding ground.

Temperature must influence chick growth. Above a certain temperature, the chicks can be continually active and growth conditions are optimal, as long as it doesn't become too hot. However low temperatures bring high heating costs. On cold and wet days, chicks that cannot keep themselves warm have to be brooded long and often, during which time they cannot feed. Eventually, this can reach the point where the chick is unable to gather enough food to enable it to grow.[39] So there is a critical lower limit below which the chicks are unable to grow. This limit differs for the different species, but for all species it becomes lower as the chicks age.[40] You would expect that chicks that grow up on the desolate tundra become energetically independent the fastest. And this is what we find when we compare chicks at the same ambient temperature of 5 °C (Figure 5.7). Chicks of the very northerly breeding red knots are already able to survive without regular brooding after a week. Redshanks and godwits need somewhat longer to become independent. Lapwing chicks would still have to be brooded for considerable periods after more than one month, which is probably why lapwings don't breed so far north.

Wader eggs may be beautifully camouflaged, but the chicks' camouflage is even better. The chicks' fluffy down looks more like vegetation than the smooth surface of the eggs. Plus chicks can walk to a suitable hiding place before quietly pressing themselves to the ground. They know when they need to hide because they can hear their parents raising the alarm when danger threatens. Very young chicks immediately press themselves to the ground. The older the chicks the further they spread out before taking cover. This makes it more and more difficult to find them. When the danger has passed, the parents entice their chicks back out by making soft calls. If the disturbance lasts for a long time the hidden chicks may get too cold and hungry. This may be why young knots will come out of hiding after a while and run off, even though the parents are still frantically making alarm calls. Perhaps at some point, when the danger has persisted for a long time, it's better to run the risk and gather food.

Some species defend a territory while they have chicks where the young can forage. Oystercatcher territories have strict boundaries that change little from year to year. Chicks that stray into a nearby territory run the risk of being pecked to death by the neighbours. It is horrible to see the adult's long red bill disappear into the back of a little fluffy chick, which then lies convulsing on the ground. Very occasionally, a chick that crosses a boundary is adopted by the neighbouring pair. Such adoptions can only be determined if the chicks have been individually marked as soon as they hatch. A successful adoption is probably most likely to occur when the chicks are still very small and the parents don't recognise their own chicks so well. This is the case in gulls, where adoption occurs regularly and has been well studied.[95] In gulls, it seems to be the chick that initiates the adoptions, as changing families increases its chance of survival, mostly by it being older and stronger than its new siblings. Having an extra mouth to feed is not usually profitable for adoptive parents or their current chicks. But geese are the exception to the rule. Goslings feed themselves and large families will beat small families in the competition for a good feeding area. Consequently, it's advantageous for a family of barnacle geese to adopt.[524]

Rather than defending a specified area, many waders defend the area around their young. Over time, the young can cover quite large distances, and so the defended area moves with them. This is the system used by red knot families that can cover many hundreds of metres per day, or sometimes even one kilometre.[879] Female curlew sandpipers take a different approach and intentionally migrate with their chicks to an area that other families also head

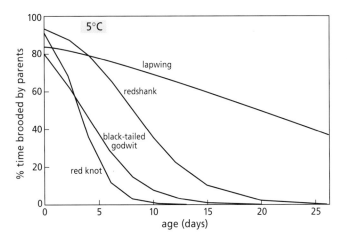

Figure 5.7 The amount of brooding a chick requires depends on its age and species. The time that a chick has to be brooded by its parents to maintain body temperature has been calculated for chicks of different ages at an ambient temperature of 5 °C, a typical Arctic climate.[879] Chicks of the highest Arctic species, the red knot, can thermoregulate well without their parents after just five days. Black-tailed godwits and redshanks are dependent on their parents for longer, while lapwings, which do not breed anywhere near as far north as knots, must rely on their parents for more than a month.

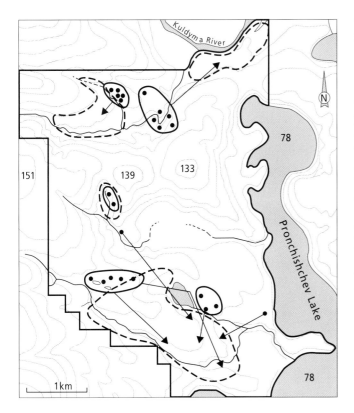

Figure 5.8 Nests and chick-rearing areas of curlew sandpipers on an area of tundra in Taimyr in 1991.[776] Nests are marked with black dots, while areas used to raise chicks are encircled with dotted lines. In places where females have taken their chicks into the same area, their nests are encircled with solid lines, and the direction they moved is shown with an arrow.

for (Figure 5.8). At this site, loose aggregations begin to form, which may also involve other species such as turnstones, little stints and sanderlings. This seems to benefit the participants through their being able to discover any threats more quickly and to raise the alarm together.[776] These family gatherings are so conspicuous in species like the whimbrel, bristle-thighed curlew (*Numenius tahitiensis*) and Hudsonian godwit (*Limosa haemastica*) that they are described as crèches.[691] These larger wader species don't just make a fuss when danger threatens, they actively attack the threat. However crèching behaviour is not limited to waders. Eiders and shelducks with young also tend to search for conspecifics. Maybe taking their offspring to crèches enables the parents to leave their young earlier.[496] Not for a few hours, like human parents that take their offspring to a crèche, but permanently.

## 5.3 The best place to breed

A bird's choice of breeding site is made on various scales. On the largest scale, the bird chooses the geographical location: Greenland, the Taimyr Peninsula or a Wadden Sea island? Within that locality the bird has to choose a suitable living area. This area may be defended against conspecifics, or conversely, the species may breed in colonies. On the smallest scale, the bird has to choose where to make its nest. This is related to what we've just discussed, the problems birds face when they incubate eggs and care for their young. So let's start with this, the smallest scale.

### 5.3.1 Choosing a nest site

Of the many dangers that threaten wader eggs and chicks, predation is undoubtedly the greatest. Waders, with their often fragile bills and long thin legs, are not built for active combat against predators. There are always strong, fast and dangerous predators around that would make short work of an egg, a chick or even a parent bird. For most species, all they can do is try to breed as discreetly as possible and hope that they are not discovered. So first of all, the plumage of the incubating parent bird must not be obvious amongst the vegetation. This is no problem for curlews and whimbrels since they wear the same cryptic brown plumage year-round. But many other waders are brightly clad in red-brown or black and white nuptial plumage during the breeding season. When you see a grey plover on the mudflat in spring, it's difficult to imagine that its conspicuous plumage would not be equally eye-catching on the breeding grounds - but it's not. This is partly because when a wader is incubating all that is usually visible is its back, not its strikingly coloured underside. But the birds must also nest in vegetation that matches their plumage. American golden plovers are hardest to spot when they nest between lichens, rather than between sedges. And the nests amongst lichens have a lower risk of predation.[116] It may be that the southern population of the Eurasian golden plover has less contrasting plumage than the northern population, because the southern birds breed in a less contrasting environment.[116]

Sometimes the nest must be left for a short time. While the shelduck's white eggs are well hidden down in its nesting cavity, white wouldn't be a good colour for eggs that lie out in the open. Wader eggs are beautifully camouflaged to fit with the vegetation near the nest and are usually light brown with darker brown spots and stripes. Oystercatcher, avocet and ringed plover eggs are rather pale and more yellow than brown, which matches the often sparsely vegetated soil around the nest. Golden plover and grey plover eggs are darker with a golden yellow gleam, mimicking the lichens amongst the stony tundra where they nest. Godwits breed in quite high lush grass, so their eggs are green with an inconspicuous brown speckle. As well as differences in egg colour between the species, eggs of different females can also look quite different. In cuckoos (which lay their eggs in other birds' nests) this variation is related to the host's eggs. Females are specialised on a particular host species and lay eggs that look like those of that species. It's possible that female waders that lay more sandy coloured eggs nest more often in the sand and females that lay more stony coloured eggs nest more often amongst rocks, but nothing is known about this.

So which partner chooses the nest site? If it's the male, it seems unlikely that he would be able to take his partner's egg colour into account. In some species, such as the ruff and the curlew sandpiper, we know that the females decide on the nest site, as the males are not involved in

# REPRODUCTION

This grey plover removes the eggshell of a newly hatched chick as quickly as possible, since the white inside of the shell creates a dangerous break in the nest's camouflage (photo top left). These chicks are even more beautifully camouflaged than the eggs that they have recently hatched out of (middle photo). Soon, the family will leave the nest, never to return.

Flooding during unusually big high tides is a real risk for birds that breed low on the salt marsh. These oystercatchers (photo left below) are lucky that their breeding territory wasn't flooded earlier, as their chick can swim well but eggs might have been washed away.
It is not known how this bar-tailed godwit has assessed the potential hazards such as flooding around his nest site near this small lake in Lapland (photo page 263). Maybe he can tell by the local vegetation, or perhaps he has been returning to the same area for many seasons and has learnt from experience.

nesting at all. In redshanks it seems to be the males that choose the nest site. Male redshanks arrive on the nesting territories before the females, and scrape out a nest site. They then perform a little welcoming ceremony to any female who comes to have a look.[138] Grey plover and golden plover females inspect the nest scrape after the males have left. As part of her inspection the female may also show nest building behaviour, such as scraping out the nest floor or covering it with pieces of lichen. Male and female oystercatchers both scrape out nests, and may take turns working on the same nest site. They sometimes appear to be holding discussions, or even negotiations, over which nest is in the best spot. Amongst species like the redshank and the golden plover you can imagine

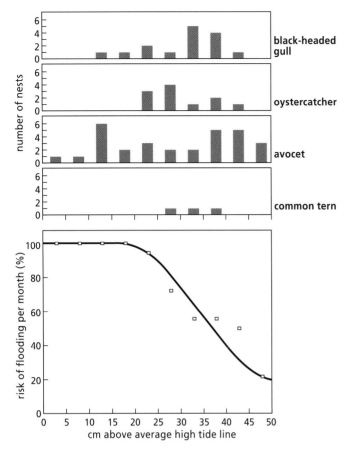

Figure 5.9 How high on the salt marsh do different bird species breed, and what are the risks of their nests flooding? The elevation above average high water was determined for the nests of four coastal bird species in salt marshes along the Frisian Wadden Sea coast near Holwerd.[960] Long-term tidal height measurements were used to calculate the chance of flooding at least once per month in the breeding season for a given elevation (lower plot).

hypothesis.[309] To test this hypothesis, chicken eggs and artificial eggs were placed next to the nest of an aggressive species, the black-tailed godwit, and also at some distance from the nest. As expected, predation on the 'unprotected' eggs was much higher than on the 'protected' eggs.[353, 854] The same experiment was carried out using grey plover nests on the tundra, with comparable results.[498] Greenshanks become very noisy at the first hint of danger,[116] and it is assumed that species breeding near them profit from the greenshanks' presence, more from their vigilance than their defence capabilities.

Flooding is a serious risk for waders that nest on salt marshes. The higher on the salt marsh the birds nest, the smaller their risk of flooding. This might lead you to expect that all birds would breed as high on the salt marsh as possible. Curiously, many birds nest so low that they run a significant risk of being washed away (Figure 5.9). It's easy to imagine that there is less protection against ground predators higher on the salt marsh. In lower areas, channels would certainly be a real barrier to hedgehogs. If so, the birds must make a trade-off between the risk of predation by ground predators and the risk of flooding. It may also be that the vegetation on the drier parts of the salt marsh is too high and too rough and therefore less suitable for nesting. For oystercatchers, the preference for breeding low on the salt marsh is related to the proximity to their feeding grounds. The heightened risk of washing away is more than compensated for by the better food provisions for the young.

So the nest site needs to be suitable for incubating the eggs, should hold few risks for the eggs and the incubating parent bird, and also offer opportunities for the chicks to find food. The best nest site is clearly not always the best place for the chicks. As a consequence, the parent birds often move away from the area with their chicks once they have hatched (Figure 5.8).

### 5.3.2 The breeding habitat

Every field biologist knows what is meant by the word habitat. In the dictionary it's defined as "the natural home of an animal or plant", so it's about the characteristics that make an area suitable for a species to survive or breed there. Some species live in the same habitat year-round. Waders usually don't do this, and for most wader species it's possible to identify a breeding habitat and a wintering habitat. These two habitats may be very different.

Anyone who's seen a curlew out on the tidal flats up to its eyes in the mud knows immediately why curlews have such long bills – it's to catch deep-living prey. If you see curlews foraging in the dunes, the need for a long curved bill is not so obvious. It has been suggested that their bills may be useful for plucking berries, but there is no evidence of this.[160] So curlews probably don't decide to breed in dune areas because they can find a lot of food for themselves there. It's much more likely that the dunes provide good nest sites where their chicks can find enough food to grow without being chomped by the first predator that comes along. The breeding habitat is probably a compromise between a site where the parents can

that the male has made a series of proposals and that the female decides which she finds the most appealing, since in the end she's the one that has to lay the eggs.

What can an incubating wader do to keep itself and its nest as safe and as inconspicuous as possible? He or she can nest as close as possible to a species that aggressively attacks predators and is reasonably successful at chasing them away. Lapwings always go for potential egg predators and make a huge uproar while doing so. When lapwings and redshanks nest in the same area, a redshank nest can often be found in a clump of grass not far from a lapwing's nest. Common and Arctic terns are also very aggressive towards possible robbers. So it's no wonder that Kentish plovers in Zeeland, in the southern part of The Netherlands, often breed amongst common terns, and that purple sandpipers in Iceland search for the protection of Arctic terns. Up on the tundra, grey plovers are one of the aggressive species, and curlew sandpipers and red knots are happy to make use of this by nesting nearby.[138] Also on the tundra, some waders nest very near possible predators like skuas, probably because they keep the even more dangerous predators at a distance.[774] The idea that breeding in the vicinity of an aggressive species provides additional protection is known as the umbrella

survive well and a site where the young can grow safely. Even from behind a desk, it's easy to think of potential conflicts that could arise. The small young can almost certainly forage profitably on prey that are too small for their parents. And vice versa, prey that are suitable for the parents would often be too big for the young, especially when chicks are still tiny. The chicks may also want good cover where they can hide from predators, while the parent birds can simply fly off. The parents can also fly to a distant pond if they are thirsty, but the young don't have this option. Sometimes there is plenty of water around, but the salinity of the water is a problem, not for the adult waterbirds that can easily handle the salt with their salt glands (Chapter 2), but for their chicks. Eider ducklings can't survive their first days of life without being able to drink fresh water.[833]

This is why tidal flats are for the most part unsuitable places to raise young, even though the breeding season coincides with the maximum food availability.[50, 982] Intertidal areas may be great for the adult waterbirds, but are largely out of reach for the flightless downy young. And if the chicks did walk out across the exposed flats, they would be easy prey for gulls and other predators. The only bird that can successfully take its downy chicks to the barren mudflats is the eider duck. The little eider ducklings can swim well, so crossing small channels is no problem for them. They can also dive well to escape from gulls. Plus the female fiercely defends her young. Often various females will gather with their young in a crèche (see Section 5.2.5). The more watching eyes, the greater the chance that a predator will be seen in time. A single female is a formidable adversary for a herring gull, but when the gull is confronted by a group of furious mother ducks it may find itself in serious danger. Eider ducks have been observed catching a gull and drowning it.

Shelducks also take their chicks down to the bare mudflats, but unlike the eider ducks, this is usually an unsuccessful venture. This may seem strange since the male shelducks help the females care for the young, and both parents are very aggressive towards potential predators, even chasing predators up into the air. And perhaps that is the problem. The shelduck male chases after the first gull and the shelduck females chases the second. Meanwhile, a third gull kills the ducklings. The whole clutch may be lost in less than an hour, bringing that breeding season to an end for the shelducks that were careless enough to try and take their chicks down to the tidal flat. The aggressive behaviour of the shelduck parents works very well in enclosed habitats such as ditches, pools and the edges of gullies, but not on the open tidal flat.

Is there any evidence in support of the idea that areas suitable for eider ducklings are less suitable for their parents, the main theme of this section? Swennen[833] studied the feeding behaviour of eider ducklings by disguising himself as mother duck out on the tidal flats with her chicks. His disguise was not particularly elaborate! Konrad Lorenz had discovered that ducklings and goslings accept the first moving object that they see after hatching as their mother. So to become an eider duck mother, Swennen only had to put his bald head in an incubator of hatching eider ducklings. From that moment on, there was a close bond between the researcher and his chicks. He learnt to tell from their peeps when they had found food or a seepage of fresh drinking water. And after a day or six the ducklings warned him when they could see a distant gull or raptor, instead of the other way around. When Swennen took his ducklings to a tidal flat in 1976, although they occasionally found a small worm it was clear that they couldn't find enough food. Only chicks that were supplementary fed during the high tides survived. In that same year hardly any wild ducklings fledged in Swennen's research area in the Wadden Sea.[835] 1988 was a much better year for both the wild and the experimental ducklings. It was obviously a much better year for intertidal benthic animals and the chicks fed well on the excess of amphipods, ragworms, lugworms, small shrimps and small crabs. However these are not the sorts of prey that would fill an adult eider's stomach. Adult eiders need to feed on mid-sized mussels and cockles.[830] Such prey reach their highest densities much lower in the tidal zone (Section 1.4). So even in years when the ducklings can find enough food on the upper tidal flat, the mothers that guard the chicks are in a bad way. For the adult females the upper tidal flat is like an empty pantry. And all this after the females have already had to incubate the eggs without any assistance from the males, causing them to endure weeks of fasting.

Oystercatchers also take their chicks down to the tidal flats. These chicks are adept at swimming and diving,

| | Coastal or Inland | Arctic desert | Tundra | Forested tundra | Taiga | Freshwater marshes/bogs* | Salt marshes and fields* |
|---|---|---|---|---|---|---|---|
| oystercatcher | C/I | | | | | | ▨ |
| avocet | C | | | | | | ▨ |
| ringed plover | C | ▨ | ▨ | | | | ▨ |
| Kentish plover | C | | | | | | ▨ |
| grey plover | I | ▨ | ▨ | | | | |
| golden plover | I | | ▨ | ▨ | ▨ | ▨ | |
| red knot | I | ▨ | ▨ | | | | |
| sanderling | I | ▨ | ▨ | | | | |
| curlew sandpiper | I | ▨ | ▨ | | | | |
| dunlin | I | | ▨ | ▨ | | ▨ | ▨ |
| little stint | I | ▨ | ▨ | | | | |
| purple sandpiper | C/I | ▨ | ▨ | | | | |
| bar-tailed godwit | I | | ▨ | ▨ | | | |
| black-tailed godwit | I | | | | | ▨ | ▨ |
| curlew | I | | | | ▨ | ▨ | |
| whimbrel | I | | ▨ | ▨ | ▨ | ▨ | |
| spotted redshank | I | | ▨ | ▨ | ▨ | | |
| redshank | C/I | | | | | ▨ | ▨ |
| greenshank | I | | | ▨ | ▨ | ▨ | |
| common sandpiper | I | | | ▨ | ▨ | ▨ | |
| turnstone | C | ▨ | ▨ | | | | |

*temperate zone

Table 5.2 Breeding habitats of various wader species.[220, 678, 691]

which can come as something of an unfortunate surprise if you're trying to catch them. However the chicks always stay near the cover provided by the salt marsh, especially when they are small. This means that only the upper edges of the tidal flats are suiting rearing sites for young oystercatchers. The same applies to avocets and redshanks that, like oystercatchers, regularly nest in salt marshes. The redshank and avocet chicks can search for insects in the salt marsh and for ragworms, amphipods and other benthic animals in the gullies, and hide in the salt marsh vegetation when danger threatens. For oystercatcher chicks the salt marshes are simply a place to hide; the parent birds must bring in all of their food from the tidal flats. This results in the adult oystercatchers usually eating smaller prey items than those they feed to the chicks.[759] This is the reverse of what happens in wader species whose young forage for themselves. Since the oystercatchers have to carry each food item individually to the chicks transport costs come into play, especially when the young grow larger and require more food. It's more efficient for them to take larger prey to their young and eat the small prey themselves.[504] This doesn't diminish the fact that the prey found near the salt marsh are usually very small from an adult oystercatcher's point of view. However in spring and summer the parents manage to live off it.[104] At this time of year, the ragworms often come up to feed and the Baltic tellins are fat and nearer the surface. It is debatable whether the oystercatchers would show much interest in these areas if they weren't raising young. They are rarely seen in these areas outside the breeding season.

There are also other waterbird species that breed along the coast. Rather than breeding on the densely vegetated salt marshes, ringed plovers and Kentish plovers nest on the sandy, very sparsely vegetated, transitional areas where mudflats meet the beach.[876] There the chicks mainly forage on insects that they find along the high tide line in the rotting plant material left by the sea. So the sea is very important to these species in the breeding season. It may even be that the freshly deposited and still wet debris contains the most insects.[876] The sea is also important for breeding turnstones, but they are not at all partial to sand and mud. In contrast, the rockier the better, and the low rocky islands in the Baltic Sea are a favoured breeding habitat.

With this, our list of European waders that breed along the tidal flats or the sea is exhausted. Other species breed in more freshwater areas such as damp fields, dunes, peat land and heather areas, and on the taiga and tundra. For example, the knot breeds on islands and peninsulas in the Arctic Ocean, but never really along the coast itself. You can't go further north than that and it's bitterly cold. These habitats have already been described in Chapter 3. Table 5.2 provides a rough overview of the breeding habitats of different wader species. Some species are found in the same habitat more often than other species. Some, like the avocet and the greenshank, are never found together. Because Table 5.2 is very generalised the differences between the species appear less dis-

Redshanks defend a territory in the breeding season. Like most territorial species, they usually limit this defence to display behaviour, but occasionally it comes to blows (photo left).
Avocet pairs often breed in colonies on the salt marsh, but each defends their own feeding territory from the edge of the salt marsh to the tidal flat. There the chicks must find their own food. Every so often they take a break from feeding and warm up under a parent's wings and belly. All four chicks have survived their first few days and now hardly fit under their parent (photo page 267).

tinct than they would from a detailed description of each species' breeding habitat. For example, in Table 5.2 it appears that there is no difference between the habitats used by red knots and sanderlings; both species breed inland on the High Arctic tundra. And it's true that both red knots and sanderlings are usually found on dry slopes not far from the coast. Their nest sites are usually on stony plains, sparsely vegetated with the low-growing Arctic willow, with clumps of mountain avens (*Dryas*) and *Saxifraga*, and almost always have a good view. However the knots tend to breed on the somewhat damper clay slopes on the undulating tundra. Perhaps access to the nearby water's edge is important for their young.[691] Ferns reviewed resource partitioning amongst the various wader species that breed on the tundra and found that the differences were related to foraging behaviour.[281] Ringed plovers like open areas with little vegetation that they can race through while visually hunting insects. However, these insects need vegetation, which is why you don't find ringed plovers on the completely barren stony desert. Turnstones like dry dense vegetation that they overturn in their characteristic way to search for any prey that might be hiding underneath. Dunlins have a long thin bill that they use to hunt by touch in small pools and amongst damp vegetation. Interestingly, Ferns didn't find any obvious differences between the species as far as areas for raising chicks were concerned.[281] This fits with the idea that the habitat requirements of parents and their young differ. Perhaps while the chick is still in the egg, the habitat needs of the parent are paramount, at least as long as the distance between the nest and an area suitable for raising young can be traversed by a young chick.

### 5.3.3 Distribution and competition

Conspecifics cause many desires and disappointments during reproduction. The desires are necessary, as without a partner of the opposite sex reproduction cannot occur. In theory, the density of a population could become so low that some individuals within the population would be unable to find partners, and so would be unable to breed. As the population density increases, the likelihood of finding a partner increases, as does the breeding success. An increase in the breeding success with increasing density is known as the Allee effect.

Professor Warder Allee was the informal leader of the Chicago Ecology Group. They believed that evolution eventually leads to better and higher forms of cooperation, between both individuals and species. So group members diligently searched for evidence that the team is more important than the individual. Allee admitted that overpopulation could have negative impacts on growth and reproduction, but thought that this did not mean that lower densities would always be better. He believed that critical densities existed for many, if not all, species. Below these critical densities conspecifics would profit from each other's presence. In other words, reproduction increases with density, as long as the density doesn't exceed the critical limit.

If the Allee effect only results from problems finding a partner, then it hardly seems of any relevance to waders. Birds are some of the most mobile organisms, and few birds are as mobile as waders. Plus, waders breed in open and easily surveyed habitats and can be very noisy when it suits them. For example, an oystercatcher could easily fly from the westernmost point to the easternmost point of the Dutch island Schiermonnikoog in half an hour. Even if there were only two oystercatchers on the whole island, they would probably be able to find each other within a day. What we really don't know is whether this one pair would be able to find the best breeding site available. Perhaps they would have more success if there were other oystercatchers around and they could profit from each other's knowledge. In fact there are thousands of very territorial oystercatcher pairs breeding on Schiermonnikoog, and in these densities the oystercatchers only seem to annoy each other. So much so, that each year several thousand adult oystercatchers are unable to breed.

So although Allee effects could occur in waders that defend a large breeding territory, they seem little more than a theoretical possibility. However it seems very likely that Allee effects occur amongst avocets. Avocets are one of the few colonial wader species, which suggests that the breeding birds profit from each other's presence in one way or another. Even if there are disadvantages to having conspecifics nearby, they must be outweighed by the advantages. One advantage could be the ability to collectively chase away potential egg predators, especially gulls and harriers, something avocets are particularly fanatic about.[410] This collective defence must be effective as the nests of avocets that breed alone run a much higher risk of predation.[411] Predation risks are lowered as soon as a handful of avocet pairs nest a few tens of metres away from each other. However avocets often breed in colonies comprising many tens of pairs whose nests lie less than 1 m from each other. Pairs are very aggressive in such colonies, spending up to 10% of their time squabbling

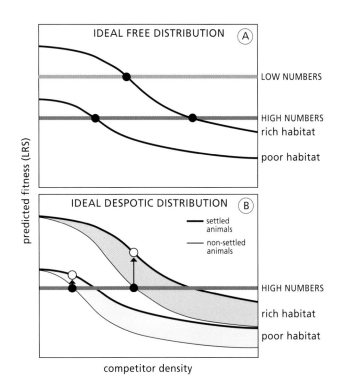

Figure 5.10 Graphical representations of the ideal free and ideal despotic models.[285, 286] Note that the shapes of the curves are arbitrary and are not based on biological data. In both cases it is assumed that fitness decreases with density. It is also assumed that there are good and poor habitats. For a given density of competitors, fitness in the good habitat is always higher. In the ideal free distribution (A) animals can settle wherever they want, without any costs. When populations are small, all the individuals can use the good habitat and have higher fitness than if they were in the poor habitat. As more individuals settle in the good habitat, the difference between the fitness in the good habitat and the poor habitat decreases, until at some point it becomes better for some individuals to settle in the poor habitat. At equilibrium, all the animals have the same fitness, regardless of whether they are in the good or the poor habitat. This is not true for the despotic distribution (B). The non-settled animals can still distribute themselves freely and thus have the same fitness expectation everywhere, just as in the ideal free distribution. But there are costs to settling in an area, as a resident bird is loathe to part with its territory without a fight. These settling costs (marked by arrows) are larger in the good habitat (where there are more competitors). Having overcome these costs, the fitness at equilibrium becomes higher for the individuals in the good habitats. The best and most practicable measure of fitness is the lifetime reproductive success (LRS), the total number of young produced during the life of an individual in a certain habitat. Often, however, the parents' survival cannot be measured, and there may be just a single season's data on the number of young raised. When the focus is on foraging animals, intake rate is often used as a surrogate for fitness.

with neighbours. Extremely small or extremely large clutches often result from nest parasitism and many clutches are deserted. As a result, the birds in the largest colonies have lower breeding success than birds in somewhat smaller colonies. So why do the avocets breed in such large, dense colonies? These sorts of colonies tend to arise on islands early in the season. The fox-free islands offer some protection against predation, as once a fox finds a mainland colony all of the nests are lost. However there is limited space on the islands, especially since other colonial species also breed there. So the last avocets to arrive at an island colony simply try to squeeze in between the other pairs, thus causing all the problems. But what alternatives do these latecomers have? If nesting with a few other pairs in area where there are foxes means your nest is less likely to succeed than if you'd squeezed into an overpopulated island colony, then it's worthwhile to use the island colony. Consequently, an avocet colony is sometimes much larger and more densely packed than is optimal for the average pair.

Avocets don't feed their young. As soon as they've hatched, the chicks must forage by themselves. There is nothing for the chicks to feed on in the colony, so as soon as the chicks hatch the family leaves the colony and goes to a feeding territory where the little ones can forage undisturbed. Having conspecifics nearby is an advantage during incubation, but a disadvantage during chick rearing.

The vast majority of wader species defends a breeding territory at some point. There are large differences between species in the activities that occur within the territory, how long the territory is defended, and which sex does the defending. Oystercatchers are extremely territorial. For the entire breeding season an oystercatcher pair will defend the area where they forage, nest and raise their young.[244] Sometimes a territory is made up of two separate areas, one for nesting, the other for foraging. Grey plovers and golden plovers are also very territorial, but in these two species it's usually the males that are aggressive. Furthermore the birds, particularly the females, often forage outside their territories, and the borders are not defended as enthusiastically once the chicks get older and start to wander into the neighbouring territories.[116] In dunlins, only the males are territorial, and the areas being defended by the males may overlap extensively prior to the eggs being laid.[443]

There are obvious disadvantages to territorial behaviour. The time and energy spent on alertness and aggression cannot be used for other behaviours. Fighting can lead to physical damage. It's even possible that conspicuous territorial behaviour makes the birds more vulnerable to predators. But territorial behaviour must also bring significant advantages. Once the costs and benefits of territorial behaviour have been accurately calculated, it's possible to predict the optimal territory size.[167] And research into the territorial behaviour of breeding waders has almost reached this stage. It seems likely that food availability plays a role in some cases, just as it does in the territorial behaviour of some waders outside the breeding season (Chapter 4). But at present we have just a working hypothesis for waders that while a bigger and more exclusive territory is better, it may also have more intruders and be costlier to defend. However credible this seems, it is still only a hypothesis.

Unsurprisingly, there are large quality differences between territories. If there are only a few animals in an area, they can each have a good territory. When many animals live in the same area, the good territories will all be occupied and some animals will be forced to make use of territories that aren't as good. This theory, known as the buffer hypothesis, was developed by Kluyver and Tinbergen[474] from their observations of breeding tits. The theory was taken further by Fretwell,[285, 286] with his publications on the ideal free distribution and the ideal despotic distribution (Figure 5.10A and B). Fretwell was both an enthusiastic field observer and an inspired theoretician. He was, in many ways, a man ahead of his time. Maynard Smith and Price[542] were first to recognise the relevance of the economists' game theory for biology. In essence this theory states that the costs and benefits of a certain behavioural choice are largely dependent on the choices made by the other animals. So an animal's best option depends on what its competitors have chosen. This may seem straightforward, but its implications quickly become very complex.[782]

Fretwell's genius was in simplifying and idealising the problem. At first, no-one understood what he meant. In his ideal free distribution, the word ideal suggests that there are such things as ideal animals. This means that each animal strives to attain the best territory and always knows exactly where this territory can be found. In the free distribution the animals can go wherever they want, remain wherever they want and settle wherever they want without incurring any extra costs. When there are only a few animals they can all settle in the best habitats (Figure 5.10A). Possible Allee effects are not considered, which means that as the density of breeding birds increases their fitness decreases. At a certain point their fitness is as low as it would be if the birds had settled in the areas that weren't so good. From that point on, some of the birds will settle in good habitat and some in poor habitat. In this situation the birds in the good habitat do not have better fitness than the birds in the poor habitat. In fact, the fitness of the birds at the two sites is the same. So under the ideal free distribution all of the birds everywhere have the same fitness. Any differences in fitness would contradict the assumption that the birds are ideal and free. Any birds that would be slightly worse off on average would be aware of this (since the birds are ideal) and could move to a better place without incurring any costs (because they are free).

Of course it's extremely unlikely that animals are absolutely free to settle wherever they want. If they are, what is territorial behaviour in aid of? So the alternative scenario is that territory owners are despots who keep the other animals out of their breeding territory by force. Hence the name despotic distribution. The animals that have no territory (usually the young ones) can acquire one, but only at a fitness cost such as a fight to the death. If you win, you're the proud owner of a beautiful territory, but if you lose, you're dead. Since these animals

REPRODUCTION

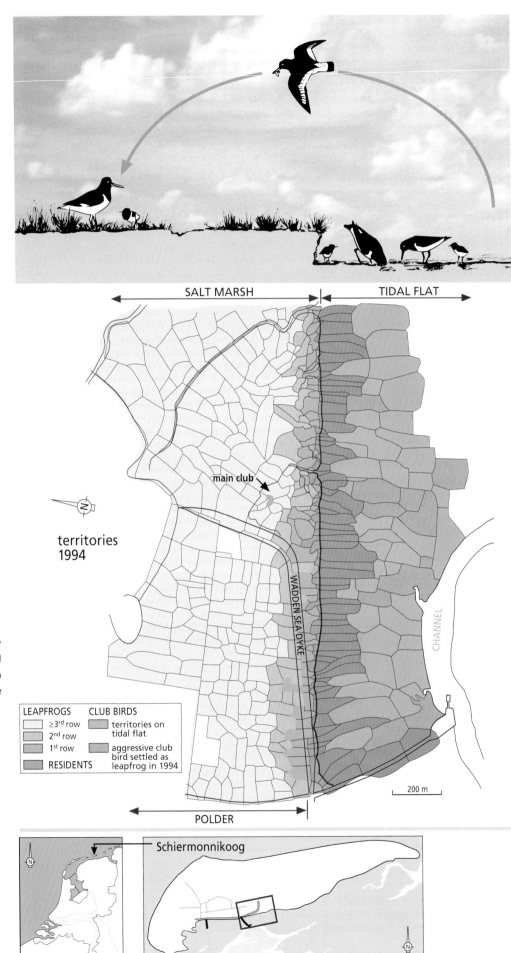

Figure 5.11 The oystercatcher society on the salt marsh and adjacent tidal flats of Eastern Schiermonnikoog in a nutshell.[388] Residents breed on the edge of the salt marsh and defend a neighbouring area of tidal flat. Leapfrogs breed further into the salt marsh and have to fly over the residents' territories to collect food to bring back to their chicks. The leapfrogs' territories can be identified by the distance to the edge of the salt marsh. They may have to defend their feeding territory on the tidal flats from territorial club birds. The club birds also regularly perform aerial 'tepiet' displays above the most attractive territories on the edge of the salt marsh. The map is based on territories from 1994.

have had to pay a fitness price, you would expect that once they have a territory their fitness is higher than it was before. This isn't really as strange as it sounds. Let's take lifetime reproductive success or LRS as a fitness measure. A territory owner lives for $L$ years on average and produces $K$ chicks per year. So his LRS is $LK$. An animal without a territory has a probability $p$ of winning the fight and a probability of $1 - p$ of losing. If the second animal wins, it will produce $LK$ chicks over the rest of its life. If it loses it will produce none. So in total $pLK + (1 - p)0 = pLK$, and $pLK$ is lower than $LK$ since $p < 1$. Animals without a territory have a serious fight on their hands, and some will not survive that fight. Territorial animals already have this fight behind them, and by definition the losers don't count anymore in this equation. Territorial animals also have further fights ahead of them, but this is accounted for in their life expectancy $L$.

In Fretwell's ideal despotic distribution, the non-territorial animals are still completely free (Figure 5.10B). This means that they have the same fitness expectation everywhere, so are both ideal and free. However the costs of acquiring a territory increase with the density of the territory owners. So in our simple example, the chance of losing $(1 - p)$ increases with the density of territorial animals. This density will be higher in better habitats, so these better habitats will have higher settlement costs.

Finally, amongst the territorial animals, those in the better habitats have a higher LRS than those in the worse habitats. So under the ideal despotic distribution we'd expect a positive relationship between the density of breeding birds and their breeding success when we compare different quality habitats.

So does this theory have any relevance to the real world? To find out we need to look for differences in habitat or territory quality. There are many ways that incubation and chick raising can fail, so territory quality can be determined by many different factors. It's clear that for oystercatchers the cost of transporting food is by far the most important factor. Their breeding success is much higher in good territories than in bad territories, just as the despotic distribution predicts. There are few wader species that are as strongly territorial during the breeding season as the oystercatcher. Both the male and the female will defend their territory against every intruder, often by storming towards the intruder as a 'tepieting' pair. Most intruders quickly flee when confronted with this superior fire power. Maybe once in a while tougher measures will need to be taken. Males may do slightly more to defend a territory, but their mates also take an active role. The female is nearly as brightly coloured as the male, and will stand her ground in a fight when she has to.

Good territories are those where the parent birds can

Oystercatchers zealously defend their breeding territory as a pair. Disputes with neighbours often arise when one pair crosses onto their neighbour's land to chase away an intruder.

safely take their young to good feeding grounds, and don't have to spend vast amounts of energy flying each tiny prey item to the chicks.[244] This can be seen very clearly on the eastern salt marsh on Schiermonnikoog (Figure 5.11). The oystercatchers use the tidal flat as a feeding area, but they cannot nest there. They nest on the salt marsh. The edge of the salt marsh erodes creating a very sharp transition between the feeding area and the nesting area, rather than a gradual transition zone, where neither foraging nor nesting would be possible. Some pairs breed on the edge of the salt marsh and defend a neighbouring piece of tidal flat. These pairs, which we call 'residents', can just step from the edge of the salt marsh onto the tidal flat. Other pairs breed further away on the salt marsh and have to fly quite far to reach the feeding territories on the tidal flat. These birds that must jump over the residents' territories are called 'leapfrogs'. Other birds that haven't yet acquired a breeding territory roost together at high tide in flocks of some tens of individuals. These flocks are called 'clubs' and the non-territorial adult oystercatchers that gather in these clubs are known as the 'club birds'.

Residents raise three times more fledglings than leapfrogs, on average.[244, 250] However there isn't a large difference in clutch size or in the likelihood that the eggs will hatch between the two groups. There are also no large differences in the survival rate of young chicks, which is very low for both resident and leapfrog families. The difference in breeding success is mainly due to differences in the survival of larger young. The more the chicks grow, the more food they need. Resident parents can simply guide their chicks to the tidal flat and feed the chicks as much food as they can find. As the chicks grow, they get bolder and will leave their hiding places on the salt marsh for longer and longer periods. So resident parents have few problems fulfilling the increasing food requirements of their growing chicks. Leapfrog parents are quite the opposite. Whether or not they are able to (see Section 5.1.5 about the costs of reproduction), leapfrog parents do not take sufficient food to their chicks, so the chicks become hungrier and fall behind in their growth. Because they are so hungry, the chicks will often come out into the open to beg their parents for food,[756] which increases their risk of predation. Malnutrition probably also makes them more susceptible to disease,[757] and the chicks may starve to death if they receive insufficient food for a long enough period. If these starving leapfrog chicks are provided with supplementary food, they begin to grow better and may even fledge (Figure 5.6).[244] This suggests that food transportation is the main problem.

So is the eastern salt marsh of Schiermonnikoog a unique situation, or is this happening to oystercatchers everywhere? It's certainly a unique site, but not aberrant. Studies at many different sites have found that oystercatchers that can take their young to the food are more successful than oystercatchers that have to take the food to their young. On the Welsh island Skokholm, one group of oystercatchers forages inland on buried crane fly larvae and caterpillars amongst the low vegetation. Another group searches for limpets at low tide, along Skokholm's rocky shore. Like the Schiermonnikoog residents, the caterpillar eaters can walk with their young to their feeding grounds. The limpet eaters, like the leapfrogs, have to fly back and forth between the top of the island where they breed, and the tidal zone where they forage. Just as you'd expect, the caterpillar eaters are far more successful at fledging young than the limpet eaters.[758]

Even so, it would be wrong to think that food transport is the only factor that determines territory quality for oystercatchers and that territory quality is the only factor that determines breeding success. Plenty of things can go wrong well before the young are big and need

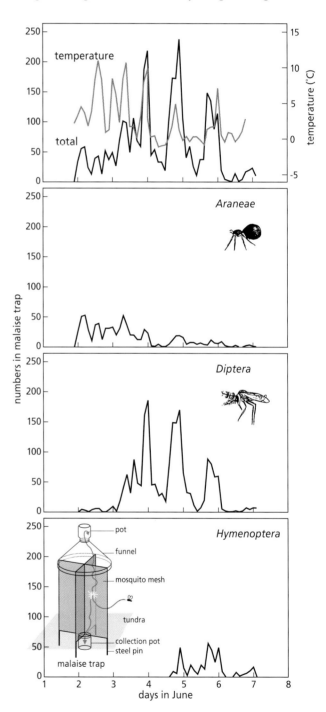

Figure 5.12 Changes in temperature and the number of spiders (Araneae) and insects (Diptera and Hymenoptera) caught in a malaise trap on the tundra of Cape Sterlegova in June 1994.[879]

more food. For example, the salt marsh may be flooded during a storm. This would affect the birds that breed on the lower parts much more than those that breed on the higher parts. Some territories provide more cover where the young can hide than others, and chicks without a good hiding place are much more likely to get munched. So the risk of flooding and the potential for hiding places also help determine territory quality. But breeding success can never be guaranteed, not even in a territory that scores well on all fronts. The owners just have to make one silly mistake and no young will fledge, even in the best territories. Pairs that are late in laying their clutch usually fledge few or no young. This holds for both residents and leapfrogs. The same individuals tend to consistently lay late, year after year, though there does seem to be some improvement as the birds age.[246, 388] So an individual's success in any particular territory depends on the quality of that territory, its own abilities, and those of its partner. It's very likely that individuals with certain characteristics have more chance of getting a good territory. Such correlations make it extremely difficult to determine whether the difference in success between two animals has to do with a difference in the quality of the individuals or a difference in the quality of the territories. We will return to this in the final section of this chapter (Section 5.7), where we'll learn that the settlement costs that are so important in the establishment of a despotic distribution actually result from increasing social status, rather than fights to the death.

## 5.4 The best time to breed

When should waders breed? The answer seems simple: they should breed at the time that provides the optimal growing conditions for their chicks. Food seems an obvious factor in this,[490, 491, 492] but day length and temperature also change a lot in the course of the year. The Arctic tundra, for example, is covered in snow and ice for much of the year, providing few feeding opportunities for waders. The tundra only becomes attractive to waders during the Arctic summer when the snow melts and, for a brief period, the plants grow and flower, causing an increase in the insect populations. Wader chicks can then feed on these insects. But the insects don't become active as soon as the temperature rises above freezing (Figure 5.12), and they must also be prepared for the early return of winter. This explains why insect availability at Barrow, Alaska, peaks before the highest summer temperatures occur.[590] Between the times when insect availability peaks and summer temperatures peak, insect (and spider) availability shows high daily variation related to the large fluctuations in temperature (Figure 5.12). There are also clear differences in the phenology of the different prey species. At Cape Sterlegova in Taimyr the spiders are the first to emerge, followed by the dipterans (mosquitoes and flies) and, at the end of the season, the hymenopterans (bees, wasps and ants). Which part of the season provides the most food for the chicks depends on which prey the chicks prefer.

So is the breeding season arranged to ensure that the chicks are out foraging on the tundra just when the availability of their food peaks? In broad terms it is, but on a smaller scale there are exceptions. There are many reasons why the timing can seem imperfect. Problems en route (such as strong headwinds) may make it impossible for a bird to arrive on the breeding grounds at the optimal time and mass (Chapter 3). Such problems will always put birds behind on their travel schedule. A bird that has especially good luck (and gets strong tailwinds) will not use its good fortune to arrive on the breeding grounds excessively early. There would be costs to the adult of arriving too early in the season. It's also true that what is best for the chicks isn't necessarily best for the adult birds. Every breeding season, the parent birds must trade their own survival (and thus their future reproductive success) against the survival of their present chicks (their current reproductive success). For the chicks, the advantages of being big early in the season may outweigh the disadvantage of encountering somewhat less food at this time.

So far, we've implicitly assumed that waterbirds have only one clutch per season, and indeed this may be all that is possible for many species with simple reproductive strategies. Even for individuals that nest early in the season, there may simply not be enough time to complete two full breeding attempts. And this can also be the case when the first nest is lost early due to predation or flooding. If a grey plover breeding in the far north loses its nest, that is almost always it for that season.[116] The same applies to the red knot, which also breeds very far north. When an Arctic fox preyed on red knot nests in Sterlegova in early July, that breeding season was finished for the knots. The females quickly departed and after a few days of aerial displaying the males also left. If the knots had immediately renested after losing their first clutch perhaps they could have managed to successfully raise a brood before winter set in, but this probably wouldn't have been profitable in the long run.[650] Species that breed further south have a better chance of renesting after losing their brood, depending on whether they have lost eggs or chicks. All oystercatchers that have laid

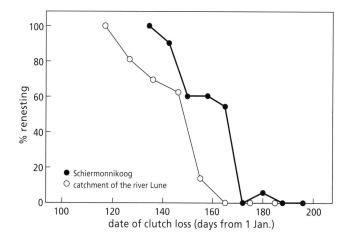

Figure 5.13 Probability of oystercatchers renesting in relation to date after losing a clutch or a brood at either the salt marsh on the Dutch island Schiermonnikoog, or the meadows and agricultural areas around the river Lune in the United Kingdom.[255]

# REPRODUCTION

Newly arrived on the tundra, a grey plover (photo above) and little stints (middle photo) search for food on the edge of the melting snow. Sometimes the thaw reveals well-preserved berries, which make a tasty meal for a whimbrel (photo below).

A newly fledged red knot has a distinctive, flecked plumage (top of page 275). Now 25 days old, the bird has grown quickly since hatching (this bird is one of the chicks shown on page 254). Its father, who raised the chick, will leave the breeding grounds in Taimyr for the Wadden Sea today, 11 August. The first juveniles will soon follow, and will arrive in the Wadden Sea in the second half of August. One such newcomer can be seen resting in the central foreground of the lower photo on page 275; in the large group of resting birds the juveniles form a very small minority.

a relatively early clutch and then immediately lost those eggs without exception produce a second clutch (Figure 5.13). The later in the season the clutch is lost, the less likely it is that the birds will renest. Oystercatchers in the Lune valley in Great Britain breed earlier than those on Schiermonnikoog, and also stop renesting earlier. Perhaps the most interesting part of Figure 5.13 is the long period over which some individuals will continue to renest after a clutch is lost, even though other birds wouldn't. Differences between individuals manifest in many aspects of reproduction and will be discussed further later in the chapter.

## 5.4.1 Geographical trends in the seasons

The further north the breeding grounds, the later spring starts and the later the birds can start breeding. This is what we'd expect, and we'd be correct. Turnstones in Southern Sweden lay their eggs in late May and early June, while in Spitsbergen they lay their eggs three weeks later. In England ringed plovers begin to lay in the second week of May on average. In Northern Norway they

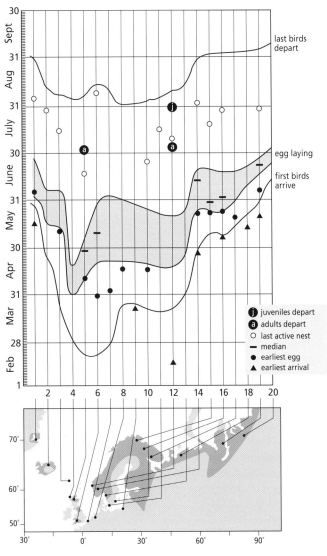

Figure 5.14 Geographical variation in the timing of the golden plover's breeding activities: arrival in the breeding area, egg laying and departure.[116]

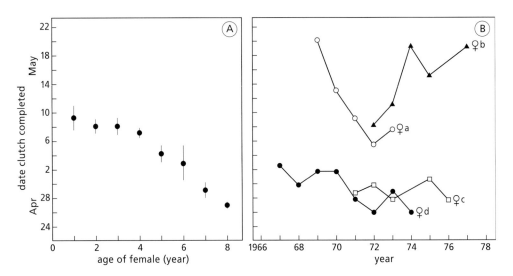

Figure 5.15 Female greenshanks tend to lay their eggs earlier in the season as they get older (A: average date for 64 individually recognisable birds, ± standard error). This does not explain all of the variation between individual females, as some females always lay relatively early whereas others continue to lay late throughout their lives (B).[852]

start to lay more than a month later. The differences in laying dates are even more extreme for the redshank. Redshanks in The Netherlands and England lay their eggs in late April to early May, but in Northern Norway they lay as much as six weeks later. One rather interesting study on the geographical variation in the timing of breeding in waders was based not on field observations, but on 7176 clutches held in museums.[894] Eggs are usually blown prior to being stored in a museum and egg collectors usually think it's unethical to blow incubated eggs (and it wouldn't really work that well either). This means that the date of collection differs little from the laying date. The study was published in 1977 and included data from 120 years of egg collecting.

We would also expect that the further north the birds breed, the shorter their breeding season. Golden plovers breed in both temperate areas and in the far north. The birds are already on their English breeding grounds in February, but only arrive on the most northerly breeding areas in June (Figure 5.14). There are two options here: the animals may have a longer breeding season in the south, or there may be little difference in the length of the southern and northern breeding seasons. The latter would mean that the northern breeding grounds would be deserted much later than the southern ones. And the golden plovers seem to make use of both options. While there is a four month difference in arrival dates, the difference in departure dates is only one month. There is little variation in the length of the period between egg laying and departure from the breeding grounds. But the golden plovers are present on the southern breeding grounds long before the first eggs are laid and the period over which they lay their eggs is quite extensive. On the far northern breeding grounds there are three weeks at most between the time that the first birds arrive and the time at which the last eggs are laid.

### 5.4.2 Arrival and egg laying

How do birds determine when they should arrive on the breeding grounds and lay their eggs? Whimbrels arrive on the Shetland Islands an average of 11 days before they begin laying.[349] This means that the females must have started making the eggs immediately after arriving on the breeding grounds, which in turn means that the timing of egg laying is determined by when they depart from the last stopover site. Whimbrels migrate across very large distances so it's questionable whether the weather at their last stopover site gives an indication of what the weather will be like on the breeding grounds. It's entirely possible that the birds rely on an internal clock for the timing of their migration to the breeding grounds.

Birds that fly to northern breeding grounds at the same time each year run the risk that these areas will often still be covered in snow and ice when they arrive. This makes it difficult for the birds to forage and build their nests, although sometimes they find ingenious solutions to the food shortages. As the snow thaws, it often reveals ripe berries from last season that had been hidden by snow and ice. Golden plovers and grey plovers eat these fruits with great relish,[116] as do whimbrels. Within a short time, the first areas will become snow-free, providing sites where the birds can nest. However many waders wait until much of the tundra is free of snow before laying their eggs. So when the thaw is late, breeding is delayed.[116] This can be risky; if the birds wait too long before nesting, their young may not be ready to migrate south by the time that winter returns. Why don't the birds immediately start nesting in the first snow-free areas? If they did they would put their clutch at a much higher risk of predation. When large areas remain snow covered, all predators would have to do is search the small snow-free areas to find the nests. An experiment using artificial nests in different snow conditions found that predation risks increased when more of the habitat was covered in snow.[114] In late spring the birds are caught between a rock and a hard place. If they breed late their young may run into problems at the end of the season, but if they breed early they run a higher predation risk. The birds do something of both: they breed later, but not so much later that they are at too high a risk.

### 5.4.3 Timing of the departure from the breeding grounds

If the birds are late to begin nesting it seems logical that they will be late departing from the breeding grounds, both the chicks that must be ready to migrate, and the parents that must care for the chicks until they are able to look after themselves. At least this must be the case unless there is some way that development can be accelerated. And there is one developmental process where this is known to happen. Dunlins that breed in the east of Taimyr moult their primaries (large wing feathers) before migrating south. In two late seasons the birds began to moult more than 10 days later than average. By moulting faster they had finished their moult on time and were able to start their migration to the wintering grounds at the normal time.[794] If development can't be accelerated, the birds have two options: depart later, or leave when they are not quite ready. Leaving before you are ready may prove very costly for the young, as has been shown for barnacle geese. Unlike wader fledglings that must find their own way to the wintering grounds, young geese migrate south with their parents. In 1986, winter arrived exceptionally early in Spitsbergen and the geese were forced to leave for their Scottish wintering grounds very early. On arrival it was clear than an unusually high number of young had not survived the journey. Since some hundreds of young were weighed and marked before they left Spitsbergen, researchers could determine that it was mainly the young with a low mass before departure that had died en route.[620]

The start of the Arctic summer may be unpredictable, but when it will end is even harder to predict. Or at least it is in Barrow, Alaska. In Barrow cold and warm days regularly alternate during the summer. However at the end of the season it drops below freezing point one day and remains there.[590] This is why there is so much variation between years in the cumulative deviation from the average temperature, especially at the end of the season (Figure 5.4C). This unpredictability is one possible explanation for why so many Arctic wader parents seem to be in such a hurry to leave the breeding grounds. Another explanation is that it is highly advantageous to return quickly to the wintering grounds. Whatever the cause, female golden and grey plovers desert their young before the chicks have fledged, leaving the males to take over the tail end of the chick raising. The males then leave the breeding grounds about two weeks before the bulk of the young begin to migrate.[116] The same pattern is found in other wader species. It suggests two conflicts in the parental care: between the sexes and between the parents and young. It is probably better for the young to be looked after for as long as possible by as many parents as possible. The parents probably want to leave the breeding grounds as soon as possible. They must make a trade-off between the fitness gains of an early departure versus the risk that the survival of their young will be reduced due to the early withdrawal of parental care. It will take quite a while before we can actually measure the costs and benefits of these sorts of decisions.

### 5.4.4 Individual differences

The date that the eggs are laid doesn't only differ from place to place and year to year. There are also differences between individuals that breed in the same year at the same site. This variation between individuals is correlated to their age. Female greenshanks lay their eggs earlier as they get older (Figure 5.15A). But this is not the whole story. Some females always lay earlier than other females (Figure 5.15B), and this individual variation is more important than the age effect. Plus, territory quality is also involved in the laying date. In territories that include a small river, greenshanks lay an average of eight days earlier than in territories where the animals must make do with a small lake.[852] Research on oystercatchers suggests similar factors are involved in their laying dates. Females

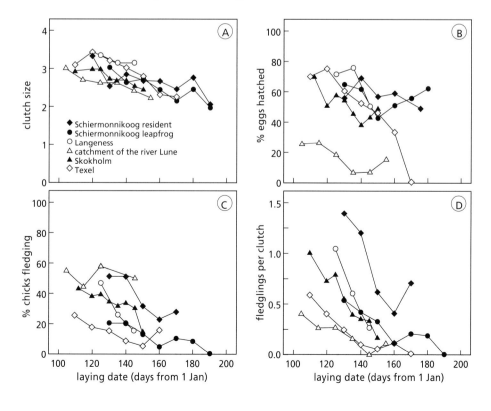

Figure 5.16 Components of breeding success as a function of laying date for oystercatchers in different areas. (A) Clutch size, (B) percentage of eggs that hatch, (C) percentage of chicks that fledge, and (D) total number of fledglings per clutch.[255]

Being beautifully red to attract females does not have to be a problem for your camouflage when incubating if, like this bar-tailed godwit in Lapland, you keep your conspicuous coloured belly well hidden in the vegetation.

start to lay earlier with age[246] but there are large and constant differences between individual females.[388] And oystercatchers in the good resident territories lay a few days earlier on average than the oystercatchers in the poorer leapfrog territories.[388]

For both oystercatchers and greenshanks an explanation for the observed correlations can be found in differences in the food supply prior to the eggs being laid. In a warm April, when the food supply has probably increased earlier, the greenshank females may lay up to 10 days earlier than normal.[852] When oystercatchers were supplementary fed prior to laying their eggs, the females did lay a few days earlier.[388] The correlations fit into the food supply hypothesis as follows. As they age, the females may become more adept at foraging; a bird that is not particularly good at finding food will improve somewhat, but not spectacularly. Greenshanks forage on the water's edge throughout the breeding season, and the best place to do this is along a river, hence the earlier laying in the river territories. Finally, although there is not a large difference in food availability between resident and leapfrog territories on the tidal flats, resident territories are exposed longer making it possible to forage longer at these sites.

So if differences in food intake explain the differences between individuals in the timing of egg laying, does this mean that food limits the egg production physiologically? No, it simply means that the whole physiological process involved in laying an egg does not start until the bird reaches a certain energy intake. This may be because laying too early increases the mortality risk; when there is not enough to eat, egg laying can lead to a negative energy balance. It may also be because the bird uses its intake rate at the beginning of the season as an indication of the time at which there will be the most food for the chicks. The birds might use the time that a certain intake rate is reached to determine not only when they should lay, but also how many eggs should be laid. This seems to be the case for kestrels.[145] Every kestrel parent must balance the following factors: (1) the harder it works for its young the greater the chance it will die in the winter, (2) the earlier in the season the young fledge the greater the chance that these young will produce their own offspring, (3) the later in the season breeding occurs, the better the food supply is. From this it follows that a bird that can hunt well and/or has a good feeding territory is better off laying many eggs early in the season. A less adept hunter is better off laying somewhat fewer eggs later in the season. And this prediction seems to be surprisingly accurate. Oystercatchers also show a clear decrease in clutch size over the course of the breeding season (Figure 5.16). That doesn't necessarily mean that the same explanation applies. Many fitness measures related to oystercatcher breeding have yet to be determined, even though it's clear that the probability that an egg will become a fledgling decreases greatly over the course of the season. This suggests that for oystercatchers an early egg has a higher reproductive value than a late egg, just as it does for kestrels.

So it seems that birds that lay early have better territories or are themselves better than the birds that lay later. This is rather odd. If the early birds do so much better than the late birds, why do late birds still exist? Why hasn't natural selection eliminated all the late birds? The answer to that question comes to light indirectly when we examine partner choice in the next section. We already know one thing: the early birds should be more attractive as potential partners.

## 5.5 Choosing a partner

### 5.5.1 Sexual selection

Sexual selection is an important subject that deserves to be dealt with in some detail. It is a form of natural selection.[20] In many ways, natural selection is concerned with the development of more efficient anatomy, morphology or behaviour enabling the individual to make better use of its environment. For example, more insulating plumage so that the bird survives the cold better, a longer bill so that the bird can catch deep-living prey more easily, more vigilant behaviour so that the bird can discover predators better. In each case the individual's chance of survival increases as these properties are enhanced. The animal may also improve its reproductive traits, increasing its fitness. For example, a well-developed brood patch enables the parent bird to incubate its eggs more efficiently. Finally, natural selection may also act on the qualities that enable the individual to compete more successfully against its conspecifics for limited resources such as food or nest sites.

Sexual selection is concerned with the qualities that influence an individual's chance of success in finding a partner.[20] These qualities may enable the animal to get more partners or better quality partners. When the males have to fight amongst themselves for access to females, there is strong selection for males to be big and to carry weapons, as is the case for deer. Some birds are more successful at getting partners if the opposite sex finds them more attractive. For example, the peacocks with the largest and most beautiful tails were probably more successful in attracting mates. However, traits that are helpful when you are trying to get a mate may cause problems in other situations. Being large may provide more fighting power, but also requires more food. Stags regularly die a miserable death by starvation, while the hinds aren't experiencing any feeding problems.[134] And peacocks probably don't find their large tails quite so handy when they trying to escape a predator. Such traits that actually seem to decrease an individual's chance of survival initially created a problem for Charles Darwin when he formulated his theory of natural selection. This eventually led him to the idea of sexual selection: these traits make the individuals more successful at finding partners and thus help the individuals to produce more offspring. However it's obvious that there must be limits to sexual selection. It's great if the opposite sex finds you attractive, but if the same qualities that make you so attractive also mean you're unlikely to survive the week, they aren't very useful. Evolution quickly eliminates the playboy that might not live to see tomorrow.

Sexual selection has played a large role in the development of birds' plumage, morphology and behaviour. All of these traits have been inherited from successful ancestors. And sexual selection continues to act; each generation must fight its own battles for partners. When selection has led to every competitor being armed with the same weapon that once gave a single ancestor a decisive advantage, this weapon no longer has any competitive advantage; it is now just the prerequisite for having any chance at all. You would expect all animals to make the most of every single advantage they have, no matter how small. Such advantages exist because individuals vary in their abilities, which can reflect genetic differences and environmental conditions during growth.

As well as their stunning plumage, male dunlins have distinctive display behaviours to help them attract females. By performing on top of a little hump in the undulating terrain, this male makes an even bigger spectacle of himself.

Let's start with the conditions while the chick is being raised. Variations in these conditions are inevitable.[623] There can be good and bad weather, years when food is abundant and years when it is scarce, parents that are inexperienced or parents that have raised many clutches, predators that might eat a competitive sibling, and diseases that may or may not strike etc. So there will always be some lucky birds that grew up in good weather, with plenty of food, with healthy parents, without siblings to compete with and without ever getting ill. Some families tend to be very lucky with their offspring, the 'silver spoon effect',[347] but this isn't always enough. Young waders must complete their first migratory journey without any assistance from their parents. So no matter how able the parents, they cannot assist their offspring if, through sheer bad luck, their young are caught in a storm en route.

In each generation, chance events give rise to differences in quality between individuals.[438, 733] How can these quality differences be capitalised on? The competitive sex can reveal their quality through being especially good at fighting, or investing more resources in external features that are attractive to the opposite sex, the so-called secondary sexual characteristics. For the peacock, it is the length of his tail. For male barn swallows, it is both the length and the symmetry of the tail, according to Møller.[576] Males show their quality by growing longer and more symmetrical tail streamers. A prudent bird will not invest all of its extra energy in just one trait, such as a longer tail, but spread its investments appropriately across the various quality traits.

So how come there are genetically 'good' individuals? While there is no lack of theories and mathematical models,[20] good empirical studies are scarce, especially in waders. Instead of ploughing through a mountain of theories, let's examine one interesting hypothesis in detail. This hyphothesis[374] starts with the observation that there is an evolutionary arms race between parasites and their hosts. Parasites are always trying to break through the defences of the host; hosts are always trying to stay ahead of the parasites in this competition. If the parasites do break through, they will multiply rapidly, at the cost of the host. The host benefits if it can prevent this happening. But this is not a race to a final end point; instead it is an ongoing process with visible changes in the genetic composition of the host and the parasite populations in even short ecological timeframes. The reason for this lies in the cyclic nature of the process. If, after a number of generations, the parasite catches up with the host, then both are in principle back to the start and everything starts all over again. But the hosts are not all equal, and at any moment there is considerable genetic variation between individual hosts (in our case individual waders) in their ability to resist parasites. Some individuals are genetically ill equipped to handle parasitism. Others are never sick and can parade this fact, by strutting around with completely moulted glossy plumage or other adornments.

There is now an intricate hypothesis for how colourful bills, eye-rings and conspicuous feathers can be reliable indicators of a bird's health.[772] It revolves around the free radicals that challenge pathogens in the body. But while free radicals are fighting pathogens (which is obviously beneficial), they can also harm the bird's own body. In response, the body must utilise antioxidants to 'catch' the roaming free radicals. It seems that many strongly sexually selected external features are made of tissues that are sensitive to free radicals, or get their colour from antioxidants such as carotenoids (red pigments) and vitamin C. When the body produces a response to an immunological challenge, it can do so in two ways, by making a specific response or a non-specific response. A specific response requires the production of a specific antibody (which the bird must have the genetic variation to do) but then the pathogen can be destroyed by a small dose of free radicals. The specific response is like a precision weapon, a single bullet. A non-specific response involves firing free radicals at the whole pathogen; this shotgun approach will also inadvertently shoot the animal in the foot, so to speak. If an animal can avoid the pathogens, or can produce the specific antibodies needed when these pathogens hit, then its oxidative stress will be lower. There will be more antioxidants left over after they have done their jobs in scavenging free radicals, and these can be incorporated into the vulnerable organs like the combs of roosters and facial warts of ruffs. These quality signals then become larger and more colourful.

That's why the male jungle fowl (the wild ancestor of domestic chickens) prances around showing off his wattles and comb. The size and colour of the comb provides information about the health of the animal – how often he has been ill and what his current state is. Many waders also have striking and highly visible fleshy bits. Some tropical lapwings have wattles below their bills, while other waders have coloured rings around their eyes. Oystercatchers possess a conspicuous orange bill, as well as a bright yellow eye-ring. But we really have no evidence about whether these traits actually relate to the health of the bird. It is known that male golden plovers with the most black in their breeding plumage on the belly tend to be dominant and can obtain the territories with the most food; these males would be very attractive to female golden plovers.[210] However the samples are small and the study area where this was found is considered by some to be rather atypical.[116] In bar-tailed godwits the intensity of the rusty-red breeding plumage is thought to be linked to the quality of the individual.[448, 670] Godwits in Africa start their moult out of the grey winter plumage into the rusty summer plumage before migration towards the breeding grounds, but interrupt the moult when they come to take off on the first flight north. In May the birds make a short stopover in the Wadden Sea. When refuelling, only some of the birds resume moult to improve their plumage further. Curiously though, it is not the poorly moulted birds that restart the process to try and catch up. Instead it is those that already have a well-moulted, attractive plumage. They also turn out to be the heavier individuals in the population. So it seems that the reddest birds really are the best of the bunch, as they have managed to do the most moult before leaving, while also presumably putting on the most fuel for the migration. And during the stopover in the Wadden Sea, they can add the final touches to their attire, while other birds are struggling just to get enough fuel on board for the next flight. So the redness of the bar-tailed godwit plumage could tell potential partners about the migratory abilities of the individual, though the breeding ground research that would tell us whether these birds are actually favoured as partners has not yet been done.

### 5.5.2 Display

Sometimes beautiful colours aren't quite enough to get a partner. Serious amounts of showing off may be required, and what better way to show off your many talents than in a display flight.

Few sounds are as melancholy as the song of a solitary golden plover out on the desolate breeding grounds of the far north. Perhaps the only comparable cry is the plaintive call of a displaying curlew making its peculiar stiff butterfly-like display flight. Hearing the lonely call of a whimbrel while you're out walking on a misty morning really sends a shiver down your spine. Although not all waders have such melancholy songs, they always have a beautiful display call. Even the oystercatcher. Oystercatchers usually make an unpleasantly loud shriek that can quickly become rather annoying. But as they perform their butterfly-like display, slowly flapping their wings, they sing beautifully. It's just a shame that oystercatchers perform their graceful display so rarely.

The red knot also has a lovely singing voice, which rings out across the silent tundra in the most northerly parts of the world. The knots fly in a figure of eight, high above the ground (on average 150 m), loudly calling "wie, poewie, poewie, poewie". They cover a large area before gracefully gliding down and landing on the tundra. A male will continue its "wie, poewie" song on the ground as it runs, holding its tail upright, after a female in the hope that she will be interested.[932] These breeding songs can also be heard in the Wadden Sea area in May and early June, just before the knots depart for the breeding grounds.[683]

Male sanderlings try a rather different tactic. They hover low (usually between 5 -10 m) above the same High Arctic tundra, drooping their tails and looking more like tiny helicopters than anything else. During the display the male holds his head back and spreads his tail out and angles it down slightly, giving him a stocky appearance. He alternately flaps his wings rapidly and glides on somewhat down-curved wings. While in flight the male regularly makes a whirring call that lasts for 1 - 2 seconds. Displaying sanderlings tend to arouse more laughter than deep emotions amongst human observers. But of course it's only the emotions that are stirred in the hearts of the sanderling females that are important, and beauty is in the eye of the beholder.

Displaying grey plovers fly higher than sanderlings, sometimes even 100 m above the ground (Figure 5.17). They look like butterflies as they slowly flap their wings, and circle above their territories singing. These flights can last for 20 minutes and end with a falcon-like dive in which the bird falls like a stone towards the ground. Just as it nears the ground, its flight changes and the bird skims along very close to the tundra.[116]

Why do birds perform such complicated displays? People used to believe that the birds put on such beautiful shows entirely for the benefit of their human observers. As knowledge of natural world increased, it was realised that birds sing and display to attract sexual partners. They also sing to demarcate the borders of their territories to keep competitors out. Their general message can be summed up in two sentences: "over here honey" and "get off my property". At least, that's the case when the males perform the displays, as they do in many species. The males are usually the first to arrive on the breeding grounds, and they wait there for the females. Some species, including the curlew sandpiper,[287] are believed to form pair bonds en route to the breeding grounds.[862] If you arrive on the breeding ground as a pair, you can begin to nest immediately rather than needing to spend time finding a mate. In some species like the phalarope and dotterel the usual sex roles are reversed. Curiously enough, neither male nor female phalaropes have an extensive repertoire of display flights and songs. You might expect something more of the females given their otherwise 'masculine' role, but they seem to have drawn the line at beautiful plumage and territorial behaviour.

The more we learn of sexual selection the more obvious it becomes that the number of partners is important

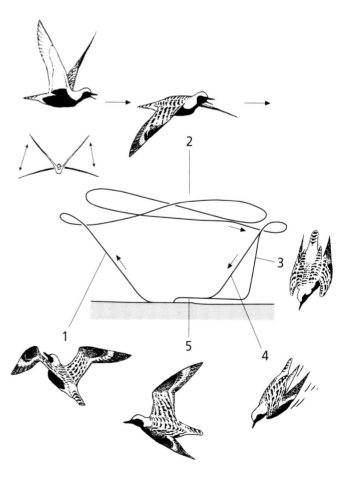

Figure 5.17 The display flight of a grey plover. The bird sets off with powerful wing beats (1) and performs a butterfly flight (2), before plummeting straight down (3) or at an angle (4) then skims along just above the tundra (5).[116]

Avocets perform elegant courtship rituals before copulating (photos above).

It was once thought that when groups of oystercatchers began 'tepieting' it was a kind of display behaviour to attract a mate. Thanks to research on individually recognisable animals we now know that these animals are actually a previously established pair trying to chase the neighbours out of their territory. It's likely that tepiet ceremonies involving more than four animals (two pairs) are due to a territory dispute on a site where three territories meet (middle photo).

Male black-headed gulls feed their partners. They usually do this by regurgitating food in the colony near the nest, but here a recently caught worm is presented to a begging female on the tidal flat (photo below). This sort of behaviour is not known to occur in waders.

A male redshank wishing to copulate with his mate must hover like a small helicopter before slowly landing on her back (photo page 283).

and that not all partners are equal. Sometimes animals have the opportunity to attract more than one partner. Through frequent and beautiful displays an individual may do an excellent job of self-promotion, thus increasing its chances of having several partners. For other species, their pair bond has an unwritten clause that stipulates that only one partner is allowed. In which case, it's very important to get the best possible partner. And to do this, the displaying bird has to prove its worth as a mate. Although little research has been done on wader display behaviour, it seems likely that the birds use different parts of the display flight to prove different aspects of their quality. The only guarantee that a display flight is an honest signal of the individual's quality is when high costs are associated with it.[348, 956] Flying slowly takes considerably more energy than flying at normal speed (Chapter 3). And a striking number of display flights include a very slow butterfly-like flight as part of the routine. As yet, the lapwing is the only wader species whose display flight has been studied in terms of its effects on mate choice.[358] Unlike grey plovers which must migrate over vast distances and have pointed wings, lapwings have rounded wings that they can use to perform acrobatic flight displays. Lapwing display flights often include alternate bouts of flying tilted to the left and tilted to the right. We can assume that it's very difficult to fly on an angle like this. So that the lapwing doesn't lose its balance, the wing that is pointing up must beat more deeply than the wing pointed down. The males that could tilt the furthest in flight had the largest territories with highest prey densities. There was often more than one female in the territory of these males, and the females in these territories laid their eggs early. So the males' strongly tilted flying correlated well with their attractiveness as a partner, just as we'd expect.

### 5.5.3 Copulation, adultery and rape

Successful display behaviour leads to pair formation, copulation and eventually a clutch. Generally, this makes both partners responsible for the parental care that is undertaken for the eggs and chicks. But before we

discuss the variation that exists in parental care, let's consider whether it is possible to reproduce without having to do the parenting. It certainly is possible. Females can lay their eggs in someone else's nest, which is called nest parasitism. This is all that cuckoos ever do. Some species parasitise the nests of their conspecifics,[381] including some ducks, geese, and colonial swallows. In the barn swallow, parasitism is most prevalent when individuals breed in very close proximity, so in years when populations are high and birds are in especially large colonies. Both partners guard the nest against nest parasites (i.e. other females). If a parasite does slip through, there is still a second line of defence: every egg found in the nest before the female has laid her own eggs will be thrown out.[576] In snow geese being a nest parasite is less successful than having your own nest. The cuckoo-like parasitism comes about when the female geese are unable to build a nest or find their clutch predated while it is still incomplete, so the birds have a couple of eggs in the system they still have to get rid of.[136] One of the few species in which intraspecific nest parasitism is really common is the redhead (*Aythya americana*), a duck related to the tufted duck (*A. fuligula*) and the pochard (*A. ferina*).[795] The redhead may be on its way to becoming a full parasite, as it also often parasites the nests of the related canvasback (*A. valisineria*).[797] But even though nest parasitism occurs regularly in ducks, geese and other waterfowl,[796] it is virtually unknown in waders. Occasionally a strange-looking egg will occur in a nest, but genetic analyses of the parents and chicks are necessary to determine whether this egg really is from another female. Of 66 chicks of the semipalmated plover (*Charadrius semipalmatus*), a species that looks a lot like the European ringed plover, not a single one was from a parasitic mother.[958] The same was true for 82 purple sandpiper chicks[649] and 65 oystercatcher chicks.[390] These three are all monogamous species. Buff-breasted sandpiper (*Tryngites subruficollis*) males behave just as ruffs do, they do not provide any parental care to their offspring at all. Of 164 chicks that were studied genetically, only one chick may have had a different mother than the female that raised it.[497] It seems that the 'cuckoo behaviour' just doesn't occur in waders. It's not hard to imagine why. Waders generally nest in open habitats, making it difficult to sneak into a neighbour's territory unnoticed. The nest has to be found before the resident female has laid all of her own eggs. And there is also no reason why the resident male should let a strange female lay in the nest. Only four eggs can be incubated properly, so it is in the male's best interests to ensure that all four eggs are his.

This brings us to males that want to produce more offspring than they want to take care of. They can manage this by copulating with a female other than their mate. There are many hypotheses as to how the females might benefit from such a liaison, though these are mostly unproven.[72] But there are also many reasons why such a mating would not be in the females' best interests. If the female doesn't wish to copulate with the male, rape is another option for the male. Some behavioural investigators strongly denounce the use of anthropomorphic terminology such as rape. Animal behaviour should be described in neutral terms, as the observed animal behaviour may not be directly comparable to human behaviour, and our human terminology is value-laden and emotionally loaded. While these arguments may be valid, there are nevertheless many advantages offered by using everyday human terms instead of the often rather artificial technical terms. The technical term for rape amongst non-human animals is forced copulation. Almost everyone who has seen a female mallard pursued by a group of males that fight amongst themselves in their efforts to copulate with her will think of the same term for this: gang rape. The behaviour of the female makes it clear that she is doing everything she can to escape her attackers. The behaviour of the males makes it clear that they have only one purpose – to copulate with the unguarded female. Most female mallards have a permanent mate who vigorously defends his partner, attacking any other male that comes too close. The female even has a special head movement to prompt her own male to chase and attack others. Shelducks also have this display, and in spring the aggressive and colourful drakes often fight (these fights may also be about territorial boundaries). In eiders only the male is conspicuously coloured; the cryptic female incubates the eggs out in the open. Before the eggs are laid, the pair stays close to each other at all times. This is advantageous for both the male and the female. The female is protected against other males and can forage undisturbed, while the male is certain of the paternity of the imminent clutch. Sometimes the male finds himself in a dilemma. While wanting to protect his own mate from other males, he may also want to seek out an unguarded female for himself, thus increasing his reproductive output. This would require him to leave his own female, which is risky from his (and her) point of view and is seldom rewarding. Any female that dares to go out on the mudflats without her partner in the breeding season will probably be attacked by any remaining unpaired males. There may even be males whose mate is already incubating, out on the tidal flats looking for the opportunity to father more ducklings.

Among birds, ducks are the most notorious rapists, and unlike most other birds, the drakes of many species have a penis. Birds generally have just a single reproductive and excretory opening in the body, the cloaca. This is where waste products like faeces and uric acid are excreted, sperm is taken in, and eggs are laid. To copulate successfully, the cloaca of the male must be pressed up against the cloaca of the female for long enough to allow sperm to be transferred. The male normally sits on top of the female, who bends her tail to one side. The male can then bend his tail downwards over the other side, enabling their cloacae to meet. Ducks copulate in the water, the male holding the neck of the female (sometimes even forcing her underwater). This makes it difficult for females to fend off unwelcome suitors. And even if the male's penis did not evolve specifically for the purpose of making it easier to perform forced copulations, it can only make it easier for the male.

Waders don't have a penis and copulation requires a delicate balancing act on the part of the male. Redshank males often seem to follow the female endlessly with their wings upstretched, calling hopefully. At a certain point the male will ascend like a little helicopter and hover above the female. Slowly descending, he tries to make a precision landing on the female's back. Having reached that step, the male has to stay balancing, flapping his wings, while trying to get his tail underneath the female's. All too often the female steps forward at the critical moment, leaving the male tumbling unsuccessfully off. Oystercatcher males also flap their wings to balance themselves, but the females usually stand still and bend forwards to help the whole process along. The fact that most waders have long legs without claws suitable for gripping certainly doesn't make it any easier for the males. They can't just grab hold of a female. All in all it's quite clear why no-one has ever seen rape among waders. If a female is unwilling, there is nothing a male can do to force her. His only chance will be with a consenting female. What the male could do in theory is defend a territory, and only let females enter it if they are willing to copulate. But why would a female breed with one male yet go and feed in the territory of another? If the territory were that good, then she would be just as likely to have settled there in the first place. This brings us back to the female's role in the breeding process. In species where the males make a significant contribution to parental care, you might expect that extrapair behaviour by the female could cause the male to neglect his parental duties. The male gains nothing by caring for another male's offspring. If the female is to stray beyond the pair bond, she should do so discretely. She will not want her helpful resident mate to notice her infidelity.

### Box 5.3 DNA fingerprinting

While a chick may have hatched in a particular nest, it's not necessarily related to the adults that made the nest and raised the chick. But we can tell who its real parents are by taking blood samples. Birds' red blood cells contain chromosomes, which are the long DNA molecules that hold the information for making proteins. These proteins are needed for the various biochemical processes essential for life. And for increasing numbers of these proteins we know exactly where on the DNA molecule their 'recipe' has been stored. But much DNA does not seem to code for proteins or have any useful purpose. When animals reproduce, they pass this 'junk' DNA on to their offspring. Some pieces of this junk DNA occur in repetitive sequences, and seem to be good at hitchhiking with the useful DNA. When the sequences are short, they are called minisatellites; when they are very short they are called microsatellites. These names refer to the length of the sequence, not to the number of times that the sequence is repeated. The number of repetitions determines how long a piece of satellite DNA is. There is a lot of variation in the length of these pieces of DNA, and that is the key to DNA fingerprinting.[105, 106] Molecular analyses can isolate the satellite DNA in the bird's blood that have certain base sequences (the order of the four components of DNA: adenine, cytosine, thymine and guanine), copy them many times, and then run the pieces down a gel where they form visible bands. Because short pieces travel along the gel faster than larger fragments, they will end up further along the gel after a given time.

A chick gets all of its genetic material from its true parents, so the bands that show in its DNA must have come from its mother or father, or both. If a chick has bands that are not found in either of its 'nest' parents' DNA, this means that it is probably the offspring of another adult. This is evident in chick C (marked with an asterisk) in the left-hand family shown in Figure B5.3.[390] In this oystercatcher family the male was clearly unrelated to chick C. The researchers studying this family had seen the female of the pair copulating with both her mate and a neighbour before egg laying. Sure enough, the neighbouring male (also marked with an asterisk) turned out to be chick's father, as he had bands that were found in chick C but not in the supposed parents (for example, the thick band below 6.4 on the y-axis).

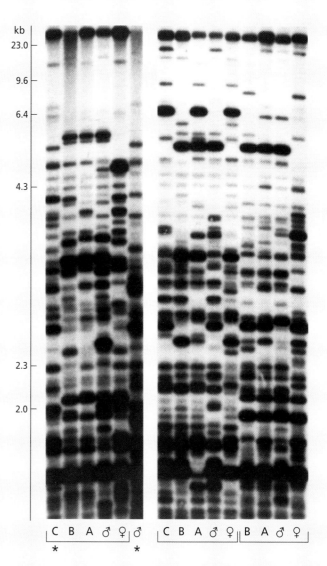

Figure B5.3 DNA fingerprinting of three oystercatcher families (chicks labelled A - C, adults with gender symbols).[390] The black bands represent DNA fragments of different sizes (kilobases, kb) found in the different individuals.

For a successful copulation the birds must push their cloacae together, as these oystercatchers are doing. The female is usually underneath the male during copulation, but sometimes it is the other way around.

Again, it's rather difficult to keep secrets in the open landscapes in which waders breed, unless the male cannot stay with the female all the time.

One of the very few paternity studies on waders is on the purple sandpiper.[649] In this species the female helps to incubate the eggs but does not take care of the young. This makes it very important for the male to be certain that he is not raising someone else's offspring. Sure enough, only one of 82 chicks (1.2%) was genetically different from its apparent father. But this isn't because the males guard the females extremely closely. In the fortnight before the eggs were laid, the pair were together (within 5 m of each other) for around 40% of the time. There was no indication that the males guarded their partners more intensely just prior to egg laying, or that they tried to minimise the risk of extramarital young by copulating with their mates very frequently. It seemed that the purple sandpiper females were just not interested in strange males. When the pair flew separately to foraging sites it was mostly the female who followed the male.

Amongst the other waders whose paternity has been studied, the frequency of extramarital young is also low, though not as low as in purple sandpipers: 4.7% (n = 85) in the semipalmated plover,[958] 1.5% (n = 65) in the oystercatcher,[390] and 4.6% (n = 44) in the Eurasian dotterel (*Charadrius morinellus*).[621] Are these wader males good at protecting their paternity, or are the females uninterested in other males? Early in the breeding season, female oystercatchers are prone to indulging in extramarital behaviour. But oystercatchers are different to other waders in that copulating is one of their main pastimes early in the season. Oystercatchers have been estimated to copulate 700 times before the first egg is laid! It seems unlikely that males can ascertain their parenthood by copulating so frequently. Instead, the copulations are probably at least partly a signal to other birds. Males and females make it clear for all to see that they have a strong pair bond and won't allow any intrusions into it. Anyone eyeing up their territory or mates can easily tell they are up against a unified cooperative pair. So when oystercatcher females are guilty of extramarital behaviour, it's probably because they have been looking for a new partner.[390] This has also been proposed for the spotted sandpiper (*Actitis macularia*).[135] Extramarital young, rather than being the objective of the copulation, are the by-products of pair-forming behaviour between the birds. This was certainly the case in the only extramarital offspring to have been found so far amongst oystercatchers (Box 5.3). This chick was fathered by a male from a neighbouring territory. After two years of dallying with this male, the female made a permanent switch and became his new partner.[228]

## 5.6 Variations on a reproductive theme

### 5.6.1 Reproduction systems: evolution and selection

Raising chicks requires more than just sexual contact and social bonds between the parent birds; the chicks need parental care. For most birds, the minimum parental investment required is to incubate the eggs, then brood and guard the young for a number of weeks. Amongst the waders, the maximum undertaken is to feed the young for many months, as oystercatchers do. What makes waders so interesting is the huge variation in the way that different species distribute the tasks involved in parental care between the sexes, and also the variation in sexual contacts and social bonds. Through long-term studies of individually marked animals our empirical knowledge of the various reproductive systems is steadily increasing. And the theoretical frameworks that we use to analyse these systems are also progressing. Each and every explanation must be in keeping with modern insights into natural selection and evolution. Many people think that animals breed to guarantee the continued existence of their species. As we discussed earlier in this chapter this is not the case. Each individual simply strives to maximise its own number of offspring. It is also important to remember that investing in offspring comes at a cost (see Section 5.1.5). But at the same time, investing more heavily in parental care will lead to better survival of the offspring. This knowledge allows us to make some important predictions about reproductive systems. First, parental care will stop at the moment that the costs of providing extra care exceed the benefits of improved survival of the young. Such costs could include decrease in the survival of the parent bird or a reduced chance of raising a second clutch that season. Secondly, animals will do everything they can to avoid raising another animal's offspring. Males guard their females to ensure she doesn't mate with anyone else, and both birds protect their nest against strange females wanting to offload an egg. Finally, there is strong pressure to allow the other parent to take on the bulk of the parental care. And since both partners would feel this way, a conflict of interests can arise.[219, 872]

Which sex is most likely to win this conflict? Female mammals must carry their young in their wombs, and then provide milk for the young for a considerable period. Each juvenile requires a large investment from the females, while the males can easily limit themselves to simply donating sperm. Unfortunately, this makes female mammals born losers in the evolutionary conflict over parental care. In many mammalian species the males make no contribution to parental care and simply fight amongst themselves for access to the females. In these species, the male's reproductive rate is mainly determined by the number of females that he can gain sole access to. This, in turn, is likely to depend on the distribution of the females. When females live in a group, it is easier for one male to defend the whole group against other males. But when the females live alone, defending a group of them becomes much more difficult if not impossible. And what does the female distribution depend on? Most likely, it depends on the distribution of limiting factors such as food. Since female mammals usually do all the parental care, they have no problems finding a male to copulate with. The females' main problem is finding enough food and keeping their young safe.[165]

The evolutionary answer to the conflict over parental care in mammals may not seem a particularly satisfactory one; the males win because the battle is already over. You could question why male mammals haven't developed functional lactation glands. But we're not going to speculate about that here. Our discussion on mammals was really just an introduction to a way of thinking: a species' reproductive system is determined by that species' ecology, and the nature of the costs and benefits of parental care and sexual behaviour for males and females of that species. Perhaps determined is a big word. It is more about demonstrating the links between ecology and reproductive behaviour. It's very possible that species that developed increased parental care over the course of evolution were consequently able to colonise habitats where the juveniles would not have been able to survive without the increased parental care.

Here we are wandering into the realms of sexual selec-

As soon as they arrive on the Taimyr tundra, these red knots begin their breeding displays. They don't seem to mind that much of the breeding grounds are still covered in snow. However egg laying will be delayed if there is still a lot of snow around later in the season.

REPRODUCTION

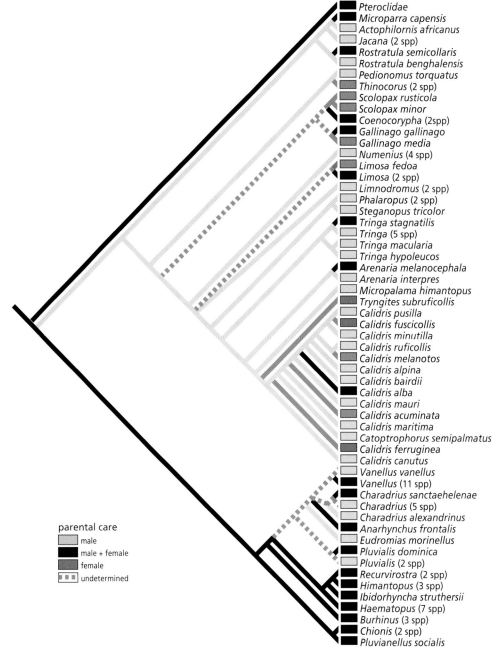

Figure 5.18 The family tree of wading birds, showing the hypothetical evolution of parental care for the various species.[846] The boxes at the end of the branches represent the current parental care strategy adopted by the species. It was not always possible to determine the parental care strategy adopted by some ancestral species, so these species are classed as 'undetermined'.

tion theory. We know that sexual selection is all about the characteristics that influence an individual's success in acquiring partners (Section 5.5.1).[20] There is an obvious connection between sexual selection and the species' reproductive system. The sex that invests more heavily in reproduction will be very selective in forming a pair bond. So the sex that does more of the parental care will limit the reproductive potential of members of the opposite sex. Members of the sex with fewer parental responsibilities will often have to engage in serious combat amongst themselves to acquire one or more partners.

In birds there is only one unavoidable difference between the sexes: only females can lay eggs. All the parental care that follows, such as incubating, guarding and feeding can be administered by both males and females. In this way female waders are in a better position than female mammals. This might also explain why male waders often take on more of the parental care than fe-

males. Székely and Reynolds[846] reviewed published studies of breeding in around 96 of the 203 species of waders. The males remained with the young until they fledged in 87% of the species. The females only did so in 56%. There were also more species of waders in which the males took on all the parental care than species in which the females did so. Incidentally, the most common system was for both the male and the female to stay with the young until they fledge.

So was this common system also used by the ancestral wader that all current waders have evolved from? This might seem like the type of question that is open to endless speculation with no reasonable answers in sight. But it isn't. Of course, it's impossible to be absolutely certain of how the ancestral wader cared for its young, but thanks to new molecular methods and statistical techniques we have some strong clues to follow. The first step is to reconstruct an evolutionary family tree: the phylogeny. To do this, you work back from the existing species searching for common ancestors. Eventually you end up with a family tree that shows how the waders we see today originated from the ancestral wader. If we had been able to travel through time and watch the species evolve, we would also have been able to determine where the dead branches occurred, i.e. the species that have become extinct in prehistoric times. We could have also seen exactly when the different branches split off, and what the common ancestors were like. Fossils might also provide us with some of this information. But unfortunately there is not an extensive collection of fossilised wader species.

So our wader family tree or phylogeny will always be hypothetical and can only show the branches that are still alive.[382] Before genetic techniques were available, people mainly used morphological and anatomical characteristics such as plumage and bones to construct phylogenies. This method assumed that species that looked very similar were more recently descended from a common ancestor, than species that looked less similar. The first problem with this is that while species may be similar in some characteristics, they can be very different in others. The different characteristics have to be weighed against each other using complex calculation methods. The second problem is that the common ancestors of the common ancestors (and so forth) have to be inferred. Mathematical calculations to enable this have also been developed. But the main problem is that species can look similar as a result of parallel or convergent evolution and not as a result of common descent. Convergent evolution means that different species have run into the same problems and adopted the same solutions to these problems independently. For example, if you don't want to be conspicuous while incubating your eggs, regardless of what colour feathers your ancestors had, you're best off to have very cryptic plumage.

Modern genetic techniques seem to offer a solution to these problems. It is assumed that some changes in the genetic material are selectively neutral. That means that they have no apparent consequences for the animal's characteristics, and that the occurrence and development of those genetic changes is mainly a matter of chance. When the probabilities of this chance process occurring are constant, we can calculate how much time it should have taken for two species to reach a certain level of genetic differences. The larger the genetic differences, the longer ago that the species diverged and the further back their common ancestor. The details of this 'molecular clock' are the subject of much discussion amongst experts, but the principle that the clock exists is widely accepted. Different assumptions about these details give rise to different family trees, so these details are not trivial. Székely and Reynolds[846] did not limit themselves to genetic data when they constructed their wader genealogy (Figure 5.18), they also considered other sources of information. In spite of this, the resulting phylogeny has been the subject of much criticism and it's very possible that the supposed wader family tree will look completely different within a few years.

As well as their family tree we also know about the division of parental care amongst the waders species around us today. This parental care data is not used to construct the family tree, or else we would end up with circular reasoning. Female curlew sandpipers do more parental care than the males, while in dunlins the males do more. So what did the ancestral waders, the plover and the sandpiper forebears, do? According to Székely and Reynolds[846] males did the bulk of the parental care in the ancestral sandpiper, while in the ancestral plover and going further back to the original wader, both sexes cared for their offspring (Figure 5.18). But how did they reach that conclusion? They have assumed that the number of evolutionary changes would be as small as possible. In the majority of *Calidris* sandpipers, the males do most of the work. So to minimise the number of evolutionary changes, we assume that the males of the ancestral sandpipers also did most of the work. It is possible that there have been more steps in the process, with switches back and forth through time, but the analysis comes up with the simplest, 'shortest' route from the ancestor to the present species. As is evident from the number of changes in parental care shown in Figure 5.18, evolution has apparently not put tight restrictions on who can undertake parental duties. According to Székely and Reynolds[846] this is related to the fact that both parents can incubate and that the young quickly become independent. This makes it possible for one of the parents to desert the brood prematurely and to try and find a new partner.

The phylogenetic tree enables us to analyse evolutionary changes in an insightful way. For example, take the question of whether increased paternal care is related to decreased maternal care. You might think that the best way to find this out is simply to plot the contribution of the male versus the contribution of the female, so that every species becomes a single dot on the graph. Such a graph can be drawn but you cannot simply test whether there is a negative relationship because the dots are not independent. Dots close together may be there because the species had a common ancestor, not because they independently evolved the same solution as a result of natural selection. But using our family tree we can now work with contrasts that get around this relatedness problem.[382] Instead of using a single species, we use two species with a common ancestor as the unit for our analysis. We can then check how this pair differs in the parental care exhibited by the males and females. The differences are called contrasts. The contrast in female care can then be plotted against the contrast in male care. When we do this for all species pairs we find that when the paternal care increases, maternal care decreases.[846]

The most interesting result of this comparative analysis is that with an increasing size difference between the sexes, the parental care of the largest sex decreases. In many wader species, the females are larger than the males, as in the curlew and the oystercatcher, but in a number of species the males are larger, as in the lapwing. Nowadays the first thought, which is not necessarily correct, is that size differences result from sexual selection

REPRODUCTION

The sanderling (photo above) and the curlew sandpiper (photo below) are both capable of incubating a nest without assistance from their partners. In some curlew sandpiper pairs the male and female each look after their own nest, but other pairs raise one brood together.

There are large differences among the waders in how parental care is divided between the sexes. Female purple sandpipers (top photo) leave the nest as soon as the eggs have hatched: the male must brood and guard the chicks alone. The female dotterel (middle photo: the male sits on the nest) does not even help to incubate the eggs; as in phalaropes, the beautiful females only come to the nest to lay their eggs. Task division is much more equal in grey plovers (lower photo). The pair takes care of eggs and chicks together, although the female is usually the first to leave the almost fully-grown chicks. You can also see how well camouflaged the grey plover chick is.

# REPRODUCTION

| breeding system | | spatial distribution of the sexes | | parental care |
|---|---|---|---|---|
| **monogamy:** one male has a pair bond with one female | **resource defence:** many songbirds, waders and seabirds | | multipurpose territory defended by male or both sexes, or only nest site defended (seabirds) | male and female |
| | **mate guarding:** some songbirds, many ducks and geese | | male guards female and sometimes also the nest site | male and female (for instance finches) or only the female (many ducks) |
| **polygyny:** one male has pair bonds with different females at the same time | **resource defence:** some songbirds and waders | | male defends large multipurpose territory in which more than one female defends nests or subterritories | male and female, or mainly female |
| | some songbirds (weavers) | | male defends a group of nests | male and female, or mainly female |
| | polyterritorial (pied flycatcher) | | male defends separate nest sites or territories each with its own female | male and female, or mainly female |
| | **harem defence:** pheasants | | male defends group of females that nest alone | female |
| **polyandry:** one female has pair bonds with different males at the same time | **cooperative:** Galapagos hawk, Tasmanian native-hen | | different males defend the territory of one female | all males can help |
| | **resource defence:** jacanas, spotted sandpiper | | female defends large multipurpose territory in which males defend exclusive subterritories | mainly or exclusively the male |
| **polygynandry:** different males have pair bonds with different females at the same time | dunnock | | different males defend the territory in which different females nest together or defend their own territories | all males can help all females |
| | acorn woodpecker | | | |
| **promiscuity:** no pair bond; contact between the sexes limited to copulation | **resource defence:** some hummingbirds | | male defends feeding territory and copulates with female that comes to feed | female |
| | **defence of display area:** manikins, birds of paradise, ptarmigans, waders, kakapo | | males defend display area where females only come for copulation. Display areas can be far apart or very close together (in arenas) | female |
| **sequential polygamy:** males or females have pair bonds with different partners in succession | **sequential polygyny:** woodcock | | male guards female until clutch is complete and then searches for new female, no territorial behaviour | female |
| | **sequential polyandry:** phalaropes, dotterel | | female guards one male until clutch is complete and then searches for a new male, no territorial behaviour | male |
| | **sequential polygyny and sequential polyandry:** Temminck's stint, little stint | | female lays eggs in male's territory, then searches for a new male; males can mate with a new female who incubates the clutch herself when the male is already incubating | male or female |
| | **simultaneous polygyny and sequential polyandry:** common rhea, tinamoes | | group of females lays eggs in the nest of a male that incubates them, while the females start searching for a new male | male |

● nest
(dashed outline) territory ♂
(shaded) territory ♀

Figure 5.19 Summary of reproductive systems in birds. After outlining the sexual and social nature of the system, examples of representative species are given. The general spatial arrangement is described, followed by which parents are involved in parental care.[165]

(see Section 5.5.1). The relationship observed is exactly what would be expected if it is assumed that fighting power and body size are positively correlated. It also seems reasonable to assume that sexual selection will act most strongly on the males of 'polygynous' species and on the females of 'polyandrous' species. 'Polygyny' means that one male can mate with more than one female, 'polyandry' means that one female can mate with more than one male, while 'monogamy' means that there is a bond between one male and one female. If the analysis is repeated for the males, without including the polygynous species, the relationship that previously existed between size differences and parental care is no longer observed. Males of monogamous species do not take on more of the parental care when their partners become relatively larger. So it appears that the transition from monogamy to polygyny, through which the males become responsible for less parental care and must fight between themselves more, seems to have caused the relationship between parental care and size differences. However, when we repeat the analysis for the females, excluding the polyandrous species, the relationship between parental care and size differences still occurs. And that isn't the only confusing result. For males an increasing migration distance is correlated to a decrease in their parental care. But this is not the case for females.

Working with contrasts may solve the problem of common ancestry making species not independent. However it does create a new problem. Absolute measures are replaced by relative measures. And it is very difficult to compare these relative measures. A slight increase in parental care by a species that already exhibits a high level of care is considered equal to a slight increase by a species that hardly does anything for its offspring. The most important conclusion of the family tree analysis still seems to be that evolution hasn't placed large restrictions on the task division. This conclusion is not only important, it's more concrete than the hypothesis about the primitive wader's parental behaviour. Evolutionary change can only have been larger than it is currently evident. And if the parental behaviour can still easily change, then measuring the costs and benefits of the current behaviour is an important step in our search for an explanation of the observed behaviour. Before we examine some of the better-known reproductive systems, let's take a look at the different breeding systems in more detail.

### 5.6.2 Monogamy, polygamy and polyandry

Figure 5.19 summarises all of the known mating systems in birds.[165] Many of these systems can be found in the waders, making them a particularly interesting group. However we shouldn't forget that the majority of the wader species discussed in this book is monogamous (Table 5.3). In these monogamous species one male and one female form a pair and care for their eggs and young together. In many of these species the female abandons her parental duties earlier than the male. In fact, red knot, purple sandpiper[648] and spotted redshank females leave the breeding grounds as soon as the eggs have hatched.

In most species the male stays with the young until they can fly, but not dunlins and greenshanks. In these two species the males abandon the young before they fledge. In all waders, the bond between the parent birds and their young is broken before the young begin their first migration to the wintering grounds. This is why for many species, females are the first to be seen in the stopover areas, followed by the males, and eventually the juveniles.

The Kentish plover is unique amongst the monogamous wader species. These plovers breed so far south that their breeding season is long enough for them to raise more than one brood. This allows the species to commit 'serial monogamy'. Which means that within one breeding season the animals can have a series of monogamous relationships with different partners. So Kentish plovers desert their broods because they want to start another clutch, rather than get to the wintering grounds as quickly as possible. We'll discuss the battle of the sexes over who gets to leave the brood first in more depth later in the chapter.

Serial monogamy is sometimes called 'serial polygamy', but that is not as accurate. In polygamy, one indi-

| | pair bond | number of parents | laying date | male parental care | female parental care |
|---|---|---|---|---|---|
| oystercatcher | M | 2 | | 7 | 7 |
| avocet | M | 2 | | 7 | 7 |
| ringed plover | M | 2 | E | 7 | 7 |
| Kentish plover | SM | 2/1 | | 7 | 4 |
| grey plover | M | 2 | E | 7 | 6 |
| golden plover | M | 2 | E | 7 | 7 |
| red knot | M | 2 | E | 7 | 3 |
| sanderling | M | 2/1 | E/L | 7 | 7 |
| curlew sandpiper | PG | 1 | E/L | 0 | 7 |
| dunlin | M | 2 | E | 6 | 4 |
| little stint | M/SM | 1 | L | 7 | 7 |
| purple sandpiper | M | 2 | E | 7 | 3 |
| bar-tailed godwit | M | | | 7 | 6 |
| black-tailed godwit | M | 2 | E | 7 | 7 |
| curlew | M | 2 | E | 7 | 7 |
| whimbrel | M | 2 | E | 7 | 7 |
| spotted redshank | M | | | 7 | 3 |
| redshank | M | 2 | E | 7 | 5 |
| greenshank | M | 2 | | 6 | 6 |
| common sandpiper | M | 2 | | 7 | 5 |
| turnstone | M | 2 | E | 7 | 5 |

Table 5.3 Characteristics of the reproductive systems of various wader species. Pair bond: M = monogamous, SM = serially monogamous, PG = polygynous. Laying date: E = relatively early in the season, L = relatively late in the season. Number of parents: how many parents contribute parental care. Parental care: expressed on a scale of 0 - 7, in which 0 = no parental care, 3 = leaves nest as soon as chicks hatch, 4 = leaves nest long before chicks fledge, 7 = takes care of chicks until they have fledged.[116, 220, 678, 691, 727, 933]

vidual has more than one partner. If the male has more than one partner it's known as polygyny. And when the female has more than one mate it's called polyandry. In principle, a polygamous individual has more than one partner at the same time. In serial monogamy, an individual has different partners consecutively.

It's usually not that difficult to determine whether a certain individual is monogamous or polygamous, if enough time is spent observing the behaviour of the individual concerned. Classifying an entire species however is more problematic. Lack[492] classified the majority of all birds as monogamous. However in many monogamous bird species, polygyny seems to occur to some extent. Polygyny was regularly identified in 39% of 122 well-studied European passerines (passerines are the song bird species).[574] In 25 of the 47 species that displayed polygyny to some extent, as many as 5% of the males were polygamous. The lapwing is another good example. Until recently this species was classified as monogamous.[138] But then individual lapwing males were repeatedly observed holding a territory in which two or more females bred.[44, 79] In one study as many as 42% of the males had two or even three partners.[358] So there are three main problems with trying to classify a species' breeding system. First, the animals must have been monitored intensively and that is lacking for many species. Secondly, there are differences between populations. Finally, there is no species in which all individuals are polygamous. The boundary between monogamous and polygamous behaviour is somewhat arbitrary at the species level.

The scientific challenge lies not in classifying the species but in finding an explanation for the observed frequency of polygamy. It's highly likely that all of the wader species described as monogamous in Table 5.3 are in fact polygamous to some extent. This suspicion has been confirmed for the oystercatcher, despite its improbably low frequency of occurrence. Less than 2% of the breeding males and 3% of breeding females on Schiermonnikoog are polygynous.[391] And because of the extensive research that has been done, we know why these bonds are so rare!

Species that don't form social bonds, and whose females visit an area where the males display (called a lek) to copulate, are something of a special case. Sometimes these lekking species are classified as polygamous and sometimes they are treated as a separate category all together. The ruff is perhaps the best known lekking species and its display behaviour has been described extensively.[41] Polygamous species are more likely to encounter sperm competition than monogamous species. This means that the sperm of one animal can be confronted with the sperm of another male. So whoever supplied the most sperm cells may win the race to fertilise the egg. It should come as no surprise that on average polygamous wader

Kentish plover parents take turns incubating their eggs. A week after the chicks hatch, the females often leave the brood.

Male and female little stints each care for their 'own' clutch of eggs and the chicks that subsequently hatch. A recent theory suggests that this is due to the species being specialised on a rich food source that becomes available relatively late in the season.

species have larger testes than monogamous species.[691] A ruff's testes may constitute 5% of its total body weight, making them even heavier than its brains.

Curlew sandpiper females, like reeves, incubate the eggs and take care of the young. However instead of showing off in a lek, the male curlew sandpipers defend a breeding territory. Perhaps their differing social system is due to the unpredictable weather in the curlew sandpiper's high northerly breeding grounds. Ruffs can be found on the same display areas year in year out. However the number of curlew sandpipers in a certain area on the tundra is very variable.[865] The animals search for the snow-free areas. In cold years when the tundra remains snow-covered they breed further south. Some curlew sandpipers arrive on the breeding grounds as a pair, which means that pair formation has occurred en route. This may be an adaptation to help them cope with the short breeding season. But why don't the male curlew sandpipers help their mates incubate the eggs? The most obvious explanation is that the males prefer to take their chances at picking up other females.[862] So maybe they're not that different from the ruffs after all.

In the Temminck's stint[91] and the little stint,[401] the male and the female can each incubate a clutch and raise a brood on their own. If the tasks were evenly divided, you would expect the male to incubate the first clutch, and the female to incubate the second. But nature is seldom that fair. The Temminck's stint males do incubate the first clutch. But while the male is incubating, the female finds a new mate; Temminck's stint females never make a second clutch with the same male. Sometimes the females then leave the area. At the same time, new females arrive from other sites, so in principle, each male can get a second partner. However some males don't achieve this. There are also some clever females that manage to catch two males and don't actually have to incubate until they've laid their third clutch.[862] As well as helping them outsmart as many males as possible, the way that the females move around might be related to the changing environment on the tundra. You would expect the females to nest wherever is most favourable at that particular time. Areas that are slow to thaw will not be suitable until late in the season.[91]

This system in which males and females each incubate their own clutch is called double clutching. Sanderlings can either use the double-clutching system or incubate a clutch normally as a monogamous pair.[862, 890] This is rather perplexing. If the pair is able to incubate eight eggs between them, then neither partner is doing itself an evolutionary favour when the pair restricts itself to four eggs. The first hypothesis that needs to be investigated is whether every pair is capable of double clutching. Differences in the extent of double clutching between different populations would be a good starting point for further research.

When one parent bird is capable of successfully incubating the eggs and raising the young alone, there is no reason why the males can't take on all the parental care. And these reproductive systems have arisen in the course of wader evolution. One example is the red-necked phalarope, which incidentally is not your average wader. The red-necked phalarope usually forages while swimming with the assistance of its webbed feet. And phalaropes are not really shorebirds; these dainty little birds winter on the open sea. While on migration they are sometimes seen in intertidal areas, not walking on the mud, but swimming in little pools. Apart from their beautiful plumage, the striking thing about red-necked phalaropes is their apparent tameness. On the open sea the birds have nothing to fear from humans, and the same probably applies to their breeding grounds on the tundra. As you'd expect of a species where the male takes on all parental responsibilities, male phalaropes have much duller plumage than females. Although each female can lay a large number of eggs, each male can only incubate four eggs. So the females must compete for the males and, as befits the competing sex, the females are first to arrive on the breeding grounds. Unpaired males may be hotly pursued by a number of females. When a pair bond is established, the partners stay in close proximity until the eggs are laid. The female has to chase away other females that want to steal her partner and the male must ensure that his partner doesn't copulate with other males. Once the eggs have been laid the female can try to find a new male. This form of serial monogamy is also sometimes called serial polyandry, as the female pairs with a number of males in succession.[931]

So some species have developed special reproductive systems whereby one individual can incubate the eggs and raise the young. One theory is that these systems arise in species dependent on food sources that become available late in the season.[933] Compared to the food sources available early in the season, the later food sources would be relatively rich, thus enabling one bird to undertake all of the parental responsibilities. There is good evidence that the one-parent species do breed later than two-parent species at the same site.[933] The species shown in Table 5.3 are only a small sample of the total number of wader species that have been studied. But even amongst this small sample the one-parent species do breed relatively late. However this doesn't prove that the food sources theory is correct; it is still just speculation. For the theory to be confirmed, at least two things have to be proven. First, that the late-breeding, one-parent species are dependent on food sources that only become available late in the season. Secondly, that these later food sources are richer than the earlier food sources.

Furthermore you would have to prove that the species in which two parents always incubate are unable to cope otherwise, which is best tested using removal experiments. This involves taking away one partner and seeing whether the remaining individual can cope with its increased workload. As yet this has only been done in oystercatchers. Both males and females have been removed during incubation,[227, 255, 378] and the clutch always suffers. The partner that has been left continues to incubate for a while, but there comes a point at which it can't carry on. Territorial fighting is so intense that only pairs can hold onto their territories. A bird that has been 'widowed' has the choice of staying single and losing the territory or quickly finding a new partner. Because there is a large pool of unattached birds without territories, it is usually possible to find a new partner. But the new partner has no interest in investing resources in raising the previous partner's offspring. So the oystercatcher's inability to incubate on its own results not from inadequate food, but from the fierce competition for a good territory. However the oystercatcher is something of an extreme case, and this result can't be generalised for other waders where the pair care for their offspring together.

### 5.6.3 The deserting plover

Not all bird names are equally well chosen. Some names are fairly clear in their meaning: Temminck's stint was probably discovered by Temminck, and Baird's sandpiper by Baird. Other names say something about the bird itself, rather than a researcher. This may be a physical characteristic (spoonbill, redshank, ruff, bar-tailed godwit), a habitat (marsh sandpiper, sanderling), a sound (hummingbird) or even a behaviour (turnstone). Most birds received their names in a time when the ornithologist's main tool was a gun. Fortunately, guns have since been replaced with telescopes and binoculars, so we can now make long-term observations of individuals. Being able to mark and recognise individuals has enormously enriched our knowledge about the behaviour of many species. If we were to name the Kentish plover now, we might well call it the deserting plover.

Kentish plovers breed on fairly barren sandy areas, but that is not what makes them so special. The interesting thing about Kentish plovers is the variation in the extent of their parental care. The female incubates the eggs during the day, while the male takes the night shift. They care for the chicks in the standard wader way: brooding them when they are cold, taking them to suitable foraging areas, raising the alarm when danger approaches, distracting or attacking unwelcome visitors. But their caretaking roles are not strictly determined. As long as both parents take part, they share the duties. This may continue until the chicks fledge and can take care of themselves. But more often than not, one parent deserts the brood, leaving the other to do the work all by itself. It's usually the female that deserts, although the male may also do so. A parent may leave as early as a week after the chicks hatch, though there is no fixed point at which desertion occurs.[507] This breeding system is extremely interesting for a number of reasons. To start with, desertion during chick rearing is unusual in birds. Desertion more often takes place as soon as the clutch is complete and genetic parenthood is guaranteed. It is also unusual to have a situation where either sex may desert.

This provides the researcher with a unique opportunity to manipulate the breeding system of the birds and discover what the costs and benefits are for the deserting parent and the abandoned family. Such research is much more difficult, if not impossible, in species where the desertion is fixed and seems to be part of genetically programmed marriage contract.

Two birds can take better care of the offspring than one can, so if the chicks could choose both parent birds would have to stay right to the end.[844] Unfortunately for the chicks, this decision is not left to them. The remaining parent bird still has options; it too could up stakes and leave. This seems like a drastic response since the young would almost certainly perish. Nevertheless, such behaviour is theoretically possible. The bird that stays behind must trade the expected success of continuing its present breeding attempt against the possibility of quickly finding a new and possibly more faithful partner.

Kentish plovers never desert while the eggs are being incubated. To find out what would happen to the clutch if one parent deserted, you could catch and remove one of the parents.[506, 845] It seems that the remaining parent is quite capable of incubating on its own. But only some of the widowed birds are prepared to finish the job. Females are especially likely to quit the clutch. This means that a male who deserts early runs a serious risk that his brood will not survive. Males also have much more trouble than females in finding a new partner after they have been deserted, unless the nest is also removed (by researchers). If a breeding attempt fails, for whatever reason, it seems that it doesn't matter greatly to the female whether she looks for a new partner, or simply starts again from scratch with her current mate. She can be back on eggs again almost immediately, one way or another. This difference between males and females is probably caused by a surplus of males on the marriage market.[842, 845]

It seems as if it's mainly the females that desert since they can easily find a new mate. Experiments in which one partner was removed directly after hatching reveal that there is even more going on.[842] Broods that are taken care of by the male have a higher chance of surviving and grow better than broods that are taken care of by the female. Males seem to be better able to defend the territory and young against unfriendly neighbours and other potential dangers.

It could be that both males and females are discretely on the lookout for new mates once the eggs have hatched. Desertion could occur whenever the opportunity arises and it is reasonably likely that the abandoned partner will remain with the nest. This scenario could explain the large natural variation that is present in Kentish plover populations. Alternative scenarios can be envisaged, in which individuals only search for a new partner after they have deserted their nest and partner. Kentish plovers have been recorded moving 170 km between breeding attempts in the same season in Europe, while the American subspecies (*nivosus*) has a record of 600 km! These observations support the idea that birds desert first, and then go looking for a new mate.

While ingenuous detective work has helped measure what once seemed immeasurable, the story is not yet complete. Ideally, we would observe both members of many Kentish plover pairs throughout the breeding season, day and night. However this remains a dream at present.

### 5.6.4 Polygyny in the monogamous oystercatcher

When the parents make a large contribution to the survival and success of their offspring, it follows that their reproductive opportunities are limited by the number of mates they can have. By having an extra mate, a bird of either sex could bring up more young than if it had just one partner. Both sexes would benefit from getting as much personal assistance from the other sex as is possible, but in a three-way relationship, one partner has to share their mate with another bird of the same sex. Polygyny (when one male has more than one female mate) is therefore favourable for males but unfavourable for females.[165, 166] We would expect that a female oystercatcher in such a relationship would bring up less young than a bird in a monogamous relationship. And that's what we find. We might also expect the male oystercatcher to profit from this set up. But that is not the case, which explains why polygyny is so rare in oystercatchers.[391]

To understand why polygyny isn't profitable for male oystercatchers, we have look deep into the pair bond of the oystercatchers and the way in which trios arise. Oystercatchers with a strong partnership can be found in their breeding territory months before the eggs are laid. They copulate frequently and defend their territory together against intruders of either sex. When an intruder arrives the partners sometimes join forces to demonstrate their solidarity, and then charge the intruder, 'tepieting' loudly. They also regularly have disputes with their neighbours. These disputes or 'tepiet ceremonies' involve the two pairs running up and down the boundary of their territory calling loudly. Between these hectic moments they feed fairly close together, or roost side by side. Once the eggs have been laid the pair share incubation duties, regularly swapping shifts. The bird that is free guards the territory, chasing away gulls and other predators, or forages. When they have chicks, one bird broods the young or stands on guard while the other brings in food. They will also fiercely attack intruders and neighbours.

Most intruders are 'club' birds. As described earlier (Figure 5.11) these are adult oystercatchers without breeding territories, which gather at high tide. Some club females manage to disrupt a previously happy marriage i.e. they succeed in chasing the original female out of the territory and form a pair bond with the male.[246] It's unclear what role the male plays in such an event. Was the relationship already on the rocks, and did he make overtures towards the new female before the final showdown? Or was the male a realist who supported his wife until it was evident that the new female would win? We don't know. What we do know is that such takeover fights don't always lead to a clear result. The territory can then be defended by a male and two females that argue between themselves. On Schiermonnikoog, the females each defended their own part of the territory in just over half of the threesomes documented.[391] This arrangement often arises in polygyny.[165, 166] The first female has no interest in any other females, and tries to keep them away by being territorial. But once she has eggs, it is more difficult to be aggressive. This makes life easier for the other female, but at the same time she is at a large disadvantage. Males will target their brood care towards the initial mate, probably because the earlier young are more successful that the late young (see Section 5.4). This means that the second female has to undertake most of the parental care herself. Some species can manage this, even though the breeding success will be lower. In the oystercatcher this is not possible: nests of females undertaking solo incubation or chick raising are inevitably lost. Unfortunately, even with the help of the oystercatcher male, the first nest is also not very productive. Aggressive trios (n = 25) raised on average only 0.04 fledglings per attempt. While the average production for monogamous pairs (n = 1729) was also fairly low, 0.23 fledglings per pair, it is still six times higher than for a trio. From the viewpoint of the male, he is much better off in a secure relationship with one female rather than having two disputing females in his territory.[391]

Curiously, in just over half the cases the females made a truce at some point. They didn't just decide to tolerate each other's presence; they became allies and worked together in territory defence. Even more surprisingly some began to copulate together! These lesbian oystercatchers caused so much fuss in the scientific community that their story made it into Nature, the most prestigious international scientific journal.[391] But the females didn't limit their relationship to copulation. They also made a communal nest, each laying their eggs in it, and finally caring jointly for the young. Could this cooperation be occurring because the females were related to each other? DNA fingerprinting showed that this was not the case. For the moment it remains a mystery as to why some females continue to fight, while others start to cooperate. Can three cooperating oystercatchers raise more young than two cooperating oystercatchers? With a production per trio (n = 20) of 0.16 fledglings, cooperative trios clearly do better than aggressive trios, but still have lower rate of fledgling production than monogamous pairs.[391] The problem is that the females regularly lay too many eggs in the communal nest. More than half of the nests had five or six eggs. We've already seen that large clutches suffer from poor incubation and have reduced hatching success (Figure 5.2). If the females could agree to each lay only two eggs, then their problems would be solved. This seems beyond the birds at this point.

All in all, it is odd that there seems to be no advantage to either the oystercatcher females or the male from polygyny. It may be that our assessment of the pros and cons is too simplistic, because we looked only at the breeding success. Perhaps a male with two females doesn't need to put as much effort into defending his territory, resulting in increased survival. Currently, the data don't suggest that this occurs although it must be

This young oystercatcher is still being fed in August (photo right below) amongst dunlins that have already finished breeding some time ago. It is still a long way from being ready to breed itself. It will be a few years before this bird even starts visiting the 'club', where oystercatchers without a breeding territory gather at high tide (photo page 298). The club birds lay claim to breeding territories by flying over them calling loudly (photo left).

admitted that the samples are small. Studies are ongoing at Schiermonnikoog, and what we currently believe may well be overturned in the future. For the moment, the long-term perspective does offer some hints about the behaviour of the females. The average club bird has a 9 - 13% chance of being promoted to a breeding bird. After a season as part of a trio, the chance is much higher, 33% - 73%. Many trios only last one season, suggesting that the females use the truce as a stepping stone towards settling themselves permanently in an area. This brings us to the final stage of this chapter about reproduction: an individual's career planning.

## 5.7 Career planning

A waterbird that wants to reproduce has to make decisions about where and when to breed, the nature of its social and sexual contacts with the opposite sex, and how much effort to put into raising offspring. By now it should be clear that these are not independent problems that can be solved separately. And the choices made in one season may have far reaching consequences. Efforts made one year don't only determine the probability of offspring in that one season; they also have consequences for later seasons. Perhaps an animal that works extremely hard for one season's offspring runs an increased mortality risk in the following winter. Reproductive costs could also affect an animal's social status or its bond with its partner. And by selecting a certain area the animal probably reduces its chance of being successful at another site. A well-chosen career path is necessary for the animal to maximise the number of offspring it produces during its life.

The first important career question is at what age the animal should start reproducing. An animal that has already bred must decide whether or not to breed in the same place with the same partner in the following season. This decision is complicated by the animal's lack of control over many factors. Its previous partner may have died during the winter, or chosen another bird in spring, or perhaps its old breeding area is already occupied when it returns. Animals are very rarely monitored intensively enough for researchers to determine the exact causes of any changes in breeding areas and/or partners.

### 5.7.1 The best age to debut as a breeding bird

Many waterbirds do not breed when they are one year old. The larger species in particular may wait another year or more before risking a breeding attempt. For the experienced evolutionary biologist this is a problem. It's obvious why human beings don't start to reproduce as one year olds. But waterbirds are physically fully grown after one year. A fully-grown bird that misses an opportunity to reproduce is making a mistake, unless reproducing at a young age comes at very high cost. So what sort of costs could be involved? Perhaps a one year old bird that attempted to breed would only have a small chance of success, but would put itself at a much greater risk of dying before the next breeding season. If so, a bird that postpones breeding for one or more years will bring up more young over the course of its life than a bird that starts to breed immediately. This is the standard explanation.[938] However it's only a satisfactory explanation if we understand why young birds are likely to be unsuccessful breeders and why they run a much bigger risk of not living to see the next breeding season. There are various ideas about why this is so.

Before we start to discuss those ideas it is good to emphasise that there are actually two important decisions to be made, rather than just one. Although the age of first breeding may sound like one single decision, it is not. First, the young bird must decide whether or not to return to the breeding grounds in spring. This decision can be made independently of what its conspecifics choose. However a successful return to the breeding grounds is not an automatic guarantee that the bird can breed. Birds must fight for territories and partners on the breeding grounds. Success is never guaranteed and it depends on the willingness of conspecifics to form a pair bond. So the final decision about whether a bird becomes established as a breeding bird depends on the decisions made by conspecifics.

Let's start with the age that birds first return to the breeding grounds. Young birds that don't go to the breeding grounds in their first year spend the summer on their intertidal wintering grounds. In June 1998 a count was conducted across the whole Banc d'Arguin, the most important wintering ground for waders in Africa.[194] Large numbers of almost all species were found. The only exception to this was the smallest species, the little stint. Less than 1% of the 40 000 little stints that had been present in the winter were counted in June.[996] The proportion of birds that remain depends not only on how old birds are when they return to the breeding grounds, but also on the population's age structure. If the birds are relatively long-lived, the percentage of first year birds will be low on average. Estimates for the annual survival of adult waterbirds are usually somewhere between 70 – 90%.[270, 727] If only the one year old birds don't go to the breeding grounds, we would expect to see around 10 – 30% of the wintering birds spend the summer in their winter quarters. These percentages become higher if some second year birds also skip a breeding season (and so on). With the exception of the little stint, the percentage of birds that had remained on the Banc d'Arguin for the summer was between 5 - 47% for the most numerous species. When we combine this information with data from the breeding grounds,[270, 678, 691] the following picture forms. The smaller wader species are already able return to the breeding grounds in their first year, but don't always do so. The larger species never do. As well as differences between species dependent on their size, there also differences between the sexes within a species. Females often return earlier than males. And finally within one sex of one species, there is variation between individuals. The best-studied example is the oystercatcher in the Exe estuary. Thousands of oystercatchers that breed in Scotland, The Netherlands and Norway winter in this estuary.[325] After spending some months during summer on the breeding grounds, the adult birds return to the Exe year in year out (Figure 5.20). First year birds have never been observed leaving for the breeding grounds. And the bulk of the second year birds also summers in the Exe. The opposite applies to the third year birds: nearly all will spend a number of weeks on the breeding grounds. This continues until the fifth year, when the birds begin to stay on the breeding grounds for as long as the adults.

How can you tell what fitness price the young bird that didn't return would have paid if it had returned? Actually you can never find this out, unless you force the animal to do things that it hadn't intended to do (maybe in the long run this would be possible using hormonal treatment). Instead, your next step is to find out whether the young birds do face extra costs and how these costs operate.

The first possibility to spring to mind is foraging behaviour. Young birds need to learn which prey are most profitable and how best to find and access these prey. There are numerous studies comparing the behaviour of young and old birds in one or more of these aspects. For example, in early spring juvenile curlew sandpipers have a lower pecking rate and catch less prey per unit time foraging than adult birds.[715] However there are few studies in which the resulting intake rate has been measured over a longer period.[126] The most important question is whether or not one winter is adequate for a young bird to learn everything that can be learnt. It is highly plausible that it takes young oystercatchers a few years to become adept at opening difficult prey like mussels.[324, 613] Strangely enough, the few young oystercatchers that continue to eat mussels through the winter do have an intake rate in spring that is comparable to that of the adult birds.[326] Nevertheless, they are socially subordinate to the more dominant adult birds and through this lose a lot of mussels.[234] It takes many years before the young oystercatchers can socially measure up to the adult birds and are assured of the same food intake rate. Likewise, while juvenile redshanks had similar intake rates in winter to adults foraging in the same habitat, the juveniles were subdominant and forced to forage in places where the risk of predation by raptors was high.[141, 142, 143] In summary, young birds are not as good at foraging as the older birds and are lower in the pecking order. It's likely that because of this a young bird is at higher risk during the long migration to the breeding grounds than an older bird. This must mean that young birds also run more risks than the older birds when they first migrate from the breeding grounds to the wintering grounds. The scarce data available confirm this suspicion.[620, 644]

A related hypothesis is based on the idea that young birds are more vulnerable to parasitic infection than older birds.[548] When the young birds get infected they may either die as a result of the infection or, after a period of illness, build up their immune systems to some degree. It is conceivable that the young birds use their first year to recover fully from any infection and build up their immune systems. This theory that young birds are more vulnerable to parasites could reflect age-related differences in foraging ability. Young animals that have problems finding enough food are more likely than replete adults to eat prey containing parasites.

Factors such as a lower foraging rate, low social status, higher risk of predation, and a higher risk of infection by parasites don't only make the journey to the breeding grounds risky, they lower the bird's chance of success should it reach the breeding grounds. However the bird has several options. Rather than choosing not to return to the breeding grounds at all that year, it may decide to migrate later in the season. During spring temperatures increase, as do the biomass and activity of benthic animals. All of which would be expected to make it easier to successfully complete a long migratory journey. Whether or not this is the real reason behind it, in a large number of species the young birds that do return to the breeding grounds arrive later than the older birds.[587, 588]

Returning to the breeding grounds is a prerequisite for breeding, but it's not the only requirement, as the oystercatcher clearly demonstrates. Oystercatchers begin to return to the breeding grounds in large numbers in their fourth year, but it can still be years before the birds actually start to breed. When young oystercatchers first return they become club members. These clubs are groups of sexually mature birds without breeding territories and also often without partners. Club members try to infiltrate the territories of the established pairs. Often, they let the owners chase them away without too much effort. It seems as if the club birds are mainly just curious. This curiosity peaks in June when there are chicks. This is understandable as the presence of chicks is an indication of the territory's quality. There are regular gatherings of these curious intruders while the chicks are present, particularly when, for example, the territory owners have a serious dispute with their neighbours. Sooner or later, the club birds also become aggressive. They display above the territories comparing their strength with that of the other club birds, or making their intentions for a certain area clear. After some years the animals conquer territory space for themselves. Some fill a situation that has become vacant and form a bond with the resident widow or widower; others chase away the previous territory owner, and some form a trio (see Section 5.6.4). Most often a pair of club birds will conquer and seize the territory of an established pair.[227, 388] The pair of club birds may chase the territory holder away or squeeze themselves in between two existing territories. Sometimes a pair of club birds already has a territory on the tidal flat and will fight their way in from there to the edge of the salt marsh, in an attempt to occupy a resident territory (Figure 5.11). The existence of these would be residents illustrates the importance of site faithfulness. Making their way to the edge of the salt marsh requires a number of seasons of fighting. At the end of one season, the fight stops, and at the beginning of the next season, the battle resumes. So not only do the animals remember each other, they also know who had been in control of each square metre of terrain in the previous season. And for each square metre of desirable land, there is probably a number two, three and four already in the queue. When the number one dies during the winter, then the number two can take over the territory in the following season. The most likely number two pair is the neighbours. And when a good resident territory on the edge of the salt marsh is concerned, as well as the pair of hopeful club birds that had been badgering the previous owners, there will also be a neighbouring pair of leapfrogs on the salt marsh. Just as we'd expect, only the leapfrogs already in a territory that directly borders the resident territory have any chance of acquiring that resident territory. To finally succeed in acquiring a territory a club bird must build up a locally dominant position over a number of seasons.[227, 228, 255, 388]

This requirement to build up local dominance has serious consequences. The bird must commit itself to a certain location, relinquishing its chances of finding a territory elsewhere. This leads to the queue hypothesis, which also explains why oystercatchers settle in the poor leapfrog territories. It's possible to designate a number of waiting club birds to each territory. A club bird that manages to acquire a resident territory has got it made. So you might think it would be best for club birds to join the queues that may lead to a resident territory. However the longer the queue, the longer wait until the club bird can establish itself. And the worst thing that might happen while you are still waiting is that you might die before you have had a chance to breed. The queue for good

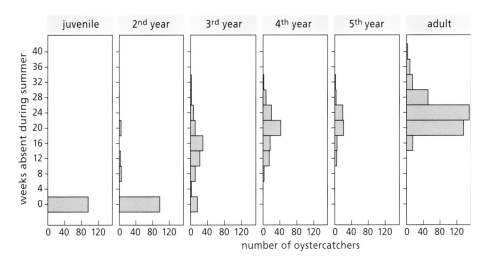

Figure 5.20 Age of first return to the breeding grounds for oystercatchers that winter in the Exe estuary. The frequency distributions show the number of weeks that different age-classes of oystercatcher were away from the Exe during the summer (presumably at the breeding grounds).[255] The figure is based on unpublished data of S.E.A. le V. dit Durell and J.D. Goss-Custard.

# REPRODUCTION

territories can become so long that the benefits of being able to settle in a good territory are completely nullified by the risk of a premature death. In which case you are better off settling in a poor territory. A number of predictions follow from this.[250] First, the better the territory, the lower the chance that a club bird that tries to get that territory will actually succeed. This prediction is difficult to test. Secondly, the queue for good territories will be longer than the queue for bad territories. And there are good indications that this is the case. Finally, club birds that settle in a poor territory will do so, on average, at a younger age than club birds that settle in a good territory. This prediction can only be tested through long-term research. Of the many hundreds of oystercatcher chicks ringed on Schiermonnikoog since 1983, a total of 40 had established themselves in territories by 1997. Sure enough, the animals that settled in the poor territories were, on average, 1.5 years younger than those in better territories (5.9 years in a leapfrog territory compared to 7.6 years in a resident territory). This might seem a small difference, but it is exactly what recent calculations had predicted.[388]

So oystercatchers generally wait until they are three years old before returning to the breeding grounds, and must then wait for at least a further three years before they begin to produce offspring. If the bird is unlucky enough to join a slow-moving queue (where none of the incumbent territory holders die), it may have to wait much longer. This, and the difference in queue length, explains why some animals don't begin to breed until they are 14 years old. By the way, our use of the word waiting is somewhat misleading. The birds don't get anywhere if they just passively wait. To become locally dominant and retain this position a bird must continually fight off its competitors. This is not a fight to the death. Competitors will respect another bird's claims as long as that bird makes regular displays of its strength. In a system reliant on competition for the best sites it's probably worth spending several years enhancing your status, rather than putting all your efforts into breeding on your first return visit. In such a system you would expect that a reduction in competition would allow waiting animals to settle and begin breeding at a younger age. Competition can be reduced dramatically if there is high mortality due to, for example, extremely harsh winter weather. It's also possible that new breeding areas can become available. And when they have the opportunity the birds do begin to breed at a younger age.[137]

### 5.7.2 Site faithfulness (or not)

The young waterbird making its first journey from the wintering grounds back to the breeding grounds can choose between returning to the site where it hatched or searching for somewhere new. A bird that already bred in a previous season faces a comparable choice: should it to return to the same site as last year or find a new site? It often happens that when adults are site faithful to the

Young terns continue to be fed by their parents until well after they have fledged. This juvenile sandwich tern (photo right) is being fed in the Dutch Delta area in October. Juvenile terns fly with their parents to the African wintering grounds and will not return to their place of birth to breed for a few years. Large gulls also don't start to breed until they are older. However the juvenile gulls (photo above) stay in the Wadden Sea.

breeding grounds they've used in previous years, the young are very likely to return to the site where they were raised.[220] Being site faithful is advantageous when key aspects of the site, such as places where food is plentiful or that are relatively safe from predators, are constant from year to year. In this case previously acquired site knowledge will come in very handy in future years.

The oystercatcher is a classic example of this. Territories on the edge of the salt marsh near the tidal flat are always better than those further into the salt marsh. There are large annual variations in the food availability, which could explain why breeding success also fluctuates annually. However the same territories consistently produce the most young. The oystercatcher's system of queuing for a territory makes it extremely important to be site faithful. In principle, an oystercatcher is free to go to a completely different site at any time, but in doing so it would waste all of the time and effort it had invested in building up local dominance.[227, 388]

Like the oystercatcher, the grey plover defends a territory where it breeds and forages. And, like the oystercatcher, the grey plover is very site faithful. Curlew sandpipers and little stints breed in the same tundra areas as grey plovers. However these two species are not remotely site faithful. The numbers of curlew sandpipers and little stints breeding in a certain area vary greatly from year to year and there is almost no chance of observing a breeding bird return to the site where it was ringed as a chick.

Purple sandpipers (photo left) and red knots (photo below) are very site faithful to their breeding grounds and nesting areas. They are also very faithful to their partners. These purple sandpipers were out on the Taimyr tundra on 15 June, when it was still covered in snow. The female knot is laying her second egg. Later, the two parents are rarely seen at the nest at the same time.

What causes this difference between the species? Perhaps the grey plover experiences less dramatic annual changes in its habitat than the other two species. It all comes down to how the animals use the area. Maybe the sandpipers experience greater difficulties with snow and ice than grey plovers, and it's mainly the structure of the area that is important to grey plovers. Or maybe each species is forced to make certain choices based on its social system. Site faithfulness makes it possible to build up local dominance, and being locally dominant increases the advantages of site faithfulness. And vice versa, when everybody searches anew for the prime sites each year, you're probably also better off looking around for that season's best site rather than returning to an area that you're familiar with. It is theoretically possible that two very different solutions have evolved in response to the same problem.

As well as differences in site faithfulness between species, there are also differences between the sexes. Territoriality and site faithfulness seem to be related. Since the males tend to invest more in territory defence, you would expect that males would be more site faithful than females. And you'd be correct, even in the oystercatcher where the female is very nearly as territorially aggressive as the male. When oystercatchers changes their partners, either due to death or divorce, the males only change territories in around 10% of the cases. But females will move in 50% of cases. The birds are often just moving to the neighbouring territory. Very few females move over a large distance.[246] The difference between the sexes is even more obvious when we examine the likelihood that a chick will return to its place of birth. Based on survival estimates for the first years of life, it seems that all or almost all of the surviving males that fledged on Schiermonnikoog return to the island. However the proportion of females that return to Schiermonnikoog is so low that the most likely explanation is that some of them have moved elsewhere. And there have been a few resightings of young (probably female) oystercatchers ringed on Schiermonnikoog that settled far from their natal area.[388]

When animals are very site faithful and the young return to their places of birth, it is possible that the various populations will start to become genetically different. The genetic makeup of the extremely site faithful oystercatcher has been studied in the Dutch Wadden Sea, using blood samples collected from Texel, Griend, Schiermonnikoog and Holwerd. There was considerable genetic variation within the various populations, but no systematic differences between populations were found. Even a very low exchange of individuals between the populations would have been enough to prevent genetic differentiation, and the young females are almost certainly responsible for this exchange.[869]

### 5.7.3 Divorce

Waders like the curlew sandpiper that choose their breeding site dependent on what seems best in that particular year, don't need to decide whether or not to team up with last year's partner. Their chance of encountering their old partner on the immense tundra is minuscule. It would be interesting to know if species that regularly change partners within one season, like the Kentish plover and Temminck's stint, are equally unfaithful between seasons. Good data on this are lacking. However there are data on the species that show a high degree of site faithfulness. If, for these species, we define divorce as the chance that both partners are alive but do not form a pair again, the observed divorce rate is low: 10.7% for the oystercatcher,[379] 7.7% for the purple sandpiper,[649] 27.9% for the dunlin[793] and 4.8% for the grey plover.[752] What do these low divorce rates mean? Is it rewarding for both birds to return to the same site, which automatically leads to the restoration of their old pair bond? Or are there some special advantages associated with remaining with the same partner year after year. And why are there any divorces at all?

As ever, there is no shortage of hypotheses.[130, 256] The only problem is the lack of evidence. If there were substantial advantages to keeping the same partner, then a number of predictions about breeding success should hold. First, breeding success should increase with the number of 'married' years. And there are indications that this is the case for oystercatchers.[246, 388] The only problem is that the results are confounded, since for every year that the oystercatchers maintain their pair bond, they are also a year older and a year more experienced. It is very difficult to correct for this properly, especially since the age of many oystercatchers is not known. Secondly, the unintentional break up of a pair (for example through the death of one of the partners) must have repercussions for the breeding success of the remaining partner. And there are some indications that this is the case,[246, 379] but again these data are difficult to analyse. The new partner will often be younger and less experienced, so it should be no surprise that breeding success is lower with the new partner than with the old.

These analysis problems arise because we're trying to understand changes in social relationships, when little is

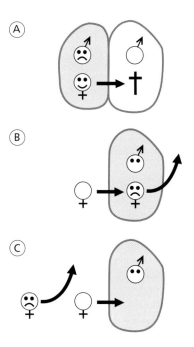

Figure 5.21 How an oystercatcher pair can become divorced. (A) The female may desert the male, (B) she may be usurped by another female, or (C) she may be pre-empted by an earlier arriving female taking her mate. Oystercatcher males usually stay in the same territory while the females move. The 'face' shows whether an individual will benefit or be disadvantaged by the moves shown by the arrows.

known of these relationships themselves. Often researchers will find nests and determine which parents belong to which nest, but nothing is known of the history of the nest and thus of the breeding pair. A more direct but very labour intensive way to approach the problem of divorce is to monitor all of the relationships between individuals from the time that the first animals return to the breeding grounds. Due to the large number of students and participants on animal ecology courses it has been possible to continually study the love lives of the Schiermonnikoog oystercatchers for a number of years from observation towers. This work has made it clear that there are two completely different paths to divorce for an oystercatcher: desertion or usurpation (Figure 5.21).[246, 255, 256] Desertion occurs when one of the pair, and it's almost always the female, leaves the original partner and forms a pair with a new partner. In many cases the new partner is the neighbour who has recently been widowed. It seems that the deserting female chooses to take up a better position that suddenly becomes available. If the neighbouring female had still been alive, the deserting female wouldn't have got into the territory without a fight, but since the neighbour has passed away, there is no need. The other way divorce can occur is when one oystercatcher, often a club bird, successfully usurps one existing member of a territorial pair. A special form of usurpation is pre-empting. It's a classic case of don't stand up or you'll lose your seat. Early in the breeding season a male whose female has not yet returned may form a pair bond with a new female in the interim. The longer the new female is in the territory before the other female returns, the smaller the chance that the previous incumbent will eventually succeed in chasing her away.

The most important conclusion from these observations is that the reasons for the divorce have a bearing on its effects on reproductive success. A female that leaves her partner when the neighbour becomes available will probably be better off. A female that is usurped from her territory is worse off. A male that is deserted by its mate also has a problem. And it's doubtful whether it's in the male's best interests to have to deal with two fighting females. Slowly enough data are being gathered on the Schiermonnikoog oystercatchers for these predictions about the reasons for divorce and breeding success to be tested.

5.7.4 In conclusion

At the beginning of this chapter we established that the breeding season is all about raising as many offspring as possible. To do so, each waterbird must select the best time, place and partner, as well as the appropriate amount of parental care to invest in its offspring. All of these choices are inter-related, both within and between seasons. The ultimate goal is not simply to raise the most offspring in one particular season, but to maximise the individual's output over its lifetime.

Now of course it's easy to say that these choices should not be answered in isolation, but we inevitably end up having to study each component one at a time. This is understandable. Scientific progress is often reliant on the ability to isolate a certain problem from other problems. This ability allows researchers to identify the critical pieces of information that are needed, and design specific experiments to address a particular question. However sooner or later we reach the point where all of the different parts of the jigsaw must be pieced together.[941] It's very easy to state that all of the breeding choices facing waterbirds must be considered in their entirety. However this is only meaningful when it stems from the theories we propose.

Although we still have to go a long way, the first theoretical cross-connections have been made. Sexual selection is linked to the extent of parental investment. So the answer to the question 'with whom?' is dependent on the answer to the question 'how much time shall I invest?' If one sex has a greater investment in parental care, that sex will quickly limit the reproductive potential of the opposite sex. Or, sexual selection is the inevitable price for not investing in parental care.

Another theoretical cross-connection is the queue hypothesis. To obtain a good territory you must wait longer, at the risk of dying before you've even had a chance to breed. So the decisions of where to breed and what age to begin breeding are closely connected. The queue hypothesis was developed based on observations of oystercatcher behaviour. These observations also highlighted that having a mate is an important part of acquiring a territory. The most obvious advantage for females in cooperative trios is that they gain access to a territory. Most other club birds will team up with another to force their way in between existing pairs. We have yet to develop a theory that connects the question 'with whom' to the questions 'where and when'.

It's very likely that this all embracing theory would start with the inescapable fact that an individual can only be in one place at one time, performing one function. An individual that wants to visit a large area can only do so by spending a short time at each distinct site. In all likelihood, this would result in that individual failing to enhance its status at any of those sites. And without status, it is unlikely to succeed in acquiring a territory or a partner. We also know that this theory will be based on game theory: the best option for one animal depends on the choices made by other animals. We are really searching for a theory that not only describes the social careers of the individuals but also the society in which they live; the frequency distribution of the various social classes and the factors that limit access to these classes. We would like to use our understanding of the factors limiting an individual to understand its society, and in turn the population. In the next and final chapter we continue our quest, but the emphasis is less on the individual and more on the population and the threats facing that population.

# 6. Looking to the future

## 6.1 Counters and twitchers

Many people enjoy watching birds. Only a handful of these people are lucky enough to do so professionally. But amongst amateur birdwatchers there are many who take their hobby as seriously as if it was their occupation. These enthusiasts can often be identified as belonging to one of two groups: the twitchers and the counters. Twitchers are characterised by their desire to excel at distinguishing the different bird species and their delight at seeing species in places where they don't normally occur. In contrast, the counters enjoy counting the birds in a particular area as accurately as possible and they delight in finding more birds in an area than usual. This chapter is especially written for the birdwatchers who hold counting birds in high esteem, with no disrespect to the twitchers. It's thanks to the twitchers that we can better recognize species that are difficult to identify in the field and are better informed about changes in species' distribution. Thanks to the counters we get a constantly improving picture of the total population sizes of different species and the areas that are important to them. Slowly but surely we are becoming better informed about changes in species abundance. The virtual sea of numbers gathered leads to many questions. Why are there so many more oystercatchers than turnstones? Why do most red knots occur on the tidal flats of Schleswig Holstein in spring, to the virtual exclusion of the rest of the Wadden Sea? Why has there been such a spectacular increase in the number of grey plovers wintering in British estuaries in recent decades?

There are no definitive answers to any of these questions. However in many cases there are hypotheses supported by varying amounts of evidence. There are different methods available to help us find the answers. Mathematical models play an important role in our quest. They help to structure the problems and force us to clearly state any assumptions made. At the same time the models can narrow our perspectives. The phrase 'carrying capacity' is a good example of this. It seems intuitively obvious that there is a limit to the number of animals that can live in a particular area. Yet views on 'carrying capacity' have become strongly associated with one particular mathematical equation: the logistic growth curve. This curve, which can be found in every biology textbook, shows how the abundance of a certain animal species increases over time when you start with just a few animals. Initially there is no competition and the number of animals can increase exponentially. As the number of animals increases in that area, their density increases. After some time, this increasing density causes problems for the animals and the rate of population increase begins to drop. Eventually the number of animals (and thus also their density) settles around an asymptotic value, usually indicated by the letter K, after which no further increase is possible. According to the mathematical model, the carrying capacity of that area has been reached. This suggests that (1) the carrying capacity of a certain area has a fixed definitive value and that (2) the numbers in an area can continue to increase as long as the carrying capacity of that area has not yet been reached. However the real world is not so simple, as has been shown by field research and some of the more advanced models.

The behaviour of individuals and the differences between them can form a useful starting point for research into population processes[248, 323, 523, 823] and so play a central role in many of the advanced models. Our knowledge of individuals is greater than our knowledge of populations, as it's much easier to run a controlled experiment on an individual rather than a population. Natural selection is usually discussed at the level of the individual. In previous chapters we've discussed the numerous evolutionary adaptations of individuals extensively. Furthermore populations are made up of individuals. Finally, it is reasonably simple to create a computer model in which individuals compete with each other for food or breeding territories. This makes it possible to calculate survival or chick production at different densities of competing conspecifics. And these are exactly the relationships we need to determine if we want to predict the population trend.

When these predictions don't stack up against the

A flock of waterbirds roosting together at high tide is an impressive sight. It also provides a good opportunity to count the birds. High tide bird counts have been carried out across the entire Wadden Sea area for many years. These counts have provided valuable insights into the status of many species and changes in their numbers.

real situation it may be a pity for the researcher, but it's usually of little consequence. In fact, philosophers of science would suggest that researchers should be delighted when their results differ from what was expected. A large discrepancy between the predicted and the actual results is an exceptionally good starting point for further research. Increasingly, mathematical models are used not only to increase our understanding of natural systems but also to make predictions about the effects of human interference – such as shellfish harvesting, disturbance or gas extraction – on waterbird abundance in a certain area.[94, 248, 322, 340, 823] In these situations erroneous predictions can have serious consequences.

Which leads us to the crux of this chapter. Is the increasing human pressure on our planet's limited resources threatening the survival of the waterbirds? Vast areas of tidal flats have been and continue to be reclaimed and many estuaries are highly polluted. Some human activities in tidal areas can also cause problems, particularly activities like commercial shellfish harvesting, hunting, disturbance-causing recreation, sand extraction, and exploration drilling. When the exploration drilling is successful, the commercial extraction that follows can cause the seafloor to sink. Finally we now face the knowledge that human activities are changing the world's climate. A worldwide temperature increase could lead to a reduction or maybe even the loss of the High Arctic tundra, endangering the waterbirds that breed in the far north. Such a temperature increase could also cause sea levels to rise, permanently covering the tidal flats. Since dykes are now ubiquitous in most European intertidal areas, in the event of rising sea levels the intertidal zone would no longer be able to move inland as it would have in the past.

An increasing number of people and organisations are aware of these problems and trying to do something about them. Environmental organisations lobby and organise protest actions to raise awareness of the various threats. There has been an explosive increase in the number of international and national treaties, conventions, directives and laws to protect migratory birds and wetlands. In principle these treaties should ensure greater protection for the waterbirds, but their application has often been rather disappointing. This often leads to legal battles, where environmental groups employ professional lawyers to try to ensure that the authorities in question uphold their own laws and regulations. Data from waterbird counts play an important role in these legal battles, as the counts can be used to demonstrate the importance of an area. Without bird counters, there wouldn't be any of these data. Which takes us back to the start. Through conducting frequent counts now, the bird counters ensure that they will still have something to count in the future.

Chapter 4 described the best methods for counting waterbirds and possible sources of error. As you read the following discussions of the processes that limit bird numbers, it's useful to remember that counting waterbirds is completely different to counting a bag full of marbles that can be removed one by one from the bag.

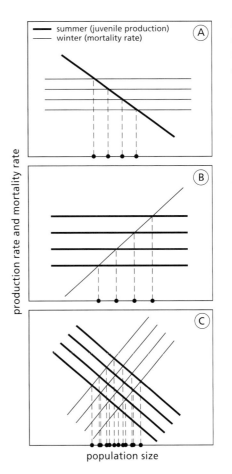

Figure 6.1 Stylised diagram of the interaction between density-dependent and density-independent population processes in summer (affecting juvenile production) and winter (affecting mortality rate). Density-independent factors include variation in food resources, weather and hunting pressure. The population is stable where the two lines intersect, as at this point reproductive success in summer compensates for winter mortality. (A) Regulation in summer and strong, density-independent variation in winter; (B) regulation in winter and strong, density-independent variation in summer; (C) regulation and variation in both seasons.

## 6.2 Where and how are numbers limited?

### 6.2.1 The problem

In the struggle for existence only animals that manage to raise more offspring than other animals remain. Why isn't the land black with oystercatchers or the tidal flats red with knots? It's because no population can increase forever. Sooner or later, numbers increase to the point where resources become limiting, causing juvenile production to decrease, mortality to increase, or both. The limiting factors are a result of the increased density of the animals, i.e. the factors are density-dependent, which is what regulates the population growth.[285, 491] The idea is simple enough, but the proof is often hard to come by. Shortages may begin to arise, but shortages of what? Perhaps food or secure nest sites. A shortage of safe nest sites can only occur in the breeding season, but food shortages aren't connected to a season. Food shortages could occur in the breeding grounds, the stopover sites, or the wintering grounds. Plus food shortages may not affect the different age-classes equally. There may also be a difference between the sexes. The shortages may not occur every year; years of abundance can alternate with years of scarcity. And large annual differences in mortality or production of young may result from processes unrelated to the population density, such as a harsh winter, prevailing headwinds on migration, or many hungry predators on the tundra breeding grounds and few lemmings to satisfy their appetite. And just because one tidal area is fully occupied it doesn't necessarily mean that other tidal areas are also full. Finally it is still possible, and perhaps even likely, that there are times of the year in which the animals compete not only with conspecifics, but also with individuals of other species. Do all these problems mean that only a masochist should attempt to research the population dynamics of waterbirds, or are they simply an excuse for our frequent incapacity to be certain of the causes of changes in animal abundance? Neither, and nor are they meant to give the impression that research on population changes is doomed to fail.

### 6.2.2 Processes in summer and winter

We have already mentioned that the logistic growth curve is too limited a perspective for viewing population changes. So what would a better perspective look like?[285, 321] The most striking feature of waterbird biology is that the birds depend on different habitats through the year (Chapter 3). Put as simply as possible, there are breeding habitats where young are produced and adult mortality is usually very low, and in the nonbreeding season there are tidal areas where young and old birds try to survive, with varying degrees of success. A waterbird species can only survive when, on average, winter mortality is completely compensated for by the production of young in summer. It is inevitable that in some years slightly more young are produced than in others and that the number of birds that survive the winter also varies. That means that rather than being constant, the abundance of a species fluctuates. However one thing it never does is continually increase. The only explanation for this is the density-dependent processes that have negative feedback on population growth. This can mean that the chance of an individual surviving the winter decreases when there is a higher density of wintering birds. Or that when the density of birds in a breeding area increases, the number of young raised by each individual decreases. So density-dependence can be expressed during winter, or summer, or both.

In the first scenario the population is regulated on the wintering grounds. It is easy to imagine that this is the case. When you walk across the Wadden Sea tidal flats at low tide they seem vast, but on a world map they are just a just a tiny dot. In comparison the tundra, where many birds breed, is very extensive even on a world map. Even so, the second scenario that the population is regulated in the summer may also have some truth in it. The tundra may seem endless, but that doesn't automatically mean that all of it is suitable breeding habitat. Maybe there are few suitable sites, so the birds breed there in high densities out of sheer necessity and thus impact on each other. We can also consider the birds that winter and breed in coastal areas. In the winter the whole tidal area is available to them, but in summer they can only use the edges. In this case the population is not regulated by one or other of the seasons, although competition could play a role in both. Perhaps the intensity of the competition differs seasonally. The different birds must compete for food in the winter and breeding territories in the summer. From the number of aggressive interactions it appears that the competition for breeding territories is more intense than that for food.

Figure 6.1 summarises these three scenarios. Each diagram depicts the juvenile production and mortality rate per head of population as a function of the total world population. The lines intersect at the equilibrium population for different values. This is only the case if we are considering the total world population of a species or subspecies. It depends on the densities in each of the wintering areas and each of the breeding areas, and the cumulative effect of what is happening at all of these sites. We know that there can be large annual differences in the production or mortality rates between the different sub-areas, even at the same densities. When the conditions in an area vary independently from those in other areas, this density-independent variance does not affect the overall outcome. Low breeding success in one area may be compensated for by higher breeding success in another. This seems to be the case in some oystercatcher breeding populations that have been the subject of long-term studies.[339] However some factors, particularly the weather, have the same effect across large areas. A harsh winter in the Wash in Britain usually means a harsh winter in the Wadden Sea. This can deal a blow to the entire population of some species. Which again means that the relationship shown in Figure 6.1 between the number of young produced (or the mortality rate) and the size of the world population can differ greatly from year to year. In conclusion, we can't expect the

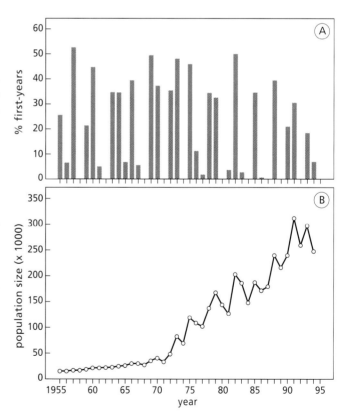

Figure 6.2 Variation in reproductive success and population size of brent geese that breed in Northern Siberia and winter mainly in France, England and The Netherlands.[206, 207] (A) Juveniles as a percentage of total population; (B) total population size.

world population of a species to be very stable. We can state the limits between which the population should vary. But these limits may be determined by density-dependent and independent processes in both summer and winter. When large numbers of birds die in harsh winters, their conspicuous demise can cause people to think that winter conditions regulate the populations. This is incorrect. Harsh winters may influence abundance, but regulation requires density-dependence and thus competition.

The brent goose is a good example of a species with a fluctuating world population. Brent geese winter in a rather limited number of areas that are relatively easy to survey, which means that the size of the world population of each subspecies can be readily determined. Adults can also be distinguished from juveniles in their first winter. This makes it possible to determine the juvenile production of the preceding breeding season each autumn. The number of young produced varies greatly from year to year. In some years there are almost no young fledged, and in other years, up to half of the autumn population consists of young birds.[206, 207] (Figure 6.2A). As we will see later, this variation is regulated by the lemming population and has nothing to do with goose density. If we only examine the years in which the geese raise young successfully, then reproductive success does seem to be density-dependent. The population of brent geese has increased more than ten-fold and, particularly in recent years, the number of young produced in good seasons seems to have decreased. So the brent goose population could be an example of a species that is regulated in the breeding season (Figure 6.1A). The overall increase in the population is due to hunting discontinuing on the wintering grounds, which has greatly increased winter survival.

### 6.2.3 From individuals to populations

Much of this book has focussed on individual waterbirds. In this chapter our focus shifts to populations and the processes that determine their distribution and sizes. Populations consist of individuals, but making an explicit connection between these two biological levels has only recently become fashionable. In the past the number of animals belonging to a species was the only variable included in population models. These models made no distinction between different ages, sizes or sexes, let alone including social interactions or spatial distribution. Nowadays there is growing awareness that the individual is a fundamental biological unit.

This push to understand population processes from a lower and more fundamental level is reminiscent of a major breakthrough in physics back in the 17th century: the discovery of Boyle and Gay-Lussac's law. Boyle's law states that there is a fixed relationship between pressure, volume and temperature for a gas in an enclosed space. When the pressure changes and the volume remains the same, it is possible to calculate, according to the law, precisely how the temperature will change. This break-

There are few species whose world population is as well-known as the northerly-breeding brent goose (shown with newly-hatched chicks on page 310). This is because brent geese winter in areas that are easy to survey. Furthermore, juveniles can be readily recognized by the pale tips on their wing feathers (photo below), allowing breeding success to be determined via counts of winter flocks. Such counts have been made for many years in the Wadden Sea area. Some individual animals are also monitored. This family of brent geese on Terschelling in the Dutch Wadden Sea (bottom photo) was ringed with numbered colour rings on Taimyr Peninsula, Russia.

through was due to the work of Stefan Boltzmann. He modelled gas not as a gas but as a collection of individual molecules, each with its own speed and direction of movement. The average speed with which the molecules move is a measure of the temperature, and the power with which the molecules collide with the wall is a measure of the pressure. By summing up the behaviour of individual molecules Boltzmann was able to deduce the gas laws.

At present it seems unrealistic to expect a similar breakthrough in biology that would allow us to deduce population processes from the behaviour of individual animals. Our models allow us to treat individual animals like the gas molecules; impose a few simple behaviour rules on them and then calculate what this means for the population. But individual animals are infinitely more complex than individual gas molecules. Unlike a molecule, an animal develops over the course of its life. This development is irreversible and inevitably ends with death. Plus, animals can adapt their behaviour from minute to minute depending on the situation. These behaviours exist as a result of the selective forces their ancestors were exposed to over the course of evolution. In essence, the problem with modelling populations based on the fate of individuals is that there is a lot of interplay between the two levels. The best choice for the individual depends on the state of the population. The state of the population results from the choices made by individuals. There are also problems with differences in scale between the two levels. We want to understand why the world population is chang-

ing, but an individual animal is only concerned with the desires and difficulties of its neighbours. The fortunes of a conspecific 1000 km away are not important to an individual. These scaling problems occur in both space and time. It is not easy to translate prey choice decisions to the likelihood that an animal will survive the winter. And we're not really that interested in whether it survives one winter; we want to know what happens after many winters.

So it seems unlikely that there will be a spectacular development in the near future, even though some researchers claim the opposite. It's striking that many of the researchers making these claims study waterbirds. This is because tidal flats provide relatively simple, accessible and easily surveyed research systems. Some of these waterbird researchers have constructed very simple models.[823] Simple can quickly become confused with generally valid. The simple models that have been constructed to date contain very specific assumptions. It's true they are simple, but they're not generally valid and have little relevance to real life. This has induced other researchers to design more complex models that allow individual oystercatchers to choose the best foraging sites from hour to hour.[337, 338] And it's no accident that we're talking about individual oystercatchers rather than individual waterbirds. The oystercatcher is the show pony of the latest research linking individuals to populations.

### 6.2.4 The oystercatcher model in the winter

A complex simulation model has been constructed from data on individually marked oystercatchers that feed on mussel beds in the Exe estuary in winter.[337, 338] In this model each oystercatcher has a number of characteristics: sex, age, foraging efficiency and fighting ability. It's assumed that individual oystercatchers differ in their foraging and fighting abilities and that these differences are only partly determined by age. So one oystercatcher is better able to find enough food than another at the same prey availability. Differences in fighting ability determine which of the two animals would win a fight if there was conflict over food. This may seem logical but it assumes that dominance is determined solely by fighting ability, which is very unlikely. In many waterbird species a variable proportion of the population defends a feeding territory outside the breeding season. This has been studied extensively in curlews,[239] whimbrels,[975] sanderlings,[589] and grey plovers[867] (see also Section 4.7.3). Within its territory the owner is lord and master and can chase everyone else away, but this is not the case outside the territory. For these birds, the location of the conflict, rather than an individual's fighting abilities, determines the outcome. The same almost certainly applies to species such as the oystercatcher[234] and the turnstone[930] that seem to defend pseudo-territories. Within the pseudo-territories some individuals are very dominant and will only tolerate conspecifics if they behave as if they are subdominant. Outside their small fixed living areas these tyrants are not so dominant. So dominance in oystercatchers and turnstones depends not only on differences in fighting abilities, but also on the location and who is locally dominant.

The complex simulation model allows each oystercatcher to repeatedly choose the best foraging sites. The attractiveness of a site is determined by food availability and competition. When one animal moves to another site, the levels of competition change locally, which may make it worthwhile for other animals to also move. This results in a dynamic equilibrium, with a reasonably stable number of oystercatchers per mussel bed. In these simulations many animals move around as time progresses. However in the real world dominant oystercatchers are very site faithful and return to the same mussel bed year after year.[332] Predictions about the distribution of the oystercatcher across the different mussel beds are more realistic.[338] Larger numbers of oystercatchers forage in the richer feeding areas. Over the winter these richer feeding grounds are depleted and other, poorer areas will be increasingly used. Eventually, some of the birds may be unable to find enough food. These will be birds that are lowest in the pecking order or less skilled at foraging i.e. the young birds. They have two options: stay or leave. Staying means certain starvation, so leaving (even though conditions may not be any better elsewhere) seems to be the best option, although not an attractive one.

We can now calculate how many birds will survive the winter for a given food availability. All these predictions are based on observations from the Exe where around 9% of the approximately 2000 oystercatchers that have wintered there over the last few years died during the winter.[344] So the values given to the model's parameters must allow exactly 91% of the 2000 birds to survive the winter. Once that has been achieved, the

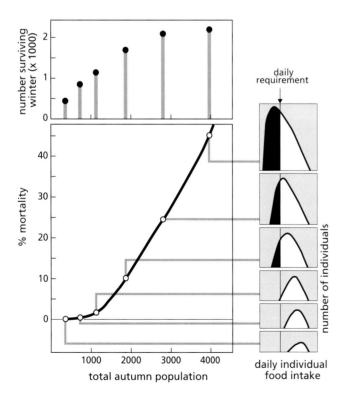

Figure 6.3 Model of density-dependent mortality in oystercatchers settling in the Exe estuary in autumn. As the number of oystercatchers increases, their spatial distribution across the feeding area changes and more individuals encounter problems finding sufficient food.[322]

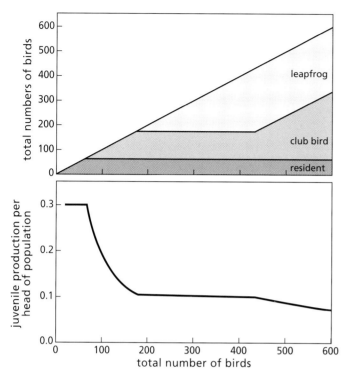

Figure 6.4 Model of density-dependent reproductive success in oystercatchers. As the potential number of breeding birds increases, a larger portion of birds will breed in the marginal ('leapfrog') territories, or queue for a territory as a non-breeding 'club' bird (with no reproductive success). Based on long-term studies on Schiermonnikoog, in the Dutch Wadden Sea.[388]

number of oystercatchers at the start of the season can be varied. If you start with 8000 birds, the model calculates that it's possible for 3000 oystercatchers to survive the winter. But then more than half have died. A lower mortality rate would require more favourable conditions, for example, no loss of body condition in shellfish or a relatively mild winter. Figure 6.3 shows how this works. As the number of oystercatchers increases, a higher proportion of the animals can't find enough food to survive, causing an increased mortality rate.

### 6.2.5 The oystercatcher model in the summer

The relationship shown in Figure 6.3 between winter mortality and bird density is just what we were looking for: a density-dependent mortality rate. We also need to find the relationship between the production rate (number of young per head of population in one year) and bird density. There are three ways that the production rate could be density-dependent. First, it may only be possible for all the animals to breed at high densities if they have smaller territories. This means that territory size decreases in proportion to the increasing competitive pressure. And of course it's more difficult to successfully raise offspring in a small territory. Secondly, as density increases a higher proportion of the animals may be forced to settle in bad territories, causing the average chick production to decrease. Finally an increasing proportion of animals may not even attempt to breed, also causing the average chick production to decrease. There are indications that all three apply to oystercatchers. During a 15-year study on the Welsh island Skokholm, the density of oystercatcher territories varied from 0.8 - 1.0 per ha. At low densities, production was 0.8 chicks per pair and at high densities production dropped to just 0.2 chicks per pair.[344] No such relationship could be found for oystercatchers breeding in the Lune catchment in England,[344] nor on the Dutch island Schiermonnikoog. There was little change in the number of territories on Schiermonnikoog from 1983 - 1994, but the juvenile production was highly variable. From 1995 - 1998 the number of pairs in bad territories halved, while the number in the good territories remained fairly constant. Over this period, there was a slight increase in juvenile production.[388] At the same time the number of club birds (birds without a breeding territory) hit an all-time low. Similar relationships between the number of club birds and the population density were also observed in the Lune valley and on the German island Mellum.

The data collected on Schiermonnikoog allow us to estimate the effects of density-dependence on fledgling production (Figure 6.4). However some assumptions must be made, the most important of which, is that territory size is fixed. Imagine that a disaster has wiped out all the oystercatchers on Schiermonnikoog. The first oystercatchers to arrive would probably settle in resident territories on the edge of the salt marsh and have a high reproductive output (Section 5.7). When all the resident territories along the edge of the salt marsh are occupied, the next wave of birds won't immediately settle in the less favourable leapfrog territories, as they can produce more young during their lives by trying to establish themselves as residents. But as the queue for resident territories gets longer, an oystercatcher's chance of acquiring such a territory decreases and eventually it becomes more profitable to settle in a leapfrog territory (Section 5.7.1). Once all the leapfrog territories are occupied, the next generation of oystercatchers must join the queues for the good and bad territories.

So now we have managed to construct the two curves of negative feedback that we needed: a density-dependent mortality rate in winter and a density-dependent production rate in summer. But have we now reached our goal? Only if we dare to extrapolate these curves to the world population. And that would be rather a risky business, as it requires making many additional assumptions. For example, we must assume that surplus birds in the Exe are doomed to die over the winter. Yet there are good indications that the Burry Inlet further north still has room available.[804] We also have to assume that birds that cannot find a breeding site on Schiermonnikoog are also unable to find a site elsewhere. Yet until recently the number of oystercatchers breeding inland was still increasing.[344] Finally we have to assume that the curves for the Exe estuary and Schiermonnikoog are representative of all the other areas, and whether this is the case is still unknown. We have much to learn about movements between different wintering areas and between different breeding areas, and the costs associated with this dispersal.

## 6.3 Effects of summer conditions on populations

This section provides an overview of the various processes known to affect the reproductive rate of the waterbirds, all of which breed during the northern summer. How many birds will attempt to breed and how successful will they be? To what extent does competition (and therefore density) regulate the reproductive rate? Do large differences between good and bad years make it impossible to measure the effects of density and competition? And on what scale are conditions good or bad? Is a good year good everywhere, or only very locally?

### 6.3.1 Lemmings and snow in spring

Up on the tundra, the weather and lemming abundance have a huge influence on the number of fledglings produced. Lemmings are small herbivorous rodents that use an extensive burrow system to run around beneath the snow in winter. So how can they affect the survival of waterbird chicks? Most people have heard the story that when there are too many lemmings around, they throw themselves into the sea en masse thus solving their overpopulation problems. Of course the idea that lemmings commit suicide is nonsense, but it is true that in some years the tundra is overrun with lemmings. In such years, the lemmings might take far more risks while searching for feeding areas than they would normally. There are plenty of predators that like to put lemmings on their menu: Arctic foxes, snowy owls and several species of skua (*Stercorarius spp.*). In years when lemmings are abundant, these predators have an easy life and raise large numbers of young. The problem is that lemming populations tend to peak one year and then collapse the next for no known reason. This leaves plenty of predators from the previous year's successful breeding season but hardly any lemmings for them to eat. The hungry Arctic foxes, skuas, snowy owls and other predators are forced to switch to eating the eggs and chicks of waders and geese. This results in poor breeding success for the local waders and geese.

Interestingly enough, this hypothesis was first proposed on the basis of the numbers of curlew sandpipers counted and caught in autumn.[745] These data showed a clear three-yearly cycle of peaks and troughs. When the 'Roselaar hypothesis' was first presented at a scientific conference it was greeted with much disbelief by goose researchers. Brent geese breed in the same areas as curlew sandpipers and patterns in the number of young produced by curlew sandpipers were mirrored by the geese.[355, 809, 812] However the researchers working on geese had strong evidence that the probability of a female brent goose returning from the breeding grounds with young depended on her condition in spring. Jokes about lemmings causing all sorts of mysterious relationships were frequent throughout the rest of the symposium. But the goose researchers have stopped laughing now

When lemmings (such as this Siberian lemming, photo on page 314, middle) are abundant on the tundra, they form the main prey of predators such as the Arctic fox (upper photo page 314) and the long-tailed skua (above). In years without lemmings the eggs of geese and waders, like those in the red knot nest (page 314, lower), must pay the price.

and no-one denies the role of the lemming cycle. First, statistical evidence was gathered for a three-yearly cycle in goose reproductive success.[355, 809, 812] This was followed by heated discussions about the relative importance of the lemming cycle and the condition of the birds on arrival.[204, 205, 888, 889] Their condition depends on the grass growth on the stopover sites in spring and the headwinds encountered en route.

Expeditions to the Siberian tundra and collaboration with Russian researchers who had already been studying tundra ecology for many years were the next crucial steps. Neither had been possible until the fall of the Iron Curtain. This resulted in an explosion of expeditions to Russia and greatly increased contact with Russian scientists. The Russian researchers had known for some time that the lemming cycle was very important, and that the precise chain of events involved could be very complex.[862] This complexity is also evident in a report on six expeditions to the Pyasina Delta in Taimyr between 1990 - 1995.[798] In this area, the brent geese mainly breed on small offshore islands, which offer them some protection against egg predation by foxes. As predicted, the researchers found that breeding success was low in two years that followed lemming peaks. But it wasn't because Arctic foxes were eating the majority of the eggs or goslings. In 1992, the first fox year, there were many foxes. That year the ice had broken up late, allowing foxes to regularly visit the island and search for eggs. During each visit the geese stopped their border disputes about breeding territories and formed one group. After many disturbances most geese gave up and left the breeding grounds without having bred. During the second fox year, in 1995, the island wasn't visited by foxes and the geese began to breed successfully. But unlike other years the geese remained on the island with their goslings. Many of the goslings disappeared, probably due to predation by Taimyr herring gulls (*Larus argentatus taimyrensis*) that were also breeding on the island. In short, in years with few lemmings and plenty of Arctic foxes the geese were not producing many young, but it wasn't necessarily because the foxes were eating the eggs and chicks.

The lemming story could only be detected on the nonbreeding grounds outside Russia because the lemming cycle is synchronous across much of the Russian tundra. If it wasn't synchronous, we would expect lemmings to peak in around one third of the tundra areas each year. This would result in the same number of chicks fledgling annually across the tundra as a whole, and a constant percentage of juveniles among northerly breeding species like the brent goose, grey plover, curlew sandpiper, sanderling and turnstone on the wintering grounds, rather than a three-yearly cycle.[888] The cycle is not absolutely synchronous,[866] as can be seen in Figure 6.5, which might explain why the cycle sometimes appears to jump when you look at wintering numbers. It might also explain why waterbirds migrate to the tundra in fox years: there's always a chance that foxes won't be a problem. There are also other explanations for why birds return to the breeding grounds in years when their chances of breeding successfully are low. The more site faithful species might visit to ensure they retain their property investment.

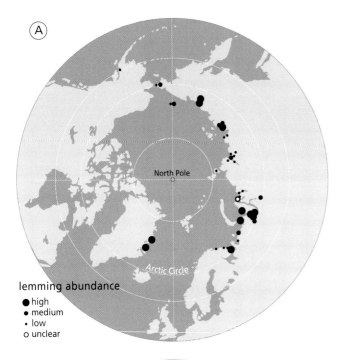

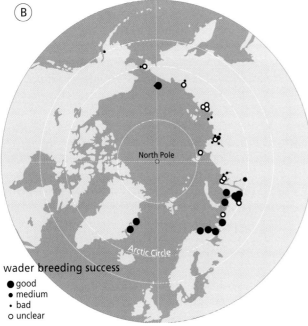

Figure 6.5 (A) Abundance of lemmings and (B) reproductive success of waders on the Arctic tundra in 1998. Overview compiled by the Russian researchers Mikhail Soloviev and Pavel Tomkovich and can be found on Internet: http://soil.msu.ru/~soloviev/arctic/

Lemmings aren't found in temperate areas, but common voles (*Microtus arvalis*) are. Common voles are the main prey of long-eared owls, kestrels and other raptors and, like the lemmings, their numbers go through cyclical changes. Raptor survival and breeding success are high in years when common voles are abundant.[203] In some areas, wader chicks provide an alternate food source for raptors in years when voles are scarce.

The abundance of lemmings and voles isn't the only density-independent factor that influences breeding success. The weather also plays a large role. In years when the snowmelt is late, the birds' breeding is delayed and they are at greater risk of predation (Chapter 5). Bitterly cold weather is detrimental to the chicks as they must spend more energy keeping warm and can't find as much food. Insects are the main food source on the tundra and they are mainly active in warm sunny weather. In more temperate regions rain can have a positive effect on food availability, since invertebrates such as earthworms are more active when the soil is damp. This might explain the positive relationship between rainfall in May and the survival of lapwing chicks.[36] But rain can also have a negative effect, and new chicks are particularly vulnerable to downpours.[36] In the Wadden Sea region most rain comes from the North Sea, which means that the wind is westerly or northwesterly. A northwesterly storm in June can be fatal for chicks hatched in the Wadden Sea salt marshes, which can become flooded (Section 5.2.3). However, the same storm will have no effect on birds that breed on salt marshes in the Wash, as in this area northwesterly winds cause water levels to drop. The effects of torrential downpours and hail are probably more localised. So the weather can influence reproductive success on varying spatial scales and with variable outcomes.

### 6.3.2 Competition on the breeding grounds

Lemming abundance and the snow conditions in spring have a dramatic effect on the breeding success of Arctic waterbirds. This may make the less dramatic effects of competition easily missed. Waders are long-lived birds, possibly due to the large amounts of selenium they ingest.[305] The longest-living species is the oystercatcher. Oystercatchers can live to 40 years of age and once they reach adulthood are difficult to kill off. The annual adult survival of most other waders is also over 80%.[220, 270] This means that the animals will have long lives, and their numbers will change little from year to year. This also means that there may be little variation in the numbers of breeding birds, even over many years. That makes it exceptionally difficult to demonstrate any effects of breeding density, even when these effects almost certainly occur. The existence of territorial behaviour means that it's likely that competition is occurring. Apparently, it is profitable to occupy one of the limited number of good sites and to defend such sites against conspecifics. The tundra is large, but the brent geese breed in close proximity on islands as a protection against Arctic foxes. Some brent geese do breed on the mainland, but even in years when lemmings are abundant they seem to suffer high levels of nest predation, although truly long-term data are lacking.[798]

The competition between oystercatchers for breeding territories of different quality was extensively discussed in Chapter 5. Section 5.7.1 describes this competition-driven queue hypothesis, which is used in Figure 6.4 to predict the role of density-dependence in reproduction. Evidence supporting this hypothesis came from events a few years prior to 1999. Over this period, a decrease was observed in the number of breeding pairs in the intensively studied populations on Schiermonnikoog and Texel in the Wadden Sea (Figure 6.6). This decrease was also noted in the rest of The Netherlands.[482] On Schiermonnikoog the decrease was mainly limited to the lower-quality leapfrog territories; the density in the better-quality resident territories remained more or less constant (Figure 6.6). This is exactly what the hypothesis would lead you to expect.

You would also expect the number of club birds to decrease dramatically. They did, but the change was much more erratic than the decrease in the number of breeding pairs (Figure 6.6). There are good explanations for this irregularity. In some years, a high proportion of the young oystercatchers on Schiermonnikoog fledge whereas in other years almost none do (this is the case in both the good quality and poor quality territories). These annual fluctuations are completely in sync with fluctuations in the numbers of chicks produced by the oystercatchers on Texel, another Wadden Sea island. This suggests that the weather, or an as yet unknown factor that causes the same effect across a large area, is responsible for that variation. These same fluctuations are reflected in the numbers of chicks banded throughout the Wadden Sea area (Figure 6.6F). How many of these chicks end up as club birds depends again on their survival during their first years. So in the end the number of club birds in any particular year depends on the numbers of 'new club members' that joined in previous years and the proportion of those club birds that acquire territories during these years or die prematurely.

## 6.4 Effects of winter conditions on populations

### 6.4.1 Severe winters and years of superabundance and scarcity

Outside the breeding season, waterbirds are often at the mercy of the weather, an unpredictable factor that can have serious implications for survival. Not much is known of the influence of weather on autumn migration. But we do know about the consequences of harsh winters on birds in the Wadden Sea. Heavy frosts always result from strong easterly winds, which bring in very cold air from Russia. These easterly winds also cause a longer exposure period on the tidal flats, allowing the Wadden Sea to quickly become covered in a layer of ice. This ice forms a barrier between the birds and their prey. The birds have two choices: wait and hope that the situation improves before they have exhausted their fat stores, or fly away in search of ice-free tidal areas. The first option tends to be

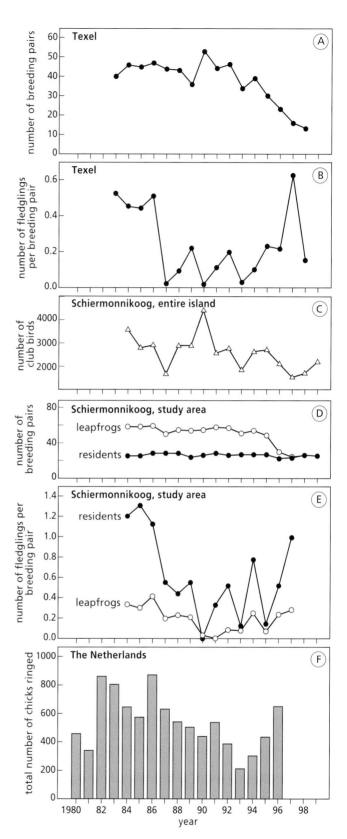

Figure 6.6 Changes in population size and breeding performance of oystercatchers in The Netherlands. (A) Number of breeding pairs at the study site on the island of Texel.[240] (B) Reproductive success on Texel.[240] (C) Total number of birds without a breeding territory (club birds) on the island of Schiermonnikoog.[388] (D) Number of breeding pairs occupying territories of different quality in the study area on Schiermonnikoog. (E) Reproductive success in the study area on Schiermonnikoog.[388] (F) Total number of chicks ringed in The Netherlands.[597]

Oystercatchers face hard times when the Wadden Sea freezes over (top photo this page). They need extra food because of the cold, but cannot reach it because of the ice that covers the tidal flats. Some birds migrate south to areas such as the bay of Mont St. Michel in France (page 319, upper). Here, there is much less risk of freezing, but a much greater risk of being shot by a hunter. Other oystercatchers winter in the Exe estuary in England (lower photo below). The Exe is not hunted and rarely freezes over. However, oystercatcher numbers seem to have reached their limit here. There are few good options for Dutch oystercatchers in harsh winters, and many perish. (lower photo page 319).

favoured by the larger species, such as the oystercatcher and the curlew (Chapter 4). This can lead to massive mortality. But the second option is not without drawbacks. Some oystercatchers migrate to the nearby North Sea coast. The coast is slow to freeze over, but there is little food available. In harsh winters, more dead oystercatchers may be found on the beach than are counted there alive in mild winters.[124] The birds may also escape to more southerly tidal areas. The first sizeable area south of the Wadden Sea is the Dutch Delta and in cold winters there can be a marked influx of birds to this area. During frosty winters more than 71% more red knots, 26% more shelducks, 22% more oystercatchers and 8% more curlews can be found in the Delta than in frost-free winters. There is no indication of such an increase in the numbers of some other species, such as the redshank, bar-tailed godwit, dunlin and grey plover.[76] This might mean that the Delta is 'full' and that these other birds fly further south or west to Britain. Oystercatchers are one of the species whose numbers do increase (Figure 6.7), but it seems that even some oystercatchers have difficulty finding room in the Delta. Of the 100 000 oystercatchers estimated to leave the Wadden Sea, only 40 000 are found in the Delta.[124] A considerable number fly further south to the French estuaries.[716] These estuaries are full of hunters. Almost all the ringed oystercatchers that have been resighted in France were shot![420] The end result is that in cold winters 25% of all oystercatchers may die. In mild winters the mortality is often just a few percent.

Harsh winters don't just cause high mortality in waterbirds. Benthic animal populations also suffer. Cold snaps may cause almost the entire population of some species, like the cockle, to die (Chapter 4). So in the following spring, little food can be found on the tidal flats. Despite the scarcity of adult benthic animals, which causes the total production of larvae to be low, a very strong spatfall usually follows. In the following autumn the tidal flats teem with small worms and young shellfish. Waders, like the red knot, that eat these small prey find themselves spoilt for choice. The oystercatchers prefer the larger size-classes, and have to wait another year before they find themselves in this happy position. The precise timing of this depends on the growth rate of the species. Red knots prefer 10 -15 mm Baltic tellins and it takes two to three years before the slow-growing Baltic tellins reach this size. Oystercatchers prefer larger Baltic tellins, so they must wait even longer (Chapter 4). Benthic animals on the upper tidal flats grow more slowly. However these details do not really change the overall picture. Each year fewer benthic animals of the strong year-class remain. In the first few years this mortality is compensated by the increase in biomass as a result of growth. But older benthic animals invest more in reproduction than in growth and slowly but surely the strong year-class dies out, leading to food shortages if there isn't another strong spatfall.

There are two very important factors to bear in mind. First, harsh winters often affect much of Western Europe. So the high mortality amongst benthic animals due to low

winter temperatures and the resulting strong spatfall are not local phenomena. Instead, these events are synchronised over a large area.[68] A bad year at one site is very likely to be a bad year at neighbouring sites too. The second important point is that the annual variation in biomass of benthic animals is much larger than the annual variation in bird abundance. These two factors mean that in some years there is so much food for the birds in the Wadden Sea that it must be an almost inexhaustible supply. The birds will only have problems in years when there is little food.[552]

Is there any evidence that the waterbird mortality is higher in years that food is scarce? The only way to determine this is to study the effects of harsh winters and food availability simultaneously. To date only oystercatcher mortality has been subjected to such scrutiny. Oystercatcher mortality has been measured in different ways: (1) via the total number of dead animals found (Figure 6.7C), (2) via the number of marked animals that don't return to the breeding grounds and (3) via resightings of ringed birds.[124, 597, 995] All the analyses lead to the same conclusion: when the additional mortality resulting from winter cold is taken into account, mortality is higher in years that food is scarce. Interestingly, young birds are most affected by food shortages.[597] Which brings us to our next topic: competition in winter.

### 6.4.2. Competition on the wintering grounds

What role does competition play in harsh winters when food is scarce? Young oystercatchers are less skilled at finding food and have lower status than older birds, so it's not surprising that they are the first winter casualties.[344, 420, 597, 837, 992] But what exactly is happening during a harsh winter? Do so many young oystercatchers die because they haven't stored as much fat and so have lower reserves to fall back on? This suggestion seems unlikely as young oystercatchers have particularly high fat stores, probably because they can be less certain about their access to food.[992] So is it because they suffer more during the cold snap itself? During the first few days of the cold snap, a few ice-free areas remain which become densely packed with foraging birds. Less dominant birds are likely to fall victim to conspecifics that steal their food and chase them away from the few feeding spots available. It seldom freezes in the Exe, but it can be very cold in the winter months. These are also the times when interference between foraging oystercatchers is at its strongest.[322] Finally it seems that young birds that move south because of the cold are at a disadvantage compared to southwards moving older birds. This is certainly the case when these southerly areas become 'full'.

Finding evidence that more birds die when food is scarce is somewhat different from proving that more birds die when bird density increases and food availability is unchanged. Although it seems reasonable to assume that if the first occurs the second does too, we still need to find a way to connect these two relationships. One possibility is to assign values to the different parameters in population models from our field data, and then (as we described earlier), construct the negative feedback loop between bird density and survival.

Until now we have assumed that competition on the wintering grounds leads directly to increased mortality. The birds may starve to death, or be forced to forage in areas where they run a higher risk of predation. The latter seems to occur in redshanks.[142] The birds may also risk ingesting prey infected with harmful parasites. Mussels on beds that are exposed for long periods contain more parasites than those that are only exposed for a short period.[335] There are indications that oystercatchers that feed on parasite-rich mussel beds carry higher parasite loads. These birds have more parasite eggs in their faeces, suggesting that they also have more parasites in their intestines.[716] Oystercatchers avoid parasite-rich mussel beds. They also avoid meadows where they risk catching a different suite of parasites.[302] Meadows are mainly used by the oystercatchers unable to find enough food at low tide (Figure 4.9).[146, 342] Once a bird is infected, the parasites are very unlikely to cause serious harm immediately. It may be quite some time before they really trouble the bird. But eventually the cost of carrying a parasite load can mean that a bird has less chance of acquiring or defending a breeding territory or successfully raising young.

Increased competition as geese prepare to migrate north in spring doesn't lead to increased mortality, but to lower chick production. This is because the geese build up their body stores before migrating north, and the more of this stored energy that remains on arrival, the more eggs the birds can lay.[187, 209] This in turn leads to more fledglings.

It's clearly very difficult to measure and/or predict the impacts of density-dependent effects. This important area of research is still largely an unexplored field.

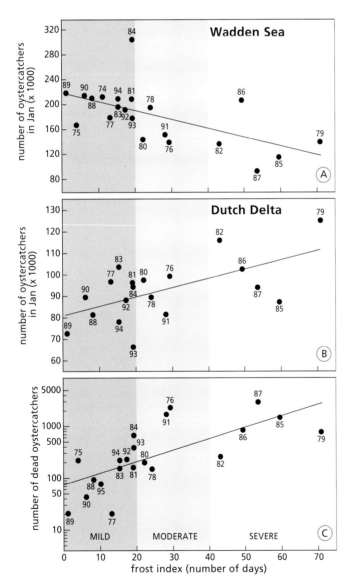

Figure 6.7 The effect of harsh winters on the distribution and mortality of oystercatchers. (A) Numbers in the Wadden Sea, (B) numbers in the Dutch Delta, and (C) numbers of dead birds found along the Dutch coast, as a function of the severity of the winter.[124]

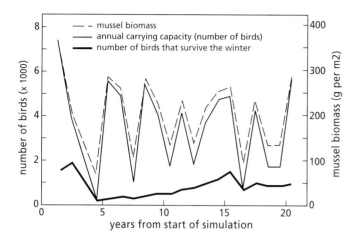

Figure 6.8 A simulation of the effects of an annually varying food supply on the oystercatcher population in the Exe estuary.[345] The model assumes that oystercatchers that survive the winter have a realistic reproductive success. It appears that oystercatcher numbers in the Exe are largely determined by years when food is scarce.

### 6.5 What is carrying capacity?

We briefly discussed carrying capacity earlier in this chapter. Now that we've examined the population processes at work in summer and winter, it's time to return to it in more detail. We'll restrict ourselves to considering the Wadden Sea's carrying capacity for passing and wintering waterbirds. Researchers can use different methods to try to determine whether the carrying capacity of their research area has been reached. One often-used method is to plot the total number of animals in the research area against the total number of animals in a much larger area, preferably the world population. When no further increase of birds in the research area occurs it is concluded that the carrying capacity has been reached.[207, 333, 584, 961, 964] Although such a levelling off is a necessary condition, it's not enough. There is no evidence that the numbers cannot increase further. This realisation has led some researchers to state that a tidal area has reached its carrying capacity when for every new bird that settles in the area, another bird must leave or die.[321]

A consequence of this definition is that the carrying capacity is higher in years when food is abundant than in years when food is scarce. The oystercatcher model based on the Exe estuary data makes it possible to predict the maximum numbers of oystercatchers that can survive the winter at different levels of food availability. The predicted annual carrying capacity closely follows the food availability (Figure 6.8). If we imagine that the Exe was the only wintering area used by oystercatchers, then all the adults would return there after the breeding season and all of the juveniles would winter there. That makes it possible to calculate long-term changes in oystercatcher abundance. In very unfavourable years, the population is dealt a harsh blow. These are the only years in which the actual number of wintering birds corresponds to that year's predicted carrying capacity (Figure 6.8). In the subsequent years, the oystercatcher population slowly increases, but before it can reach the carrying capacity of an average winter, the population is knocked back again by food shortages. So the long-term carrying capacity is mainly determined by the carrying capacity in unfavourable years. It seems likely that if food availability was more consistent, the oystercatcher numbers would more closely resemble the average carrying capacity. Even though annual fluctuations in prey availability on the Banc d'Arguin haven't been measured directly, it's very possible that the Banc d'Arguin provides a more stable food supply than the Wadden Sea or other West European estuaries. It is almost certain that the density of waterbirds on the Banc d'Arguin is much higher than that in the Wadden Sea, although the average biomass of benthic animals is much lower.[17, 950, 983] Perhaps this is due to the supposedly more stable food supply of the Banc d'Arguin.[946] However, the giant bloody cockle (*Anadara senilis*) doesn't fit the picture, and it is a favourite oystercatcher food. From 1980 to 1997 the density of giant bloody cockles in a number of permanent quadrats fell from 56 per m² to 7 per m², due to the lack of a strong spatfall. These cockles, which were 29 years old in 1997, only grew 13 mm over this time (from 72 to 85 mm), so their decrease in abundance was nowhere near compensated by an increase in biomass. As a consequence, the number oystercatchers wintering on the Banc d'Arguin halved from 1980 to 1997.[996] Despite recent increases in the giant bloody cockle population, oystercatcher numbers have continued to decline.

Of course the Exe is not the only estuary in the world, which is one problem with the calculations shown in Figure 6.8. Birds that struggle to find food there can always try to find more elsewhere. Such a quest is probably risky, but is undoubtedly preferable to certain death. It only makes sense to leave if the other tidal areas are not equally overcrowded, and this certainly seems to be the case. The Exe may be full, but there is evidence that the Burry inlet further can still take more oystercatchers.[610, 804] The animals most at risk of starvation stand to gain the most by finding a new wintering area. Amongst oystercatchers, these are the young birds. Sure enough, when there was a large cockle spatfall in the mouth of the Ribble after years of little food, most of the many oystercatchers that discovered this feast were young birds.[821]

So what does this mean for carrying capacities? It means that bird abundance in a particular area is usually largely determined by the carrying capacity in unfavourable years rather than by the average carrying capacity. It also means that we shouldn't look at individual tidal areas in isolation. It is theoretically possible that a situation could exist where the world population cannot expand further, even though some wintering sites haven't reached the carrying capacities of even the unfavourable years.

### 6.6 Natural selection and population dynamics

In our discussion of processes that affect bird abundance, we used existing behavioural adaptations as our starting point. In the earlier chapters, we approached from the other direction: we started with the bird abundance and used this to understand birds' adaptations, such as migration patterns. Ideally, we should be taking both factors into account at the same time. Some thought has already been put into this line of investigation.[9]

When the amount of suitable wintering habitat is the limiting factor, then population regulation (when most animals die) will mainly occur in winter. Conversely, the superabundance of breeding habitat will mean that many young are produced in summer, and clutches will be large as a result of natural selection. In this book, we describe these species as 'winter-restricted'.

When the amount of the breeding habitat is the limiting factor, population regulation will mainly take place in summer. That means that the birds have fewer problems surviving the non-breeding season than the winter-restricted species. It also means that there will be strong selection for characteristics that aid the acquisition of breeding habitat. When the struggle for breeding habitat is fierce, there may be benefits to not taking part in it in the first year of life. We refer to these species as 'breeding-restricted'.

Winter-restricted species are characterised by a short life expectancy, large clutches and strong fluctuations in abundance. Breeding-restricted species are characterised by a long life expectancy, small clutches, a later age of first breeding and a gradual change in abundance.

Let's look at this in more detail for a typical breeding-restricted species, the oystercatcher. Why do oystercatchers defend their territories so zealously in the breeding season? It seems that good territories are very scarce, and the few lucky owners of good territories must continually defend their properties against competitors. Of course, scarcity is relative. It depends on the total surface area of

Grey plovers mainly fed on ragworms and have recently shown an increase in wintering numbers (photo left). They do not really compete for food with shellfish-eating birds such as oystercatchers (below). Red knots feed on shellfish, but only on individuals that are too small for oystercatchers. These species can happily forage near each other (page 323, on a tidal flat near Texel, in the Dutch Wadden Sea). But there may still be indirect competition, if the knots eat so many small shellfish that few survive to grow large enough for the oystercatchers to eat.

good quality sites, the size of a good territory and the number of contestants. Territories wouldn't be so scarce if there were fewer oystercatcher, more suitable habitat, or if the oystercatchers were content with smaller territories. So how big should a good breeding territory be? This is a very difficult question to answer, but it's worth noting that they are often less than 1 ha in size.[244, 250] The large number of competitors for territories is partly due to the oystercatcher's specialised feeding techniques. The oystercatcher is generally believed to be the only bird able to open cockles and mussels with its bill and extract the flesh (apart from a little-known duck in New Zealand).[1008] Much of the biomass of benthic fauna in European tidal areas comprises cockles and mussels and a few other shellfish species that are also food for the oystercatcher.[51, 982]

This rich food source is available year-round to the non-breeding oystercatchers (provided that the stocks are not diminished by commercial shellfish harvesting). Breeding oystercatchers however do have a problem. The richest mussel and cockle beds are situated some distance from the coast. Oystercatchers continue to feed their young well after they have fledged. A parent can transport food over some distance, but as it holds the prey in its bill tip, it can only carry one food item at a time. Elsewhere in the bird world this transport problem has been cleverly solved. Many seabirds store food in their stomachs and can regurgitate this for their young when they return to the nest. Puffins don't regurgitate food, instead they return to their nest burrows with a bill full of fishes.

So they must be able to catch new fish while already carrying some fish in their beaks. Oystercatchers have yet to discover such a clever solution. Carrying only one prey item at a time is only profitable when the prey are large and the distances are small.[504] Oystercatchers will usually only carry food a few hundred metres. The record is 2 km, but this is exceptional. As a consequence, oystercatchers breeding in the Wadden Sea area are confined to the narrow edge of the tidal zone. This means that in the breeding season they are restricted to a small fraction of the available food resource. The area may provide enough habitat of sufficient quality for the parents to find enough food to survive, but limited high quality habitat close to the nesting sites where the birds can reproduce. This results in intense competition for the good feeding territories in the breeding season. This can mean that young birds cannot acquire a territory until later in life, exactly what we would expect of a breeding-restricted species.[9]

According to Meltofte[568] nearly all tidal flat species can be categorised as breeding-restricted. They do not breed until after their first year and have small clutches. The adult birds leave the breeding grounds early in the season, before they have moulted and before the juveniles are able to migrate. So many species are only on the breeding grounds for a few months, and spend the remaining 9 – 10 months of the year in tidal areas. This would explain why despite intense competition for good breeding territories, these birds are highly adapted physically and behaviourally to life on the tidal flats.

## 6.7 Competition between species

So far we have only considered the competition between conspecifics, and not that with other species. It is true that in general two birds of the same species will compete more than two birds of different species. We saw in Chapter 4 how little overlap there is between the different species in their food choice and distribution on the tidal flats. However it is very unlikely there is no competition between the different waterbird species. There are many ways competition can occur.

To start with, species can steal food off each other. Stealing is a very direct type of food competition. This occurs, but as we saw in Chapter 4 such theft tends to be the exception rather than the rule. Another possibility is that one species may lower the food availability for other nearby species, if they disturb their prey which then try to hide. For example, ragworms retreat into their burrows after a disturbance (Chapter 4). There are indications that this type of competition does occur, but there are very few observations that suggest it really is a big problem. There is also a third more indirect form of competition for food. Red knots feed on small shellfish, while oystercatchers eat the larger individuals of the same species. It is theoretically possible that red knots could eat so many small shellfish that few would remain to grow big enough for the oystercatchers to eat.[977] In a reverse scenario, the oystercatchers could eat so many large shellfish that few would be left to produce new small shellfish. This type of indirect food competition has never been measured in the field and developing adequate methods to do so would probably be very complex. However, we do know something of the processes involved, such as the prey size selection of the different bird species and shellfish recruitment. This information has been used to create a simple model that describes the competition between red knots and oystercatchers for cockles.[552] This model assumes that cockles are the sole food source for both bird species, and that the stock of old cockles has no influence on the spatfall of new cockles. The model predicts that when there is little variation in the annual spatfall, the number of red knots increases markedly and the knots would leave very little for the oystercatchers to eat. However when the annual spatfall is highly variable, the number of red knots remains low because of food shortages in years with a poor spatfall. This allows the number of oystercatchers to increase. According to this model, red knots can have a large effect on oystercatchers, but the oystercatchers don't affect the knots. Of course, if it came to a physical fight a red knot would have no chance against an oystercatcher, but in this case the small guy wins every time.

Cockle beds are often made up of just one year-class, which means oystercatchers and knots would very rarely encounter each other on the tidal flat even if they do compete indirectly. This demonstrates just how difficult it is to make conclusions about interspecific competition from data about overlap, or the lack of overlap, or prey choice and spatial distribution. So it's no surprise that different researchers have come to different conclusions about the extent of overlap in the wintering grounds of different wader species. Amongst the genus *Calidris*, morphologically similar species overlapped to a large extent in their breeding grounds, but not in their wintering grounds. From this, it was concluded that these species mainly compete with each other in the winter.[864] Many species winter along the west coasts of Europe and Africa in completely separate wintering grounds. However wader researchers have yet to reach a consensus[568] as to whether southerly wintering species and populations are forced south (and over the more attractive northerly wintering areas) by direct competition with the more northerly wintering species and populations. These southbound animals must fly over the northerly wintering grounds while these grounds are still unoccupied! Wintering in either the south or the north has its advantages and disadvantages. Still you would expect these advantages and disadvantages to be related in some way to competition and the numbers of other animals choosing each option.

## 6.8 Are we leaving the birds enough room?

Now we have a picture of the processes that determine waterbird abundance we can ask ourselves what effects

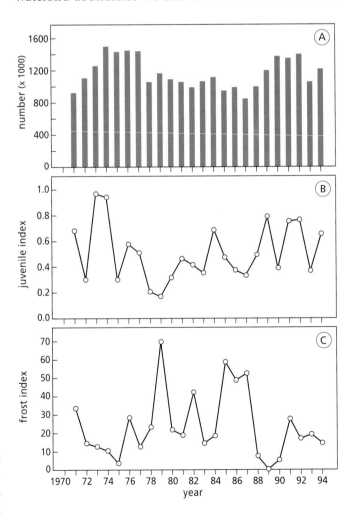

Figure 6.9 Dunlin populations in Western Europe:[748] (A) wintering numbers, (B) juvenile index with maximum value set at 1 and (C) frost index, indicating winter harshness.[124]

human activities are having. There are two ways to tackle this problem. We can record the changes in bird numbers and see whether these changes can be explained by human activities. Or we can monitor a human activity and measure its impacts on the birds.

### 6.8.1 Waders and humans: a brief history

When you're thinking of waterbird species that have been adversely affected by people, the grey plover is one of the first to spring to mind. Grey plovers breed on the most northerly tundra, an area that few people visit recreationally but where some go to fish, hunt or drill for oil. None of these activities benefit the plovers. Plus if global warming becomes a reality, the High Arctic tundra may be one of the first habitats to disappear. Outside the breeding season grey plovers are exclusively found in tidal areas (unlike their close relatives, golden plovers, who prefer to go inland). The total area of available tidal flats has decreased dramatically over the years, while human use has increased. Despite this, the number of grey plovers wintering in British estuaries has increased spectacularly since the 1960s. In the winter of 1970 - 71 there were 5570 grey plovers. Twenty years later, there were 47 620 grey plovers, an almost fourteen-fold increase. How was this possible? After extensive analysis, Moser[584] concluded that this increase resulted from an increase in the world population. Presumably, more grey plovers were breeding on the High Arctic tundra than in previous decades. The scarce data available from the tundra do not indicate such an increase. If the world population had increased, you would also expect an increase in the number of animals passing through fixed count stations on migration between the breeding and wintering grounds. There are more data available on this, but these don't suggest an increase either. An alternative explanation is that there hasn't been an overall increase; the birds have simply changed their distribution across the different wintering grounds. The British wintering grounds may have become more attractive, or the other wintering grounds less attractive. The prohibition of hunting in British estuaries may have encouraged more grey plovers to stay in Britain for the winter, rather than heading south as soon as they'd finished their moult in autumn.[116]

The dunlin, with a population of more than two million, is by far the most abundant waterbird using the East Atlantic flyway. More than half of these two million birds winter in Western Europe each year. Changes in the numbers of dunlin have been analysed extensively.[328, 748] In the 1970s numbers were high, in the 1980s low and in the early 1990s they seemed to be back to their old levels (Figure 6.9A). What was going on? Perhaps there were fewer young produced in the 1980s. Juvenile dunlins are easily distinguished from adults in autumn, so we can compare the percentages of juveniles in different counts and catches. It is not possible to determine the percentage of juveniles for the population as a whole since juveniles, unlike adults, mainly stay in marginal areas.[383] The best we can do is use these data to construct an index

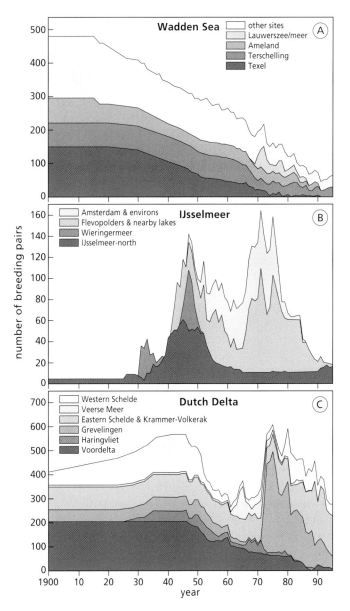

Figure 6.10 Reconstruction of the breeding populations of Kentish plovers in three areas in The Netherlands: (A) Wadden Sea, (B) IJsselmeer, and (C) Dutch Delta.[555]

(Figure 6.9B). This index revealed that dunlins are influenced by the lemming cycle, although not as strongly as the more northerly breeding species.[748] In the early 1980s the juvenile index was lower on average than in the years before or after. An additional explanation for the decrease in the 1980s is higher winter mortality due to very cold weather. There certainly were many severe winters in the 1980s (Figure 6.9C). It's difficult to determine the relative importance of each hypothesis, but together they form a possible alternative to a third hypothesis. This hypothesis was that the spread of the cordgrass *Spartina anglica* through the key English feeding grounds, where half a million dunlins winter, was the primary cause of the decrease observed in early eighties.[328] An alternative to this hypothesis is now required since the number of dunlins has increased again, even though *Spartina* remains prevalent.[748]

Meininger and Arts[555] described the Kentish plover as the typical breeding bird of Dutch coastal areas. The species prefers to nest near the water's edge on small, sparsely vegetated beaches. The authors reconstructed the changes in numbers of breeding pairs in different parts of The Netherlands, based on an extensive literature search. Although too much emphasis shouldn't be put on the exact numbers since birds have only been counted properly in recent decades, the impacts of humans couldn't be clearer (Figure 6.10). In the Wadden Sea region, the number of breeding pairs has steadily decreased due to this area's ever increasing recreational use. Tourists and Kentish plovers have similar tastes when it comes to beaches, and it's clear who is losing out.[876] Human activities can also have positive spin-offs. Dyke construction in the Lauwerszee (creating the Lauwersmeer), in the Dutch Wadden Sea, led to a temporary local increase in numbers, as tidal flats once inundated every high tide were transformed into permanently exposed barren sand flats. However once dense vegetation grew on these flats, as it inevitably did, the Kentish plovers no longer felt at home. The same phenomenon explains the increase in Kentish plovers in the IJsselmeer region in the Northern Netherlands after a 30-km long dyke (the Afsluitdijk) was constructed, transforming the Zuiderzee into Lake IJsselmeer. Although these drainage and reclamation projects may be beneficial to Kentish plovers in the short-term, without the sea's influence the sandy areas quickly become overgrown making them unsuitable in the long-term. Large construction projects that once provided temporary habitats, like the Dutch Delta Project in Zeeland, are now less common and the Kentish plover's future in Western Europe is looking bleak.

### 6.8.2 Land reclamation and the Delta Project

Dyke construction and land reclamation may bring temporary benefits to a few waterbird species, but they are a disaster for the birds that are depend on the intertidal areas. From one day to the next they lose their feeding grounds.

Around the Wadden Sea, this encroachment into tidal areas has steadily progressed seaward from the inland areas. Wicker dams were constructed on tidal flats in front of the salt marshes, to encourage sedimentation. Once the salt marshes had built up high enough, they were reclaimed. Eventually, many of very silty tidal areas in the Wadden Sea were lost. The continued existence of many marshes on the mainland now depends on the reclamation works. But land reclamation stopped when the Wadden Sea was declared a nature reserve. Salt marshes in areas like Wierum in the north of Fryslan,

Reclamation has removed large areas of tidal flats. When inlets are dammed by a flood barrier, parts of the flats remain available as feeding grounds for birds. Here, red knots forage on tidal flats inland of a flood barrier on the Eider River, in Germany.

Land reclamation in salt marshes destroys the feeding and roost sites of many species, such as these 'shanks' (redshanks, spotted redshanks and greenshanks) that use the salt marshes during high tide (upper photo). Parts of the Wadden Sea are still threatened by reclamation, like the estuary of the River Eems (lower photo).

where the reclamation works are no longer maintained, are now eroding. The continued existence of some salt marshes, like those along the Groninger coast, depends on the maintenance of the wicker dams.[201]

Land reclamation began hundreds of years ago, before the bird counts began. This makes it difficult to say just how severe the impacts on waterbirds have been. But even in areas where the reclamation occurred more recently, impacts may still be difficult to determine. Various European tidal areas have been subject to reclamation in recent decades. The surface area of tidal flats in the Tees estuary was 2400 ha in the 19th century; by 1973 it was down to 400 ha. But only the impacts of the most recent reclamation in 1974 which left 140 ha of tidal flats have been studied.[266] Around half of the original 2284 ha of tidal flats in the Firth of Forth estuary have been sacrificed to agriculture, harbours and industry. Within Valleyfield Bay in the Firth, 20% of the waterbird feeding areas have disappeared.[101, 550] In the Dutch Wadden Sea, the Lauwerszee has been closed off. The Hojer reclamation in Danish Wadden Sea resulted in the loss of 1100 ha of tidal flat and salt marsh.[502] Many tidal flats

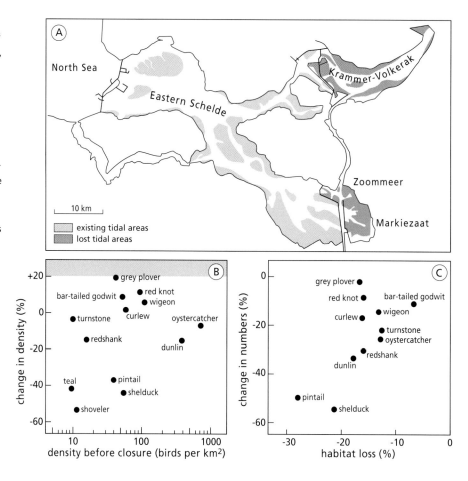

Figure 6.11 The effect of tidal flat loss on waterbird populations in the Eastern Schelde estuary.[775] Dykes were used to close off the Krammervolkerak, Zoommeer and Markiezaat from the main estuary, while a moveable flood barrier was built between the Eastern Schelde and the North Sea. (A) Map of the area, indicating dykes and flood barrier (bold lines), and tidal flats (grey areas). (B) Change in foraging bird densities after tidal flat closures. The change is expressed as the post-closure density as a percentage of the pre-closure density. Only species with wintering numbers exceeding 2000 individuals are shown. (C) Changes in wintering bird numbers after tidal flat closures in relation to the percent of relevant habitat lost for each species. The change is expressed as the post-closure population as a percentage of the pre-closure population.

have also ceased to exist in the German Wadden Sea: 1600 ha from Voslapperwatt, north of Jadebusum in 1974; 15 000 ha in 1969 - 1979 due to various dyke reconstructions along the Elbe downstream of Hamburg; 1280 ha in 1972 when the Eider was dammed; 4800 ha at Meldorferbucht from 1973 – 1978; 850 ha at Heringsand in 1973; 750 ha at Leybucht in 1993; and 3310 ha at the Nordstranderbucht.[189]

So what were the consequences of so much habitat loss? Usually, bird numbers in these areas declined, as you would expect. But this wasn't always the case. The number of grey plovers in the Tees estuary increased after the reclamation. However, as we've discussed numbers of wintering grey plovers have increased throughout Britain for reasons that are not well understood. So this increase doesn't necessarily mean that the reclamation of the Tees was beneficial for grey plovers. We can only conclude that the number of grey plovers in the Tees depends not only on the local situation but also on what happens in other estuaries and on the breeding grounds.

The most dramatic loss of European tidal areas took place in the Dutch Delta area (Figure 6.11A). Whole estuaries were closed as a result of dyke reinforcements following the Dutch Delta law, which aimed not to reclaim land but to provide better flood protection. The Delta law was passed after a disastrous flood in 1953 breached the dykes in the southwest of The Netherlands and many people were drowned. Fifty years later, the extensive Delta Project was completed. When the Grevelingen estuary was shut off in 1971, a sudden increase in the numbers of some waterbird species was observed in the western part of the Eastern Schelde, south of the Grevelingen.[494] It seems likely that some of the Grevelingen birds took refuge in the Eastern Schelde, but there is not enough evidence to be certain.

More is known about the impacts of the storm flood protection works in the Eastern Schelde.[558] Between 1982 and 1987 around 33% of the 170 km² intertidal area in the Eastern Schelde was lost.[775] If the Krammer-Volkerak area of the Eastern Schelde is not included, 17% of the tidal flats were lost. Where did the waterbirds that used that 17% go? If they had spread over the rest of the Eastern Schelde, then densities at these other sites would have increased by 20%. Yet of the common waterbirds, only the grey plover showed this sort of increase (Figure 6.11). As was described above, the numbers of this species in many English estuaries have also increased. In spite of the increase in grey plovers at these other sites, they haven't really increased in the Eastern Schelde. The observed increase is simply an assimilation of the refugees. Densities of most other species in the Eastern Schelde have declined since the reduction in tidal habitat (Figure 6.11). These changes have more to do with the type of habitat that was lost - soft silt areas on the edge of salt marshes - than population changes on a European scale. Species like the shelduck, pintail, shoveler and redshank, which depend on this soft silty habitat, showed the greatest declines (Figure 6.11). Species like the bar-tailed godwit that use sand flats in the western parts of the Eastern Schelde were less affected.

### 6.8.3 Poisoning

If reclamation is like an enemy that fights out in the open, poison is like an assassin. Suddenly there is a mass die-off of animals for no apparent reason. Poisons can also act more subtly; instead of dying immediately, the animals may slowly but surely disappear from the ecosystem as they become ill or produce fewer offspring. Once poisons reach a certain concentration, they act by disabling one or more of the animal's basic life processes. An animal's body is an incredibly complex piece of machinery. To ensure that this machinery does not seize up, signals are continually being sent around the body. Many poisons act by disrupting these signals. The potential for an entire animal to be immobilised by a tiny amount of poison has not gone unnoticed by evolution. Poisonous snakes, poison dart frogs, poisonous mushrooms and poisonous berries are living proof that all things natural aren't necessarily good for our health. Evolution has also supplied the potential victims with appropriate responses, such as evasive behaviour or metabolic adaptations to break down poisons.

The problem with evolving an appropriate response is that it's usually a very slow process and some animals may never achieve this. The best way to conserve our nature heritage is to prevent poisons from entering the environment. And we should bear in mind that humans are animals too and most substances that are poisonous for birds and other animals are poisonous for us as well. People in The Netherlands had a wake-up call in the 1960s when the number of sandwich terns on the inhabited Wadden Sea island Griend underwent a dramatic decrease from 20 000 pairs to less than 2000 pairs.[895] The terns had been poisoned by the insecticides telodrin and dieldrin which had reached the Wadden Sea via the Rhine.[477] Since that time, much needed environmental laws have been brought in, both in The Netherlands and elsewhere in Europe. This legislation seems to have taken effect. Programmes to clean up polluted sediments have also begun in The Netherlands. In 1994 sediments in a canal that flows to the Wadden Sea near Delfzijl that were contaminated with hexachlorobenzene (HCB) were rehabilitated.[30] HCB is a very hazardous chemical, like polychlorinated biphenyls (PCBs) it disrupts mammalian reproductive systems,[726] but it is also carcinogenic. The AKZO company in Delfzijl had been dumping this toxin from 1969 till 1985. Stopping the discharges and rehabilitating the contaminated sediment seems to have had an effect: in the Dollard River near Delfzijl the HCB concentrations in mussels, and in the eggs of oystercatchers that eat the mussels, have steadily declined (Figure 6.12).

So does this mean our pollution problems are over? Unfortunately not. Many toxic chemicals continue to damage our environment. For decades ships have been smeared with the anti-fouling chemical tributyl tin (TBT) to prevent the growth of barnacles and other marine life. Even in very low concentrations TBT disrupts the gonad development of marine animals such as the common whelk (*Buccinum undatum*). It seems likely that this whelk, which has disappeared from the Wadden Sea, cannot return due to TBT levels. The International Maritime Organization (IMO) passed a resolution prohibiting the use of TBT in 2003, but this resolution will not be fully implemented until 2008. Moreover, industries continue to develop and produce new chemical processes and substances with breath-taking speed. It is not difficult to compile a list of requirements that new substances must adhere to, but it is impossible to predict all of their potential ecotoxicological effects in advance. Some substances only become poisonous after exposure to the environment. Through sheer necessity, only substances currently known to be poisonous can be monitored.[35] The real problem is the as yet unknown assassins. Regularly counting the birds in tidal areas and monitoring their breeding success is a good way to monitor the health of tidal ecosystems.[482, 855]

### 6.8.4 Effluent discharges in the Dollard estuary

Eutrophication is a very different type of pollution. It is defined as an increase in the amount of organic material entering an ecosystem.[603] Unlike toxic chemicals that weaken or kill plants and animals, eutrophication supplies them with extra nutrients. This might sound beneficial, but it is not always. Usually one particular group of plants or animals profits spectacularly from this sudden bounty, causing problems for the other organisms in the ecosystem.

Eutrophication can occur in two ways: directly, such as a point source discharge of effluent, or indirectly through an increased input of the nutrients nitrogen and phosphate. In coastal ecosystems the limited amounts of these nutrients usually limit the production of pelagic and benthic algae.[393] The growing use of artificial fertilisers and larger cattle stocks have increased the amounts of these nutrients being applied to the land.[603] Sooner or

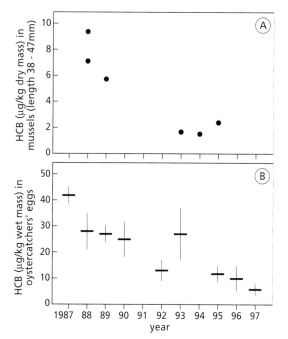

Figure 6.12 Hexachlorobenzene (HCB) concentrations in (A) common mussels (*Mytilus edulis*) and (B) oystercatcher eggs (averages and 95% confidence intervals are shown).[30]

Fish-eating birds, like these little terns (photo above), are at the end of a long food chain that can accumulate poisonous substances. Mussel beds (photo opposite) are very important for waterbirds. The birds feed not only on the mussels, but also on the many small animals that live amongst the mussels.

later these nutrients are washed into coastal waters.

You might expect that more algae or bacteria would automatically mean that the benthic animals have more to eat, which in turn means more food for the birds. But it's no simple task finding the evidence. There are a few long-term studies of tidal areas where an increase in the biomass of benthic animals happened to be correlated to an increase in nutrients,[65, 431, 874] but evidence for a causal relationship is lacking.[263] Many different and complex processes are taking place at the same time.

Direct eutrophication seems easier to understand. An increase in the available organic material leads to increased activity by aerobic bacteria, which, like humans, use oxygen for the metabolism. In extreme cases, this can lead to oxygen shortages. No higher organisms can survive long without oxygen. But it's when there is a shortage of oxygen on the tidal flat (in other words it becomes anaerobic) and there is plenty of organic material to provide food for bacteria, that the real problems begin. Under these conditions the anaerobic bacteria that produce hydrogen sulphide ($H_2S$) thrive. Hydrogen sulphide is an extremely poisonous gas, well-known for its characteristic rotten egg smell. At varying depths below the surface, mudflats are always anaerobic, stinking and black due to the limited available oxygen. The benthic animals that live at this depth take oxygenated water from the surface for respiration. So the walls of their little burrows are coloured rusty brown rather than black. Eventually, a very high input of organic material will cause even the upper layer of the tidal flat to become anaerobic. This produces stinking black spots on the tidal flats, where all of the benthic animals have died. Not so long ago, black spots of a considerable size were found at various sites in the Wadden Sea.[442]

The annual discharge of effluent into the Dollard, an estuary in The Netherlands is a well-researched case of eutrophication. The effluent came from potato processing and straw carton industries in Groningen, in the Northeastern Netherlands. The canals that transported this effluent to the sea had a terrible stench and each year an anaerobic area developed at the discharge site where many benthic animals died (Figure 6.13A). Further from the discharge site the effects of the effluent were less dramatic, but there was no doubt that the whole estuary was affected. From 1971 onwards treatment of the effluent within the factories improved. This treatment became so effective that by the 1990s there was almost no pollution occurring (Figure 6.13B).

Thanks to an extensive research programme, the reaction of the Dollard's benthic fauna to the declining nutrient input is very well known.[261] In the world of benthic animals ragworms are the tough guys; they are

LOOKING TO THE FUTURE

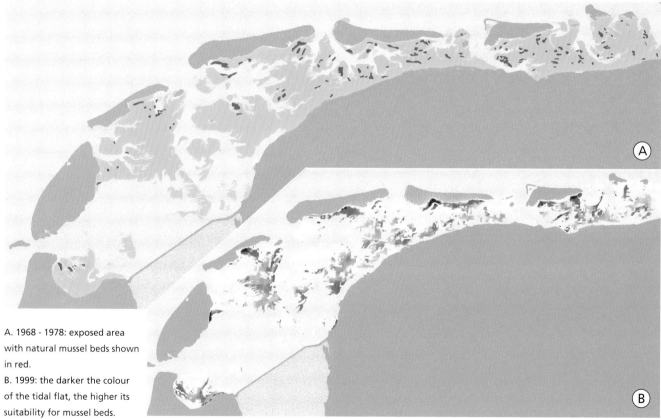

A. 1968 - 1978: exposed area with natural mussel beds shown in red.
B. 1999: the darker the colour of the tidal flat, the higher its suitability for mussel beds.

Figure 6.14 The mussel (*Mytilus edulis*) beds that were present in the Dutch Wadden Sea in the 1970s[202] have been almost eliminated. Research has pinpointed the locations where managed recovery of mussel beds is most likely to succeed.[93]

highly resistant to large fluctuations in salinity. In brackish areas with a high silt content, like the former Ventjagersplaten[961] in Zeeland in the Southern Netherlands and the nutrient-enriched Dollard, ragworms were the main food of a large number of waterbirds. Each bird species has its own favoured catching technique and size preferences when feeding on ragworms.[961] As effluent discharge to the Dollard decreased, so did the ragworm biomass (Figure 6.13C). We can use count data and publications on food requirements to estimate the annual food consumption of waterbirds in the Dollard.[709] Over the years, their total consumption has halved (Figure 6.13F). This is due to a decrease in abundance of many waterbirds species. Since the abundance of most species has increased rather than decreased across Europe as whole, it seems likely that the declining bird numbers were connected to the improved water quality in the Dollard. The only species that didn't decline was the spot-

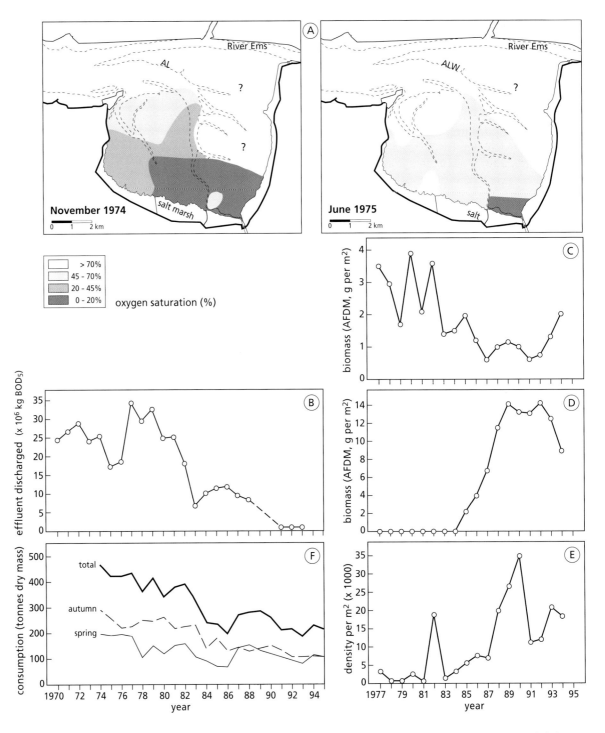

Figure 6.13 Consequences of lowered effluent discharges in the Dollard for benthic fauna and birds. (A) Oxygen saturation (%) during high tide in November 1974 and in June 1975 (effluent was still being discharged at these time). (B) Annual water discharge water (millions of kg $BOD_5$). (C – E) Abundance of selected invertebrates in the central Heringplaat: (C) biomass of the ragworm *Nereis diversicolor*, (D) biomass of the exotic worm *Marenzellaria*, and (E) density of the amphipod *Corophium volutator*. (F) Annual food consumption by birds, broken down into spring and autumn components.[260, 709]

ted redshank. However spotted redshanks feed on shrimps and the amphipod *Corophium*, rather than relying on ragworms. Unlike most other species, amphipod density has increased with the decrease in nutrients (Figure 6.13E).

So far the story may seem simple, but it's about to become more complicated. In the mid 1980s a new worm species appeared on the scene, *Marenzellaria* cf. *wireni*, from North America.[260] This worm quickly became the dominant benthic animal in the Dollard (Figure 6.13D). As a result, rather than decreasing, the total biomass and production of benthic animals in the Dollard are higher than ever! Ecologically this worm is very similar to the ragworm, so competition between the two species cannot be ruled out.[260] Nevertheless, the combined biomass of the two worm species is higher since the discharges stopped. Presumably the decrease in bird numbers is because the birds that normally ate ragworms didn't switch to feeding on *Marenzellaria*. Unfortunately good data are lacking. However ragworms aren't the only prey species, and when we are considering bird food we shouldn't exclusively focus on ragworm biomass.[709] When *Marenzellaria* is excluded, the production of useful benthic animals for birds was estimated to be 48 g per m$^2$ in the 1970s and 24 g per m$^2$ in the 1990s. Across the 77 km$^2$ of exposed tidal flat, the birds were calculated to have consumed only 4.9 and 2.6 g per m$^2$ respectively. So even without taking *Marenzellaria* into account, the birds in the Dollard were estimated to have eaten just 10% of the total macrobenthic production.

Perhaps the most important effect of the discharges for the birds was their effect on prey availability. The discharges always occurred in autumn, the season when the total amount of food consumed by waterbirds declined the most over time (Figure 6.13F). On a per species basis, much of the recent decline in bird abundance in the Dollard is due to lower numbers in autumn. Being tough, ragworms are fairly resistant to anaerobic conditions, but every so often even they would be forced to come up to the surface for a breath of fresh air. They may also have surfaced more frequently to grab dead or weakened benthic animals with their jaws, or to graze on the enriched sediment surface. Whatever the mechanism, it is likely that the large numbers of waterbirds were attracted by the increase in ragworm availability caused by the discharges each year. So the relationship between the input of organic matter and the available biomass for birds at the Dollard was far from straightforward. Knowledge of other biological processes regulating food availability for birds, such as prey behaviour, is needed if we are to understand the resulting changes in bird abundance.

### 6.8.5 Commercial shellfish harvesting

Hidden amongst the mangrove forests in the Sine Saloum Delta in Senegal, stand small villages built on hills composed entirely of shells from giant bloody cockles. Local fishermen may have been dumping the empty shells at these same sites for centuries. On some of these islands, giant baobab trees flourish. When you sit in the shade under one of these trees it is easy to imagine that hu-

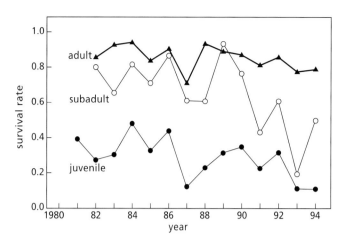

Figure 6.15 Annual survival rates of adult, immature and juvenile oystercatchers in the Dutch Wadden Sea, based on resightings of ringed individuals.[597]

mans have eaten shellfish from the moment that the first people set foot on the tidal flats. The result is that humans and waterbirds often compete directly for the same food resource. Eider ducks and oystercatchers live off the same large cockles and mussels that people like to eat. As long as the people collected the shellfish more or less with their bare hands, as in Senegal, this doesn't appear to have been a large problem. However we can't be completely sure of this. The enormous giant bloody cockles found on the mudflats of the Banc d'Arguin are decades old (Chapter 4). But in the more southerly tidal areas of Senegal and Guinea-Bissau, the giant bloody cockles are much smaller. This size difference could easily be due to differing levels of human exploitation. Nowadays, the inhabitants of the Banc d'Arguin, the Imraguen, have little interest in harvesting shellfish: they much prefer fish.

Before we examine the relationship between birds and commercial shellfisheries in depth, it is good to remember that there isn't always a conflict of interests. It's in the best interests of both groups to have clean intertidal areas that will not be reclaimed. It's even questionable whether conservationists could have succeeded in opposing plans to completely close off the Eastern Schelde if it hadn't been for the commercial shellfishermen joining the battle.

In the latter part of the 20$^{th}$ century, the equipment used to harvest shellfish in Europe became more and more highly developed. Commercial shellfish farming using wild spat was also perfected. Human use of wild shellfish became incredibly intensive and has pushed our competition with waterbirds to extremes. Unfortunately, the birds are the losers in this. Oystercatchers in the Burry Inlet were held responsible for a decline in the number of cockles and, after pressure from commercial shellfisheries, many thousands of oystercatchers were shot: 7000 in the winter of 1972 - 1973 and, despite a storm of protests, another 3000 in the following winter.[21, 429, 703] It's unlikely that permission for such slaughter would be given nowadays. But that doesn't mean that the problems have been solved. Since the 1990s a huge debate has raged about the effects of large-scale mechanical harvests on mussels and cockles, particularly those in the Wadden Sea. The fishery now uses large vessels able to cruise into very shallow waters. The mussel boats carry large dredge nets that are used to collect wild mussels and transport them to mussel farms in subtidal areas of the Wadden Sea. Cockle ships have large suction-dredges that are used to suck up cockles and layers of sediments ploughed loose from the tidal flats.

In The Netherlands the debate really heated up in 1990 when fishermen won a legal battle and fished away virtually all the natural mussel beds.[487] Natural mussel beds are a key intertidal habitat and play an important role in preserving intertidal biodiversity. Bird densities are much higher on and around mussel beds than elsewhere on the tidal flats (Chapter 4: Figure 4.38). Mussel beds are very rich habitats because the spaces between the mussels provide cover for other animals, such as small crabs. Oystercatchers can feed on the mussels, while curlews eat the crabs. In the late 1970s around 4100 ha of exposed mussel beds remained in the Dutch Wadden Sea.[202, 247] Since then, the beds have been decimated by harvesting. Now just hundreds of hectares remain at most.[400] Large-scale recovery of the mussel beds has not occurred, largely because of sediment disturbance by fisheries that still continue.[672, 1009] Recovery is also slowed by the lack of mature mussel beds that could otherwise form a foundation for new mussel brood. On old beds, the young mussels can easily attach themselves to their elders and hide between them as a protection

Curlews are hunted in Denmark (photo above), which partly explains their very wary nature. In general, foraging waterbirds try to keep a large distance between themselves and people, even when the people are school children on a field excursion (shown here in the Mokbaai, Texel, in the Dutch Wadden Sea; photo below). At low tide the birds have plenty of space and low-intensity disturbances do not have serious consequences. At high tide it is a different matter. There are only a limited number of quiet and safe roost sites where high concentrations of waterbirds can gather to avoid disturbance. Many of these high tide roosts are in protected areas such as Griend, an uninhabited island in the Dutch Wadden Sea (page 335).

against predators such as crabs.[549] Mussel spat can also settle in other places. But new mussel beds, where the young mussels mainly anchor onto each other, are much less resistant to storms or ice drift than mature beds that have built up on dead shells of a variety of year-classes.

What unequivocal evidence do we have that damage has been caused by commercial shellfish harvest? We know that the mussel fisheries did enormous damage when they greatly diminished mussel beds in the Dutch Wadden Sea. It is also certain that since 1990 these fisheries have caused food shortages for the birds, resulting in an increased mortality in eider ducks and oystercatchers.[122, 124, 487, 597] We can also deduce that the oystercatchers had great difficulties finding enough food in 1990 from the exceptionally high mortality of prey like the Baltic tellin and sand gaper. These prey are usually less attractive to oystercatchers, but that year the oystercatchers on the Balgzand were forced to feed on them.[62] But even that hasn't been enough to help the oystercatchers. There has been a clear decline in oystercatcher survival in the Wadden Sea (Figure 6.15). Analysis has shown that when winter temperature is taken into account, lower food availability leads to lower survival. The loss of the mussels beds means that food availability for the oystercatchers is now continually on the low side, causing higher mortality, especially in young oystercatchers. There is little doubt that this explains the decline in the numbers of oystercatchers breeding around the Wadden Sea (Figure 6.6).

That the mussel fisheries have caused the almost complete disappearance of a very important habitat in an internationally renowned natural area is remarkable to say the least. So in 1993 the Dutch government took action. By permanently closing 25% of the tidal flats to harvesting and reserving enough shellfish to cover an estimated 60% of the energy requirements of two shellfish-eating species (oystercatchers and eiders) in poor shellfish years, the government hoped to meet the needs of the birds and allow the mussel beds to return. Neither aim has been achieved to date. The lack of mature mussel beds continues to cause regular food shortages for the oystercatchers. The restrictions on cockle harvesting, which occur independently of the mussels, have also failed to have the desired effect.

Although commercial cockle fisheries did not contribute to the disappearance of the mussel beds in the 1980s, they probably impede their reestablishment. In a natural system, mussel spat can attach to cockleshells leading to the eventual development of a new mussel bed. Cockle beds are also ecologically important in their own right. In terms of bird density, cockle beds are the only areas that compare to mussel beds (Figure 4.38).

Recent data also emphasise the importance of cockle beds as a distinctive habitat with a high biodiversity. The extent to which the tidal flats suitable for cockle harvesting overlapped with those suitable for birds in a large, intensively-studied area in the Western Wadden Sea in 1998 was determined (Figure 6.16). This study particu-

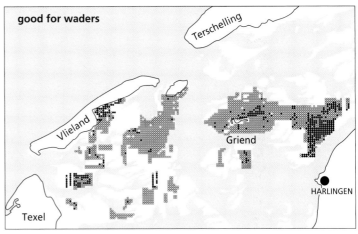

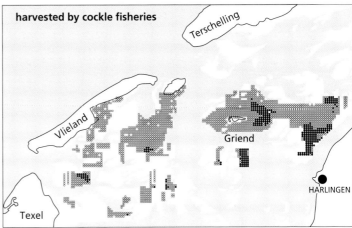

Figure 6.16 Overlap between shellfisheries and food resources for birds in the Western Wadden Sea. The tidal flats exposed during an average low tide are shown on the maps. The upper map shows the points sampled for benthic fauna by NIOZ researchers in August 1998 as circles, with sites suitable for birds indicated with black dots. These held cockles (*Cerastoderma edule*), Baltic tellins (*Macoma balthica*), sand gapers (*Mya arenaria*) or ragworms (*Nereis diversicolor*) at densities of 250 per m² or higher. In the subsequent shellfishing period in September - November 1998, automatic recorders on board cockle-dredging boats logged the fishing intensity (time and location) over the same areas. In the lower map, black dots indicate the sites dredged for cockles, open circles the points not fished.

larly focused on birds that don't live on cockles. From September to November 1998 the cockle fishermen carried 'black boxes' which registered the areas where they had been fishing. The fishery maps produced were made public after intervention by a judge, and can be compared to the results of a detailed benthic sampling programme that took place in the same area in August 1998 just prior to the fisheries (Figure 6.16). The harvested areas initially contained an average of 160 cockles per m², three times the initial densities of the non-harvested areas. These areas of high cockle densities were also rich in other benthic animals. Before harvesting, those areas contained twice the number of Baltic tellins, sand gapers, and razor clams (*Ensis* sp.) than the non-fished areas. The densities of common shore crabs, shrimps, ragworms and the polychaete worms *Nephtys* and *Heteromastus* were also two to three times higher in the harvested than non-harvested areas, before the fishing began. Adding insult to injury, the only sites where mussels had been found in August, were harvested by the cockle fishers that autumn. Only one species occurred in high densities in the areas that were not dredged later in the season, a polychaete worm, *Scoloplos*. This small worm species is common in impoverished and sandy tidal areas. The main question now is how badly do cockle fisheries damage species other than cockles and how long will it take them to recover? This probably depends on the fishing intensity.[495]

The main threat posed by mechanical shellfish harvesting is its potential to harm entire intertidal ecosystems. Mechanised harvesting creates coarser sediment beds in three ways:[483, 672] (1) ploughing the sediment causes fine sediment particles to be washed away, especially during heavy storms; (2) mussel beds shelter the sediment, when they are lost the edges of the flats become lower allowing current velocities to increase, which decreases silt deposition; (3) mussel and cockle beds filter seawater, producing silt-rich (pseudo) faeces that build up the sediment around them, an important process that is halted when the beds are destroyed. These three impacts can be depicted as a negative bio-deposition spiral (Figure 6.17). The spiral shows how ploughing, the loss of protective natural structures (such as mussel beds), and the loss of the silt-producing shellfish can disrupt a silt-rich sediment system. Thus one intense period of shellfish dredging can set off a chain reaction that causes the sediment to become increasingly coarse, leading to a further loss of shellfish that no longer produce and/or shelter silt, causing fewer shellfish to settle, and so it continues. The result of this vicious cycle is a barren, sandy tidal flat devoid of food. Although this is a hypothetical model, it is based on data collected near Griend, an island in the Dutch Wadden Sea.[672] Additional data are now available from the Wash in Britain and several other Wadden Sea sites.[1009, 1010] These new data attest to the validity of the model, making it clear that mechanical shellfish harvesting and a healthy coastal environment cannot co-exist. We must choose whether these natural areas are to be conserved or allowed to become barren wastelands.

### 6.8.6. Hunting and other disturbances outside the breeding season

As well as desiring their foods, humans also hunger for the birds themselves. Hunters like to bring home a trophy. Some species are hunted for food. Necessity was probably the main motivation for our ancestors to hunt. But today some species are hunted simply because they are difficult to kill. To the birds it makes little difference: dead is dead. On the population level it is mainly the magnitude of the hunt that is important and the way in which increased mortality due to hunting is or is not compensated by reduced mortality due to other factors.[600] It's reasonable to assume that such compensation occurs when there is intense competition for food. Hunting results in fewer competitors and through this a decline in mortality due to food shortages.

There is no doubt that hunting can have a large impact on populations. The dodo, passenger pigeon and Eskimo curlew did not survive excessive hunting, although the introduction of pigs (the dodo) and habitat loss (the passenger pigeon and Eskimo curlew) probably also played a role. The Eskimo curlew may not yet be extinct, but the millions of birds that once made stopovers along the east coast of the United States have been reduced to a handful of birds at most.[691] Further evidence for the impacts of hunting come from what happens when the hunting stops. Hunting of brent geese was prohibited in France in the late 1960s and in Denmark in the early 1970s. Since then the population has increased fourteen-fold (Figure 6.2).[207] In The Netherlands almost no wader species can be legally hunted, but they are less well protected in some other European countries. In a harsh winter, oystercatchers that abandon the frozen Dutch Wadden Sea may unfortunately go to France, where there are many hunters. Recoveries of bird rings in France are almost without exception from birds that were shot.[420]

It's clear that waterbirds are disturbed by hunters. But although hunting is popular in some countries, it's always a minority of the human population involved. In affluent West European countries people tend to visit natural areas armed with their binoculars rather than a gun. They certainly don't intend to disturb the wildlife. Sadly, people walking on the tidal flats or salt marshes can cause a lot of disturbance among the birds.[156] Why is that? Birds on the tidal flats have nothing to fear from people going for a walk, so there's no need for them to take flight. When other large and innocuous animals such as cows walk across the salt marsh, the birds aren't disturbed. Nor do cars cause as much disturbance as people on foot. The most likely explanation is that the birds cannot distinguish between dangerous hunters and other people. If this is the case, you might imagine that if hunting was completely abolished the birds would become very tame. This is a long way off. And if the hunting did finally stop, who knows how many waterbird generations it would take before the waterbirds became completely tame.

Until then we must accept that our human presence disturbs the birds. And to properly protect the waterbirds from this disturbance we need to know what impacts it has. A bird that takes flight because it is disturbed uses time and energy. Exactly how much time and energy is lost has been estimated in different studies.[42, 735] The problem lies in trying to relate this increase in energy expenditure to a reduction in survival rates. Another study assessed the costs associated with taking flight due to a variety of sources of disturbance.[799] There are striking differences between different species and sites. Larger species take flight at greater distances than the smaller species (Figure 6.18A). Across all species, the disturbance distance is smaller in the areas more frequently visited by people. For example, the Mokbaai, a bay on the Wadden Sea island Texel, is surrounded by roads and is often visited by the armed forces, people collecting lugworms, fishers and tourists. Waterbirds in the Mokbaai allow people to approach to a much closer range than the waterbirds along the Groninger coast, an area visited far less often. A comparison of disturbance distances from the Eastern Schelde and the Wadden Sea is also very interesting. The Eastern Schelde has always been more frequently visited by people than the Wadden Sea. Disturbance distances in the Eastern Schelde are shorter than those in the Wadden Sea and have shown little change over time. Human use of the Wadden Sea tidal areas has increased in recent years and the disturbance distances have become progressively shorter, although they are still larger than those in the Eastern Schelde (Figure 6.18B).

Disturbance distances, like time and energy loss, are difficult to relate to survival rates. Do shorter disturbance

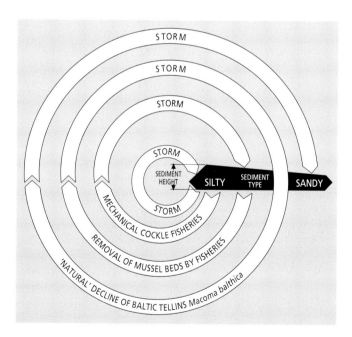

Figure 6.17 Negative bio-deposition spiral model. According this model, silty tidal flats can lose equilibrium and become coarser due to mechanical ploughing of the sediment, the disappearance of protective mussel beds, and the disappearance of silt-producing animals. One substantial push by a cockle harvester can cause a chain reaction: the sediment suffers an initial increase in coarse particles, leading to decreases in shellfish numbers, which can therefore no longer produce or retain silt. As a consequence, fewer shellfish can settle in the area and the cycle goes on.

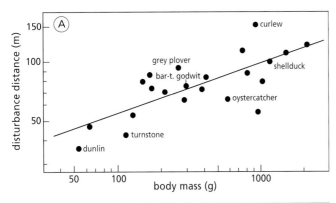
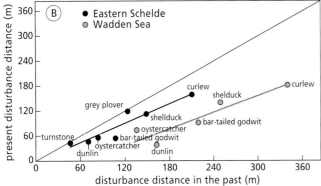

Figure 6.18. Disturbance distances of various waterbirds in The Netherlands. (A) Disturbance distance in the Wadden Sea (as measured in 1995) in relation to body mass. (B) Past and present disturbance distances in the Wadden Sea. Present data are from 1995 for both sites; past data are from 1982 for the Wadden Sea, and 1984 for the Eastern Schelde.[799]

distances mean that the birds are in a worse condition and prepared to run a higher risk of predation? (Perhaps frequently visited areas like the Mokbaai are populated with the misfits of the bird world). Or do shorter disturbance distances mean that the birds have somehow realised that people don't really pose a high predation risk? Without long-term research on individually marked animals, this question cannot be answered. What we can is draw a circle around each source of disturbance on the basis of the measured disturbance distance. This circle represents the minimum disturbance area. The birds must concentrate in the undisturbed areas and when this leads to increased interference lower intake rates may result. So at least we can determine the reduction in tidal areas available for the birds as a result of disturbance.

More information on the effects of disturbance has been gathered through monitoring the effects of disturbances created near individually marked animals in the field,[892] and experimentally varying the foraging time available to captive birds.[841] These foraging experiments were carried out on captive oystercatchers and the birds did forage more quickly when the available time decreased. By coincidence the available time was so short that the birds reached their digestive limits (Figure 4.7 and Section 4.3.4).[462] Faced with a digestive bottleneck, the birds were no longer able to maintain their energy balance. An opportunity for a rather unique field experiment arose when the movable flood barrier in the Eastern Schelde was temporarily closed over several tides because of the very high water levels caused by a fierce storm.[562] As a consequence, the oystercatchers were unable to forage over a number of low tides. It was predicted that when the tidal flats became available again, the birds would increase their food intake rates to compensate for their accumulated energy shortfalls. This did not happen. So there are situations in which lost energy stores cannot just be augmented. However the true costs of having low energy stores are only paid in severe winters.

The work that has provided the greatest insights into the effects of disturbance has been on pink-footed geese (*Anser brachyrhynchus*) in Northern Norway, during their last stopover on migration to the northern breeding grounds.[532, 1011] Some farmers don't like the geese and continually chase them off their land. These chased geese are less successful at building up the fuel stores necessary for their final flight north. This has negative impacts on their breeding success: individually recognisable geese that had often been disturbed and had increased in mass less quickly, returned from the breeding season with fewer juveniles than the undisturbed geese.

An area subject to continual disturbances is permanently lost to the birds. The food supplies in that area are of no use. It's been found that more beet foliage remains uneaten by foraging pink-footed geese closer to the source of disturbance, in this case a road.[298] This is only a problem for the pink-footed goose population if competition in the winter is a limiting factor and this competition is caused by depleting food sources. In principle, when an area of tidal flat is not continually visited by tourists or hunters, the birds have the option of consuming the food there during quiet periods. Unfortunately food seldom sits and waits to be consumed. Brent geese excluded by hunters from exploiting a large area of seagrass in Denmark immediately visited that seagrass when the hunting season finished in late autumn. But it seemed that they weren't able to get as much food there as they had from a comparable area that had been closed to hunting for the entire season. Storms and other causes had led to a decrease in quality for some of the seagrass while other parts had disappeared.[531]

### 6.8.7 Disturbance on the breeding grounds

The breeding season is a time of plenty when birds can forage for themselves and invest time and energy in raising their young. During the breeding season disturbance is not as much a problem for the adult birds as it is for the eggs and young. Chapter 5 maintains that human visits to nests do not necessarily attract predators. But that does not exclude the possibility that a human presence makes it impossible for the birds to incubate their eggs or care for their young. Regular visits by people can reduce the likelihood that eggs will hatch and young will survive.

Surely sites that are intensively visited by humans would be less attractive to breeding waterbirds than sites that are rarely visited. To investigate this a well planned study was carried out on Vlieland, an island in the Dutch Wadden Sea.[742] A number of areas were alternately open or closed to the public during different breeding seasons. It appeared that densities of breeding birds, like the oystercatcher, were higher in the years when the paths were closed to the public. The cause of the changing bird densities in this study was not known. Did potential breeding birds avoid the areas visited by people, or did they decide to move after a failed breeding season? A study on a population of individually marked ringed plovers on a beach in England provided the answers.[513] Human visitation to this beach varied strongly from place to place. Each year, the same places were visited intensively by people. The ringed plovers were already established in their territories before the tourists arrived on the beach en masse. It was usually the young inexperienced ringed plovers that settled in more disturbed areas. The sudden increase in human visitors sometimes caused them to abandon their territories before they had even laid their eggs. In the following year, these birds often tried to obtain territories in an undisturbed area. It seems that the ringed plovers must learn by experience which potential breeding areas are often disturbed and best avoided.

It is clear that a frequent human presence on salt marshes and dunes make these sites unsuitable as breeding areas for waders. Humans and breeding waders do not mix well, so a choice must be made. In dunes and salt marshes the choice often favours the birds. But birds that need sandy beaches, like the Kentish plover, are less fortunate. It seems there is little that relaxes people more than lying practically or completely naked in the sun on the beach. Sandy beaches are increasingly becoming the domain of human holidaymakers and less and less remains for the Kentish plovers.

## 6.8.8 Grazing in salt marshes

Salt marshes grazed by cattle look very different from salt marshes without cattle.[31] Ungrazed European salt marshes quickly develop a monoculture of quackgrass (*Elymus repens*) on the higher parts of the marsh. Small wild grazers such as caterpillars, hares and geese are unable to stop this process, which is unfortunate for them as it is detrimental in the long run. Brent geese prefer to graze on juicy salt-tolerant plants that only grow low on the salt marsh in the absence of cattle, such as sea plantain (*Plantago maritima*), seaside arrow-grass (*Triglochin maritima*) and common saltmarsh-grass (*Puccinellia maritima*).[710] Cattle grazing removes the quackgrass from the upper salt marsh, allowing the salt-tolerant plants to re-establish themselves further up, and making the upper salt marsh attractive to the geese once more.[618]

Waders that breed on salt marshes grazed by cattle run the risk of being trampled. But at the same time, the grazing results in more varied vegetation improving the quality of these salt marshes as breeding habitat. We know this because of an extensive British study on grazing management.[611] The highest densities of breeding redshanks were found not on the ungrazed salt marshes, but on salt marshes with low grazing pressure. Intense grazing pressure, especially by sheep, is unfavourable. Under intense grazing a kind of 'billiard table' effect develops with too few tall tussocks for the redshanks to hide their nests amongst. Intensive grazing pressure also increases the risk of trampling. From 1985 - 1996 the number of breeding redshanks on salt marshes in Great Britain decreased by 23%. This is assumed to result from an increase in sheep grazing on the salt marshes.[611]

## 6.8.9 Climate change: melting ice caps and rising sea levels

The waterbird species we see today have already survived various ice ages and interglacial periods. At the start of an ice age the vegetated zone moves south and sea levels fall. During an interglacial period the vegetated zone moves north again and sea levels rise again, sometimes by more than 100 m. The locations and surface areas of the different tidal areas must have changed a lot over these times, as must those of the birds' breeding grounds.

If waterbirds have survived all these changes, is there any need for us to worry about the effects of global warming and the associated sea level rise on waterbirds?[229] Of course what happens to the waterbirds is unlikely to be our main concern when climate change occur. It's clear that humans are able to cause climate change, as other organisms have done so before us. The present oxygen-rich atmosphere and its associated climate makes life possible, but only exist by grace of that life.[525] And the question of whether humans are changing the climate is the subject of less and less discussion. Nowadays scientific debate mainly centres on how quickly changes will occur and what exactly will happen. For example, it's possible that the warm Gulf Stream in the East Atlantic Ocean will stop flowing. At least this would make some Dutch skaters happy: each winter they would be able to ice skate outside! Most likely politicians won't come up with a solution until the climate has already changed dramatically, assuming that there would still be organised politics in the chaos. For example, a changed climate would mean that completely different crops need to be grown. And a higher sea level would cause huge problems with drainage in land below sea level.

Waterbirds (and other animal species) will be confronted with two problems that they didn't face in earlier periods of climate change. First, the changes may happen more quickly than ever before, giving the animals less time to adapt. Secondly artificial constraints, such as the construction of seawalls, have restricted the freedom and space of natural areas to follow large 'natural' changes on the same large scale. An important part of the earth's surface is now being cultivated by humans and nature is, especially in the temperate regions, mostly limited to a few humble reserves. For many species, climate change will cause the habitat in the reserves to deteriorate. It won't be difficult for plants and animals to die out in their reserves. The challenge will be to find a reserve where, as a result of the climate change, the conditions have become more favourable. Thanks to their wings, most birds belong to the group of animals that have a good chance of finding new areas. And as we have discussed at length in this book, most waterbirds belong to the group of birds that can fly spectacularly well and are capable of making extremely long flights. But even so, in the event of global warming waterbirds could get into trouble. First, as temperatures rise much of the High Arctic tundra will disappear;[516] when the North Pole melts it will not expose any habitable land, just water. Secondly, tidal areas can no longer move inland as they would have in the past. At many sites extensive dykes have been built. There is no doubt among the experts that if the sea level rises at a rate of 1 m per century the tidal flats in the Wadden Sea will be permanently inundated.[619] However, as the tidal difference is about 2 m and rising sea levels will cause sedimentation to increase, it might take several centuries before the last of the exposed tidal flats disappear.

### 6.8.10 Subsidence and other volume changes in the Wadden Sea

With the risk of the Wadden Sea sinking due to an accelerated sea level rise, it seems a shame to assist that process by letting the seafloor subside or extracting sediment. The Dutch gas and oil drilling company's (NAM) plans to extract gas from the Wadden Sea have met with much resistance from environmentalists. As a result of this criticism, NAM carried out a study on the effects of subsidence and gas extraction on the geomorphology and wildlife of the Wadden Sea.[619]

Predicting such effects is very difficult, but not impossible. First, the expected amount of seafloor subsidence due to gas extraction from 'new' gas fields must be estimated. Gas extraction already occurs in the Dutch Wadden Sea, on the island of Ameland and near Griend. These gas fields are exploited from the land but extend beneath the Wadden Sea in some areas. Extracting the gas causes the seafloor to drop, which in turn increases the volume of the Wadden Sea itself. The continued existence of the Wadden Sea in its present form would require old seafloor levels to be reinstated, and one way to do this is by infilling with sediment. So this volume increase can be viewed as sediment shortage.

It was predicted that by 2000 existing gas drilling would have resulted in a sediment requirement of 27 000 000 $m^3$ (Table 6.1). Once the subsidence stopped, sometime between 2000 and 2050, another 32 000 000 $m^3$ of sediment was to be added. If new fields were exploited, NAM predicted that an additional 12 000 000 $m^3$ of sediment would be required. Many sedimentologists expect that natural sedimentation processes will fully compensate for the sediment shortages caused by gas extraction.[619] More than that, they predict that even with the present rates of sea level rise (20 cm per century), existing and proposed gas extraction and other human activities will not result in any major problems for the system apart from increasing erosion along the coasts of the Wadden Sea islands (the sediment must come from somewhere). However the intertidal area won't return to its old state of equilibrium overnight, this will take quite some time. For example, when the Lauwerszee in the north of The Netherlands was closed off, the channels that had transported water into and out of the Lauwerszee were suddenly much too large. It has taken decades for these channels to fill up and reach a new smaller equilibrium level.

What will happen to the waterbirds that forage on the exposed flats while the system slowly recovers from the subsidence caused by the 'new' gas extraction? Subsidence will cause some tidal flats to disappear completely, while others will have a shorter exposure period. Of all the species, we only have enough data to make tentative predictions about the effects of subsidence on the carrying capacity of the oystercatcher.[94] The effects appear to be minimal. We can also use habitat correlations to predict what might happen to the other species of which less is known. We know the foraging density on different types of tidal flat for a large number of waterbird species from counts made in small plots at low tide. For bird abundance, the two most important characteristics are the tidal flat's silt content and exposure time. A bell-shaped curve can be produced for each waterbird species, showing the density of foraging birds for various combinations of silt content and exposure time. Using maps that show the height of the flats and their sediment composition it is possible to predict how many birds of each species would occur in the Wadden Sea as a whole. And for a satisfactory number of species these predictions are consistent with the actual numbers of birds counted in the Wadden Sea.[569] These accurate predictions were set as 100% and the predicted changes in abundance due to rising water levels (as a result of subsidence or sea level rise due to other causes) were then calculated.[94] As expected, the numbers of all species would decline, apart from the greenshank (Figure 6.19). According to the data, greenshanks prefer relatively sandy tidal flats that are exposed for only a short time. And initially this type of flat will increase. But an extra metre of water would have dire consequences for waterbird populations. Such an increase in water levels could only occur with an extreme acceleration in the rates of

| Process or activity | Period | Million $m^3$/year | Million $m^3$ |
|---|---|---|---|
| Sea level rise (20 cm per century) | 2000 - 2050 | 4 | 200 |
| Existing gas extraction | 1960 - 2000 | 0.7 | 27 |
| Existing gas extraction | 2000 - 2050 | 0.6 | 32 |
| Proposed gas extraction |  | 0.2 | 12 |
| Shell extraction | 1997 | 0.14 |  |
| Sand extraction | 1997 | 0.6 |  |
| Lauwerszee closure | 1970 - 1990 | 2 | 36 |
| Loss of mussel beds | 1960 - 1980 | 0.5 - 1 | 10 - 20 |

Table 6.1 The amount of sediment required to restore the geomorphological equilibrium of the Dutch Wadden Sea in response to different factors that increase its volume. For each factor the annual sediment requirement and the total requirement over the estimated period of impact are given. All gas extractions refer to the activities of the Dutch gas drilling company NAM.[535] Elf-Petroland's extractions have not been included. For proposed gas fields the most likely scenario is given.

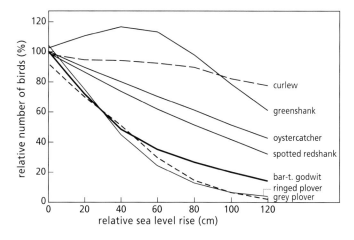

Figure 6.19 Effect of rising sea levels on waterbirds in the Dutch Wadden Sea.[233] Habitat correlations were used to predict the numbers of birds that would occur in late summer. Where the prediction matched the number counted, the predicted number (y-axis) was set to 100%. Calculations were then made of how numbers of each species would change with rising sea levels.

sea level rise. Subsidence would only cause water levels to change by centimetres. According to NAM's predictions the maximum possible subsidence that could occur in the centre of the gas field would be 15 cm. And 8 cm of subsidence in the centre is thought to be much more realistic. Natural sedimentation processes would partly compensate for this 8 cm of subsidence, so the seafloor would only be 5 cm lower. But even this 5 cm is expected to cause a decline in the abundance of most bird species. However it seems very unlikely that this decline in abundance would be detectable. The numbers of waterbirds counted throughout the Wadden Sea fluctuate by more than 10% annually.[94, 569] Species are only predicted to decline by a few percent as a result of subsidence and that decline is expected to be temporary.

Moreover, it is important to realise that these predictions were based on the assumption that the decrease in the bird abundance would be linearly related to the loss of habitat. That appears a good starting point. But although a loss of wintering habitat leads almost without exception to a decrease in abundance in the different models that exist, this decrease is not linearly related to the loss of habitat.[345] For the oystercatcher, this could lead to an overestimate of the predicted decrease.

Making these predictions requires many different steps and no one researcher has sufficient knowledge predict exactly what would happen at every turn. But it does seem that gas extraction from unexploited beds in the Wadden Sea would be unlikely to have serious consequences for tidal flat birds. At least, as long as these consequences are mainly due to subsidence. And that of course is questionable. In any case, gas extraction can only be started after exploratory drillings have occurred, which would require the use of drilling rigs. The rigs would stick out like sore thumbs in a wide-open landscape like the Wadden Sea. By the way, you could be forgiven for thinking that drilling rigs, like jet fighters, do not belong in nature reserves.

## 6.9 From species protection to habitat protection

For many years coastal birds in Western Europe were treated like herrings and mussels, natural resources to be harvested at will. When the harvest was regulated this was simply to safeguard the interests of the rightful claimants. Nobody was concerned about the animals' best interests. This began to change early in the 20th century. In The Netherlands, the Dutch Society for the Protection of Birds (now Birdlife Netherlands) was established in 1899 in reaction to the widespread slaughter of terns for the millinery trade. Feathers, or even whole stuffed birds, were seen as fashionable hat accessories! Now all West European countries protect their birds or the birds' habitats, but in different ways.

The protection of the Wadden Sea is an international affair. Since 1978 Denmark, Germany and The Netherlands have made joint agreements on scientific research, fisheries, recreation, water quality and management. However each country has retained its own rules on some contentious issues. In Denmark eider ducks may still be shot during winter, while the Dutch Government has prohibited this hunting but condemns the same eider ducks to death by starvation by permitting commercial shellfisheries to remove their food.

In future the protection of birds and natural areas will be increasingly considered in a European context, rather than from just one country's perspective. The European Union's (EU) Wild Birds Directive and Habitats and Species Directive focus on the protection of habitats and birds. These directives will play important roles in the future of the Wadden Sea and its wildlife.

## 6.10 The convention circus

### 6.10.1 Some birdwatchers and the origins of conventions

Of all animals, birds really capture our imagination. So it's not surprising that birds were the first group of animals to receive legal protection. Their worldwide migratory flights led to the first international cooperation between bird protection agencies. In 1922 the International Committee (or Council since 1950) for Bird Preservation was founded. Then, in 1948, the collaboration of various non-governmental environmental organisations led to the foundation of the International Union for Protection of Nature, renamed the International Union for Conservation of Nature (IUCN) in 1956. The IUCN remains an active force in international conservation.

In 1963 a group of waterbird enthusiasts founded the International Waterfowl Research Bureau (IWRB) as a non-profit organisation. 1971 was probably the right time politically for an international agreement on waterbird and wetland protection, and during an IWRB meeting in the town of Ramsar in Iran a wetland convention was agreed on. This Convention on Wetlands of International Importance Especially as Waterfowl Habitat later became known as the Ramsar Convention. A very enthusiastic bird watcher, Mike Smart, was involved in this convention from the start. Although the realisation, growth and acceptance of the Ramsar Convention was due to the work of many people, all agree that that the dedication, linguistic abilities and the political competence of Mike Smart made the convention what it is today: the most widely known and perhaps the most respected international nature conservation treaty.

The Ramsar Convention was followed a year later by a UNESCO convention designed to protect the world's cultural and natural heritage, known as the World Heritage Convention (Table 6.2). Only a few waterbird sites were given the recognition they deserved (such as the Florida Everglades, Lake Ichkeul in Tunisia and the Djoudj area downstream of the Senegal River). According to Mike Smart other Rolls-Royces from the ranks of the world's wetlands, such as the International Wadden Sea and the Banc d'Arguin, also deserve to be considered World Heritage sites![784] To date this hasn't happened.

In 1973, the Convention on International Trade in Endangered Species of Wild Fauna and Flora, better known as CITES, was initiated in Washington, USA. As there is no significant trade in threatened waterbird species, this very important and successful convention is not so relevant to waterbirds. The 1979 Bonn Convention, the Convention on the Conservation of Migratory Species of Wild Animals, however was very relevant. Almost 20 years later, under the auspices of the Bonn Convention, Gerard Boere succeeded in creating the African-Eurasian Waterbird Agreement. Countries that signed this agreement were obligated to undertake more species-focussed management than under the site-led Ramsar Convention.[85] Finally, in 1992 the Rio Convention resulted from a conference in which world leaders collectively aimed to preserve the environment, climate and biodiversity (Table 6.2).

Throughout these developments, as the treaties became more binding, the input of field biologists during meetings decreased and that of lawyers and politicians expanded. Currently, 115 countries have signed the Wetlands Convention, but at recent Ramsar meetings biologists have been very thin on the ground. Although there were still plenty of biologists at the 1983 Ramsar meeting in Groningen in The Netherlands! In every case, as these treaties have been taken more seriously the focus has drifted away the reason it all began, the birds. In the 1980s the name of the successful and influential IWRB was changed to the International Waterfowl and Wetland Research Bureau. In the Americas, the success of the Western Hemisphere Shorebird Reserve Network led to the foundation of Wetlands for the Americas (WA), and in Southeast Asia a working group called Interwader became the Asian Wetland Bureau (AWB). Since 1996, the IWRB, WA and AWB have worked together under the name Wetlands International. Although you may notice

| Area | Name of treaty or convention | From |
|---|---|---|
| Worldwide | Ramsar Convention | 1971 |
| | World Heritage Convention | 1972 |
| | CITES | 1973 |
| | Bonn Convention | 1979 |
| | Convention on Biological Diversity (Rio Convention) | 1992 |
| Europe/Africa/West Asia | African Convention | 1968 |
| | Berne Convention | 1979 |
| | EU Wild Birds Directive | 1979 |
| | EU Habitats and Species Directive | 1992 |
| | African-Eurasian Waterbird Agreement (AEWA) | 1995 |
| East Asia/Australasia | Asia-Pacific Migratory Waterbird Conservation Strategy | 1996 |
| | East Asia-Australasia Shorebird Reserve Network | 1996 |
| | Various bilateral treaties between the USA, Japan, China, Australia, India and Russia, for example the Japan-Australia Migratory Birds Agreement (JAMBA) | |
| North and South America | Protection of Migratory Birds Convention | 1916 |
| | Protection of Migratory Birds & Game Mammals Convention | 1936 |
| | Western Hemisphere Convention | 1940 |
| | US-Japan Migratory Birds Convention | 1976 |
| | US-USSR Migratory Birds Convention | 1976 |
| | Western Hemisphere Shorebird Reserve Network (WHSRN) | 1985 |
| | North American Waterfowl Management Plan (NAWMP) | 1986 |
| | United States Shorebird Conservation Plan | 2000 |
| | Canadian Shorebird Conservation Plan | 2000 |

Table 6.2 International treaties, conventions and key action plans dealing with the protection of waterbirds and their habitats around the world.[159]

a waterbird in this organisation's emblem, waterbirds now only play a ceremonial role. Wetlands International and Birdlife International are jointly based in an office block in the city of Wageningen in The Netherlands. Administration of the Ramsar Convention has moved from the relatively modest accommodation of the IWRB at the Wildfowl Trust (later called the Wildfowl and Wetlands Trust) in the small village of Slimbridge in England to a new, official Ramsar office in Switzerland. The administrative headquarters of the Bonn Convention are housed in, yes you guessed it, Bonn, in very elegant buildings located near the Rhine.

As well as the international conventions, many bilateral treaties (Table 6.2) and regional treaties have sprung up, such as the EU's Wild Birds Directive and Habitat and Species Directive. These two directives are particularly important, as they are more binding. If the directives are not followed a national government can be prosecuted. Following a conviction, the EU can implement very firm sanctions (although legal proceeding may take many years before this point is reached). For example, in 1993 Birdlife Netherlands notified the EU that the Dutch government had not obeyed the Wild Birds Directive and the Habitat and Species Directive: the government had allowed the commercial shellfisheries too much space in the Dutch Wadden Sea. In March 1998 the EU took this to the European Court of Justice. In April 1999 a report was produced on The Netherlands' performance in terms of the international legislation. The report concluded that although various plans had been produced, their implementation was unsatisfactory. In short, the report stated that The Netherlands needed to implement the Ramsar Convention, the Wild Birds Directive and the Habitat and Species Directive earlier, more systematically and with greater care.

### 6.10.2 Effectiveness of the conventions

So what have all these good intentions, management advice, and environmental designations actually meant for the tidal flats and the waterbirds? International recognition! The sustainable protection of the Dutch Wadden Sea receives a lot of attention in every management document produced by the Dutch government (whether it concerns permission for gas drilling, harvesting sand or shells or mechanical cockle dredging). There are frequent international meetings about the Wadden Sea and the state of its wetlands and waterbirds. One important product of all this activity is an overview of the current status of these waterbirds.

Estimates of population size are only available for about two thirds of the 340 known populations of plovers and sandpipers (Charadriidae and Scolopacidae).[744] And for half of these populations it is known whether they are increasing, stable or decreasing. In half of these cases the populations are decreasing! Between 1980 and 1997 there seem to have been considerable declines in the numbers of various waders that winter in West Africa.[996] There are no indications of far-reaching ecological changes on the wintering grounds in Mauritania and Guinea-Bissau or on the Arctic breeding grounds, so it seems we must search for the cause on the West European stopover sites.[663] The number of red knots on the West African wintering grounds has halved since the 1980s. During their autumn migration, these red knots are completely dependent on small shellfish usually found in the Wadden Sea.[686, 986] This knot population may have fallen victim to the decreasing shellfish stocks in the Dutch Wadden Sea.

In spite of all the treaties and conventions, waterbird protection often comes second when commercial interests are at stake, as was recently demonstrated in Delaware Bay in the United States. Adult horseshoe crabs, particularly the fertile females that come to the beach to bury their billions of eggs in the sand, make attractive bait for eel fishers who are prepared to pay up to about US$1 per crab. For the last 15 years, large numbers of these crabs have been collected. In 1997 it was noticed that the numbers of horseshoe crabs were undergoing a dramatic decline, as were the waders that feed on the horseshoe crab eggs in May. Harvesting regulations have now been put in place, but it is debatable whether these will be enough to halt the declines. The red knot populations that pass through Delaware Bay on migration from the wintering grounds in South America to the breeding grounds in Northern Canada are completely dependent on the horseshoe crab eggs. In the last 20 years, the abundance of this red knot population has halved. Like the Dutch Wadden Sea, Delaware Bay is a very well protected area on paper.

## 6.11 Scientific research and monitoring

### 6.11.1 What is needed most?

The fate and protection of waterbirds that migrate across vast distances, and that of all specialised organisms that live in vulnerable habitats, depend on the key attributes of these organisms. When we succeed in transforming knowledge about these attributes into effective protection measures, these vulnerable systems have a chance. The Swedish biologist Staffan Ulfstrand points out that the evolutionary pressure on organisms due to environmental changes is now many times greater than it has been in the past.[884] If a species cannot adapt to the negative changes caused by an exploding human population and the ever increasing ecological footprint of each person on the planet, the species has little future. Unless of course, we are aware of its vulnerable characteristics and are able and willing to take these into account. If we want to protect a bird species and its habitats, we must collect critical information. There is a lot of work to do.

According to Dugan,[198] research into the ecology of some species, particularly in Third World countries, often misses the point. Although it is important to identify the north - south connections through research on migratory flyways (and with this the willingness of the more affluent north to support the efforts of the less affluent south), it would be even more useful to investigate how habitat protection in Third World countries can benefit the local people. Only the development of sustainable fisheries, grazing and regulated hunting in wetlands will guarantee the continued existence of many species in heavily populated areas.

### 6.11.2 The importance of in-depth independent research

For a long time, scientists were the subject of much criticism for removing themselves from society and retreating to their ivory towers to try and extend the limits of human knowledge. Through the 1980s and 1990s it became increasingly fashionable for universities to undertake research at the request of public interest groups or industry. At this time, privatisation was often seen as the cure for many woes. This belief led to an increasing number of research institutes being forced to earn their own money. As a result, some of these institutions began 'selling themselves' with the promise that the 'customer is always right'. Financial dependence on a commissioning body with very specific interests carries significant risks for the independence and quality of the research.[476] One does not have to be a rocket scientist to realise that the gas drilling company has no interest in funding research that shows that sediment subsidence or gas drilling are harmful. And the commercial shellfisheries have nothing to gain from research that shows shellfish harvesting damages natural systems. Or, as the chairman of the Federation of Dutch Fisheries, B. Daalder, said "As fishermen we are prepared to co-finance research. But we want to have a say in the goal and the design of the research and co-think about what has to happen with the research results". Environmental organisations have a particular interest in research that demonstrates harmful effects. No-one cannot blame either of these groups for pursuing their own interests. However you can blame governments for not being aware of the dangers of privatising research. You may even wonder how independent the research institutions are that directly work for the government.

Protection of waterbirds and their habitats requires more than the vigilance of environmental organisations who, armed with the fine words of their and other governments, must fight the many commercial interests threatening natural areas. Protection also requires independent research. It needs researchers driven by curiosity and the desire to find out how the tidal systems work. These people don't just maintain an alert presence in tidal areas and provide a stream of stories that keep waterbirds and their habitat in the public eye. They are also necessary when conflicts of interest threaten honest policy evaluations. Scientists should be seen as the 'ivory lighthouse keepers'. They form the front guard in the fight to save and protect the tidal flats.

### 6.11.3 Looking to the future: waterbirds in trouble!

In recent years, numbers of many waterbird species have shown an increase, by as much as fourteen-fold in the case of the brent geese. Such seemingly positive statements can be made for few forest or wetland birds. It is much easier to drain wetlands, confine rivers, and cut down forests, than to reclaim intertidal areas or make Arctic tundras unsuitable as breeding grounds. But if we're being realistic we need to prepare ourselves for the worst that might happen to the waterbirds and their habitat in the future. The climate is changing and the ever-increasing number of people limits the potential for natural habitats to adapt to these changes. Despite the large progress that has been made in international treaties for the protection of waterbirds and their ecosystems, conservation, like charity, must begin at home. When even affluent countries such as The Netherlands and the United States do not succeed in undertaking thorough research into the consequences of intensive but financially rewarding exploitation of tidal areas (shellfish dredging and horseshoe crab harvest respectively) prior to and during the exploitation, then what should we expect of less affluent countries? Tidal areas in developing countries are next in the queue for transforming natural resources into cold hard cash.

The good news is that we will be able to monitor the fluctuations in waterbird populations with increasing accuracy. The density of enthusiastic bird counters is still increasing, which means that we will be better informed about numbers. It is still relatively easy and cheap to travel with the migratory birds (or at least it is for interested rich Westerners), which can greatly increase our knowledge of waterbird migration. And our technical abilities are still increasing; in recent years it became possible to map the migration of the eastern curlew (*Numenius madagascariensis*) between Australia and China and Japan using tiny satellite transmitters. In 1999 these transmitters weighed 20 g and were only suitable for the very large waders such as the curlew. But there is no doubt that such transmitters will get smaller in the future, and eventually we will be able to tell exactly when a certain bar-tailed godwit has left the Wadden Sea and where and when he or she has arrived in the breeding grounds in Taimyr. Or we will be able to follow the flight of a red knot between the Wadden Sea and West Africa from behind our computer screens through an Internet site.

As waterbird researchers we will do our best to come with fascinating new information about waterbirds and tidal areas. The application of new techniques such as satellite transmitters and the Internet will hopefully help to keep the attention of the public. We can only hope that through this the tidal flats and waterbirds will be valued enough by the wider community, that they will be happy to make the small public sacrifices that are sometimes necessary for the birds' wellbeing. Good research must do more than underpin sound management, it has to ensure that the waterbirds continue to capture our imagination and touch our hearts.

# References

1. Able, K.P. & M.A. Able. 1996. The flexible migratory orientation system of the Savannah Sparrow (*Passerculus sandwichensis*). J. exp. Biol. 199: 3-8.
2. Åkesson, S. 1990. Animal orientation in relation to the geomagnetic field. Introductory Paper 59, Department of Ecology, Lund University, Lund.
3. Alerstam, T. 1981. The course and timing of bird migration. In: D.J. Aidley (ed.). Animal migration: 9-54. Cambridge University Press, Cambridge.
4. Alerstam, T. 1985. Strategies of migratory flight as illustrated by Arctic and Common Terns, *Sterna paradisaea* and *Sterna hirundo*. Contr. Mar. Sci. 27, Suppl.: 580-603.
5. Alerstam, T. 1990a. Bird migration. Cambridge University Press, Cambridge.
6. Alerstam, T. 1990b. Ecological causes and consequences of bird orientation. Experientia 46: 405-415.
7. Alerstam, T. 1996. The geographical scale factor in orientation of migrating birds. J. exp. Biol. 199: 9-19.
8. Alerstam, T. & G. Högstedt. 1980. Spring predictability and leap-frog migration. Ornis Scand. 11: 196-200.
9. Alerstam, T. & G. Högstedt. 1982. Bird migration and reproduction in relation to habitats for survival and breeding. Ornis Scand. 13: 25-37.
10. Alerstam, T. & G. Högstedt. 1985. Leap-frog arguments: reply to Pienkowski, Evans and Townshend. Ornis Scand. 16: 71-74.
11. Alerstam, T. & Å. Lindström. 1990. Optimal bird migration: the relative importance of time, energy, and safety. In: E. Gwinner (ed.). Bird migration. Physiology and ecophysiology: 331-351. Springer-Verlag, Berlin.
12. Alerstam, T. & S.-G. Petterson. 1991. Orientation along great circles by migrating birds using a sun compass. J. theor. Biol. 152: 191-202.
13. Alerstam, T., C.-A. Bauer & G. Roos. 1974. Spring migration of Eiders *Somateria mollissima* in southern Scandinavia. Ibis 116: 194-210.
14. Alerstam, T., G.A. Gudmundsson, P.E. Jönsson, J. Karlsson & Å. Lindström. 1990. Orientation, migration routes and flight behaviour of Knots, Turnstones and Brent Geese departing from Iceland in spring. Arctic 43: 201-214.
15. Atkinson-Willes, G.L., D.A. Scott & A.J. Prater. 1982. Criteria for selecting wetlands of international importance. Ric. Biol. Selvaggina 8, Suppl.: 1017-1042.
16. Altenburg, W. & J. van der Kamp. 1991. Ornithological importance of coastal wetlands in Guinea. ICBP study report 47 / WIWO Report 35, Cambridge / Zeist.
17. Altenburg, W., M. Engelmoer, R. Mes & T. Piersma. 1982. Wintering waders on the Banc d'Arguin, Mauritania. Stichting Veth tot steun aan Waddenonderzoek, Leiden.
18. Alvarez Laó, C.M. 1995. La Ria de Avilez: un enclave de interes para las Limícolas. Airo 6: 24-28.
19. Anderson, M. 1978. Optimal egg shape in waders. Ornis Fenn. 55: 105-109.
20. Andersson, M.B. 1994. Sexual selection. Princeton University Press, Princeton.
21. Andrews, J. 1974. Death on the sands. Birds 15: 32-33.
22. Åstrand, P.-O. & K. Rodahl. 1986. Textbook of work physiology. Physiological bases of exercise. Third edition. MacGraw-Hill, New York.
23. Avise, J.C. 1994. Molecular markers, natural history and evolution. Chapman & Hall, New York.
24. Badgerow, J.P. & F.R. Hainsworth. 1981. Energy savings through formation flight? A re-examination of the Vee formation. J. theor. Biol. 93: 41-52.
25. Baird, R.H. 1966. Factors affecting the growth and condition of Mussels (*Mytilus edulis* L.). Fish. Invest. Lond. (Ser. II) 25: 1-33.
26. Baker, A.J. 1992. Molecular genetics of *Calidris*, with special reference to Knots. Wader Study Group Bull. 64, Suppl.: 29-35.
27. Baker, A.J. & H.D. Marshall. 1997. Mitochondrial control region sequences as tools for understanding evolution. In: D.P. Mindell (ed.). Avian molecular evolution and systematics: 51-82. Academic Press, San Diego.
28. Baker, A.J., T. Piersma & L. Rosenmeier. 1994. Unraveling the intraspecific phylogeography of Knots *Calidris canutus*: progress report on the search for genetic markers. J. Ornithol. 135: 599-608.
29. Baker, R.R. 1984. Bird navigation: the solution of a mystery? Hodder & Stoughton, London.
30. Bakker, J.F. & V.N. de Jonge. 1998. Hoe veilig is de Eems-Dollard? Ontwikkelingen in enkele belangrijke verontreinigde stoffen. In: K. Essink & P. Esselink (eds.) Het Eems-Dollard estuarium: interacties tussen menselijke beïnvloeding en natuurlijke dynamiek: 47-60. Report RIKZ-98.020, Haren.
31. Bakker, J.P. 1989. Nature management by cutting and grazing. Junk, Dordrecht.
32. Barnard, C.J. & H. Stephens. 1981. Prey size selection by lapwings in Lapwing/gull associations. Behaviour 77: 1-22.
33. Battley, P.F. 1997. The northward migration of arctic waders in New Zealand: departure behaviour, timing and possible migration routes of Red Knots and Bar-tailed Godwits from Farewell Spit, north-west Nelson. Emu 97: 108-120.
34. Battley, P.F. & T. Piersma. 1997. Body composition of Lesser Knots (*Calidris canutus rogersi*) preparing for take-off on migration from northern New Zealand. Notornis 44: 137-150.
35. Becker, P.H., S. Thyen, S. Mickstein, U. Sommer & K. R. Schmieder. 1998. Monitoring pollutants in coastal bird eggs in the Wadden Sea: Final report of the pilot study. 1996-1997. Wadden Sea Ecosystem 8: 59-101.
36. Beintema, A.J. 1991. Fledging success of meadow bird (Charadriiformes) chicks, estimated from ringing data. In: A.J. Beintema. Breeding ecology of meadow birds (Charadriiformes); implications for conservation and management: 115-127. PhD thesis, University of Groningen, Groningen.
37. Beintema, A. 1992. Mayfield moet: oefeningen in het berekenen van uitkomstsucces. Limosa 65: 155-162.
38. Beintema, A.J. & G.J.D.M. Müskens. 1987. Nesting success of birds breeding in Dutch agricultural grasslands. J. appl. Ecol. 24: 743-758.
39. Beintema, A.J. & G.H. Visser. 1989a. Growth parameters in chicks of charadriiform birds. Ardea 77: 169-180.
40. Beintema, A.J. & G.H. Visser. 1989b. The effect of weather on the time budgets and development of chicks of meadow birds. Ardea 77: 181-192.
41. Beintema, A.J., O. Moedt & D. Ellinger. 1995. Ecologische atlas van de Nederlandse weidevogels. Schuyt & Co, Haarlem.
42. Bélanger, L. & J. Bédard. 1989. Responses of staging Greater Snow Geese to human disturbance. J. Wildlife Management 53: 713-719.
43. Bennett, A.T.D. 1996. Do animals have cognitive maps? J. exp. Biol. 199: 219-224.
44. Berg, Å. 1993. Habitat selection by monogamous and polygamous Lapwings on farmland: the importance of foraging habitats and suitable nest sites. Ardea 81: 99-105.
45. van den Berg, R. 1998. Zelfzuchtig bij elkaar. Waarom in een vogelzwerm de cohesie gehandhaafd blijft. NRC-Handelsblad, 12 december. 1998, Wetenschap & Onderwijs, p. 5.
46. Berthold, P. (ed.). 1991. Orientation in birds. Birkhäuser Verlag, Basel.
47. Berthold, P. 1996. Control of bird migration. Chapman & Hall, London.
48. Berthold, P., A.J. Helbig, G. Mohr & U. Querner. 1992. Rapid microevolution of migratory behaviour in a wild bird species. Nature 360: 668-670.
49. Beukema, J.J., unpublished.
50. Beukema, J.J. 1974. Seasonal changes in the biomass of the macro-benthos of a tidal flat area in the Dutch Wadden Sea. Neth. J. Sea Res. 8: 94-107.
51. Beukema, J.J. 1976. Biomass and species richness of the macrobenthic animals living on the tidal flats of the Dutch Wadden Sea. Neth. J. Sea Res. 10: 236-261.

52. Beukema, J.J. 1979. Biomass and species richness of the macrobenthic animals living on a tidal flat area in the Dutch Wadden Sea: effects of a severe winter. Neth. J. Sea Res. 13: 203-223.
53. Beukema, J.J. 1980. Calcimass and carbonate production by molluscs on the tidal flats in the Dutch Wadden Sea: I. The tellinid bivalve *Macoma balthica*. Neth. J. Sea Res. 14: 323-338.
54. Beukema, J.J. 1981. The role of the larger invertebrates in the Wadden Sea ecosystem. In: N. Dankers, H. Kühl & W.J. Wolff (eds.). Invertebrates of the Wadden Sea: 211-221. Balkema, Rotterdam.
55. Beukema, J.J. 1982a. Calcimass and carbonate production by molluscs on the tidal flats in the Dutch Wadden Sea: II. The Edible Cockle, *Cerastoderma edule*. Neth. J. Sea Res. 15: 391-405.
56. Beukema, J.J. 1982b. Annual variation in reproductive success and biomass of the major macrozoobenthic species living in a tidal flat area of the Wadden Sea. Neth. J. Sea Res. 16: 37-45.
57. Beukema, J.J. 1985. Zoobenthos survival during severe winters on high and low tidal flats in the Dutch Wadden Sea. In: J.S. Gray & M.E. Christiansen (eds.). Marine biology of polar regions and effects of stress on marine organisms: 351-361. John Wiley, Chichester.
58. Beukema, J.J. 1989a. Long-term changes in macrozoobenthic abundance on the tidal flats of the western part of the Dutch Wadden Sea. Helgoländer Meeresunters. 43: 405-415.
59. Beukema, J.J. 1989b. Bias in estimates of maximum life span, with an example of the edible cockle *Cerastoderma edule*. Neth. J. Zool. 39: 79-85.
60. Beukema, J.J. 1991. The abundance of Shore Crabs *Carcinus maenas* (L.) on a tidal flat in the Wadden Sea after cold and mild winters. J. exp. mar. Biol. Ecol. 153: 97-113.
61. Beukema, J.J. 1992. Dynamics of juvenile shrimp *Crangon crangon* in a tidal-flat nursery of the Wadden Sea after mild and cold winters. Mar. Ecol. Prog. Ser. 83: 157-165.
62. Beukema, J.J. 1993a. Increased mortality in alternative bivalve prey during a period when the tidal flats of the Dutch Wadden Sea were devoid of Mussels. Neth. J. Sea Res. 31: 395-406.
63. Beukema, J.J. 1993b. Successive changes in distribution patterns as an adaptive strategy in the bivalve *Macoma balthica* (L.) in the Wadden Sea. Helgoländer Meeresunters. 47: 287-304.
64. Beukema, J.J. 1997. Caloric values of marine invertebrates with an emphasis on the soft parts of marine bivalves. Ocean. Mar. Biol. Ann. Rev. 35: 387-414.
65. Beukema, J.J. & G.C. Cadée. 1986. Zoobenthos responses to eutrophication of the Dutch Wadden Sea. Ophelia 26: 55-64.
66. Beukema, J.J. & E.C. Flach. 1995. Factors controlling the upper and lower limits of the intertidal distribution of two *Corophium* species in the Wadden Sea. Mar. Ecol. Prog. Ser. 125: 117-126.
67. Beukema, J.J. & J. de Vlas. 1979. Population parameters of the Lugworm, *Arenicola marina*, living on tidal flats in the Dutch Wadden Sea. Neth. J. Sea Res. 13: 331-353.
68. Beukema, J.J., K. Essink, H. Michaelis & L. Zwarts. 1993. Year-to-year variability in the biomass of macrobenthic animals on tidal flats of the Wadden Sea: how predictable is this food source for birds? Neth. J. Sea Res. 31: 319-330.
69. Beukema, J.J., G.C.Cadée & R. Dekker. 1998. How two large-scale "experiments" illustrate the importance of enrichment and fishery for the functioning of the Wadden Sea ecosystem. Senckenbergiana maritima 29: 37-44.
70. Biebach, H. 1998. Phenotypic organ flexibility in Garden Warblers *Sylvia borin* during long-distance migration. J. Avian Biol. 29: 529-535.
71. Binkley, S. 1997. Biological clocks. Your owner's manual. Harwood Academic Publishing, Amsterdam.
72. Birkhead, T.R. & A.P. Møller. 1992. Sperm competition in birds: evolutionary causes and consequences. Academic Press, London.
73. Bloksma, J., B.J. Ens & M. de Vries. 1979. Voedseloecologie van de Wulp op het Friese wad. Students report Zoological Laboratory, University of Groningen, Groningen / Rijksdienst voor de IJsselmeerpolders (RIZA), Lelystad.
74. Blomert, A-M. 1985. De Taag: het grootste waddengebied van Zuid-west Europa. Waddenbulletin 20: 187-190.
75. Blomert, A-M. & M. Engelmoer, unpublished.
76. Blomert, A-M. & P.M. Meininger. 1998. Watervogels in het Deltagebied: wintersterfte en draagkracht. Report RIKZ / University of Groningen, Middelburg / Groningen.
77. Blomert, A-M. & L. Zwarts, unpublished.
78. Blomert, A-M., M. Engelmoer & D. Logemann. 1983. Voedseloecologie van de scholekster op het Friese wad. Students report Zoological Laboratory, University of Groningen, Groningen / Rijksdienst voor de IJsselmeerpolders (RIZA), Lelystad.
79. Blomqvist, D. & C.O. Johansson. 1994. Double clutches and uniparental care in Lapwing *Vanellus vanellus*, with a comment on the evolution of double-clutching. J. Avian Biol. 25: 77-79.
80. Boddeke, R. & P. Hagel. 1991. Eutrophication, a blessing in disguise. ICES C.M./1991/E:7.
81. Bodsworth, F. 1984. De laatste wulp. Kaal Boek, Amsterdam.
82. de Boer, T. 1996. Trekbaan van Nederlandse Lepelaars. Action Report Vogelbescherming Nederland. 10, Zeist.
83. Boere, G.C. 1974. Het Waddenzeegebied als kruispunt van vogeltrekwegen. Waddenbulletin 12: 324-329.
84. Boere, G.C. 1976. The significance of the Dutch Waddenzee in the annual life cycle of arctic, subarctic and boreal waders. Part. 1. The function as a moulting area. Ardea 64: 210-291.
85. Boere, G.C. & B. Lenten. 1997. The African-Eurasian Migratory Waterbird Agreement; a technical agreement under the Bonn Convention. In: P. Straw (ed.). Shorebird conservation in the Asia-Pacific region: 67-71. Birds Australia, Hawthorn East, Victoria.
86. Boere, G.C. & B. Lenten. 1998. The African-Eurasian Waterbird Agreement: a technical agreement under the Bonn Convention. International Wader Studies. 10: 45-50.
87. Boates, J.S. & J.D. Goss-Custard. 1989. Foraging behaviour of Oystercatchers *Haematopus ostralegus* during a diet switch from worms *Nereis diversicolor* to clams *Scrobicularia plana*. Can. J. Zool. 67: 2225-2231.
88. Boates J.S. & P.C. Smith. 1979. Length-weight relationships, energy content and the effects of predation on *Corophium volutator* (Pallas) (Crustacea: Amphipoda). Proc. Nova Scotia Inst. Sci. 29: 489-499.
89. Boates, J.S. & P.C. Smith. 1989. Crawling behaviour of the amphipod *Corophium volutator* and foraging by Semipalmated Sandpipers, *Calidris pusilla*. Can. J. Zool. 67: 457-462.
90. Bredin, D. & A. Doumeret. 1987. Importance du littoral centre-ouest Atlantique pour la migration des limicoles côtiers. Rev. Ecol. (Terre Vie), Suppl. 4: 221-229.
91. Breiehagen, T. 1989. Nesting biology and mating system in an alpine population of Temminck's Stint *Calidris temminckii*. Ibis 131: 389-402.
92. Brey, T., H. Rumohr & S. Ankar. 1988. Energy content of macrobenthic invertebrates: general conversion factors from weight to energy. J. exp. mar. Biol. Ecol. 117: 271-278.
93. Brinkman, A.G. & M. van Stralen. 1999. Toelichting habitatkaart stabiele mosselbanken Waddenzee. Internal Report IBN-DLO (Alterra) & RIVO-DLO.
94. Brinkman, A.G. & B.J. Ens. 1998. Effecten van bodemdaling in de Waddenzee op wadvogels. IBN (Alterra) Report 371. Den Burg, Texel.
95. Brown, K.M. 1998. Proximate and ultimate causes of adoption in Ring-billed Gulls. Anim. Behav. 56: 1529-1543.
96. Brown, R.A. & R.J. O'Connor. 1974. Some observations on the relationships between Oystercatchers *Haematopus ostralegus* L. and Cockles *Cardium edule* L. in Strangford Lough. Irish Naturalist. 18: 73-80.
97. Browne, S. & H. Mead. 1999. Wader Study Group Colour-Marking Register. A report for. 1997 and. 1998. Wader Study Group Bull. 89: 21-23.
98. Bruinzeel, L.W. & T. Piersma. 1998. Cost reduction in the cold: heat generated by terrestrial locomotion partly substitutes for thermoregulation costs in Knot *Calidris canutus*. Ibis 140: 323-328.
99. Brunckhorst, H. 1996. Ökologie und Energetik der Pfeifente (*Anas penelope* L. 1758) im schleswig-holsteinischen Wattenmeer. Verlag Dr. Kovac, Hamburg.
100. Bryant, D.M. 1979. Effects of prey density and site character on estuary usage by overwintering waders (*Charadrii*). Estuar. Cstl. Mar. Sci. 9: 369-384.
101. Bryant, D.M. 1987. Wading birds and wildfowl of the estuary Firth of Forth, Scotland. Proc. Roy. Soc., Edinburgh 93B: 509-520.
102. Bryant, D.M. & J. Leng. 1975. Feeding distribution and behaviour of Shelduck in relation to food supply. Wildfowl 26: 20-30.
103. Bryant, D.M. & P. Tatner. 1991. Intraspecific variation in avian energy expenditure: correlates and constraints. Ibis 133: 236-245.
104. Bunskoeke E.J., B.J. Ens, J.B. Hulscher & S.J.de Vlas. 1996. Why do Oystercatcher *Haematopus ostralegus* switch from feeding on Baltic Tellin *Macoma balthica* to feeding on the Ragworm *Nereis diversicolor* during the breeding season? Ardea 84A: 91-104
105. Burke, T. & M.W. Bruford. 1987. DNA fingerprinting in birds. Nature 327: 149-152.
106. Burke, T., N.B. Davies, M.W. Bruford & B.J. Hatchwell. 1989. Parental care and mating behaviour of polyandrous dunnocks *Prunella modularis* related to paternity by DNA fingerprinting. Nature 338: 249-251.
107. Burley, N. 1985. Leg-band colour and mortality patterns in captive breeding populations of zebra finches. Auk 102: 647-651.

108. Burley, N. 1986. Sexual selection for aesthetic traits in species with biparental care. Am. Nat. 127: 415-445.
109. Burley, N., G. Krantzberg & P. Radman. 1982. Influence of colour-banding on the conspecific preferences of zebra finches. Anim. Behav. 27: 686-698.
110. Burton, P.J.K. 1974. Feeding and the feeding apparatus in waders. A study of anatomy and adaptations in the Charadrii. British Museum of Natural History, London.
111. Burton, R. 1990. Bird flight. Facts on File, New York.
112. Burton, R. 1992. Vogeltrek. De Toorts, Haarlem.
113. Bijlsma, R.G. 1990. Predation by large falcons on wintering waders on the Banc d'Arguin, Mauritania. Ardea 78: 75-82.
114. Byrkjedal, I. 1980. Nest predation in relation to snow-cover - a possible factor influencing the start of breeding in shorebirds. Ornis Scand. 11: 249-252.
115. Byrkjedal, I. 1987. Antipredator behavior and breeding success in Greater Golden Plover and Eurasian Dotterel. Condor 89: 40-47.
116. Byrkjedal, I. & D.B.A. Thompson. 1998. Tundra Plovers: The Eurasian, Pacific and American Golden Plovers and Grey Plover. T. & A.D. Poyser, London.
117. Cadée G.C. & J. Hegeman. 2002. Phytoplankton in the Marsdiep at the end of the 20$^{th}$ century; 30 years monitoring biomass, primary production, and *Phaeocystis* blooms. J. of Sea Res. 48: 97-110.
118. Cadée, G.C. 1988. Levende wadslakjes in Bergeend faeces. Correspondentieblad Ned. Malacol. Ver. 243/4: 443-444.
119. Cadée, G.C. & J. Hegeman. 1993. Persisting high levels of primary production at declining phosphate concentration in the Dutch coastal zone (Marsdiep). Neth. J. Sea Res. 31: 147-152.
120. Cadée, N., T. Piersma & S. Daan. 1996. Endogenous circannual rhythmicity in a non-passerine migrant, the Knot *Calidris canutus*. Ardea 84: 75-84.
121. Campfield, L.A., F.J. Smith & P. Burn. 1998. Strategies and potential molecular targets for obesity treatment. Science 280: 1381-1387.
122. Camphuysen, C.J. 1996. Ecologisch profiel van de Eidereend *Somateria mollissima*. RIKZ werkdocument 96.146X: 1-124. Texel.
123. Camphuysen, C.J. & J. van Dijk. 1983. Zee- en kustvogels langs de Nederlandse kust, 1974-1979. Limosa 56: 81-230.
124. Camphuysen, C.J., B.J. Ens, D. Heg, J.B. Hulscher, J. van der Meer & C.J. Smit. 1996. Oystercatcher *Haematopus ostralegus* winter mortality in The Netherlands: the effect of severe weather and food supply. Ardea 84A: 469-492.
125. Castro, G., N. Stoyan & J.P. Myers. 1989. Assimilation efficiency in birds: a function of taxon or food type? Comp. Biochem. Physiol. 92A: 271-278.
126. Cayford J.T. 1988b. The foraging behaviour of Oystercatcher (*Haematopus ostralegus*) feeding on Mussels (*Mytilus edulis*). PhD thesis, University of Exeter, Exeter.
127. Cayford, J.T. & J.D. Goss-Custard. 1990. Seasonal changes in the size selection of Mussels, *Mytilus edulis*, by Oystercatchers, *Haematopus ostralegus*: an optimality approach. Anim. Behav. 40: 609-624.
128. Chambers, M.R. & H. Milne. 1979. Seasonal variation in the condition of some intertidal invertebrates of the Ythan estuary, Scotland. Estuar. cstl. mar. Sci. 8: 411-419.
129. Chapman, T., L.F. Liddle, J.M. Kalb, M.F. Wolfner & L. Partridge. 1995. Cost of mating in *Drosophila melanogaster* is mediated by male accessory-gland products. Nature 373: 241-244.
130. Choudhury, S. 1995. Divorce in birds: a review of the hypotheses. Anim. Behav. 50: 413-429.
131. Clark, N. 1996. Trouble brewing on the Wash? Declining shellfish stocks pose problem. BTO News 206: 1.
132. Clarke, R.T. & J.D. Goss-Custard. 1996. Appendix. 1: The Exe estuary Oystercatcher-mussel model. In: J.D. Goss-Custard (ed.), The Oystercatcher: from individuals to populations: 390-393. Oxford University Press, Oxford.
133. Clutton-Brock, T.H. (ed.) 1988. Reproductive success: studies of individual variation in contrasting breeding systems. The University of Chicago Press, Chicago.
134. Clutton-Brock, T.H., F.E. Guinness & S.D. Albon. 1982. Red Deer: Behavior and Eeology of two Sexes. Edinburgh University Press, Edinburgh.
135. Colwell, M.A. & L.W. Oring. 1989. Extra-pair mating in the Spotted Sandpiper: a female mate acquisition tactic. Anim. Behav. 38: 675-684.
136. Cooke, F., E.F. Rockwell & D.B. Lank. 1995. The Snow Geese of La Pérouse Bay. Oxford University Press, Oxford.
137. Coulson, J.C., N. Duncan & C. Thomas. 1982. Changes in the breeding history of the Herring Gull (*Larus argentatus*) induced by reduction in the size and density of the colony. J. Anim. Ecol. 51: 739-756.
138. Cramp, S. & K.E.L. Simmons (eds.). 1983. The birds of the western Palaearctic, Vol. III. Oxford University Press, Oxford.
139. Cramp, S. & K.E.L. Simmons (eds.). 1985. The birds of the western Palaearctic, Volume IV. Oxford University Press, Oxford.
140. Cremer, J. & R. Hupkes. 1984. Voedselterritoria bij de wulp: lange en korte termijn voordeel. Students report Zoological Laboratory, University of Groningen / Rijksdienst voor de IJsselmeerpolders (RIZA), Lelystad.
141. Cresswell, W. 1994. Flocking is an effective anti-predation strategy in Redshank, *Tringa totanus*. Anim. Behav. 47: 433-442.
142. Creswell, W. 1994b. Age-dependent choice of Redshank (*Tringa totanus*) feeding location: profitability or risk? J. Anim. Ecol. 63: 589-600.
143. Cresswell, W. & D.P. Whitfield. 1994. The effects of raptor predation on wintering wader populations at Tyningham estuary, southeast Scotland. Ibis 136: 223-232.
144. Cutts, C.J. & J.R. Speakman. 1994. Energy savings in formation flight of Pink-footed Geese. J. exp. Biol. 189: 251-261.
145. Daan, S., C. Dijkstra & J.M. Tinbergen. 1990. Family planning in the Kestrel (*Falco tinnunculus*): the ultimate control of covariance of laying date and clutch size. Behaviour 114: 83-116.
146. Daan, S. & P. Koene. 1981. On the timing of foraging flights by Oystercatchers, *Haematopus ostralegus*, on tidal flats. Neth. J. Sea Res. 15: 1-22.
147. Dankers, N. & J.J. Beukema. 1984. Distributional patterns of macro-zoo-benthic species in relation to some environmental factors. In: W.J. Wolff (ed.). Ecology of the Wadden Sea, Vol. 1, part 4: 69-103. Balkema, Rotterdam.
148. Dann, P. 1987. The feeding behaviour and ecology of shorebirds. In: B.A. Lane (ed.). Shorebirds in Australia: 10-20. Nelson, Melbourne.
149. Dansgaard, W., S.J. Jonsen, H.B. Clausen, D. Dahl-Jensen, N.S. Gundestrup, C.U. Hammer, C.S. Hvidberg, J.P. Steffensen, A.E. Sveinbjornsdottir, J. Jouzel & G. Bond. 1993. Evidence for general instability of past climate from a 250-kyr ice-core record. Nature 364: 218-220.
150. Darwin, C. 1859. The origin of species by means of natural selection. John Murray, London.
151. Davidson, I. 1997. The successes and challenges of establishing a shorebird conservation program in the Americas. In: P. Straw (ed.). Shorebird conservation in the Asia-Pacific region: 63-66. Birds Australia, Hawthorn East, Victoria.
152. Davidson, N.C. 1981. Survival of shorebirds (Charadrii) during severe weather: the role of nutritional reserves. In: N.V. Jones & W.J. Wolff (eds.). Feeding and survival strategies of estuarine organisms: 231-249. Plenum Press, New York.
153. Davidson, N.C. 1984. How valid are flight range estimates for waders? Ring. & Migr. 5: 49-64.
154. Davidson, N.C. & M.W. Pienkowski (eds.). 1987. The conservation of international flyway populations of waders. Wader Study Group Bull. 49, Suppl. / IWRB Spec. Publ. 7.
155. Davidson, N.C. & T. Piersma. 1992. The migration of Knots: conservation needs and implications. Wader Study Group Bull. 64, Suppl.: 198-209.
156. Davidson, N.C. & P. Rothwell (eds.). 1993. Disturbance to waterfowl on estuaries. Wader Study Group Bulletin 68.
157. Davidson, N.C. & J.R. Wilson. 1992. The migration system of European-wintering Knots *Calidris canutus islandica*. Wader Study Group Bull. 64, Suppl.: 39-51.
158. Davidson, N.C., P.I. Rothwell & M.W. Pienkowski. 1995. Towards a flyway conservation strategy for waders. Wader Study Group Bull. 77: 70-81.
159. Davidson, N.C., D.A. Stroud, P.I. Rothwell & M.W. Pienkowski. 1998. Towards a flyway conservation strategy for waders. International Wader Studies. 10: 24-38.
160. Davidson, N.C., D.J. Townshend, M.W. Pienkowski & J.R. Speakman. 1986a. Why do Curlews have decurved bills? Bird Study 33: 61-69.
161. Davidson, N.C., J.D. Uttley & P.R. Evans. 1986b. Geographic variation in the lean mass of Dunlins wintering in Britain. Ardea 74: 191-198.
162. Davidson, N.C., D. d'A Laffoley, J.F. Doody, L.S. Way, J. Gordon, R. Key, C.M. Drake, M.W. Pienkowski, R. Mitchell & K.L. Duff. 1991. Nature conservation and estuaries in Great Britain. Nature Conservancy Council, Peterborough.
163. Davidson, P.E. 1967. A study of the Oystercatcher (*Haematopus ostralegus* L.) in relation to the fishery for Cockles (*Cardium edule* L.) in the Burry Inlet, South Wales. Fish. Invest. Lond. (Ser. II) 25: 1-28.
164. Davies, N.B. 1976. Food, flocking and territorial behaviour of the Pied Wagtrail (*Motacilla alba*) in winter. J. Anim. Ecol. 48: 235-253.
165. Davies, N.B. 1991. Mating systems. In: J.R. Krebs & N.B. Davies (eds.). Behavioural ecology: an evolutionary approach, Third edition: 263-294. Blackwell Scientific Publications, Oxford.
166. Davies, N.B. 1992. Dunnock Behaviour and Social Evolution. Oxford University Press, Oxford.

167. Davies, N.B. & A.I. Houston. 1984. Territory economics. In: J.R. Krebs & N.B. Davies (eds.). Behavioural ecology: an evolutionary approach. Second edition: 148-169. Blackwell Scientific Publications, Oxford.
168. Davis, J.M. 1980. The coordinated aerobatics of Dunlin flocks. Anim. Behav. 28: 668-673.
169. Dawkins, R. 1976. The selfish gene. Oxford University Press, Oxford.
170. Dawkins, R. 1986. The blind watchmaker. Longman Scientific & Technical, Essex.
171. Dekinga, A. & T. Piersma. 1993. Reconstructing diet composition on the basis of faeces in a mollusc-eating wader, the Knot *Calidris canutus*. Bird Study 40: 144-156.
172. Dekkers, H. & B.S. Ebbinge. 1997. Ganzen, grazers op trek langs de vorstgrens. Schuyt & Co, Haarlem.
173. Dennett, D.C. 1995. Darwin's dangerous idea: evolution and the meanings of life. Penguin Books, London.
174. Denton, G.H. & T.J. Hughes (eds.). 1981. The last great ice sheets. Wiley, New York.
175. Deutsches Hydrographisches Institut. 1966. Gezeitentafeln für das Jahr 1967. Hamburg.
176. Dick, W.J.A. (ed.). 1975. Oxford and Cambridge Mauritanian Expedition 1973. Report, Cambridge.
177. Dick, W.J.A. & M.W. Pienkowski. 1979. Autumn and early winter weights of waders in North-West Africa. Ornis Scand. 10: 117-123.
178. Dick, W.J.A., M.W. Pienkowski, M.Waltner & C.D.T. Minton. 1976. Distribution and geographical origins of Knot *Calidris canutus* in Europe and Africa. Ardea 64: 22-47.
179. Dick, W.J.A., T. Piersma & P. Prokosch. 1987. Spring migration of the Siberian Knots *Calidris canutus canutus*: results of a co-operative Wader Study Group project. Ornis Scand. 18: 5-16.
180. Dierschke, V. 1996. Unterschiedliches Zugverhalten alter und junger Alpenstrandläufer *Calidris alpina*: Ökologische Untersuchungen an Rastplätzen der Ostsee, des Wattenmeeres und auf Helgoland. PhD thesis, University of Göttingen, Göttingen.
181. Dierschke, V. & J. Kube. 1999 Feeding ecology of Dunlins *Calidris alpina* staging in the southern Baltic Sea: 1. Habitat use and food selection. J. Sea Res. 42: 49-64.
182. Dietz, M.W., A. Dekinga, T. Piersma & S. Verhulst. 1999. Estimating organ size in small migrating shorebirds with ultrasonography: an intercalibration exercise. Physiol. Biochem. Zool. 72: 28-37.
183. Dolman, P. & W.J. Sutherland. 1995. The response of bird populations to habitat loss. Ibis 137, Suppl.: 38-46.
184. Dorst, J. 1962. The migrations of birds. Heinemann, London.
185. Drent, R.H. 1967. Functional aspects of incubation in the Herring Gull (*Larus argentatus* Pont.). Behaviour Suppl. XVII: 1-132.
186. Drent, R. 1975. Incubation. In: D.S. Farner & J.R. King (eds.). Avian biology, Vol 5: 333-420. Academic Press, New York.
187. Drent, R.H. & S. Daan. 1980. The prudent parent: energetic adjustments in avian breeding. Ardea 68: 225-252.
188. Drent, R. & T. Piersma. 1990. An exploration of the energetics of leap-frog migration in arctic breeding waders. In: E. Gwinner (ed.). Bird migration. Physiology and ecophysiology: 399-412. Springer-Verlag, Berlin.
189. Drijver, C.A. 1983 Coastal protection and land reclamation in the Wadden Sea area. In: M. F. Mörzer Bruyns & W.J. Wolff (eds.). Nature conservation, nature management and physical planning in the Wadden Sea area: 130-144. Balkema, Rotterdam.
190. Drinnan, R.E. 1957. The winter feeding of the Oystercatcher (*Haematopus ostralegus*) on the Edible Cockle (*Cardium edule*). J. Anim. Ecol. 26: 441-469.
191. Drinnan, R.E. 1958a. The winter feeding of the oystercatcher (*Haematopus ostralegus*) on the edible mussel (*Mytilus edulis*) in the Conway estuary, North Wales. Fish. Invest. Lond. 22: 1-15.
192. Drinnan, R.E. 1958b. Observations on the feeding of the Oystercatcher in captivity. Brit. Birds 51: 139-149.
193. Dijksen, L. 1980. Enige gegevens over broedseizoen en broedsukses bij Scholeksters (Haematopus ostralegus) in de duinen. Watervogels 5: 3-7.
194. van Dijk, A.J., K. van Dijk, L.J. Dijksen, T.M. van Spanje & E. Wijmenga. 1986. Wintering waders and waterfowl in the Gulf van Gabès, Tunesia, January-March. 1984. WIWO Report. 11, Zeist.
195. van Dijk, A.J., F.E. de Roder, E.C.L. Marteijn & H. Spiekman. 1990. Summering waders on the Banc d'Arguin, Mauritania: a census in June. 1988. Ardea 78: 145-156.
196. Dugan, P.J. 1981. The importance of nocturnal feeding in shorebirds: a consequence of increased invertebrate prey activity. In: N.V. Jones & W.J. Wolff (eds.). Feeding and survival strategies of estuarine organisms: 251-260. Plenum Press, New York.
197. Dugan, P.J. 1982. Seasonal changes in patch use by a territorial Grey Plover: weather-dependent adjustment in foraging behaviour. J. Anim. Ecol. 51: 849-857.
198. Dugan, P.J. 1987. Socio-economic considerations in protecting shorebird sites in the developing world: some priorities and implications for the direction of future research. Wader Study Group Bull. 49, Suppl. / IWRB Special Publ. 7: 146-148.
199. Dugan, P.J., P.R. Evans, L.R. Goodyear & N.C. Davidson. 1981. Winter fat reserves in shorebirds: disturbance of regulated levels by severe weather conditions. Ibis 123: 359-363.
200. Durell, S.E.A. le V. dit, J.D. Goss-Custard & R.W.G. Caldow. 1993. Sex-related differences in diet and feeding method in the Oystercatcher *Haematopus ostralegus*. J. Anim. Ecol. 62: 205-215.
201. Dijkema, K.S. 1983. The salt-marsh vegetation of the mainland coast, estuaries and Halligen. In: K.S. Dijkema & W.J. Wolff (eds.), Flora and vegetation of the Wadden Sea islands and coastal areas: 185-220. Balkema, Rotterdam.
202. Dijkema, K.S. 1989. Habitats of the Netherlands, German and Danish Wadden Sea. Veth Foundation, Leiden.
203. Dijkstra, C. 1988. Reproductive tactics in the Kestrel *Falco tinnunculus*. PhD thesis, University of Groningen, Groningen.
204. Ebbinge, B.S. 1989. A multifactorial explanation for variation in breeding performance of Brent Geese *Branta bernicla*. Ibis 131: 196-204.
205. Ebbinge, B.S. 1990. Reply. Ibis 132: 481-482.
206. Ebbinge, B.S. 1992a. Population limitation in arctic-breeding geese. PhD thesis, University of Groningen, Groningen.
207. Ebbinge, B.S. 1992b. Regulation of numbers of Dark-bellied Brent Geese on spring staging areas. Ardea 80: 203-228.
208. Ebbinge, B.S., A. St. Joseph, P. Prokosch & B. Spaans. 1982. The importance of spring staging areas for arctic-breeding geese, wintering in western Europe. Aquila 89: 249-258.
209. Ebbinge, B.S. & B. Spaans. 1995. The importance of body reserves accumulated in spring staging areas in the temperate zone for breeding in Darkbellied Brent Geese *Branta b. bernicla* in the high Arctic. J. Avian Biol. 26: 105-113.
210. Edwards, P.J. 1982. Plumage variation, territoriality and breeding displays of the Golden Plover *Pluvialis apricaria* in southwest Scotland. Ibis 124: 88-96.
211. van Eerden, M. 1977. Vorstvlucht van watervogels door het oostelijk deel van de Nederlandse Waddenzee op 30 december. 1976. Watervogels 2: 11-14.
212. van Eerden, M.R. 1997. Patchwork. Patch use, habitat exploitation and carrying capacity for water birds in Dutch freshwater wetlands. RIZA report Van Zee tot Land 65, Lelystad / PhD thesis, University of Groningen, Groningen.
213. Eisma, D. 1998. Intertidal deposits: river mouths, tidal flats and coastal lagoons. CRC Press, Boca Raton.
214. Elkins, N. 1983. Weather and bird behaviour. Poyser, Calton.
215. Elliott, C.C.H., M. Waltner, L.G. Underhill, J.S. Pringle & W.J.A. Dick. 1976. The migration system of the Curlew Sandpiper *Calidris ferruginea* in Africa. Ostrich 47: 191-213.
216. Emlen, J.M. 1966. The role of time and energy in food preference. Am. Nat. 109: 427-435.
217. Emlen, S.T. 1975. Migration: orientation and navigation. In: D.S. Farner & J.R. King (eds.). Avian Biology, Vol. V: 129-219. Academic Press, New York.
218. Emlen, S.T. & J.T. Emlen. 1966. A technique for recording migratory orientation of captive birds. Auk 83: 361-367.
219. Emlen, S.T. & L.W. Oring. 1977. Ecology, sexual selection and the evolution of mating systems. Science 197: 215-223.
220. Engelmoer, M. & C.S.Roselaar. 1998. Geographical variation in waders. Kluwer Academic Publishers, Dordrecht.
221. Engelmoer, M., T. Piersma, W. Altenburg & R. Mes. 1994. The Banc d'Arguin (Mauritania). In: P.R. Evans, J.D. Goss-Custard & W.G. Hale (eds.). Coastal waders and wildfowl in winter: 293-310. Cambridge University Press, Cambridge.
222. Engelmoer, M., R. Kuipers, H. Gartner & A. Ferwerda. 1996. Vogeltellingen langs de Friese Waddenkust. 1990-1995. Report Wadvogelwerkgroep Fryske Feriening foar Fjildbiology, Ferwerd.
223. Ens, B.J., unpublished.
224. Ens, B.J. 1981. Identifying colour-ringed Oystercatchers *Haematopus ostralegus*. Wader Study Group Bull. 31: 28-29.
225. Ens, B.J. 1985. Tussen Sahara en Siberië. Stichting WIWO, Ewijk.
226. Ens, B.J. 1991. Guarding your mate and losing the egg: an Oystercatcher's dilemma. Wader Study Group Bull. 61, Suppl.: 69-70.
227. Ens, B.J. 1992. The social prisoner: causes and consequences of natural variaton in reproductive success of the oystercatcher. PhD thesis, University of Groningen, Groningen.

228. Ens, B.J. 1994. De carrière-planning van de Scholekster *Haematopus ostralegus*. Limosa 67: 53-67.
229. Ens, B.J. 1996. Climate change: consequences for migratory birds. Change 32: 12-14.
230. Ens, B.J. 1997. Climate change and bird migration: neat theory versus unruly reality. Change 37: 1-4.
231. Ens, B.J. & D. Alting. 1996a. Prey selection of a captive Oystercatcher *Haematopus ostralegus* hammering on Mussels *Mytilus edulis*. Ardea 84A: 215-219
232. Ens, B.J. & D. Alting. 1996b. The effect of an experimentally created musselbed on bird densities and food intake of the Oystercatcher *Haematopus ostralegus*. Ardea 84A: 193-507
233. Ens, B.J. & A.G. Brinkman. 1998. De reactie van vogels op relatieve zeespiegelstijging. In: A.P. Oost, B.J. Ens, A.G. Brinkman, K.S. Dijkema, W.D. Eysink, J.J. Beukema, H.J. Gussinklo, B.M.J. Verboom & J.J. Verburgh (eds.). Integrale bodemdalingstudie Waddenzee: 301-308. NAM, Assen. ISBN 90-804791-4-4
234. Ens B.J. & J. Cayford. 1996. Feeding with other Oystercatchers. In: J.D. Goss-Custard (ed.). Behaviour and ecology of the Oystercatcher: from individuals to populations: 77-104. Oxford University Press, Oxford.
235. Ens, B.J. & J.D. Goss-Custard. 1984. Interference among Oystercatchers, *Haematopus ostralegus*, feeding on mussels, *Mytilus edulis*, on the Exe estuary. J. Anim. Ecol. 53: 217-231.
236. Ens, B. & M. Klaassen. 1998. Logistiek van de trekvogel. Natuur & Techniek 66 (4): 10-20.
237. Ens, B.J. & R. de Vries. 1983. Voedseloecologie van de Wulp op het Friese wad, deel II. Students report Zoological Laboratory, University of Groningen, Groningen / Rijksdienst voor de IJsselmeerpolders (RIZA), Lelystad.
238. Ens. B.J. & L. Zwarts. 1980a. Wulpen op het wad van Moddergat. Watervogels 5: 108-120.
239. Ens, B.J. & L. Zwarts. 1980b. Territoriaal gedrag bij Wulpen buiten het broedgebied. Watervogels 5: 155-169.
240. Ens, B.J., C.J. Smit & M. de Jong., unpublished.
241. Ens, B.J., P. Duiven & T.M. van Spanje. 1990a. Spring migration of turnstones *Arenaria interpres* from the Banc d'Arguin in Mauritania. Ardea 78: 301-314.
242. Ens, B.J., P. Esselink & L. Zwarts. 1990b. Kleptoparasitism as a problem of prey choice: a study on mudflat-feeding Curlews, *Numenius arquata*. Anim. Behav. 39: 219-230.
243. Ens, B.J., T. Piersma, W.J. Wolff & L. Zwarts. 1990c. Homeward bound: problems waders face when migrating from the Banc d' Arguin, Mauritania, to their northern breeding grounds in spring. Ardea 78: 1-16.
244. Ens B.J., M. Kersten, A. Brenninkmeijer & J. B. Hulscher. 1992. Territory quality, parental effort and reproductive success of Oystercatcher (*Haematopus ostralegus*). J. Anim. Ecol. 61: 703-715.
245. Ens, B.J., M. Klaassen & L. Zwarts. 1993a. Flocking and feeding in the Fiddler Crab (*Uca tangeri*): prey availability as risk-taking behaviour. Neth. J. Sea Res. 31: 477-494.
246. Ens, B.J., U.N. Safriel & M.P. Harris. 1993b. Divorce in the long-lived and monogamous Oystercatcher, *Haematopus ostralegus*: incompatibility or choosing the better option? Anim. Behav. 45: 1190-1217.
247. Ens, B.J., G.J.M. Wintermans & C.J. Smit. 1993c. Verspreiding van overwinterende wadvogels in de Nederlandse Waddenzee. Limosa 66: 137-144.
248. Ens, B.J., T. Piersma & J.M. Tinbergen. 1994a. Towards predictive models of bird migration schedules: theoretical and empirical bottlenecks. NIOZ Report. 1994-5, Texel.
249. Ens, B.J., T. Piersma & R.H. Drent. 1994b. The dependence of waders and waterfowl migrating along the East Atlantic Flyway on their coastal food supplies: what is the most profitable research program? Ophelia Suppl. 6: 127-151.
250. Ens, B.J., F.J. Weissing & R.H. Drent. 1995a. The despotic distribution and deferred maturity: two sides of the same coin. Am. Nat. 146: 625-650.
251. Ens, B.J., J.D. Goss-Custard & T.P. Weber. 1995b. Effects of climate change on bird migration strategies along the East Atlantic Flyway. Dutch National Research Programme on Global Air Pollution and Climate Change Report no. 410. 100 075.
252. Ens, B.J., E.J. Bunskoeke, R. Hoekstra, J.B. Hulscher & S.J. de Vlas. 1996a. Prey choice and search speed: why simple optimality fails to explain the prey choice of Oystercatchers *Haematopus ostralegus* feeding on *Macoma balthica* and *Nereis diversicolor*. Ardea 84A: 73-89.
253. Ens, B.J., S. Dirksen, C.J. Smit & E.J. Bunskoeke. 1996b. Seasonal changes in size selection and intake rate of Oystercatchers *Haematopus ostralegus* feeding on the bivalves *Mytilus edulis* and *Cerastoderma edule*. Ardea 84A: 159-176.
254. Ens, B.J., T. Merck, C.J. Smit & E.J. Bunskoeke. 1996c. Functional and numerical response of Oystercatchers *Haematopus ostralegus* on shellfish populations. Ardea 84A: 441-452.
255. Ens, B.J., K. Briggs, U.N. Safriel & C.J. Smit. 1996d. Life history decisions during the breeding season. In: J.D. Goss-Custard (ed.). The Oystercatcher: from individuals to populations: 186-218. Oxford University Press, Oxford.
256. Ens, B.J., S. Choudhury & J.M. Black. 1996e. Mate fidelity and divorce in monogamous birds. In: J.M. Black (ed.). Partnerships in birds: the study of monogamy: 344-401. Oxford Univ. Press, Oxford.
257. Esselink, P. 1982. De verspreiding van Wulpen (*Numenius arquata*) tijdens laagwater, foeragerend op Zeeduizendpoten (*Nereis diversicolor*). Students report Zoological Laboratory University of Groningen, Groningen / Rijksdienst voor de IJsselmeerpolders (RIZA), Lelystad.
258. Esselink, P. & J. van Belkum. 1986. De verspreiding van de Zeeduizendpoot *Nereis diversicolor* en de Kluut *Recurvirostra avosetta* in de Dollard in relatie tot een verminderde afvalwaterlozing. Internal Report van RIKZ/RWS GWAO-86.155, Haren.
259. Esselink, P. & L. Zwarts. 1989. Seasonal trend in burrow depth and tidal variation in feeding activity of *Nereis diversicolor*. Mar. Ecol. Prog. Ser. 56: 243-254.
260. Essink, K. 1998. Het effect van de sanering van de lozingen van veenkoloniaal afvalwater op de bodemfauna van de Dollard. In: K. Essink & P. Esselink. (eds.). 1998. Het Eems-Dollard estuarium: interacties tussen menselijke beïnvloeding en natuurlijke dynamiek: 101-127. Report RIKZ-98.020, Haren.
261. Essink, K. & P. Esselink (eds.). 1998. Het Eems-Dollard estuarium: interacties tussen menselijke beïnvloeding en natuurlijke dynamiek. Report RIKZ-98.020, Haren.
262. Essink, K., J. Eppinga & R. Dekker. 1998a. Long-term changes (1977-1994) in intertidal macrozoobenthos of the Dollard (Ems estuary) and effects of introduction of the North American spionid polychaete *Marenzellaria* cf. *wireni*. Senckenbergiana maritima 28: 211-225.
263. Essink, K., J.J. Beukema, P.B. Madsen, H. Michaelis & G.R. Vedel. 1998b. Long-term development of biomass of intertidal macrozoobenthos in different parts of the Wadden Sea governed by nutrient loads? Senckenbergiana maritima 29: 25-35.
264. Evans, A. 1987. Relatively availability of the prey of wading birds by day and by night. Mar. Ecol. Progr. Ser. 37: 103-107.
265. Evans, P.R. 1976. Energy balance and optimal foraging strategies in shorebirds: some implications for their distributions and movements in the non-breeding season. Ardea 64: 117-139.
266. Evans, P.R. 1981. Reclamation of intertidal land: some effects on Shelduck and wader populations in the Tees estuary. Verh. Orn. Ges. Bayern 23: 147-168.
267. Evans, P.R. 1979. Adaptations shown by foraging shorebirds to cyclical variations in the activity and availability of their intertidal invertebrate prey. In: E. Naylor & R.G. Hartnall (eds.). Cyclic phenomena in marine plants and animals: 357-366. Pergamon Press, New York.
268. Evans, P.R. 1988. Predation of intertidal fauna by shorebirds in relation to time of the day, tide and year. In: G. Chelazzi & M. Vannini (eds.). Behavioral adaptation to intertidal life: 65-78. Plenum Press, New York.
269. Evans, P.R. & N.C. Davidson. 1990. Migration strategies and tactics of waders breeding in arctic and north temperate latitudes. In: E. Gwinner (ed.). Bird migration. Physiology and ecophysiology: 388-398. Springer-Verlag, Berlin.
270. Evans, P.R. & Pienkowski, M.W. 1984. Population dynamics of shorebirds. Behavior of marine animals 5: 83-123.
271. Evans, P.R. & D.J. Townshend. 1988. Site faithfulness of waders away from the breeding grounds: how individual migration patterns are established. Acta XIX Congr. Intern. Ornithol. (Ottawa): 594-603.
272. Evans, P.R., D.M. Herdson, D.J. Knight & M.W. Pienkowski. 1979. Short-term effects of reclamation of part of Seal sands, Teesmouth, on wintering waders and Shelduck. Oecologia (Berl.) 41: 183-206.
273. Evans, P.R., N.C. Davidson, J.D. Uttley & R.D. Evans. 1992. Premigratory hypertrophy of flight muscles: an ultrastructural study. Ornis Scand. 23: 238-243.
274. Exo, K-M., G. Scheiffarth & U. Haesihus. 1996. The application of motion sensitive transmitters to record activity and foraging patterns of Oystercatchers *Haematopus ostralegus*. Ardea 84A: 29-38.
275. Eyssink, W.D., R.J. Fokker, Z.B. Wange, M. Buijsman & M.J.F. Stive. 1998. Effecten van bodemdaling door gaswinning in en rond de Waddenzee: morfologische, infrastructurele en economische aspecten. WL/Delft Hydraulics.
276. Fairbanks, R.G. 1989. A. 17,000-year glacio-eustatic sea level record: influence of glacial melting on the Younger Drias event and deep-ocean circulation. Nature 342: 637-642.

277. Farke, H. & E.M. Berghuis. 1979. Spawning, larval development and migration of *Arenicola marina* under field conditions in the western Wadden Sea. Neth. J. Sea Res. 13: 529-535.
278. Farke, H., P.A.W.J. de Wilde & E.M. Berghuis. 1979. Distribution of juvenile and adult *Arenicola marina* on a tidal mudflat and the importance of nearshore areas for recruitment. Neth. J. Sea Res. 13: 354-361.
279. Fasola, M. & L. Canova. 1993. Diel activity of resident and immigrant waterbirds at Lake Turkana, Kenya. Ibis 135: 442-450.
280. Fenschel, T. 1978. The ecology of micro- and meiobenthos. Ann. Rev. Ecol. Syst. 9: 99-121.
281. Ferns, P.N. 1979. Resource partitioning amongst breeding waders in NE Greenland. Wader Study Group Bull. 26: 24.
282. Fitzgerald, L. 1991. Overtraining increases the susceptibility to infection. Int. J. Sports Med. 12: S5-S8.
283. Flach, E.C. 1992. The influence of four macrozoobenthic species on the abundance of the amphipod *Corophium volutator* on tidal flats in the Wadden Sea. Neth. J. Sea Res. 29: 379-394.
284. Frenzel, B., M. Pésci & A.A. Velichko. 1992. Atlas of paleoclimates and paleoenvironments of the northern hemisphere; late Pleistocene-Holocene. Gustav Fischer Verlag, Stuttgart.
285. Fretwell, S.D. 1972. Populations in seasonal environment. Princeton University Press, Princeton.
286. Fretwell, S.D. & H.L. Lucas, Jr. 1970. On territorial behavior and other factors influencing habitat distribution in birds. I. Theoretical development. Acta Biotheoretica XIX: 16-36.
287. Frodin, P., F. Haas. & Å. Lindström. 1994. Mate guarding by Curlew Sandpipers *Calidris ferruginea* during spring migration in North Siberia. Arctic 47: 142-144.
288. Fuchs, E. 1973. Durchzug und Überwinterung der Alpenstrandläufers *Calidris alpina* in der Camargue. Ornithol. Beob. 70: 113-134.
289. van Gaalen, P. & J. van Gelderen. 1995. De wereld van de Zwarte Stern. Van Reemst Publisher, 's-Hertogenbosch.
290. Gendron, R.P. 1986. Searching for cryptic prey: evidence for optimal search rates and the formation of search images in Quail. Anim. Behav. 34: 898-912.
291. Gendron, R.P. & J.E.R. Staddon. 1983. Searching for cryptic prey: the effect of search rate. Am. Nat. 121: 172-186.
292. Gerrets, D.A. 1999. Evidence of political centralization in Westergo: The excavations at Wijnaldum in a (supra-)regional perspective. In: T. Dickinson & D. Griffiths (eds.). The Making of Kingdoms: Papers from the 47th Sachsensymposium, York, September. 1996. Anglo-Saxon Studies in Archaeology and History. 10: 119-126. Oxford University Committee for Archaeology, Oxford.
293. Gerritsen, A.F.C. 1988. Feeding techniques and the anatomy of the bill in sandpipers (*Calidris*). PhD thesis, University of Leiden, Leiden.
294. Gerritsen, A.F.G. 1988. General and specific patterns in foraging acts of sandpipers (*Calidris*). In: A.F.G. Gerritsen. Feeding techniques and the anatomy of the bill in sandpipers (*Calidris*): 91-137. PhD thesis, University of Leiden, Leiden.
295. Gerritsen, A.F.C. & A. Meijboom. 1986. The role of touch in prey density estimation by *Calidris alba*. Neth. J. Zool. 36: 530-562.
296. Getty, T. & H.R. Pulliam. 1991. Random prey detection with pause-travel search. Am. Nat. 138: 1459-1477.
297. Gill, J.A. 1996. Habitat choice in Pink-footed Geese: quantifying the constraints determining winter site use. J. appl. Ecol. 33: 884-892.
298. Gill, J., W.J. Sutherland & A.R. Watkinson. 1996. A method to quantify the effect of human disturbance on animal populations. J. appl. Ecol. 33: 786-792.
299. Gill, R.E., Jr., R.W. Butler, P.S. Tomkovich, T. Mundkur & C.M. Handel. 1994. Conservation of North Pacific shorebirds. Trans. 59th North Am. Wildl. & Natur. Resour. Conf.: 63-78.
300. Gill, R.E., Jr., P. Canevari & E.H. Iversen. 1998. Eskimo Curlew (*Numenius borealis*). In: A. Poole & F. Gill (eds.). The birds of North America, No. 347. The Birds of North America, Inc., Philadelphia.
301. Ginn, H.B. & D.S. Melville. 1983. Moult in birds. BTO Guide. 19, Tring.
302. Goater, C. 1988. Patterns of helminth parasitism in the Oystercatcher, *Haematopus ostralegus*, from the Exe estuary, England. PhD thesis, University of Exeter, Exeter.
303. Gochfeld, M. 1984. Antipredator behavior: aggressive and distraction displays of shorebirds. Behavior of marine animals 5: 289-377.
304. Godard, R. 1991. Long-term memory of individual neighbours in a migratory songbird. Nature 350: 228-229.
305. Goede, A.A. 1993. Variation in the energy intake of captive Oystercatchers *Haematopus ostralegus*. Ardea 81: 89-97.
306. Goede, A.A., E. Nieboer & P.M. Zegers. 1990. Body mass increase, migration pattern and breeding grounds of Dunlins, *Calidris a. alpina*, staging in the Dutch Wadden Sea in spring. Ardea 78: 135-144.
307. de Glopper, R.J. 1967. Over de bodemgesteldheid van het waddengebied. RIZA Report Van Zee tot Land no. 43. Tjeenk Willink, Zwolle.
308. Gonzalez, P.M., T. Piersma & Y. Verkuil. 1996. Food, feeding and refuelling of Red Knots during northward migration at San Antonio Oeste, Rio Negro, Argentina. J. Field Ornithol. 67: 575-591.
309. Göransson, G., J. Karlsson, S.G. Nilsson & S. Ulfstrand. 1975. Predation on birds' nests in relation to antipredator aggression and nest density: an experimental study. Oikos 26: 117-120.
310. Goss-Custard, J.D. 1969. The winter feeding ecology of the Redshank *Tringa totanus*. Ibis 111: 338-356.
311. Goss-Custard, J.D. 1970a. The responses of Redshank (*Tringa totanus*) to spatial variations in the density of their prey. J. Anim. Ecol. 39: 91-113.
312. Goss-Custard, J.D. 1970b. Feeding dispersion in some overwintering wading birds. In: J.H. Crook (ed.). Social behaviour in birds and mammals: 3-35. Academic Press, London.
313. Goss-Custard, J.D. 1976. Variation in the dispersion of Redshank *Tringa totanus* on their winter feeding grounds. Ibis 118: 257-263.
314. Goss-Custard, J.D. 1977a. The energetics of prey selection by Redshank, *Tringa totanus* (L.), in relation to prey density. J. Anim. Ecol. 46: 1-19.
315. Goss-Custard, J.D. 1977b. Predator responses and prey mortality in Redshank, *Tringa totanus* (L.), and a preferred prey, *Corophium volutator* (Pallas). J. Anim. Ecol. 46: 21-35.
316. Goss-Custard, J.D. 1977c. Responses of Redshank, *Tringa totanus* (L.), to the absolute and relative densities of two prey species. J. Anim. Ecol. 46: 867-874.
317. Goss-Custard, J.D. 1977d. Optimal foraging and the size selection of worms by Redshank, *Tringa totanus*, in the field. Anim. Behav. 25: 10-29.
318. Goss-Custard, J.D. 1977e. The ecology of the Wash III. Density-related behaviour and the possible effects of a loss of feeding grounds on wading birds (Charadrii). J. appl. Ecol. 14: 721-739.
319. Goss-Custard, J.D. 1980. Competition for food and interference among waders. Ardea 68: 31-52.
320. Goss-Custard, J.D. 1984. Intake rate and food supply in migrating and wintering shorebirds. In: J. Burger & B.L. Olla (eds.). Shorebirds. Breeding behavior and populations: 233-270. Plenum Press, New York.
321. Goss-Custard, J.D. 1993. The effect of migration and scale on the study of bird populations: 1991 Witherby Lecture. Bird Study 40: 81-96.
322. Goss-Custard, J.D. (ed.). 1996a. The Oystercatcher: from individuals to populations. Oxford University Press, Oxford.
323. Goss-Custard, J.D. 1996b. Conclusions: From individuals to populations: progress in Oystercatchers. In: J.D. Goss-Custard (ed.). The Oystercatcher: from individuals to populations: 384-389. Oxford University Press, Oxford.
324. Goss-Custard, J.D. & S.E.A. le V. dit Durell. 1983. Individual and age differences in the feeding ecology of Oystercatchers *Haematopus ostralegus* wintering on the Exe estuary. Ibis 125: 155-171.
325. Goss-Custard, J.D. & S.E.A le V. dit Durell. 1984. Feeding ecology, winter mortality and the population dynamics of Oystercatchers on the Exe estuary. In: P.R. Evans, J.D. Goss-Custard, W.G. Hale (eds.). Coastal waders and wildfowl in winter: 190-208. Cambridge University Press, Cambridge.
326. Goss-Custard, J.D. & S.E.A. le V. dit Durell. 1987. Age-related effects in Oystercatchers, *Haematopus ostralegus*, feeding on Mussels, *Mytilus edulis*. I. Foraging efficiency and interference. J. Anim. Ecol. 56: 521-536.
327. Goss-Custard, J.D. & S.E.A. le V. dit Durell. 1988. The effect of dominance and feeding method on the intake rates of Oystercatchers, *Haematopus ostralegus*, feeding on Mussels. J. Anim. Ecol. 57: 827-844.
328. Goss-Custard, J.D. & M.E. Moser. 1988. Rates of change in the number of Dunlin, *Calidris alpina*, wintering in British estuaries in relation to the spread of *Spartina anglica*. J. appl. Ecol. 25: 95-109.
329. Goss-Custard, J.D., R.E. Jones & P.E. Newbery. 1977a. The ecology of the Wash I. Distribution and diet of wading birds (Charadrii). J. appl. Ecol. 14: 681-700.
330. Goss-Custard, J.D., R.A. Jenyon, R.E. Jones, P.E. Newbery & R. Le B. Williams. 1977b. The ecology of the Wash II. Seasonal variation in the feeding conditions of wading birds (Charadrii). J. appl. Ecol. 14: 701-719.
331. Goss-Custard, J.D., D.G. Kay & R.M. Blindell. 1977c. The density of migratory and overwintering Redshank (*Tringa totanus*) and Curlew (*Numenius arquata*) in relation to the density of their prey in south-east England. Estuar. Cstl. Mar. Sci. 5: 497-510.
332. Goss-Custard, J.D., S.E.A. le V. dit Durell & B.J. Ens. 1982. Individual differences in aggressiveness and food stealing among wintering Oystercatchers, *Haematopus ostralegus*. Anim. Behav. 30: 917-928.

333. Goss-Custard, J.D., S.E.A. le V. dit Durell, S. McGrorty & C.J. Reading. 1982b. Use of Mussel, *Mytilus edulis* beds by Oystercatchers *Haematopus ostralegus* according to age and population size. J. Anim. Ecol. 51: 543-554.

334. Goss-Custard, J.D., R.T. Clarke & S.E.A. le V. dit Durell. 1984. Rates of food intake and aggression of Oystercatchers *Haematopus ostralegus* on the most and least preferred Mussel *Mytilus edulis* beds of the Exe estuary. J. Anim. Ecol. 53: 233-245.

335. Goss-Custard, J.D., A.D. West & S.E.A. le V. dit Durell. 1993. The availability and quality of the Mussel prey (*Mytilus edulis*) of Oystercatchers (*Haematopus ostralegus*). Neth. J. Sea Res. 31: 419-439.

336. Goss-Custard, J.D., E.W.G. Caldow, R.T. Clarke, S.E.A. le V. dit Durell, J. Urfi & A.D. West. 1994. Consequences of habitat loss and change to populations of wintering migratory birds: predicting the local and global effects from studies of individuals. Ibis: S56-S66.

337. Goss-Custard, J.D., E.W.G. Caldow, R.T. Clarke, S.E.A. le V. Durell & W.J. Sutherland. 1995a. Deriving population parameters from individual variations in foraging behaviour. I. Empirical distribution model of Oystercatchers *Haematopus ostralegus* feeding on Mussels *Mytilus edulis*. J. Anim. Ecol. 64: 265-276.

338. Goss-Custard, J.D., R.W.G. Caldow, R.T. Clarke & A.D. West. 1995b. Deriving population parameters from individual variations in foraging behaviour. II. Model tests and population parameters. J. Anim. Ecol. 64: 277-289.

339. Goss-Custard, J.D., R.T. Clarke, K.B. Briggs, B.J. Ens., K.-M. Exo, C. Smit, A.J. Beintema, R.W.G. Caldow, D.C. Catt, N. Clark, S.E.A. le V. dit Durell, M.P. Harris, J.B. Hulscher, P.L. Meininger, N. Picozzi, R. Prys-Jones, U. Safriel & A.D. West. 1995c. Population consequences of winter habitat loss in a migratory shorebird. I. estimating model parameters. J. appl. Ecol. 32: 320-336.

340. Goss-Custard, J.D., R.T. Clarke, S.E.A. le V. dit Durell, R.W.G. Caldow & B.J. Ens. 1995d. Population consequences of winter habitat loss in a migratory shorebird. II. Model predictions. J. appl. Ecol. 32: 337-351.

341. Goss-Custard, J.D., S. McGrorty & S.E.A. le V. dit Durell. 1996a. The effect of Oystercatchers *Haematopus ostralegus* on shellfish populations. Ardea 84A: 453-468.

342. Goss-Custard, J.D., S.E.A le V. dit Durell, C.P. Goater, J.B. Hulscher, R.H.D. Lambeck, P.L. Meininger & J. Urfi. 1996b. How Oystercatchers survive the winter. In: J.D. Goss-Custard (ed.). The Oystercatcher: from individuals to populations: 133-154. Oxford University Press, Oxford.

343. Goss-Custard, J.D., A.D. West & W.J. Sutherland. 1996c. Where to feed. In: J.D. Goss-Custard (ed.). The Oystercatcher: from individuals to populations: 105-132. Oxford University Press, Oxford.

344. Goss-Custard, J.D., S.E.A. le V. dit Durell, R.T. Clarke, A.J. Beintema, R.W.G. Caldow, P.L. Meininger & C.J. Smit. 1996d. Population dynamics: predicting the consequences of habitat change at the continental scale. In: J.D. Goss-Custard (ed.). The Oystercatcher: from individuals to populations: 352-383. Oxford University Press, Oxford.

345. Goss-Custard, J.D., A.D. West, R.T. Clarke, R.W.G. Caldow & S.E.A. le V. dit Durell. 1996e. The carrying capacity of coastal habitats for Oystercatchers. In: J.D. Goss-Custard (ed.). The Oystercatcher: from individuals to populations: 327-351. Oxford University Press, Oxford.

346. Götmark, F. 1992. The effects of investigator disturbance on nesting birds. Current Ornithology 9: 63-103.

347. Grafen, A. 1988. On the uses of data on lifetime reproductive success. In: T.H. Clutton-Brock (ed.). Reproductive success: studies of individual variation in contrasting breeding systems: 454-471. University of Chicago Press, Chicago.

348. Grafen, A. 1990. Biological signals as handicaps. J. theor. Biol. 102: 549-567.

349. Grant, M.C. 1991. Relationshiop between egg size, chick size at hatching, and chick survival in the Whimbrel *Numenius phaeopus*. Ibis 133: 127-133.

350. Gratto, G.W., L.H. Thomas & C.L. Gratto. 1984. Some aspects of the foraging ecology of migrant juvenile sandpipers in the outer Bay of Fundy. Can. J. Zool. 62: 1889-1892.

351. Graveland, J. 1995. The quest for calcium: calcium limitation in the reproduction of forest passerines in relation to snail abundance and soil acidification. PhD thesis, University of Groningen, Groningen.

352. Green, G.H., J.J.D. Greenwood & C.S. Lloyd. 1977. The influence of snow conditions on the date of breeding of wading birds in north-east Greenland. J. Zool. Lond. 183: 311-328.

353. Green, R.E., G.J.M. Hirons & J.S. Kirby. 1990. The effectiveness of nest defence by Black-tailed Godwits *Limosa limosa*. Ardea 78: 405-413.

354. Greenwood, J.G. 1986. Geographical variation and taxonomy of the Dunlin *Calidris alpina*. Bull. Brit. Ornithol. Club. 106: 43-56.

355. Greenwood, J.J.D. 1987. Three-year cycles of lemmings and Arctic geese explained. Nature 328: 577.

356. Gretton, A. 1991. The ecology and conservation of the Slender-billed Curlew (*Numenius tenuirostris*). International Council for Bird Preservation, Cambridge.

357. Grönlund, E. & O. Melander (eds.). 1995. Swedish-Russian Tundra Ecology-Expedition-94, a cruise report. Swedish Polar Research Secretariat, Stockholm.

358. Grønstøl, G.B. 1996. Aerobatic components in the song-flight display of male Lapwings *Vanellus vanellus* as cues in female choice. Ardea 84: 45-55.

359. Groves, S. 1984. Chick growth, sibling rivalry, and chick production in American Black Oystercatchers. Auk 101: 525-531.

360. Gudmundsson, G.A. 1988. Intraspecific variation in bird migration patterns. Introductory Paper 49, Department of Animal Ecology, Lund University, Lund.

361. Gudmundsson, G.A. 1993. The spring migration pattern of arctic birds in southwest Iceland, as recorded by radar. Ibis 135: 166-176.

362. Gudmundsson, G.A. 1994. Spring migration of the Knot *Calidris c. canutus* over southern Scandinavia, as recorded by radar. J. Avian Biol. 25: 15-26.

363. Gudmundsson, G.A. & T. Alerstam. 1998a. Why is there no transpolar migration? J. Avian Biol. 29: 93-96.

364. Gudmundsson, G.A. & T. Alerstam. 1998b. Optimal map projections for analysing long-distance migration routes. J. Avian Biol. 29: 597-605.

365. Gudmundsson, G.A., Å. Lindström & T. Alerstam. 1991. Optimal fat loads and long-distance flights by migrating Knots *Calidris canutus*, Sanderlings *C. alba* and Turnstones *Arenaria interpres*. Ibis 133: 140-152.

366. Gudmundsson, G.A., S. Benvenuti, T. Alerstam, F. Papi, K. Lilliendahl & S. Åkesson. 1995. Examining the limits of flight and orientation performance: satellite tracking of Brent Geese migrating across the Greenland ice-cap. Proc. R. Soc. Lond. B 261: 73-79.

367. Gwinner, E. 1977. Circannual rhythms in bird migration. Ann. Rev. Ecol. Syst. 8: 381-405,

368. Gwinner, E. 1986. Circannual rhythms. Endogenous annual clocks in the organization of seasonal processes. Springer-Verlag, Berlin.

369. Gwinner, E. 1995. Circannual clocks in avian reproduction and migration. Ibis 138: 47-63.

370. Gwinner, E. 1996. Circadian and circannual programmes in avian migration. J. exp. Biol. 199: 39-48.

371. Gwinner, E & W. Wiltschko. 1978. Endogenously controlled changes in migratory direction of the Garden Warbler, *Sylvia borin*. J. Comp. Physiol. 125: 267-273.

372. Hale, W.G. 1980. Waders. Collins, London.

373. Hamilton, W.D. 1971. Geometry of the selfish herd. J. theor. Biol. 31: 295-311.

374. Hamilton, W.D. & M. Zuk. 1982. Heritable true fitness and bright birds: a role for parasites? Science 218: 384-387.

375. Hamilton, W.D. & M. Zuk. 1989. Parasites and sexual selection: a reply. Nature 341: 289-290.

376. Hammond, K.A. & J. Diamond. 1997. Maximum sustained energy budgets in humans and animals. Nature 386: 457-462.

377. Harrington, B.A. 1996. The flight of the Red Knot. A natural history account of a small bird's annual migration from the Arctic Circle to the tip of South America and back. Norton, New York.

378. Harris, M.P. 1970. Territory limiting the size of the breeding population of the Oystercatcher (*Haematopus ostralegus*) - a removal experiment. J. Anim. Ecol. 39: 707-713.

379. Harris, M.P., U.N. Safriel, M. de L. Brooke & C.K. Britton. 1987. The pair bond and divorce among Oystercatchers *Haematopus ostralegus* on Skokholm Island, Wales. Ibis 129: 45-57.

380. Harrison, J.A., D.G. Allen, L.G. Underhill, M. Herremans, A.J. Tree, V. Parker & C.J. Brown (eds.). 1977. The Atlas of southern African birds. Vol. 1: Non-passerines. BirdLife South Africa, Johannesburg.

381. Hartley, I.R., M. Shepherd, T. Robson & T. Burke. 1993. Reproductive success of polygynous male Corn Buntings (*Miliaria calandra*) as confirmed by DNA fingerprinting. Behav. Ecol. 4: 310-317.

382. Harvey, P.H. & M.D. Pagel. 1991. The comparative method in evolutionary biology. Oxford University Press, Oxford.

383. van der Have, T., E. Nieboer & G.C. Boere. 1984. Age-related distribution of Dunlin in the Dutch Wadden Sea. In: P.R. Evans, G.D. Goss-Custard & W.G. Hale (eds.). Coastal waders and wildfowl in winter: 160-189. Cambridge University Press, Cambridge.

384. Hedenström, A. 1993. Migration by soaring or flapping flight in birds: the relative importance of energy cost and speed. Phil. Trans. R. Soc. Lond. B 342: 353-361.

385. Hedenström, A. & T. Alerstam. 1992. Climbing performance of migrating birds as a basis for estimating limits for fuel-carrying capacity and muscle work. J. exp. Biol. 164: 19-38.

386. Hedenström, A. & T. Alerstam. 1994. Optimal climbing flight in migrating birds: predictions and observations on Knots and Turnstones. Anim. Behav. 48: 47-54.

387. Hedenström, A. & T. Alerstam. 1995. Optimal flight speeds of birds. Phil. Trans. R. Soc. Lond. B 348: 471-487.

388. Heg, D. 1999. Life history decisions in Oystercatchers. PhD thesis, University of Groningen, Groningen.

389. Heg, D. 1999. Timing of breeding in Oystercatchers: II. Fitness consequences. In: Heg, D. 1999. Life history decisions in Oystercatchers: 177-204. PhD thesis, University of Groningen, Groningen.

390. Heg, D., B.J. Ens, T. Burke, L. Jenkins. & J.P. Kruijt. 1993. Why does the typically monogamous Oystercatcher (*Haematopus ostralegus*) engage in extra-pair copulations? Behaviour 126: 247-289.

391. Heg, D. & R. van Treuren. 1998. Female-female cooperation in polygynous oystercatchers. Nature 391: 687-691.

392. Heg, D. & M. van der Velde 2001. Effects of territory quality, food availability and sibling competition on the fledging success of oystercatchers (*Haematopus ostralegus*). Behav. Ecol. Sociobiol. 49 (2-3): 157-169.

393. Heip, C. 1995. Eutrophication and zoobenthos dynamics. Ophelia 41: 113-136.

394. Helbig, A.J. 1990. Are orientation mechanisms among migratory birds species-specific? Trends Ecol. Evol. 5: 365-367.

395. Helbig, A.J. 1996. Genetic basis, mode of inheritance and evolutionary changes of migratory directions in Palearctic warblers (Aves: Sylviidae). J. exp. Biol. 199: 49-55.

396. Hepburn, I.R. 1987. Conservation of wader habitats in coastal West Africa. Wader Study Group Bull. 49, Suppl. / IWRB Special Publ. 7: 125-127.

397. Heppleston, P.B. 1971. The feeding ecology of Oystercatchers (*Haematopus ostralegus*) in winter in northern Scotland. J. Anim. Ecol. 40: 651-672.

398. Heppner, F. 1997. Three-dimensional structure and dynamics of bird flocks. In: J.K. Parrish & W.M. Hamner (eds.). Animal groups in three dimensions: 68-89. Cambridge University Press, Cambridge.

399. Hicklin, P.W. & P.C. Smith. 1984. Selection of foraging sites and invertebrates prey by migrant Semipalmated Sandpipers *Calidris pusilla* (Pallas) in Minas Bay, Bay of Fundy. Can. J. Zool. 62: 2201-2210.

400. Higler, B., N. Dankers, A. Smaal & V. de Jonge. 1998. Evaluatie van de ecologische effecten van het reguleren van schelpdiervisserij in Waddenzee en Delta op bodemorganismen en vogels. Annex V in LNV report: Structuurnota Zee- en kustvisserij: Evaluatie van de maatregelen in de kustvisserij gedurende de eerste fase (1993-1997). Ministry of Agriculture, Nature Conservation & Fisheries (LNV), Den Haag.

401. Hildén, O. 1983. Mating system and breeding biology of the Little Stint *Calidris minuta*. Wader Study Group Bull. 39: 47.

402. Hockey, P.A.R., R.A. Navarro, B. Kalejta & C.R. Vélasques. 1992. The riddle of the sands: why are shorebird densities so high in southern estuaries? Am. Nat. 140: 961-979.

403. van 't Hoff, J. 1998. Wadvogeltellingen Groninger Noordkust. Report Wadvogelwerkgroep Avifauna Groningen, Groningen.

404. Hoffman-Goetz, L. & B.K. Pedersen. 1994. Exercise and the immune system: a model of the stress response? Immunology Today. 15: 382-387.

405. Hofstee, J. 1983. Explanatory memorandum on methods of analysis for soil, organic matter, water and soil moisture. Rijksdienst voor de IJsselmeerpolders (RIZA), Lelystad.

406. Holling, C.S. 1966. The functional response of invertebrate predators to prey density. Mem. Entomol. Soc. Can. 48: 1-86.

407. Holthuijzen, Y.A. 1979. Het voedsel van de Zwarte Ruiter *Tringa erythropus* in de Dollard. Limosa 52: 22-33.

408. Hötker, H. 1995. Aktivitätsrhythmus von Brandgänsen (*Tadorna tadorna*) und Watvögeln (Charadrii) an der Nordseeküste. J. Orn. 136: 105-126.

409. Hötker, H. 1998. Intraspecific variation in length of incubation period in Avocets *Recurvirostra avosetta*. Ardea 86: 33-41.

410. Hötker, H. 2000. Conspecific nest parasistism in the Pied Avocet *Recurvirostra avosetta*. Ibis 142: 280-288.

411. Hötker, H. 2000. Intraspecific variation in size and density of Avocet colonies: effects of nest-distances on hatching and breeding success. J. Avian Biol. 31: 387-398.

412. Hötker, H. & A. Segebade. 2000. The effects of predation and weather on the breeding success of Avocets *Recurvirostra avosetta*. Bird Study 47: 91-101.

413. Houston, A.I. 1996. Estimating Oystercatcher mortality from foraging behaviour. Trends Ecol. Evol. 11: 108-109.

414. del Hoyo, J., A. Elliott & J. Sargatal (eds.). 1996. Handbook of the birds of the world, Vol. 3. Hoatzin to auks. Lynx Edicions, Barcelona.

415. Hughes, R.N. 1980. Optimal foraging theory in the marine context. Oceanogr. Mar. Biol. Ann. Rev. 18: 423-481.

416. Hulscher, J.B. 1976. Localisation of Cockles (*Cardium edule* L.) by the Oystercatcher (*Haematopus ostralegus* L.) in darkness and daylight. Ardea 64: 292-310.

417. Hulscher, J.B. 1982. The Oystercatcher *Haematopus ostralegus* as a predator of the bivalve *Macoma balthica* in the Dutch Wadden Sea. Ardea 70: 89-152.

418. Hulscher, J.B. 1985. Growth and abrasion of the Oystercatcher bill in relation to dietary switches. Neth. J. Zool. 35: 124-154.

419. Hulscher, J.B. 1988. Mossel doodt Scholekster *Haematopus ostralegus*. Limosa 61: 42-45.

420. Hulscher, J.B. 1989. Sterfte en overleving van Scholeksters *Haematopus ostralegus* bij strenge vorst. Limosa 62: 177-181.

421. Hulscher, J.B. 1990. Survival of Oystercatchers during hard winter weather. Ring. 13: 167-172.

422. Hulscher, J.B. 1996. Food and feeding behaviour. In: J.D. Goss-Custard (ed.). The Oystercatcher: from individual to populations: 7-29. Oxford University Press, Oxford.

423. Hulscher, J.B. & B.J. Ens. 1991. Somatic modifications of feeding system structures due to feeding on different foods with emphasis on changes in bill shape in Oystercatchers. Acta XX Congr. intern. Ornith., Christchurch: 889-896.

424. Hulscher, J.B., E.J. Bunskoeke, D. Alting & B.J. Ens. 1996a. Subtle differences between male and female Oystercatcher *Haematopus ostralegus* in locating and handling the bivalve mollusc *Macoma balthica*. Ardea 84A: 117-130.

425. Hulscher, J.B., K.-M. Exo & N.A. Clark. 1996b. Why do Oystercatchers migrate? In: J.D. Goss-Custard (ed.). The Oystercatcher: from individual to populations: 155-185. Oxford University Press, Oxford.

426. Hume, I.D. & H. Biebach. 1996. Digestive tract function in the long-distance migratory Garden Warbler, *Sylvia borin*. J. Comp. Physiol. B. 166: 388-395.

427. Hummel, D. 1973. Die Leistungsersparnis beim Verbandsflug. J. Orn. 114: 259-282.

428. Hummel, H. 1985. Food intake of *Macoma balthica* (Mollusca) in relation to seasonal changes in its potential food on a tidal flat in the Dutch Wadden Sea. Neth. J. Sea Res. 19: 52-76.

429. Hurford, C. & P. Lansdown. 1995. Birds of Glamorgan. D. Brown & Sons Ltd., Glamorgan.

430. Hustings, F. & E. van Winden. 1998. Slechtvalken terug uit een diep dal. SOVON-News 11 (4): 14-16.

431. van Impe, J. 1985. Estuarine pollution as a probable cause of increase of estuarine birds. Mar. Pollut. Bull. 16: 271-276.

432. Jacob, J. & J. Poltz. 1973. Chemotaxonomische Untersuchungen an Limikolen. Die Zusammensetzung des Bürzeldrüsen sekretes von Austernfischer, Rotschenkel, Knutt und Alpenstrandläufer. Biochem. Syst. Ecol. *1*: 169-172.

433. Jehl, J.R., Jr. 1997. Fat loads and flightlessness in Wilson's Phalaropes. Condor 99: 538-543.

434. Jenni-Eiermann, S. & L. Jenni. 1992. High plasma triglyceride levels in small birds during migratory flight: A new pathway for fuel supply during endurance locomotion at very high mass-specific metabolic rates? Physiol. Zool. 65: 112-123.

435. Jenni-Eiermann, S. & L. Jenni. 1998. What to take on board: metabolic constraints and adaptations of fuel supply in migrating birds. J. Avian Biol. 29: 521-528.

436. Jensen, K.T. 1985. The presence of the bivalve *Cerastoderma edule* affects migration, survival and reproduction of the amphipod *Corophium volutator*. Mar. Ecol. Prog. Ser. 25: 269-277.

437. Jenssen, B.M., M. Ekker & C. Bech. 1989. Thermoregulation in winter acclimatized Common Eiders (*Somateria mollissima*) in air and water. Can. J. Zool. 67: 669-673.

438. van der Jeugd, H.P. & K. Larsson. 1998. Pre-breeding survival of Barnacle Geese *Branta leucopsis* in relation to fledgling characteristics. J. Anim. Ecol. 7: 953-966.

439. Johnson, C. 1985. Patterns of seasonal weight variation in waders on the Wash. Ring. & Migr. 6: 19-32.

440. Johnson, D.H. 1979. Estimating nest success: the Mayfield method and an alternative. Auk 96: 651-661.

441. Jones, A.M. 1979. Structure and growth of a high-level population of *Cerastoderma edule* (Lamellibranchiata). J. mar. biol. Ass. U.K. 59: 277-287.

442. de Jonge, V., J. Bakker & D. Stoppelenburg. 1996. Zwarte vlekken in de Duitse Waddenzee. Waddenbulletin 96-4: 24.

443. Jönsson, P.E. 1988. Ecology of the Southern Dunlin *Calidris alpina schinzii*. PhD thesis, University of Lund, Lund.

444. Jukema, J. 1987a. Was de Kleine Goudplevier (Pluvialis fulva) eens een talrijke doortrekker in Friesland? Vanellus 40: 85-99

445. Jukema, J. 1987b. Were Lesser Golden Plovers Pluvialis fulva regular winter visitors to Friesland, The Netherlands, in the first half of the 20th century? Wader Study Group Bull. 51: 56-58.

446. Jukema, J. 1988. Over de Aziatische Kleine Goudplevier Pluvialis fulva, als doortrekker in Friesland. Limosa 61: 189-190.

447. Jukema, J. & T. Piersma. 1987. Special moult of breast and belly feathers during breeding in Golden Plovers Pluvialis apricaria. Ornis Scand. 18: 157-162.

448. Jukema, J. & T. Piersma. 1993. Hoe rosser de grutto, hoe beter ie trekt. Limosa 66: 32-34.

449. Kaiser, J. & S. Lutter. 1998. Do we have the right strategies to combat eutrophication in the Wadden Sea? – A critical review of current policies. Senckenbergiana maritima 29: 17-24.

450. Kålås, J.A. & L. Løfaldli. 1987. Clutch size in the Dotterel Charadrius morinellus: an adaptation to parental incubation behaviour. Ornis Scand. 18: 316-319.

451. Kalejta, B. 1992. Time budget and predatory impact at the Berg River estuary, South Africa. Ardea 80: 327-342.

452. Kalejta, B. & P.A.R. Hockey. 1994. Distribution of shorebirds at the Berg River estuary, South Africa, in relation to foraging modes, food supply and environmental features. Ibis 136: 233-239.

453. Kamermans, P. 1992. Growth limitation in intertidal bivalves of the Dutch Wadden Sea. PhD thesis, University of Groningen, Groningen.

454. Karasov, W.H. 1990. Digestion in birds: chemical and physiological determinants and ecological implications. Stud. Avian Biol. 13: 391-415.

455. Kerlinger, P. 1989. Flight strategies of migrating hawks. University of Chicago Press, Chicago.

456. Kerlinger, P. & F.R. Moore. 1989. Atmospheric structure and avian migration. Curr. Ornithol. 6: 109-142.

457. Kersten, M. 1996. Time and energy budgets of Oystercatchers Haematopus ostralegus occupying territories of different quality. Ardea 84A: 291-310

458. Kersten, M. 1997. Living leisurely should last longer: energetic aspects of reproduction in the Oystercatcher. PhD thesis, University of Groningen, Groningen.

459. Kersten M. & A. Brenninkmeijer. 1995. Growth, fledging success and post-fledging survival of juvenile Oystercatchers Haematopus ostralegus. Ibis 137: 396-404.

460. Kersten, M. & T. Piersma. 1987. High levels of energy expenditure in shorebirds; metabolic adaptations to an energetically expensive way of life. Ardea 75: 175-187.

461. Kersten, M. & C.J. Smit. 1984. The Atlantic coast of Morocco. In: P.R. Evans, J.D. Goss-Custard & W.G. Hale (eds.). Coastal waders and wildfowl in winter: 276-292. Cambridge University Press, Cambridge.

462. Kersten, M. & W. Visser. 1996a. The rate of food processing in Oystercatchers: food intake and energy expenditure constrained by a digestive bottleneck. Funct. Ecol. 10: 440-448.

463. Kersten, M. & W. Visser. 1996b. Food intake of Oystercatcher by day and night measured with an electronic nest balance. Ardea 84A: 57-72.

464. Kersten, M., T. Piersma, C.J. Smit & P.M. Zegers. 1983. Wader migration along the Atlantic coast of Morocco, March. 1981. Report of the Netherlands Morocco Expedition. 1981. RIN (Alterra) Report 83/20, Texel.

465. Kersten, M., C. Rappoldt & K. van Scharenburg. 1997. Wadvogels op Ameland. In: M. Versluys, D. Blok & R. van der Wal (eds.). Vogels van Ameland: 55-87. Friese Pers Boekerij, Leeuwarden.

466. Kersten, M., L.W. Bruinzeel, P. Wiersma & T. Piersma. 1998. Reduced basal metabolic rate of migratory waders wintering in coastal Africa. Ardea 86: 71-80.

467. Ketterson, E.D. & V. Nolan, Jr. 1983. The evolution of differential bird migration. Curr. Ornithol. 1: 357-402.

468. Kiepenheuer, J. 1984. The magnetic compass mechanism of birds and its possible association with the shifting course directions of migrants. Behav. Ecol. Sociobiol. 14: 81-99.

469. Kirkwood, J.K. 1983. A limit to metabolisable energy intake in mammals and birds. Comp. Biochem. Physiol. 75A: 1-3.

470. Klaassen, M. 1996. Metabolic constraints on long-distance migration in birds. J. exp. Biol. 199: 57-64.

471. Klaassen, M. & B.J. Ens. 1990. Is salt stress a problem for waders wintering on the Banc d'Arguin, Mauritania? Ardea 78: 67-74.

472. Klaassen, M., M. Kersten & B.J. Ens. 1990. Energetic requirements for maintenance and premigratory body mass gain of waders wintering in Africa. Ardea 78: 209-220.

473. Klaassen, M., Å. Lindström & R. Zijlstra. 1997. Composition of fuel stores and digestive limitations to fuel deposition rate in the long-distance migratory Thrush Nightingale Luscinia luscinia. Physiol. Zool. 70: 125-133.

474. Kluyver, H.N. & L. Tinbergen. 1953. Territory and regulation of density in titmice. Arch. Neerl. Zool. 10: 265-289.

475. Knol, E., W. Prummel, H.T. Uytterschaut, M.L.P. Hoogland, W.A. Casparie, G.J. de Langen, E. Kramer & J. Schelvis. 1996. The early medieval cemetry of Oosterbeintum (Friesland). Palaeohistoria 37/38: 245-416.

476. Köbben, A.J.F. & H. Tromp. 1999. De onwelkome boodschap, of hoe de vrijheid van wetenschap bedreigd wordt. Publisher Jan Mets, Amsterdam.

477. Koeman, J.H., A.A.G. Oskamp, J. Veen, E. Brouwer, J. Rooth, P. Zwart, E. van den Broek & H. van Genderen. 1967. Insecticides as a factor in the mortality of the Sandwich Tern (Sterna sandvicensis). A preliminary communication. - Med. fac. Landbouwwet. Gent 32: 841-854.

478. Koene, P. 1978. De Scholekster: aantalseffecten op de voedselopname. Students report Zoological Laboratory, University of Groningen.

479. Koivula, K. & Rönkä. 1998. Habitat deterioration and efficiency of antipredator strategy of a meadow-breeding wader, Temminck's stint (Calidris temminckii). Oecologia 116: 348-355.

480. Kokke-Smits, M.E. & J.W.M. Osse. 1968. Van der Klaauw en van Oordt's Technische Termen ten gebruike van het Zoölogische en Anatomische Onderwijs aan Nederlandsche Universiteiten. Eigth edition. Brill, Leiden.

481. Koks, B. 1998. Slechtvalken in de Nederlandse Waddenzee. Slechtvalk Newsletter 4 (2): 13-15.

482. Koks, B. & F. Hustings. 1998. Broedvogelmonitoring in het Nederlandse Waddengebied in. 1995 en. 1996. SOVON-monitoring Report 1998/05. SOVON, Beek-Ubbergen.

483. Koolhaas, A., T. Piersma & J. van den Broek. 1998. Kokkel- en mosselvisserij beschadigen het wadleven. De Levende Natuur 99: 254-260.

484. Kramer, G. 1950. Orientierte Zugaktivität gekäfigter Singvögel. Naturwissenschaften 37: 188.

485. Kramer, G. 1953. Wird die Sonnehohe bei der Heimfindorientierung verwertet? J. Orn. 94: 201-219.

486. Krebs, J.R. & A. Kacelnik. 1991. Decision-making. In: J.R. Krebs & N.B. Davies (eds.). Behavioural ecology: an evolutionary approach: 105-136. Blackwell, Oxford.

487. van de Kuip, C. 1991. Wanbeleid in de Waddenzee kost duizenden vogels het leven. Vogels 66: 230-235.

488. Kvist, A., M. Klaassen & Å. Lindström. 1998. Energy expenditure in relation to flight speed: what is the power of mass loss estimates? J. Avian Biol. 29: 485-498.

489. Lack, D. 1947. The significance of clutch size. Ibis 89: 302-352.

490. Lack, D. 1950. The breeding season of European birds. Ibis 92: 288-316.

491. Lack, D. 1954. The natural regulation of animal numbers. Clarendon Press, Oxford.

492. Lack, D. 1968a. Ecological adaptation for breeding in birds. Methuen, London.

493. Lack, D. 1968b. Bird migration and natural selection. Oikos. 19: 1-9.

494. Lambeck, R.H.D., A.J.J. Sandee & L. de Wolf. 1989. Long-term patterns in the wader usage of an intertidal flat in the Oosterschelde (SW Netherlands) and the impact of the closure of an adjacent estuary. J. appl. Ecol. 26: 419-431.

495. Lambeck, R.H.D., J.D. Goss-Custard & P. Triplet. 1996. Oystercatchers and man in the coastal zone. In: J.D. Goss-Custard (ed.). The Oystercatcher: from individual to populations: 289-326. Oxford University Press, Oxford.

496. Lanctot, R.B., R.E. Gill Jr., T.L. Tibbitts & C.M. Handel. 1995. Brood amalgamation in the Bristle-thighed Curlew Numenius tahitiensis: process and function. Ibis 137: 559-569.

497. Lanctot, R.B., K.T. Scribner, B. Kempenaers & P.J. Weatherhead. 1997. Lekking without a paradox in the Buff-breasted Sandpiper. Am. Nat. 149: 1051-1070.

498. Larsen, T. & S. Grundetjern. 1997. Optimal choice of neighbour: predator protection among tundra birds. J. Avian Biol. 28: 303-308.

499. Larsson, K. & P. Forslund. 1994. Population dynamics of the Barnacle Goose Branta leucopsis population in the Baltic area: density-dependent effects on reproduction. J. Anim. Ecol. 63: 954-962.

500. Larsson, K., P. Forslund, L. Gustafsson & B.S. Ebbinge. 1988. From the High Arctic to the Baltic: the successful establishment of a Barnacle Goose population on Gotland Sweden. Ornis Scand. 19: 182-189.

501. Lauckner, G. 1984. Parasiten- ihr Einfluss im Ökosystem Wattenmeer. In: J.L. Lozán, E. Rachor, B. Watermann & H. von Westernhagen (eds.). Warnsignale aus der Nordsee: 219-230. Parey, Berlin.

502. Laursen, K., I. Gram & L.J. Alberto. 1983. Short-term effect of reclamation on numbers and distribution of waterfowl at Hojer, Danish Wadden Sea. Proc. Third Nordic Congr. Ornithol. 1981: 97-118.

503. LeFebvre, E.A. 1964. The use of $D_2O^{18}$ for measuring energy metabolism in Columbia livia at rest and in flight. Auk 81: 403-416.

504. Leopold, M.F., J.F. van Elk & Y.M. van Heezik. 1996. Central place foraging in Oystercatchers *Haematopus ostralegus*: can parents that transport Mussels *Mytilus edulis* to their young profit from size selection? Ardea 84A: 311-325.
505. Leslie, R. & C.M. Lessells. 1978. The migration of Dunlin *Calidris alpina* through northern Scandinavia. Ornis Scand. 9: 84-86.
506. Lessells, C.M. 1983. The mating system of Kentish Plovers *Charadrius alexandrinus*: some observations and experiments. Wader Study Group Bull. 39: 43.
507. Lessells, C.M. 1984. The mating system of Kentish Plovers *Charadrius alexandrinus*. Ibis 126: 474-483.
508. Lessells, C.M. 1991. The evolution of life histories. In: J.R. Krebs & N.B. Davies (eds.). Behavioural ecology: an evolutionary aqpproach, Third Edition: 32-68. Blackwell Scientific Publications, Oxford.
509. Lessells, C.M. 1999. Sexual conflict in animals. In: L. Keller (ed.). Levels of selection in evolution: 75-99. Princeton University Press.
510. Lewin, R. 1998. Evolutiepatronen. De nieuwste moleculaire inzichten. Natuur & Techniek/Segment, Beek.
511. Lifjeld, J. 1984a. Prey selection in relation to body size and bill length of five species of waders feeding in the same habitat. Ornis Scand. 15: 217-226.
512. Lifjeld, J. 1984b. Prey and grit taken by five species of waders at an autumn migration staging post in N Norway. Fauna norv. C. Cinclus 7: 28-36.
513. Liley, D.C. 1998. Predicting the population consequences of disturbance. Ostrich 69: 202.
514. Lindstedt, S. & W.A. Calder, III. 1981. Body size, physiological time, and longevity of homeothermic animals. Q. Rev. Biol. 56: 1-16.
515. Lindström, Å. 1997. Basal metabolic rates of migrating waders in the European Arctic. J. Avian Biol. 28: 87-92.
516. Lindström, Å. & J. Agrell. 1999. Global change and possible effects on the migration and reproduction of Arctic-breeding waders. Ecol. Bull. 47: 145-159.
517. Lindström, Å. and T. Piersma. 1993. Mass changes in migrating birds: the evidence for fat and protein storage re-examined. Ibis 135: 70-78.
518. Lindström, Å. & A. Kvist. 1995. Maximum energy intake rate is proportional to basal metabolic rate in passerine birds. Proc. R. Soc. Lond. B 261: 337-343.
519. Lindström, Å., G.H. Visser & T. Piersma. 1999. Flight studies with Knots in the Lund wind tunnel. Annual Report. 1998 of the Netherlands Institute for Sea Research (NIOZ), Texel.
520. Linnaeus, C. 1759. Systema Naturae. Reprinted in. 1964, Wheldon & Wesley, New York.
521. Lissaman, P.B.S. & C.A. Shollenberger. 1970. Formation flight of birds. Science 168: 1003-1005.
522. LNV. 1998. Structuurnota Zee- en kustvisserij: Evaluatie van de maatregelen in de kustvisserij gedurende de eerste fase (1993-1997). Ministerie van Landbouw, Natuurbeheer & Visserij, Den Haag.
523. Lomnicki, A. 1988. Population ecology of individuals. Princeton University Press, Princeton.
524. Loonen, M.J.J.E. 1997. Goose breeding ecology: overcoming successive hurdles to raise goslings. PhD thesis, University of Groningen, Groningen.
525. Lovelock, J. 1988. The ages of Gaia: a biography of our living earth. Oxford University Press, Oxford.
526. Lozán, J.L., E. Rachor, K. Reise, H. von Westernhagen & W. Lenz. 1994. Warnsignale aus dem Wattenmeer. Blackwell Wissenschaftsverlag, Berlijn.
527. MacArhur, R.H. & E.R. Pianka. 1966. On optimal use of a patchy environment. Am. Nat. 100: 603-609.
528. MacIvor, L.H., S.M. Melvin & C.R. Griffin. 1990. Effects of research activity on Piping Plover nest predation. J. Wildl. Managem. 54: 443-447.
529. MacLean, S.F., Jr. 1974. Lemming bones as a source of calcium for Arctic Sandpipers (*Calidris* spp.). Ibis 116: 552-557.
530. Madsen, J. 1985. Impact of human disturbance of field utilization of Pink-footed Geese in West Jutland, Denmark. Biol. Conserv. 33: 53-63.
531. Madsen, J. 1988. Autumn feeding ecology of herbivorous wildfowl in the Danish Wadden Sea, and impact of food supplies and shooting on movements. Dan. Rev. Game Biol. 13: 1-32.
532. Madsen, J. 1994. Impact of disturbance on migratory waterfowl. Ibis 137: S67-S74.
533. Major, P.F. & L.M. Dill. 1978. The three-dimensional structure of airborne bird flocks. Behav. Ecol. Sociobiol. 4: 111-122.
534. Mangel, M. & C.W. Clark. 1988. Dynamic modeling in behavioral ecology. Princeton University Press, Princeton.
535. Marquenie, J.M. & H.J. Gussinklo. 1998. Integrale bodemdalingstudie Waddenzee. Report: abstract. Nederlandse Aardolie Maatschappij B.V., Assen.
536. Martin, T.E. 1993. Nest predation and nest sites. Bioscience 43: 523-532.
537. Masman, D. & M. Klaassen. 1987. Energy expenditure during free flight in trained and free-living Kestrels (*Falco tinnunculus*). Auk 104: 603-616.
538. Masman, D., S. Daan & H.J.A. Beldhuis. 1988. Ecological energetics of the Kestrel: daily energy expenditure throughout the year based on time-energy budget, food intake and doubly labelled water methods. Ardea 76: 64-81.
539. Masman, D., S. Daan & M.W. Dietz. 1989. Heat increment of feeding in the Kestrel, *Falco tinnunculus*, and its natural seasonal variation. In: C. Bech & R.E. Reinertsen (eds.). Physiology of cold adaptation in birds: 123-136. Plenum Press, New York.
540. Matthiessen, P. 1994. The wind birds. Shorebirds of North America. Chapters Publishing, Shelburne, Vermont.
541. Maynard Smith, J. 1958. The theory of evolution. Penguinbooks, Harmondsworth.
542. Maynard Smith, J. & G.R. Price. 1973. The logic of animal conflict. Nature 246: 15-18.
543. Mayfield, H. 1961. Nesting success calculated from exposure. Wilson Bull. 73: 255-261.
544. Mayfield, H. 1975. Suggestions for calculating nest success. Wilson Bull. 87: 456-466.
545. Mayr, E. & W. Meise. 1930. Theoretisches zur Geschichte des Vogelzuges. Vogelzug. 1: 149-172.
546. McNeil, R. 1970. Hivernage et estivage d'oiseaux aquatiques nord-américains dans le nord-est du Vénézuéla (Mue, accumulation de graisse, capacité de vol et routes de migration). Oiseaux et R.F.O. 40: 185-302.
547. McNeil, R., P. Drapeau & J.D. Goss-Custard. 1992. The occurrence and adaptive significance of nocturnal habits in waterfowl. Biol. Rev. 67: 381-419.
548. McNeil, R., M.T. Diaz & A. Villeneuve. 1994. The mystery of shorebird over-summering: a new hypothesis. Ardea 82: 143-152.
549. McGrorty, S., R.T. Clarke, C.J. Reading & J.D. Goss-Custard. 1990. Population dynamics of the Mussel *Mytilus edulis* density changes and regulation of the population in the Exe estuary, Devon. Mar. Ecol. Prog. Ser. 67: 157-169.
550. McLusky, D.S., D.M. Bryant & M. Elliott. 1992. The impact of land claim on macrobenthos, fish and shorebirds on the Forth Estuary, eastern Scotland. Aquatic Conservation: marine and freshwater ecosystems 2: 211-222.
551. Mead, C. 1983. Bird migration. Country Life books, Feltham-Middlesex.
552. van der Meer, J. 1997. A handful of feathers. Studies on the estimation and modelling of temporal and spatial fluctuations in the numbers of birds and their prey. PhD thesis, University of Groningen, Groningen.
553. van der Meer, J. & B.J. Ens. 1997. Models of interference and their consequences for the spatial distribution of ideal and free predators. J. Anim. Ecol. 66: 333-356.
554. van der Meer, J. & T. Piersma. 1994. Physiologically inspired regression models for estimating and predicting nutrient stores and their composition in birds. Physiol. Zool. 67: 305-329.
555. Meininger, P.L. & F.A. Arts. 1997. De Strandplevier *Charadrius alexandrinus* als broedvogel in Nederland in de 20e eeuw. Limosa 70: 41-60.
556. Meininger, P.L., A-M. Blomert & E.C.L. Marteijn. 1991. Watervogelsterfte in het Deltagebied, ZW-Nederland, gedurende de drie koude winters van. 1985, 1986 en. 1987. Limosa 64: 89-102.
557. Meire, P.M. 1991. Effects of a substantial reduction in intertidal area on numbers and densities of waders. Proceed. intern. Orn. Congr. 20: 2219-2227.
558. Meire, P.M. 1993a. Wader populations and macrozoobenthos in a changing estuary: the Oosterschelde (The Netherlands). PhD thesis, University of Gent, Gent.
559. Meire, O. 1993b. The impact of bird predation on marine and estuarine bivalve populations: a selective review of patterns and inderlying causes. In: R.F. Dame (ed.). Bivalve filter feeders in estuarine and coastal ecosystem process: 197-243. Springer verlag, Heidelberg.
560. Meire, P.M. 1996a. Distribution of Oystercatchers *Haematopus ostralegus* over a tidal flat in relation to their main prey species, Cockles *Cerastoderma edule* and Mussels *Mytilus edulis*: did it change after a substantial habitat loss? Ardea 84A: 525-538.
561. Meire P.M. 1996b. Using optimal foraging theory to determine the density of mussels *Mytilus edulis* that can be harvested by hammering Oystercatchers *Haematopus ostralegus*. Ardea 84A: 141-152.
562. Meire, P.M. 1996c. Feeding behaviour of Oystercatchers *Haematopus ostralegus* during a period of tidal manipulations. Ardea 84A: 509-524.
563. Meire, P.M. & A. Ervynck. 1986. Are Oystercatchers (*Haematopus ostralegus*) selecting the most profitable Mussels (*Mytilus edulis*)? Anim. Behav. 34: 1427-1435.
564. Meire, P.M. & E. Kuyken. 1984. Relations between the distribution of waders and the intertidal benthic fauna of the Oosterschelde, the Netherlands. In: P.R. Evans, J.D. Goss-Custard & W.G. Hale (eds.). Coastal waders and wildfowl in winter: 57-68. Cambridge University Press, Cambridge.

565. Meire P.M., H. Schekkerman & P.L. Meininger. 1994. Consumption of benthic invertebrates by waterbirds in the Oosterschelde estuary, SW Netherlands. Hydrobiologia 282/283: 525-546.
566. Metcalfe N.B. 1985. Prey detection by intertidally feeding Lapwing. Z. Tierpsychol. 67: 45-57.
567. Metcalfe, N.B. & R.W. Furness. 1984. Changing priorities: the effect of premigratory fattening on the trade-off between foraging and vigilance. Behav. Ecol. Sociobiol. 15: 203-206.
568. Meltofte, H. 1996. Are African wintering waders really forced south by competition from northerly wintering conspecifics? Benefits and constraints of northern versus southern wintering and breeding in waders. Ardea 84: 31-44.
569. Meltofte, H., J. Blew, J. Frikke, H-U. Rösner & C.J. Smit. 1994. Numbers and distribution of waterbirds in the Wadden Sea. Results and evaluation of 36 simultaneous counts in the Dutch-German-Danish Wadden Sea. 1980-1991. – IWRB Publication 34 / Wader Study Group Bull. 74, Special issue.
570. Merkel, F.W. & W. Wiltschko. 1965. Magnetismus und Richtungsfinden zugunruhiger Rotkehlchen (*Erithacus rubecula*). Vogelwarte 23: 71-77.
571. Mes, R., unpublished.
572. Millington, R. 1997. Separation of Black Brant, Dark-bellied Brent Goose and Pale-bellied Brent Goose. Birding World 10: 11-15.
573. Mindell, D.P. (ed.). 1997. Avian molecular evolution and systematics. Academic Press, San Diego.
574. Møller, A.P. 1986. Mating systems among European passerines: a review. Ibis 128: 234-245.
575. Møller, A.P. 1994. Sexual selection and the Barn Swallow. Oxford University Press, Oxford.
576. Møller, A.P. 1994. Phenotype-dependent arrival time and its consequences in a migratory bird. Behav. Ecol. Sociobiol. 35: 115-122.
577. Mouritsen, K.N. 1994. Day and night feeding in Dunlins *Calidris alpina*: choice of habitat, foraging technique and prey. J. Avian Biol. 25: 55-62.
578. Moody, A.L., W.A. Thompson, B. de Bruijn, A.I. Houston & J.D. Goss-Custard. 1997. The analysis of spacing of animals, with an example based on oystercatchers during the tidal cycle. J. Anim. Ecol. 66: 615-628.
579. Moore, F.R. 1987. Sunset and the orientation behaviour of migrating birds. Biol. Rev. 62: 65-86.
580. Moreira, F. 1995. Diet of Blak-headed Gulls *Larus ridibundus* on emerged intertidal areas in the Tagus estuary (Portugal): predation or grazing? J. Avian Biol. 26: 27-282.
581. Morrison, R.I.G. 1977. Migration of arctic waders wintering in Europe. Polar Record. 18: 475-486.
582. Morrison, R.I.G. 1984. Migration systems of some New World shorebirds. In: J. Burger & B.L. Olla (eds.). Shorebirds. Migration and foraging behavior: 125-202. Plenum Press, New York.
583. Morrison, R.I.G. & J.P. Myers. 1987. Wader migration systems in the New World. Wader Study Group Bull. 49, Suppl. / IWRB Spec. Publ. 7: 57-69.
584. Moser, M.E. 1988. Limits to the numbers of Grey Plovers *Pluvialis squatarola* wintering on British estuaries: an analysis of long-term population trends. J. appl. Ecol. 25: 473-485.
585. Müller, M.J. 1980. Handbuch ausgewählter Klimastationen der Erde, Zweite Ed. Gerold Richter, Trier.
586. Murton, R.K. & N.J. Westwood. 1977. Avian breeding cycles. Clarendon Press, Oxford.
587. Myers, J.P. 1981a. Cross-seasonal interactions in the evolution of sandpiper social systems. Behav. Ecol. Sociobiol. 8: 195-202.
588. Myers, J.P. 1981b. A test of three hypotheses for latitudinal segregation of the sexes in wintering birds. Can. J. Zool. 59: 1527-1534.
589. Myers, J.P. 1984. Spacing behavior of nonbreeding shorebirds. In: J. Burger & B.L. Olla (eds). Shorebirds. Breeding behavior and populations: 271-321. Plenum Press, New York.
590. Myers, J.P. & F.A. Pitelka. 1979. Variations in summer temperature patterns near Barrow, Alaska: analysis and ecological interpretation. Arc. Alp. Res. 11: 131-144.
591. Myers, J.P., P.G. Connors & F.A. Pitelka. 1979. Territoriality in non-breeding shorebirds. Studies Avian Biol. 2: 231-246.
592. Myers, J.P., J.L. Maron & M. Sallaberry. 1985. Going to extremes: why do Sanderlings migrate to the Neotropics? Ornithol. Monogr. 36: 520-535.
593. Myers, J.P., R.I.G. Morrison, P.Z. Antas, B.A. Harrington, T.E. Lovejoy, M. Sallaberry, S.E. Senner & A. Tarak. 1987. Conservation strategies for migratory species. Amer. Sci 75: 19-26.
594. Mylius, S.D. & O. Diekmann. 1995. On evolutionarily stable life histories, optimization and the need to be specific about density dependence. Oikos 74: 218-224.
595. Nachtigall, W. 1990. Wind tunnel measurements of long-time flights in relation to the energetics and water economy of migrating birds. In: E. Gwinner (ed.). Bird migration. Physiology and ecophysiology: 319-327. Springer-Verlag, Berlin.
596. Nehls, G., N. Kempf & M. Thiel. 1992. Numbers and distribution of moulting Shelduck (*Tadorna tadorna*) in the German Wadden Sea. Vogelwarte 36: 221-232.
597. Nève, G. & A. van Noordwijk. 1997. De overleving van Scholeksters in de Waddenzee. 1980-1994: De effecten van leeftijd, voedselaanbod en vorst. Interim Report NIOO-CTO.
598. Newton, I. 1969. Winter fattening in the Bullfinch. Physiol. Zool. 42: 96-107.
599. Newton, I. 1989. Lifetime reproduction in Birds. Academic Press, London.
600. Nichols, J.D. 1991. Responses of North American duck populations to exploitation. In: C.M. Perrins, J.-D. Lebreton & G.J.M. Hirons (eds.). Bird population studies: relevance to conservation and management: 498-525. Oxford University Press, Oxford.
601. Nicole, M., M. Egnankou Wada & M. Schmidt (eds.). 1994. A preliminary inventory of coastal wetlands of Côte d' Ivoire. IUCN, Gland.
602. Nieman, D.C. 1994. Exercise, infection and immunity. Int. J. Sports Med. 15: S131-S141.
603. Nixon, S.W. 1995. Coastal marine eutrophication: a definition, social causes, and future concerns. Ophelia 41: 199-219.
604. Noer, H. 1979. Speeds of migrating waders Charadriidae. Dansk orn. Foren. Tidsskr. 73: 215-224.
605. Nolan, V. Jr. 1963. Reproductive success of birds in a deciduous scrub habitat. Ecology 44: 305-313.
606. Noordhuis, R. 1989. Patronen in slagpenrui: oecofysiologische aanpassingen. Limosa 62: 35-45.
607. van Noordwijk, A.J. 1998. The EURING swallow project: amateur ringers in population studies. Ostrich 69: 15.
608. van Noordwijk, A.J. & G. de Jong. 1986. Acquisition and allocation of resources: their influence on variation in life history tactics. Am. Nat. 128: 137-142.
609. Norberg, U. 1990. Vertebrate flight. Mechanics, physiology, morphology, ecology and evolution. Springer-Verlag, Berlin.
610. Norris, K. & I. Johnstone. 1998. Interference competition and the functional response of Oystercatchers searching for cockles by touch. Anim. Behav. 56: 639-650.
611. Norris, K., E. Brindley, T. Cook, S. Babbs, C. Forster Brown & R. Yaxley. 1998. Is the density of Redshank *Tringa totanus* on saltmarshes in Great Britain declining due to changes in grazing management? J. appl. Ecol. 35: 621-634.
612. Norton-Griffiths, M. 1967. Some ecological aspects of the feeding behaviour of the Oystercatcher *Haematopus ostralegus* on the Edible Mussel *Mytilus edulis*. Ibis 109: 412-424.
613. Norton-Griffiths, M. 1969. The organisation, control and development of parental feeding in the Oystercatcher (*Haematopus ostralegus*). Behaviour 34: 55-114.
614. Ntiamoa-Baidu, Y. 1993. Trends in the use of Ghanaian coastal wetlands by migrating Knots *Calidris canutus*. Ardea 81: 71-79.
615. Ntiamoa-Baidu, Y., T. Piersma, P. Wiersma, M. Poot, P.F. Battley & C. Gordon. 1998. Water depth selection, daily feeding routines and diets of waterbirds in coastal lagoons in Ghana. Ibis 140: 89-103.
616. O'Connor, R. & A. Cawthorne. 1982. How Britain's birds survived the winter. New Scientist 93: 786-788.
617. Odum, E.P., D.T. Rogers & D.L. Hicks. 1964. Homeostasis of the non-fat components of migrating birds. Science 143: 1037-1039.
618. Olff, H. 1992. On the mechanisms of vegetation succession. PhD thesis, University of Groningen.
619. Oost, A.P., B.J. Ens, A.G. Brinkman, K.S. Dijkema, W.D. Eysink, J.J. Beukema, H.J. Gussinklo, B.M.J. Verboom & J.J. Verburgh. 1998. Integrale Bodemdalingstudie Waddenzee. NAM, Assen. ISBN 90-804791-4-4
620. Owen, M. & J.M. Black. 1989. Factors affecting the survival of Barnacle Geese on migration from the breeding grounds. J. Anim. Ecol. 58: 603-617.
621. Owens, I.P.F., A. Dixon, T. Burke, & D.B.A. Thompson. 1995. Strategic paternity assurance in the sex-role reversed Eurasian Dotterel (*Charadrius morinellus*): behavioral and genetic evidence. Behav. Ecol. 6: 14-21.
622. van Paassen, A.G., D.H. Veldman & A.J.Beintema. 1984. A simple device for determination of incubation stages in eggs. Wildfowl 35: 173-178.
623. Parker, G.A. 1982. Phenotype-limited evolutionary stable strategies. In: King's College Sociobiology Group (ed.). Current problems in sociobiology: 173-201. Cambridge University Press, Cambridge.
624. Parish, D., B. Lane, P. Sagar & P. Tomkovich. 1987. Wader migration systems in East Asia and Australasia. Wader Study Group Bull. 49, Suppl. / IWRB Spec. Publ. 7: 4-14.

625. Parker, G.A. & W.J. Sutherland. 1986. Ideal free distributions when individuals differ in competitive ability: phenotype limited ideal free models. Anim. Behav. 34: 1222-1242.
626. Peaker, M. & J.L. Linzell. 1975. Salt glands in birds and reptiles. Cambridge University Press, Cambridge.
627. Pennycuick, C.J. 1968. Power requirements for horizontal flight in the pigeon *Columba livia*. J. exp. Biol. 49: 527-555.
628. Pennycuick, C.J. 1975. Mechanics of flight. In: D.S. Farner & J.R. King (eds.). Avian biology, Vol. 5: 1-75. Academic Press, New York.
629. Pennycuick, C.J. 1978. Fifteen testable predictions about bird flight. Oikos 30: 165-176.
630. Pennycuick, C.J. 1989. Bird flight. A practical calculation manual. Oxford University Press, Oxford.
631. Pennycuick, C.J. 1998. Computer simulation of fat and muscle burn in long-distance bird migration. J. Theor. Biol. 191: 47-61.
632. Pennycuick, C.J., T. Alerstam & A. Hedenström. 1997. A new low-turbulence wind tunnel for bird flight experiments at Lund University, Sweden. J. exp. Biol. 200: 1441-1449.
633. Pennycuick, C.J., M. Klaassen, A. Kvist & Å. Lindström. 1996. Wingbeat frequency and the body drag anomaly: wind-tunnel observations on a Thrush Nightingale (*Luscinia luscinia*) and a Teal (*Anas crecca*). J. exp. Biol. 199: 2757-2765.
634. Perdeck, A.C. 1958. Two types of orientation in migrating Starlings, *Sturnus vulgaris* L. and Chaffinches, *Fringilla coelebs* L., as revealed by displacement experiments. Ardea 46: 1-37.
635. Perdeck, A.C. 1974. An experiment on the orientation of juvenile Starlings during spring migration. Ardea 62: 190-195.
636. Perez-Hurtado, A. 1995. Ecologia alimentaria de Limícolas invernantes en la bahia de Cadiz. Airo 6: 15-23.
637. Peterson, K.S., K.L. Rasmussen, J. Heinemeier & N. Nud. 1992. Clams before Columbus? Nature 359: 679.
638. Pettigrew, J.D. & B.J. Frost. 1985. A tactile fovea in the Scolopacidae? Brain Behav. Evol. 26: 185-195.
639. Pienkowski, M.W. 1977. Differences in habitat requirements and distribution pattern of plovers and sandpipers as investigated by studies of feeding behaviour. Verh. orn. Ges. Bayern 23:-105-124.
640. Pienkowski, M.W. 1983a. The effects of environmental conditions on feeding rates and prey-selection of shore plovers. Ornis Scand. 14: 227-238.
641. Pienkowski, M.W. 1983b. Surface activity of some intertidal invertebrates in relation to temperature and the foraging behaviour of their shorebird predators. Mar. Ecol. Prog. Ser. 11: 141-150.
642. Pienkowski, M.W. 1984. Breeding biology and population dynamics of Ringed Plovers *Charadrius hiaticula* in Britain and Greenland: nest-predation as a possible factor limiting distribution and timing of breeding. J. Zool. 202: 83-114.
643. Pienkowski, M.W. & P.R. Evans. 1984. Migratory behavior of shorebirds in the Western Palearctic. 1984. In: J. Burger & B.L. Olla (eds.). Shorebirds. Migration and foraging behavior: 73-123. Plenum Press, New York.
644. Pienkowski, M.W. & P.R. Evans. 1985. The role of migration in the population dynamics of birds. In: R.M. Sibly & R.H. Smith (eds.). Behavioural ecology: ecological consequences of adaptive behaviour: 331-352. Blackwell Scientific Publications, Oxford.
645. Pienkowski, M.W. & A.E. Pienkowski. 1983. WSG project on the movements of wader populations in western Europe: eighth progress report. Wader Study Group Bull. 38: 13-22.
646. Pienkowski, M.W., C.S. Lloyd & C.D.T. Minton. 1979. Seasonal and migrational weight changes in Dunlins. Bird Study 26: 134-148.
647. Pienkowski, M.W., P.R. Evans & D.J. Townshend. 1985. Leap-frog and other migration patterns of waders: a critique of the Alerstam and Högstedt hypothesis, and some alternatives. Ornis Scand. 16: 61-70.
648. Pierce, E.P. 1997. Sex roles in the monogamous Purple Sandpiper *Calidris maritima* in Svalbard. Ibis 139: 159-169.
649. Pierce, E.P. & J.T. Lifjeld. 1998. High paternity without paternity-assurance behavior in the Purple Sandpiper, a species with high parental investment. Auk 115: 602-612.
650. Piersma, T., unpublished.
651. Piersma, T. 1982. Banc d' Arguin, een warm waddengebied. Waddenbulletin 17: 109-113.
652. Piersma, T. 1983. Investigations of the benthic macrofauna. In: M. Kersten, T. Piersma, C.J. Smit & P.M. Zegers (eds.). Wader migration along the Atlantic coast of Morocco, March. 1981: 27-42. RIN (Alterra) Report 83/20, Texel.
653. Piersma, T. 1986. Breeding waders in Europe: a review of population size estimates and a bibliography of information sources. Wader Study Group Bull. 48, Suppl.: 1-116.
654. Piersma, T. 1987. Hink, stap of sprong? Reisbeperkingen van arctische steltlopers door voedselzoeken, vetopbouw en vliegsnelheid. Limosa 60: 185-191.
655. Piersma, T. 1989a. Resightings and recoveries. In: B.J. Ens, T. Piersma, W.J. Wolff & L. Zwarts (eds.). Report of the Dutch-Mauritanian project Banc d'Arguin. 1985-1986: 87-92. WIWO Report 25, Zeist.
656. Piersma, T. 1989b. Knot (*Calidris canutus*). In: B.J. Ens, T. Piersma, W.J. Wolff & L. Zwarts (eds.). Report of the Dutch-Mauritanian project Banc d'Arguin. 1985-1986: 264-275. WIWO Report 25, Zeist.
657. Piersma, T. 1990. Pre-migratory "fattening" usually involves more than the deposition of fat alone. Ring. & Migr. 11: 113-115.
658. Piersma, T. 1994. Close to the edge: energetic bottlenecks and the evolution of migratory pathways in Knots. Publisher Het Open Boek, Den Burg.
659. Piersma, T. 1996. Energetic constraints on the non-breeding distribution of coastal shorebirds. International Wader Studies 8: 122-135.
660. Piersma, T. 1997a. The biology of migratory shorebirds. In: P. Straw (ed.). Shorebird conservation in the Asia-Pacific region: 2-12. Birds Australia, Hawthorn East.
661. Piersma, T. 1997b. Do global patterns of habitat use and migration strategies co-evolve with relative investments in immunocompetence due to spatial variation in parasite pressure? Oikos 80: 623-636.
662. Piersma, T. 1998. Phenotypic flexibility during migration: optimization of organ size contingent on the risks and rewards of fueling and flight? J. Avian Biol. 29: 521-520.
663. Piersma, T. & A.J. Baker. 2000. Life history characteristics and the conservation of migratory shorebirds. In: L.M. Gosling & W.J. Sutherland (eds.). Behaviour and conservation. Cambridge University Press, Cambridge: 105-124.
664. Piersma, T. & G.C. Boere. 1983. Wetlands in West-Afrika: belangrijk voor Nederland. Vogels 3: 220-222.
665. Piersma, T. & N.E. van Brederode. 1990. The estimation of fat reserves in coastal waders before their departure from northwest Africa in spring. Ardea 78: 221-236.
666. Piersma, T. & N.C. Davidson. 1992. The migrations and annual cycles of five subspecies of Knots in perspective. Wader Study Group Bull. 64, Suppl.: 187-197.
667. Piersma, T. & B.J. Ens. 1992. Optimal migration schedules: reserve dynamics as constraint. Wader Study Group Bull. 64: 17-18.
668. Piersma, T. & R.E. Gill Jr. 1998. Guts don't fly: small digestive organs in obese Bar-tailed Godwits. Auk 115: 196-203.
669. Piersma, T. & J. Jukema. 1990. Budgeting the flight of a long-distance migrant: change in the nutrient reserve levels of Bar-tailed Godwits at successive spring staging sites. Ardea 78: 315-337.
670. Piersma, T. & J. Jukema. 1993. Red breasts as honest signals of migratory quality in a long-distance migrant, the Bar-tailed Godwit. Condor 95: 163-177.
671. Piersma, T. & M. Klaassen. 1999. Methods of studying the functional ecology of protein and organ dynamics in birds. In: N. Adams & R. Slotow (eds.). Proc. 22 Int. Ornithol. Congr., University of Natal, Durban.
672. Piersma, T. & A. Koolhaas. 1997. Shorebirds, shellfish(eries) and sediments around Griend, western Wadden Sea, 1988-1996. NIOZ Report. 1997-7, Texel.
673. Piersma, T. & Å. Lindström. 1997. Rapid reversible changes in organ size as a component of adaptive behaviour. Trends Ecol. Evol. 12: 134-138.
674. Piersma, T. & R.I.G. Morrison. 1994. Energy expenditure and water turnover of incubating Ruddy Turnstones: high costs under high arctic climatic conditions. Auk 111: 366-376.
675. Piersma, T. & Y. Ntiamoa-Baidu. 1995. Waterbird ecology and the management of coastal wetlands in Ghana. NIOZ Report 95/6, Texel.
676. Piersma, T. & M. Poot. 1993. Where waders may parallel penguins: spontaneous increase in locomotor activity triggered by fat depletion in a voluntarily fasting Knot. Ardea 81: 1-8.
677. Piersma, T. & S. van de Sant. 1992. Pattern and predictability of potential wind assistance for waders and geese migrating from West Africa and the Wadden Sea to Siberia. Ornis Svecica 2: 55-66.
678. Piersma, T. & P. Wiersma. 1996. Family Charadriidae (plovers). In: J. del Hoyo, A. Elliott & J. Sargatal (eds.). Handbook of the Birds of the World, Vol. 3. Hoatzin to Auks: 384-442. Lynx Edicions, Barcelona.
679. Piersma, T., A.J. Beintema, N.C. Davidson, OAG Münster & M.W. Pienkowski. 1987. Wader migration systems in the East Atlantic. Wader Study Group Bull. 49, Suppl. / IWRB Spec. Publ. 7: 35-56.
680. Piersma, T., M. Klaassen, J.H. Bruggemann, A.-M. Blomert, A. Gueye, Y. Ntiamoa-Baidu & N.E. van Brederode. 1990a. Seasonal timing of the spring departure of waders from the Banc d'Arguin, Mauritania. Ardea 78: 123-134.

681. Piersma, T., L. Zwarts & J.H. Bruggemann. 1990b. Behavioural aspects of the departure of waders before long-distance flights: flocking, vocalizations, flight paths and diurnal timing. Ardea 78: 157-184.

682. Piersma, T., R. Drent & P. Wiersma. 1991a. Temperate versus tropical wintering in the world's northernmost breeder, the Knot: metabolic scope and resource levels restrict subspecific options. Acta XX Congr. Intern. Ornithol. (Christchurch) II: 761-772.

683. Piersma, T., I. Tulp, Y. Verkuil, P. Wiersma, G.A. Gudmundsson & Å. Lindström. 1991b. Arctic sounds on temperate shores: the occurrence of song and ground display in Knots *Calidris canutus* at spring staging sites. Ornis Scand. 22: 404-407.

684. Piersma, T., P. Prokosch & D. Bredin. 1992. The migration system of Afro-Siberian Knots *Calidris canutus canutus*. Wader Study Group Bull. 64, Suppl. 52-63.

685. Piersma, T., P. de Goeij & I. Tulp. 1993a. An evaluation of intertidal feeding habitats from a shorebird perspective: towards comparisons between temperate and tropical mudflats. Neth. J. Sea Res. 31: 503-512.

686. Piersma, T., R. Hoekstra, A. Dekinga, A. Koolhaas, P. Wolf, P. Battley & P. Wiersma. 1993b. Scale and intensity of intertidal habitat use by Knots *Calidris canutus* in the western Wadden Sea in relation to food, friends and foes. Neth. J. Sea Res. 31: 331-357.

687. Piersma, T., A. Koolhaas & A. Dekinga. 1993c. Interactions between stomach structure and diet choice in shorebirds. Auk 110: 552-564.

688. Piersma, T., Y. Verkuil & I. Tulp. 1994. Resources for long-distance migration of Knots *Calidris canutus islandica* and *C. c. canutus*: how broad is the temporal exploitation window of benthic prey in the western and eastern Wadden Sea. Oikos 71: 393-407.

689. Piersma, T., N. Cadée & S. Daan. 1995a. Seasonality in basal metabolic rate and thermal conductance in a long-distance migrant shorebird, the Knot (*Calidris canutus*). J. Comp. Physiol. B 165: 37-45.

690. Piersma, T., J. van Gils, P. de Goeij & J. van der Meer. 1995b. Holling's functional response model as a tool to link the food-finding mechanism of a probing shorebird with its spatial distribution. J. Anim. Ecol. 64: 493-504.

691. Piersma, T., J. van Gils & P. Wiersma. 1996a. Family Scolopacidae (sandpipers, snipes and phalaropes). In: J. del Hoyo, A. Elliott & J. Sargital (eds.). Handbook of the Birds of the World, Vol. 3: 444-533. Lynx Edicions, Barcelona.

692. Piersma, T., L. Bruinzeel, R. Drent, M. Kersten, J. van der Meer & P. Wiersma. 1996b. Variability in basal metabolic rate of a long-distance migrant shorebird (Red Knot *Calidris canutus*) reflects shifts in organ sizes. Physiol. Zool. 69: 191-217.

693. Piersma, T., J.M. Everaarts & J. Jukema. 1996c. Build-up of red blood cells in refueling Bar-tailed Godwits in relation to individual migratory quality. Condor 98: 363-370.

694. Piersma, T., G.A. Gudmundsson, N.C. Davidson & R.I.G. Morrison. 1996d. Do arctic-breeding Red Knots (*Calidris canutus*) accumulate skeletal calcium before egg laying? Can. J. Zool. 74: 2257-2261.

695. Piersma, T., A. Hedenström & J.H. Bruggemann. 1997. Climb and flight speeds of shorebirds embarking on an intercontinental flight: do they achieve the predicted optimal behaviour? Ibis 139: 299-304.

696. Piersma, T., R. van Aelst, K. Kurk, H. Berkhoudt & L.R.M. Maas. 1998. A new pressure sensory mechanism for prey detection in birds: the use of principles of seabed dynamics? Proc. Royal Soc. Lond. B 265: 1377-1383.

697. Piersma, T., G.A. Gudmundsson & K. Lilliendahl. 1999. Rapid changes in the size of different functional organ and muscle groups during refueling in a long-distance migrating shorebird. Phys. Biochem. Zool. 72: 405-415.

698. Pitelka, F.A., R.T. Holmes & S.F.MacLean. 1974. Ecology and evolution of social organization in arctic sandpipers. Amer. Zool. 14: 195-204.

699. Ploeger, P.L. 1968. Geographical differentiation in arctic Anatidae as a result of isolation during the last glaciation. Ardea 56: 1-159.

700. Pomeroy, H. & F. Heppner. 1992. Structure of turning in airborne Rock Dove (*Columba livia*) flocks. Auk 109: 256-261.

701. Postma, H. 1981. Exchange of materials between the North Sea and the Wadden Sea. Mar. geol. 40: 199-213.

702. Potts, W.K. 1984. The chorus-line hypothesis, of manoeuvre coordination in avian flocks. Nature 309: 344-345.

703. Prater, A.J. 1974. Oystercatchers v. cockles. BTO News 64: 1-2.

704. Prater, A.J. 1981. Estuary birds of Britain and Ireland. Poyser, Calton.

705. Prins, H.H.T. 1987. The buffalo of Manyara: the individual in the context of herd life in a seasonal environment of East Africa. PhD thesis, University of Groningen, Groningen.

706. Prinzinger, R. 1990. Temperaturregulation bei Vögeln. I. Thermoregulatorische Verhaltensweisen. Luscinia 46: 255-302.

707. Prokosch, P. 1988. Das Schleswich-Holsteinische Wattenmeer als Frühjahrs-Aufenthaltsgebiet artktischer Watvogel-Populationen am Beispiel von Kiebitzregenpfeifer (*Pluvialis squatarola*, L. 1785), Knutt (*Calidris canutus*, L. 1758) und Pfuhlschnepfe (*Limosa lapponica*, L. 1758). Corax 12: 274-442.

708. Prokosch, P. & H. Hötker (eds.). 1995. Faunistik und Naturschutz auf Taimyr. Expeditionen. 1989-1991. Corax 16, Sonderheft: 1-264.

709. Prop, J. 1998. Effecten van afvalwaterlozingen op trekvogels in de Dollard: een analyse van tellingen uit de periode. 1974-1995. In: K. Essink & P. Esselink (eds.). Het Eems-Dollard estuarium: interacties tussen menselijke beïnvloeding en natuurlijke dynamiek: 145-168. Report RIKZ-98.020, Haren.

710. Prop, J. & C. Deerenberg. 1991. Spring staging in Brent Geese *Branta bernicla*: feeding constraints and the impact of diet on on the accumulation of body reserves. Oecologia 87: 19-28.

711. Prummel, W. 1999. Animals as grave gifts in the early medieval cremation ritual in the north of The Netherlands. In: H. Sarfatij, W.J.H. Verwers & P.J. Woltering (eds.). In discussion with the past; archaeological studies presented to W.A. van Es: 205-212. Foundation for Promoting Archaeology, Zwolle.

712. Prummel, W. & E. Knol. 1991. Strandlopers op de brandstapel. Paleo-Aktueel 2: 92-96.

713. Prummel, W. & T. Piersma 2000. Sandpipers as grave gifts in the early Middle Ages. Wader Study Group Bull. 92: 38-41.

714. Puttick, G.M. 1977. Spatial and temporal variations in inter-tidal animal distribution at Langebaan Lagoon, South Africa. Trans. roy. Soc. S. Afr. 42: 403-440.

715. Puttick, G.M. 1979. Foraging behaviour and activity budgets of Curlew Sandpipers. Ardea 67: 111-122.

716. Quénec'hdu, S. 1998. Paramètres influençant la répartition spatiale des limicoles: sédiment et parasites. PhD thesis, University of Rennes, Rennes.

717. Rappoldt, C. & M. Kersten. 1996. Telfouten en de interpretatie van wadvogeltellingen. Limosa 69: 134-137.

718. Rappoldt, C., M. Kersten & C.J. Smit. 1985. Errors in large-scale shorebird counts. Ardea 73: 13-24.

719. Rayner, J.M.V. 1990. The mechanics of flight and bird migration performance. In: E. Gwinner (ed.). Bird migration. Physiology and ecophysiology: 283-299. Springer-Verlag, Berlin.

720. Reading C.J. & S. McGrorty. 1978. Seasonal variations in the burying depth of *Macoma balthica* (L.) and its accessibility to wading birds. Estuar. cstl. mar. Sci. 6: 135-144.

721. Rehfisch, M.M., N.A. Clark, R.H.W. Langston & J.J.D. Greenwood. 1996. A guide to the provision of refuges for waders: an analysis of 30 years of ringing data from the Wash, England. J. appl. Ecol. 33: 673-687.

722. Reise, K. 1977. Predator exclusion experiments in an intertidal mud flat. Helgoländer Meeresunters. 30: 263-271.

723. Reise, K. 1978. Experiments on epibenthic predation in the Wadden Sea. Helgoländer Meeresunters. 31: 55-101.

724. Reise, K. 1979. Moderate predation on meiofauna by the macrobenthos of the Wadden Sea. Helgoländer Meeresunters. 32: 453-465.

725. Reise, K. 1985. Tidal flat ecology. Springer-Verlag, Berlin.

726. Reijnders, P.J.H. 1986. Reproductive failure in Common Seals feeding on fish from polluted coastal waters. Nature 324: 456-457.

727. Reynolds, J.D. & T. Székely. 1997. The evolution of parental care in shorebirds: life histories, ecology and sexual selection. Behav. Ecol. 8: 126-134.

728. van Rhijn, J.G. 1977. Processes in feathers caused by bathing in water. Ardea 65: 126-147.

729. van Rhijn, J.G. 1985. A scenario for the evolution of social organization in Ruffs *Philomachus pugnax* and other Charadriiform birds. Ardea 73: 25-37.

730. Richardson, K. 1998. 10 years of Wadden and North Sea protection: evaluation and perspectives. Senckenbergiana maritima 29: 7-11.

731. Richardson, W.J. 1976. Autumn migration over Puerto Rico and the western Atlantic: a radar study. Ibis 118: 309-332.

732. Richardson, W.J. 1979. Southeastward shorebird migration over Nova Scotia and New Brunswick in autumn: a radar study. Can. J. Zool. 57: 107-124.

733. Richner, H. 1989. Habitat-specific growth and fitness in Carrion Crows (*Corvus corone corone*). J. Anim. Ecol. 58: 427-440.

734. Ricklefs, R.E. 1969. An analysis of nesting mortality in birds. Smithson. Contr. Zool. 9: 1-48.

735. Riddington, R., M. Hassall, S.J. Lane, P.A. Turner & R. Walters. 1996. The impact of disturbance on the behaviour and energy budgets of Brent Geese *Branta b. bernicla*. Bird Study 43: 269-279.

736. Ridley, M. 1993. The red queen. Sex and the evolution of human nature. Viking, London.

737. Robertson, A.L. 1979. The relationship between annual production: biomass ratios and lifespan for marine macrobenthos. Oecologia (Berl.) 38: 193-202.
738. Robinson, J.A. & S.E. Warnock. 1997. The staging paradigm and wetland conservation in arid environments: shorebirds and wetlands of the North American Great Basin. International Wader Studies 9: 37-44.
739. Roff, D.A. 1992. The evolution of life histories: theory and analysis. Chapman & Hall, New York.
740. Rojas de Azuaje, L.M., S. Tai & R. McNeil. 1993. Comparison of the rod/cone ration in three species of shorebirds having different nocturnal foraging strategies. Auk 110: 141-145.
741. Rompré, G. & R. McNeil. 1994. Seasonal changes in day and night foraging of Willets in northeastern Venezuela. Condor 96: 734-738.
742. de Roos, G. Th. 1972. De invloed van rekreatie en andere verontrustingen op de broed- en trekvogels in het Staatsnatuurreservaat 'Kroonspolders' op het eiland Vlieland. Report. 186 Nature Conservation Department, Wageningen University, Wageningen.
743. Rose, P.M & D.A. Scott. 1994. Waterfowl population estimates. IWRB Publication 29, Slimbridge.
744. Rose, P.M. & D.A. Scott. 1997. Waterfowl population estimates, Second edition. Wetlands International Publication 44, Wageningen.
745. Roselaar, C.S. 1979. Fluctuaties in aantallen Krombekstrandlopers *Calidris ferruginea*. Watervogels 4: 202-210.
746. Roselaar, C.S. 1983. Subspecies recognition in Knot *Calidris canutus* and the occurrence of races in Western Europe. Beaufortia 33: 97-109.
747. Rösner, H.-U. 1990. Sind Zugmuster und Rastplatzansiedlung des Alpenstrandläufers (*Calidris alpina alpina*) abhängig vom Alter? J. Ornithol. 131: 121-139.
748. Rösner, H-U. 1997. Strategien von Zug und Rast des Alpenstrandläufers (*Calidris alpina*) im Wattenmeer und auf dem Ostatlantischen Zugweg. Shaker Verlag, Aachen.
749. Rothe, H.J. & W. Nachtigall. 1987. Pigeon flight in a wind tunnel. I. Aspects of wind tunnel design, training methods and flight behaviour of different pigeon races. J. Comp. Physiol. B 157: 91-98.
750. Roukema, B. 1984. Exploitatie van *Arenicola* door de Wulp: het probleem van prooibeschikbaarheid. Students report Zoological Laboratory University of Groningen / Rijksdienst voor de IJsselmeerpolders (RIZA), Lelystad.
751. Rufino, R., P. Mirando, J.P. Pina & A. Araujo. 1985. Limicolas invernantes na Ria do Faro. Dados sobre a sua distribuçao e desponibilidades alimentares. Actas do Colóquio das Zonas Riberheirinhas: 207-223.
752. Ryabitsev, V.K. 1993. Territorial relations and community dynamics of birds in the subarctic [in Russian]. Nauka, Ekaterinburg.
753. Sæther, B-E., J.A. Kålås, L. Løfaldi & R. Andersen. 1986. Sexual size dimorphism and reproductive ecology in relation to mating systems in waders. Biol. J. Linn. Soc. 28: 273-284.
754. Safriel, U. 1967. Population and food study of the Oystercatcher. PhD thesis, University of Oxford, Oxford.
755. Safriel, U.N. 1975. On the significance of clutch size in nidifugous birds. Ecology 56: 703-708.
756. Safriel, U.N. 1981. Social hierarchy among siblings in broods of the Oystercatcher *Haematopus ostralegus*. Beh. Ecol. Sociobiol. 9: 59-63.
757. Safriel, U.N. 1982. Effects of disease on social hierarchy of young Oystercatchers. British Birds 75: 365-369.
758. Safriel, U.N. 1985. "Diet dimorphism" within an Oystercatcher *Haematopus ostralegus* population - adaptive significance and effects on recent distribution dynamics. Ibis 127: 287-305.
759. Safriel, U.N., B.J. Ens & A. Kaiser. 1996. Rearing to independence. In: J.D. Goss-Custard (ed.). The Oystercatcher: from individual to populations: 219-250. Oxford University Press, Oxford.
760. Safriel, U.N., M.P. Harris, M. de L. Brooke & C.K. Britton. 1984. Survival of breeding Oystercatchers *Haematopus ostralegus*. J. Anim. Ecol. 53: 867-877.
761. Salomonson, F. 1955. The evolutionary significance of bird migration. Dan. Biol. Medd. 22 (6): 1-61.
762. Salvig, J.C., S. Asbirk, J.P. Kjeldsen & P.A.F. Rasmussen. 1994. Wintering waders in the Bijagos Archipelago, Guinea-Bissau. 1992-1993. Ardea 82: 137-142.
763. Sanchez-Salazar, M.E., C.L. Griffiths & R. Seed. 1987a. The effect of size and temperature on the predation of Cockles *Cerastoderma edule* (L.) by the Shore Crab *Carcinus maenas* (L.). J. exp. mar. Biol. Ecol. 111: 181-193.
764. Sanchez-Salazar, M.E., C.L. Griffiths & R. Seed. 1987b. The interactive roles of predation and tidal elevation in structuring populations of the Edible Cockle, *Cerastoderma edule*. Estuar. Cstl. Shelf Sci. 25: 245-260.
765. Sandberg, R. & G.A. Gudmundsson. 1996. Orientation cage experiments with Dunlins during autumn migration in Iceland. J. Avian Biol. 27: 183-188.
766. Sandberg, R. & B. Holmquist. 1998. Orientation and long-distance migration routes: an attempt to evaluate compass cue limitations and required precision. J. Avian Biol. 29: 626-636.
767. Sandberg, R., J. Petterson & T. Alerstam. 1988. Why do migrating Robins, *Erithacus rubecula*, captured at two nearby stop-over sites orient differently? Anim. Behav. 36: 865-876.
768. Sangster, G., C.J. Hazevoet, A.B. van den Berg & C.S. Roselaar. 1997. Dutch avifaunal list: taxonomic changes in. 1977-97. Dutch Birding 19: 21-28.
769. Sangster, G., C.J. Hazevoet, A.B. van den Berg & C.S. Roselaar. 1998. Dutch avifaunal list: species concepts, taxonomic instability, and taxonomic changes in. 1998. Dutch Birding 20: 22-32.
770. Sage, B. 1986. The arctic and its wildlife. Croom Helm, London.
771. Sauer, E.G.F. 1963. Migration habits of Golden Plovers. Proc. 13th Intern. Ornithol Congr., Ithaca: 454-467.
772. von Schantz, T., S. Bensch, M. Grahn, D. Hasselquist & H. Witzell. 1999. Good genes, oxidative stress and condition-dependent sexual signals. Proc. R. Soc. Lond. B 266: 1-2.
773. Scheiffart, G. & G. Nehls. 1997. Competition of benthic fauna by carnivorous birds in the Wadden Sea. Helgoländer Meeresunters. 51: 272-287.
774. Schekkerman, H., unpublished.
775. Schekkerman, H., P. Meininger & P.M. Meire. 1994. Changes in the waterbird populations of the Oosterschelde (SW Netherlands) as a result of large-scale coastal engineering works. Hydrobiologia 282/283: 509-524.
776. Schekkerman, H., M.W.J. van Roomen & L.G. Underhill. 1998. Growth, behaviour of broods, and weather-related variation in breeding productivity of Curlew Sandpipers *Calidris ferruginea*. Ardea 86: 153-168.
777. Schepers, F.J., G.O. Keijl, P.L. Meiningcr & J.B. Rigoulot. 1998. Oiseaux d'eau dans le Delta du Sine-Saloum et Petit Côte, Sénégal Janvier. 1997. WIWO Report 63, Zeist.
778. Scott, D.K. 1988. Breeding success in Bewick's Swans. In: T.H. Clutton-Brock (ed.). Reproductive success: studies of individual variation in contrasting breeding systems: 220-236. The University of Chicago Press, Chicago.
779. Scott, D.A. & P.M. Rose. 1996. Atlas of the Anatidae populations in Africa and Western Eurasia. Wetland International Publication 41, Wageningen.
780. Selous, E. 1931. Thought-transference (or what?) in birds. Constable, London.
781. Selman, J. & J.D. Goss-Custard. 1988. Interference between foraging Redshank *Tringa totanus*. Anim. Behav. 36: 1542-1544.
782. Sigmund, K. 1993. Games of life: explorations in ecology, evolution and behaviour. Oxford University Press, Oxford.
783. Sigvallius, B. 1994. Funeral pyres; Iron Age cremations in North Spånga. PhD thesis, and papers in Osteology 1, Stockholm.
784. Smart, M. 1987. International conventions. Wader Study Group Bull. 49, Suppl. / IWRB Special Publ. 7: 114-117.
785. Smit, C. 1982. Wader and waterfowl counts in the international Wadden Sea area: the results of the 1981-1982 season. Wader Study Group Bull. 24: 135-139.
786. Smit, C. 1984a. The importance of the Wadden Sea for birds. In: W.J. Wolff (ed.). Ecology of the Wadden Sea, Vol. 2, part 6: 280-289. Balkema, Rotterdam.
787. Smit, C. 1984b. Production of invertebrates and consumption in the Dutch Wadden Sea. In: W.J. Wolff (ed.). Ecology of the Wadden Sea, Vol. 2, part 6: 290-301. Balkema, Rotterdam.
788. Smit, C.J. & T. Piersma. 1989. Numbers, midwinter distribution, and migration of wader populations using the East Atlantic flyway. In: H. Boyd & J.-Y. Pirot (eds.). Flyways and reserve networks for water birds. IWRB Spec. Publ. 9: 24-63, Slimbridge.
789. Smit, C.J. & W.J. Wolff (eds.). 1982. Birds of the Wadden Sea. Balkema, Rotterdam.
790. Smit, C.J., N. Dankers, B.J. Ens & A. Meijboom. 1998. Birds, mussels, cockles and shellfish fishery in the Dutch Wadden Sea: how to deal with low food stocks for Eiders and Oystercatchers. Senckenbergiana maritima 29: 141-153.
791. Smith, J.N.M. 1974. The food searching behaviour of two European thrushes I. Description and analysis of search paths. Behaviour 48: 276-302.
792. Smith, J.N.M. 1974. The food searching behaviour of two European thrushes I. Description and analysis of search paths. Behaviour 49: 1-61.
793. Soikkeli, M. 1967. Breeding cycle and population dynamics in the Dunlin (*Calidris alpina*). Annales zoologici 4: 158-198.
794. Soloviev, M.Y. & T.A. Pronin. 1998. Biometrics and primary moult of Dunlin *Calidris alpina* from Taimyr, Siberia. Ostrich 69: 412-413.
795. Sorensen, M.D. 1991. The functional significance of parasitic egg laying and typical nesting in Redhead Ducks: an analysis of individual behaviour. Anim. Behav. 42: 771-796.

796. Sorensen, M.D. 1992. Comment: why is conspecific nest parasitism more frequent in waterfowl than in other birds? Can. J. Zool. 70: 1856-1858.
797. Sorensen, M.D. 1997. Effects of intra- en interspecific brood parasitism on a precocial host, the Canvasback, *Aythya valisineria*. Behav. Ecol. 8: 153-161.
798. Spaans, B., J. Blijleven, I. Popov, M.E. Rykhlikova & B.S. Ebbinge. 1998. Dark-bellied Brent Geese *Branta bernicla bernicla* forgo breeding when Arctic Foxes *Alopex lagopus* are present during nest initiation. Ardea 86: 11-20.
799. Spaans, B., L. Bruinzeel & C.J. Smit. *1996.* Effecten van verstoring door mensen op wadvogels in de Waddenzee en de Oosterschelde. IBN (Alterra) Report 202, Wageningen.
800. Staaland, H. 1967. Anatomical and physiological adaptations of the nasal glands in Charadriiformes birds. Comp. Biochem. Physiol. 23: 933-944.
801. Stearns, S.C. 1992. The evolution of life histories. Oxford University Press, Oxford.
802. Stenzel, L.E., J.C. Warriner, J.S. Warriner, K.S.,Wilson, Bidstrup & G.W. Page. 1994. Long-distance breeding dispersal of Snowy Plovers in western North America. J. Anim. Ecol. 63: 887-902.
803. Steltloperringgroep FFF. 1983. Decreases in the weight of Dunlins and Curlews in the Dutch Wadden Sea during a cold spell. Wader Study Group Bull. 38: 11-12.
804. Stillman R.A., J.D. Goss-Custard, S. McGrorty, A.D. West, S.E.A. le V. dit Durell, R.T. Clarke, R.W.G. Caldow, K.J. Norris, I.G. Johnstone, B.J. Ens, E.J. Bunskoeke, A. v.d. Merwe, J. van der Meer, P. Triplet, N. Odoni, R. Swinfen & J.T. Cayford. 1996. Models of shellfish populations and shorebirds: final report. Report to Commission of the European Communities, Directorate-General for Fisheries. CEC contract PEM/93/03. ITE project T08057O5.
805. Stillman, R.A., J.D. Goss-Custard & R.W.G. Caldow. 1997. Modelling interference from basic foraging behaviour. J. Anim. Ecol. 66: 692-703.
806. Stoddard, P.K., J.E. Marsden & T.C. Williams. 1983. Computer simulation of autumnal bird migration over the western North Atlantic. Anim. Behav. 31: 173-180.
807. van Straaten, L.M.J.U. & Ph. H. Kuenen. 1957. Accumulation of fine grained sediments in the Dutch Wadden Sea. Geol. Mijnb. 19: 329-354.
808. Strang, C.A. 1980. Incidence of avian predators near people searching for waterfowl nests. J. Wildl. Manage. 44: 220-222.
809. Summers, R.W. 1986. Breeding production of Dark-bellied Brent Geese *Branta b. bernicla* in relation to lemming cycles. Bird Study 33: 105-108.
810. Summers, R.W. & M.M. Smith. 1990. An age-related difference in the size of the nasal glands of Brent Geese *Branta bernicla*. Wildfowl 41: 35-37.
811. Summers, R.W. & M. Waltner. 1979. Seasonal variations in the mass of waders in southern Africa, with special reference to migration. Ostrich 50: 21-37.
812. Summers, R.W. & L.G. Underhill. 1987. Factors related to breeding production of Brent Geese *Branta b. bernicla* and waders (Charadrii) on the Taimyr Peninsula. Bird Study 34: 161-171.
813. Summers, R.W., L.G. Underhill, D.J. Pearson & D.A. Scott. 1987a. Wader migration systems in southern and eastern Africa and western Asia. Wader Study Group Bull. 49, Suppl. / IWRB Spec. Publ. 7: 15-34.
814. Summers, R.W., L.G. Underhill, M. Waltner & D.A. Whitelaw. 1987b. Population, biometrics and movements of the Sanderling *Calidris alba* in southern Africa. Ostrich 58: 24-39.
815. Summers, R.W., L.G. Underhill, C.F. Clinning & M. Nicoll. 1989. Populations, migrations, biometrics and moult of the Turnstone *Arenaria i. interpres* on the East Atlantic coastline, with special reference to the Siberian population. Ardea 77: 145-168.
816. Summers, R.W., S. Smith, M. Nicoll & N.K. Atkinson. 1990. Tidal and sexual differences in the diet of Purple Sandpipers *Calidris maritima* in Scotland. Bird Study 37: 187-194.
817. Summers, R.W., L.G. Underhill, M. Nicoll, R. Rae & T. Piersma. 1992. Seasonal, size- and age-related patterns in body mass and composition of Purple Sandpipers *Calidris maritima* in Britain. Ibis 134: 346-354.
818. Summers, R.W., T. Piersma, K.-B. Strann & P. Wiersma. 1998. How do Purple Sandpipers *Calidris maritima* survive the winter north of the Arctic Circle? Ardea 86: 51-58.
819. Sutherland, W.J. 1982a. Spatial variation in the predation of Cockles by Oystercatchers at Traeth Melynog, Anglesy, II. The pattern of mortality. J. Anim. Ecol. 51: 481-500.
820. Sutherland, W.J. 1982b. Do Oystercatchers select the most profitable Cockles? Anim. Behav. 30: 857-861.
821. Sutherland, W.J. 1982c.Food supply and dispersal in the determination of wintering population levels of Oystercatchers, *Haematopus ostralegus*. Estuar. Cstl. Shelf Sci. 14: 223-229.
822. Sutherland, W.J. 1992. Game theory models of functional and aggregative responses. Oecologia 90: 150-152.
823. Sutherland, W.J. 1996. From individual behaviour to population ecology. Oxford University Press, Oxford.
824. Sutherland, W.J. 1998. Evidence for flexibility and constraint in migration systems. J. Avian Biol. 29: 441-448.
825. Sutherland W.J. & P. Koene. 1982. Field estimates of the strength of interference between Oystercatcher *Haematopus ostralegus*. Oecologia 55: 108-109.
826. Sutherland, W.J. & G.A. Parker. 1985. Distribution of unequal competitors. In: R.M. Sibly & R.H. Smith (eds.). Behavioural ecology: ecological consequences of adaptive behaviour: 255-273. Blackwell Scientific Publications, Oxford.
827. Svoboda, J. & B. Freedman (eds.). 1994. Ecology of a polar oasis. Alexandra Fjord, Ellesmere Island, Canada. Captus University Publications, Toronto.
828. Swann, R.L. & B. Etheridge. 1989. Variation in mid-winter weights of Moray Basin waders in relation to temperature. Ring & Migr. 10: 1-8.
829. Swennen, C. 1971. Het voedsel van de Groenpootruiter *Tringa nebularia* tijdens het verblijf in het Nederlandse Waddenzeegebied. Limosa 44: 71-83.
830. Swennen, C. 1976a. Populatie-structuur en voedsel van de Eidereend *Somateria m. mollissima* in de Nederlandse Waddenzee. Ardea 64: 311-371.
831. Swennen, C. 1976b. Wadden Seas are rare, hospitable and productive. In: M. Smart (ed.). Proceedings International Conference on the Conservation of wetlands and waterfowl, Heiligenhafen, Germany: 184-198. IWRB, Slimbridge.
832. Swennen, C. 1981. Vogelsterfte door parasieten, een laat effect van de afsluiting van de Zuiderzee. Vogeljaar 29: 120-124.
833. Swennen, C. 1989. Gull predation of Eiders *Somateria mollissima* ducklings: destruction or elimination of the unfit? Ardea 77: 21-45.
834. Swennen, C. 1990. Oystercatchers feeding on Giant Bloody Cockles on the Banc d'Arguin. Ardea 78: 53-62.
835. Swennen, C. 1991. Fledgling production of Eiders *Somateria mollissima* in The Netherlands. J. Orn. 132: 427-437.
836. Swennen, C. & E. van den Broek. 1960. *Polymorphus botulus* als parasiet bij de eidereenden in de Waddenzee. Ardea 48: 90-97.
837. Swennen, C. & P. Duiven. 1983. Characteristics of Oystercatchers killed by cold-stress in the Dutch wadden Sea area. Ardea 71: 155-159.
838. Swennen, C. & T. Mulder. 1995. Ruiende Bergeenden *Tadorna tadorna* in de Nederlandse Waddenzee. Limosa 68: 15-20.
839. Swennen, C., H.J.L. Heessen & A.W.M. Höcker. 1979. Occurrence and biology of the trematodes *Cotylurus* (*Ichthyocotylurus*) *erraticus*, *C.* (*I.*) *variegatus* and *C.* (*I.*) *platycephalus* (Digenea: Strigeidae) in The Netherlands. Neth. J. Sea Res. 13: 161-191.
840. Swennen, C., L.L.M. de Bruijn, P. Duiven, M.F. Leopold & E.C.L. Marteijn. 1983. Differences in bill form of the Oystercatcher *Haematopus ostralegus*; a dynamic adaptation to specific foraging techniques. Neth. J. Sea Res. 17: 57-83.
841. Swennen, C., M.F. Leopold & L.L.M. de Bruijn. 1989. Time-stressed Oystercatchers, *Haematopus ostralegus*, can increase their intake rate. Anim. Behav. 38: 8-22.
842. Székely, T. 1996. Brood desertion in Kentish Plover *Charadrius alexandrinus*: an experimental test of parental quality and remating opportunities. Ibis 138: 749-755.
843. Székely, T. & Z. Bamberger. 1992. Predation of waders (Charadrii) on prey populations. J. Anim. Ecol. 61: 447-456.
844. Székely, T. & I.C. Cuthill. 1999. Brood desertion in Kentish plover: the value of parental care. Behav. Ecol. 10: 191-197.
845. Székely, T., I.C. Cuthill & J. Kis. 1999. Brood desertion in Kentish plover: sex differences in remating opportunities. Behav. Ecol. 10: 185-190.
846. Székely, T. & J.D. Reynolds. 1995. Evolutionary transitions in parental care in shorebirds. Proc. R. Soc. Lond. B. 262: 57-64.
847. Székely, T., I. Karsal & T.D. Williams. 1994. Determination of clutch-size in the Kentish Plover *Charadrius alexandrinus*. Ibis 136: 341-348.
848. Tamisier, A. 1974. Etho-ecological studies of Teal wintering in the Camargue (Rhone Delta, France). Wildfowl 25: 123-133.
849. Tennekes, H. 1992. De wetten van de vliegkunst. Over stijgen, dalen, vliegen en zweven. Aramith Publishers, Bloemendaal.
850. Thompson, A.B. 1985. *Profilicollis botulus* (Acantocephala) abundance in the Eider duck (*Somateria mollissima*) on the Ythan estuary, Aberdeenshire. Parasitology 91: 563-575.
851. Thompson, D.B.A. & C.J. Barnard. 1984. Prey selection by plovers: optimal foraging in mixed species groups. Anim. Behav. 32: 554-563.
852. Thompson, D.B.A., P.S. Thompson & D. Nethersole-Thompson. 1986. Timing of breeding and breeding performance in a population of Greenshanks (*Tringa nebularia*). J. Anim. Ecol. 55: 181-199.

853. Thorup, O. 1995. The influence of nest controls, catching and ringing on the breeding success of Baltic Dunlin *Calidris alpina*. Wader Study Group Bull. 78: 26-30.

854. Thuman, K. 1998. Bird nest density and species constellations are important factors determining nest predation intensity. Studenrapport, Uppsala University, Uppsala.

855. Thyen, S., P.H. Becker, S. Mickstein, U. Sommer & K.R. Schmieder. 1998. Monitoring breeding success of coastal birds: final report of the pilot atudy. 1996-1997. Wadden Sea Ecosystem 8: 7-57.

856. Tinbergen, J.M. & S. Daan. 1990. Family planning in the Great Tit (*Parus major*): optimal clutch size as integration of parent and offspring fitness. Behaviour 114: 161-190.

857. Tinbergen, L. 1949. Vogels onderweg. Vogeltrek over Nederland in samenhang met landschap, weer en wind. Scheltema & Holkema, Amsterdam.

858. Tinbergen, L. 1967. Vogels in hun domein. Over de vogelbevolking van Nederland in haar samenhang met landschap, plantengroei en dierenwereld. Thieme, Zutphen.

859. Tinbergen, N. 1951. The study of instinct. Oxford University Press, Oxford.

860. Tinbergen, N., M. Impekoven & D. Franck. 1967. An experiment on spacing out as a defence against predation. Behaviour 28: 307-321.

861. Tjallingii, S.T. 1969. Habitatkeuze en -gebruik van de Kluut. Students report Zoological Laboratory, University of Groningen, Groningen.

862. Tomkovich, P.S., unpublished.

863. Tomkovich, P.S. 1992. Breeding-range and population changes of waders in the former Soviet Union. British Birds 85: 344-365.

864. Tomkovich, P.S. 1998. Differences in competition intensity of *Calidris* sandpipers on wintering versus breeding grounds: indirect evidence. Ostrich 69: 311.

865. Tomkovich, P.S. & M. Soloviev. 1994. Site fidelity in high arctic breeding waders. Ostrich 65: 174-180.

866. Tomkovich, P.S. & Y.V. Zharikov. 1998. Wader breeding conditions in the Russian tundras in. 1997. Wader Study Group Bull. 87: 30-42.

867. Townshend, D.J. 1985. Decisions for a lifetime: establishment of spatial defence and movement patterns of juvenile Grey Plovers (*Pluvialis squatarola*). J. Anim. Ecol. 54: 267-274.

868. Townshend, D.J., P.J. Dugan & M.W. Pienkowski. 1984. The unsociable plover - use of intertidal areas by Grey Plovers. In: P.R. Evans, J.D. Goss-Custard & W.G. Hale (eds.). Coastal waders and wildfowl in winter: 140-159. Cambridge University Press, Cambridge.

869. van Treuren, R., R. Bijlsma, J.M. Tinbergen, D. Heg, & L. van de Zande. 1999. Genetic analysis of population structure of socially organized Oystercatchers (*Haematopus ostralegus*) using microsatellites. Molecular Ecol. 8: 181-187.

870. Triplet, P. 1987. Comparaison entre deux stratégie de recherche alimentaire de l'Huîtrier-pie *Haematopus ostralegus* en Baie de Somme. Influence des facteurs de l'environment. Unpublished PhD thesis. University of Paris, Paris.

871. Triplet, P. 1989b. Sélectivité alimentaire liée à l'age chez l'hûtrier-pie (*Haematopus ostralegus*) consommateur de *Nereis diversicolor* en Baie de Somme. Gibier Fauna Sauvage 6: 427-436.

872. Trivers, R.L. 1972. Parental investment and sexual selection. In: B. Campbell (ed.). Sexual selection and the descent of man. 1871-1971: 136-179. Aldine, Chicago.

873. Troy, D., unpublished.

874. Tubbs, C.R. 1977. Wildfowl and waders in Langstone Harbour. British Birds 70: 100-109.

875. Tucker, V.A. 1968. Respiratory exchange and evaporative water loss in the flying Budgerigar. J. exp. Biol. 48: 67-87.

876. Tulp, I. 1998. Reproduktie van Strandplevieren *Charadrius alexandrinus* en Bontbekplevieren *Charadrius hiaticula* op Terschelling, Griend en Vlieland in. 1997. Limosa 71: 109-120.

877. Tulp, I. & P. de Goeij. 1994. Evaluating wader habitats in Roebuck Bay (north-western Australia) as a springboard for northbound migration in waders, with a focus on Great Knots. Emu 94: 78-95.

878. Tulp, I., S. McChesney & P. de Goeij. 1994. Migratory departures of waders from north-western Australia: behaviour, timing and possible migration routes. Ardea 82: 201-221.

879. Tulp, I., H. Schekkerman, T. Piersma, J. Jukema, P. de Goeij & J. van de Kam. 1998. Breeding waders at Cape Sterlegova, northern Taimyr, in. 1994. WIWO Report 61, Zeist.

880. Tulp, I., H. Schekkerman, P. Chylarecki, P. Tomkovich, M. Soloviev, L. Bruinzeel, K. van Dijk, O. Hildén, H. Hötker, W. Kania, M. van Roomen, A. Sikora & R. Summers. 1998. Body mass patterns of Little Stints at different latitudes during incubation and chick-rearing. Ibis 144: 122-134.

881. Tulp, I & H. Schekkerman. Time allocation between feeding and incubation in uniparental arctic-breeding shorebirds: energy reserves provide leeway in a tight schedule. Avian Science (in press).

882. Turpie, J.K. & P.A.R. Hockey. 1993. Comparative diurnal and nocturnal foraging behaviour and energy intake of premigratory Grey Plovers *Pluvialis squatarola* and Whimbrels *Numenius phaeopus* in South Africa. Ibis 135: 156-165.

883. Tye, A. 1987. Identifying major wintering grounds of palearctic waders along the Atlantic coast of Africa from marine charts. Wader Study Group Bull. 49: 20-27.

884. Ulfstrand, S. 1996. Behavioural ecology as a tool in conservation biology: introduction. Oikos 77: 183.

885. Underhill, L.G. 1987. Waders (Charadrii) and other waterbirds at Langebaan Lagoon, South Africa, 1975-1986. Ostrich 58: 145-155.

886. Underhill, L.G. 1995a. The relationship between breeding and nonbreeding localities of waders: the Curlew Sandpiper *Calidris ferruginea* as an extreme example. Ostrich 66: 41-45.

887. Underhill, L.G. 1995b. Migratory birds. In: G.I. Cowan (ed.). Wetlands of South Africa: 163-177. Department of Environmental Affairs and Tourism, Pretoria.

888. Underhill, L.G. & R. W. Summers. 1990a. Multivariate analyses of breeding performance in Dark-bellied Brent Geese *Branta b. bernicla*; comment. Ibis 132: 477-480.

889. Underhill, L.G. & R.W. Summers. 1990b. Rejoinder. Ibis 132: 482.

890. Underhill, L.G., R.P. Prys-Jones, E.E. Syroechkovski, Jr., N.M. Groen, V. Karpov, H G. Lappo, M.W.J. van Roomen, A. Rybkin, H. Schekkerman, H. Spiekman & R.W. Summers. 1993. Breeding of waders (Charadrii) and Brent Geese *Branta bernicla bernicla* at Pronchishcheva Lake, northeastern Taimyr, Russia, in a peak and a decreasing lemming year. Ibis 135: 277-292.

891. Underhill, L.G., R.A. Earlé, T. Piersma, I. Tulp & A. Verster. 1994. Knots (*Calidris canutus*) from Germany and South Africa parasitised by trematode *Cyclocoelum mutabile*. J. Ornithol. 135: 236-239.

892. Urfi, A.J., J.D. Goss-Custard & S.E.A. le V. dit Durell. 1996. The ability of Oystercatchers *Haematopus ostralegus* to compensate for lost feeding time. Field studies of individually marked birds. J. appl. Ecol. 33: 873-883.

893. Vader, W.J.M. 1964. A preliminary investigation into the reactions of the infauna of the tidal flats to tidal fluctuations in water level. Neth. J. Sea Res. 2: 189-222.

894. Väisänen, R.A. 1977. Geographic variation in timing of breeding and egg size in eight European species of waders. Annales Zoologici Fennici 14: 1-25.

895. Veen, J. 1977. Functional and causal aspects of nest distribution in colonies of the Sandwich Tern (*Sterna s. sandvicensis* Lath.). Behaviour (Suppl.) 20: 1-193.

896. Verkuil, Y., A. Koolhaas & J. van der Winden. 1993. Wind effects on prey availability: how northward migrating waders use brackish and hypersaline lagoons in the Sivash, Ukraine. Neth. J. Sea Res. 31: 359-374.

897. Verwey, J. 1952. On the ecology and distribution of Cockle and Mussel in the Dutch Waddensea, their role in sedimentation and the source of their food supply. With a short review of the feeding behaviour of bivalve molluscs. Arch. Néerl. Zool. 10: 171-239.

898. Visser, G.H. 1991. Development of metabolism and temperature regulation in precocial birds: patterns in shorebirds (Charadriiformes) and the domestic fowl (*Gallus domesticus*). PhD thesis, University of Utrecht, Utrecht.

899. de Vlas, J. 1979a. Annual food intake by Plaice and Flounder in a tidal flat area in the Dutch Wadden Sea, with special reference to consumption of regenerating parts of macrobenthic prey. Neth. J. Sea Res. 13: 117-153.

900. de Vlas, J. 1979b. Secondary production by tail regeneration in a tidal flat population of Lugworms (*Arenicola marina*), cropped by flatfish. Neth. J. Sea Res. 19: 362-393.

901. de Vlas, J. 1987. Effecten van de kokkelvisserij in de Waddenzee. Rijksinstituut voor Natuurbeheer (RIN, now: Alterra) 87/18, Texel.

902. Vogel, S. 1992. Vital circuits. On pumps, pipes, and the workings of circulatory systems. Oxford University Press, New York.

903. Voous, K.H. 1960. Atlas van de Europese vogels. Elsevier, Amsterdam.

904. Voous, K.H. 1977. List of recent Holarctic bird species. British Ornithologists' Union, London.

905. Walcott, C. 1996. Pigeon homing: observations, experiments and confusions. J. exp. Biol. 199: 21-27.

906. Waldvogel, J.A. 1990. The bird's eye view. Amer. Scientist 78: 342-353.

907. Wallraff, H.G. 1996. Seven theses on pigeon homing deduced from empirical findings. J. exp. Biol. 199: 105-111.

908. Walters, J.R. 1984. The evolution of parental behavior and clutch size in shorebirds. Behavior of marine animals 5: 243-287.

909. Wanink, J. & L. Zwarts. 1985. Does an optimally foraging Oystercatcher obey the functional response? Oecologia (Berl.) 67: 98-106.

910. Wanink, J.H. & L. Zwarts. 1996. Can food specialization by individual Oystercatchers be explained by differences in prey specific handling efficiencies? Ardea 84A: 177-198.

911. Waterman, T.H. 1989. Animal navigation. American Scientific Library, New York.

912. Watkins, D. 1997. East Asian-Australasian shorebird reserve network. In: P. Straw (ed.). Shorebird conservation in the Asia-Pacific region: 132-137 Birds Australia, Hawthorn East, Victoria.

913. Weber, T.P. & T. Piersma. 1996. Basal metabolic rate and the mass of tissues differing in metabolic scope: migration-related covariation between individual Knots *Calidris canutus*. J. Avian Biol. 27: 215-224.

914. Weber, T.P. & A.I. Houston. 1997. Flight costs, flight range and the stopover ecology of migrating birds. J. Anim. Ecol. 66: 297-306.

915. Weber, T.P., B.J. Ens & A.I. Houston. 1998. Optimal avian migration: a dynamic model of fuel stores and site use. Evol. Ecol. 12: 377-402.

916. Wehner, R. 1989. Neurobiology of polarization vision. Trends in NeuroSciences. 12: 353-359.

917. Weiner, J. 1992. Physiological limits to sustainable energy budgets in birds and mammals: ecological implications. Trends Ecol. Evol. 7: 384-388.

918. Wenink, P.W. 1994. Mitochondrial DNA sequence evolution in shorebird populations. PhD thesis, Wageningen University, Wageningen.

919. Wenink, P.W., C.J. Smit, M.G.J. Tilanus, W.B. van Muiswinkel & A.J. Baker. 1992. DNA-analyse: achter de grenzen van de biometrie. Limosa 65: 109-115.

920. Wenink, P.W., A.J. Baker & M.G.J. Tilanus. 1993. Hypervariable controlregion sequences reveal global population structuring in a long-distance migrant shorebird, the Dunlin (*Calidris alpina*). Proc. Natl. Acad. Sci. U.S.A. 90: 94-98.

921. Wenink, P.W., A.J. Baker & M.G.J. Tilanus. 1994. Mitochondrial control region sequences in two shorebird species, the Turnstone and the Dunlin, and their utility in population genetic studies. Mol. Biol. Evol. 11: 22-31.

922. Wenink, P.W., A.J. Baker & M.G.J. Tilanus. 1996. Global mitochondrial DNA phylogeography of holarctic breeding Dunlins (*Calidris alpina*). Evolution 50: 318-330.

923. Werner, B. 1956. Über die Winterwanderung von *Arenicola marina* L. (Polychaeta sedentaria). Helgoländer Meeresunters. 5: 353-378.

924. Westendorp, R.G.J. & T.B.L. Kirkwood. 1998. Human longevity at the cost of reproductive success. Nature 396: 743-746.

925. Westerterp, K.R., W.H.M. Saris, M. van Es & F. ten Hoor. 1986. Use of doubly labelled water technique in humans during heavy sustained exercise. J. appl. Physiol. 61: 2162-2167.

926. Westerterp, K.R. & W.H.M. Saris. 1991. Limits of energy turnover in relation to physical performance, achievement of energy balance on a daily basis. J. Sports Sci. 9: 1-15.

927. Westmoreland, D. & L.B. Best. 1985. The effect of disturbance of Mourning Dove nesting success. Auk 102: 774-780.

928. Whitfield, D.P. 1985. Raptor predation on wintering waders in southeast Scotland. Ibis 127: 544-558.

929. Whitfield, D.P. 1986. Plumage variability and territoriality in breeding turnstone *Arenaria interpres*: status signalling or individual recognition? Anim. Behav. 34: 1471-1482.

930. Whitfield, D.P. 1988. The social significance of plumage variability in wintering turnstone *Arenaria interpres*. Anim. Behav. 36: 408-415.

931. Whitfield, D.P. 1990. Male choice and sperm competition as constraints on polyandry in the red-necked phalarope *Phalaropus lobatus*. Beh. Ecol. Sociobiol. 27: 247-254.

932. Whitfield, D.P. & J.J. Brade. 1991. The breeding behaviour of the Knot *Calidris canutus*. Ibis 133: 246-255.

933. Whitfield, D.P. & P.S. Tomkovich. 1996. Mating system and timing of breeding in Holarctic waders. Biol. J. of the Linnean Soc. 57: 277-290.

934. Wickelgren, I. 1998. Obesity: how big a problem? Science 280: 1364-1367.

935. Wiersma, P. & T. Piersma. 1994. Effects of microhabitat, flocking, climate and migratory goal on energy expenditure in the annual cycle of Red Knots. Condor 96: 257-279.

936. Wiersma, P., L. Bruinzeel & T. Piersma. 1993. Energiebesparing bij wadvogels: over de kieren van de Kanoet. Limosa 66: 41-52.

937. Wiley, R.H., L. Steadman, L. Chadwick & L. Wollerman. 1997. Social inertia in white-throated sparrows results from recognition of opponents. Anim. Behav. 57: 453-463.

938. Williams, G.C. 1966. Adaptation and Natural Selection: A Critique of some Current Evolutionary Thought. Princeton University Press, Princeton.

939. Williams, T.C. 1985. Autumnal bird migration over the windward Caribbean islands. Auk 102: 163-167.

940. Willis, E.O. 1994. Are *Actitis* sandpipers inverted flying fishes? Auk 111: 190-191.

941. Wilson., E.O. 1998. Consilience: toward the unity of knowledge. Little, Brown Company, London.

942. Wilson, J.R. & M.A. Barter. 1998. Identification of potentially important staging areas for 'long jump' migrant waders in the east Asian-Australasian flyway during northward migration. Stilt 32: 16-27.

943. Wilson, J.R., M.A. Czajkowski & M.W. Pienkowski. 1980. The migration through Europe and wintering in West Africa of the Curlew Sandpiper. Wildfowl 31: 107-122.

944. Wiltschko, W. & R. Wiltschko. 1987. Cognitive maps and navigation in homing pigeons. In: P. Ellen & C. Thinus-Blanc (eds.). Cognitive processes and spatial orientation in animals and men: 201-216. Martinus Nijhoff, Den Haag.

945. Wiltschko, W. & R. Wiltschko. 1988. Magnetic orientation in birds. Curr. Ornithol. 5: 67-121.

946. Wolff, W.J. 1991. The interaction of benthic macrofauna and birds in tidal flat estuaries: a comparison of the Banc d'Arguin, Mauritania, and some estuaries in the Netherlands. In: M. Elliott & J-P. Ducrotoy (eds.). Estuaries and coasts: spatial and temporal intercomparisons: 299-306. Olsen & Olsen, Fredensborg.

947. Wolff, W.J. (ed.). 1998. The end of the East-Atlantic Flyway: Waders in Guinea-Bissau. WIWO Report 39, Zeist.

948. Wolff, W.J. & C.J. Smit. 1990. The Banc d' Arguin, Mauritania, as an environment for coastal birds. Ardea 78: 17-38.

949. Wolff, W.J., A. Gueye, A. Meijboom, T. Piersma & M.A. Sall. 1987. Distribution, biomass, recruitment and productivity of *Anadara senilis* (L.) (Mollusca: Bivalvia) on the Banc d'Arguin, Mauritania. Neth. J. Sea Res. 21: 243-253.

950. Wolff, W.J., A.G. Duiven, P. Duiven, P. Esselink, A. Gueye, A. Meijboom, G. Moerland & J. Zegers. 1993a. Biomass of macrobenthic tidal flat fauna of the Banc d' Arguin, Mauritania. Hydrobiologia 258: 151-164.

951. Wolff, W.J., J. van der Land, P.H. Nienhuis & P.A.W.J. de Wilde (eds.). 1993b. Ecological studies in the coastal waters of Mauritania. Kluwer Academic Publishers, Dordrecht.

952. Woods, S.C., R.J. Seeley, D. Porte, Jr. & M.W. Schwartz. 1998. Signals that regulate food intake and energy homeostasis. Science 280: 1378-1383.

953. Worrall, D.H. 1984. Diet of the Dunlin *Calidris alpina* in the Severn estuary. Bird Study 31: 203-212.

954. Wymenga, E., M. Engelmoer, C.J. Smit & T.M. van Spanje. 1990. Geographical breeding origin and migration of waders wintering in west Africa. Ardea 78: 83-112.

955. Yates, M.G., J.D. Goss-Custard, S. McGrorty, K.H. Lakhani, S.E.A. le V. dit Durell, R.T. Clarke, W.E. Rispin, I. Moy, R.A. Plant & A.J. Frost. 1993. Sediment characteristics, invertebrate densities and shorebird densities on the inner banks of the Wash. J. appl. Ecol. 30: 599-614.

956. Zahavi, A. 1977. The cost of honesty (further remarks on the handicap principle). J. theor. Biol. 67: 603-605.

957. Zegers, P.M. 1977. Het gebruik van de. 1% norm. Watervogels 2: 174-178.

958. Zharikov, Y. 1998. Mating behaviour and paternity of socially monogamous Semipalmated Plovers *Charadrius semipalmatus* breeding in the subarctic. Students report (MsC), Trent University, Canada.

959. Zwaan, B.J., R. Bijlsma & R.F. Hoekstra. 1995. Direct selection of life span in *Drosophila melanogaster*. Evolution 49: 649-659.

960. Zwarts, L., unpublished.

961. Zwarts, L. 1974a. Vogels van het brakke getijdebied. Nederlandse Jeugdbond voor Natuurstudie, Amsterdam.

962. Zwarts, L. 1974b. Voedselopname en foerageergedrag van Tureluurs op het wad onder Schiermonnikoog. Schierboek 5: 183-227.

963. Zwarts, L. 1976. Density-related processes in feeding dispersion and feeding activity of Teal (*Anas crecca*). Ardea 64: 192-209.

964. Zwarts, L. 1981. Habitat selection and competition in wading birds. In: C.J. Smit & W.J. Wolff (eds.). Birds of the Wadden Sea: 271-279. Balkema, Rotterdam.

965. Zwarts, L. 1985. The winter exploitation of Fiddler Crabs *Uca tangeri* by waders in Guinea-Bissau. Ardea 73: 3-12.

966. Zwarts, L. 1986. Burying depth of the benthic bivalve *Scrobicularia plana* (da Costa) in relation to siphon-cropping. J. exp. mar. Biol. Ecol. 101: 25-39.

967. Zwarts, L. 1988a. Numbers and distribution of coastal waders in Guinea-Bissau. Ardea 76: 42-55.

968. Zwarts, L. 1988b. De bodemfauna van de Fries-Groningse waddenkust. Flevobericht 294, Rijksdienst voor de IJsselmeerpolders (RIZA), Lelystad.

969. Zwarts, L. 1990. Increased prey availability drives premigration hyperphagia in Whimbrels and allows them to leave the Banc d'Arguin, Mauritania, in time. Ardea 78: 279-300.

970. Zwarts, L. 1991. Seasonal variation in body condition of the bivalves *Macoma balthica, Scrobicularia plana, Mya arenaria* and *Cerastoderma edule* in the Dutch Wadden Sea. Neth. J. Sea Res. 28: 231-245.

971. Zwarts, L. 1996a. Met hoogwatertellen alleen kom je er niet. Limosa 69: 142-145.

972. Zwarts, L. 1996b. Waders and their estuarine food supplies. RIZA Report Van Zee tot Land 60: 1-386, Lelystad.

973. Zwarts, L. & A-M. Blomert. 1990. Selectivity of Whimbrels feeding on Fiddler Crabs explained by component specific digestibilities. Ardea 78: 193-208.

974. Zwarts, L. & A-M. Blomert. 1992. Why Knot *Calidris canutus* take medium-sized *Macoma balthica* when six prey species are available. Mar. Ecol. Prog. Ser. 83: 113-128.

975. Zwarts, L. & S. Dirksen. 1990. Digestive bottleneck limits the increase in food intake of Whimbrels preparing for spring migration from the Banc d'Arguin, Mauritania. Ardea 78: 257-278.

976. Zwarts, L. & R.H. Drent. 1981. Prey depletion and the regulation of predator density: Oystercatchers (*Haematopus ostralegus*) feeding on Mussels (*Mytilus edulis*). In: N.V. Jones & W.J. Wolff (eds.). Feeding and survival strategies of estuarine organisms: 193-216. Plenum Press, New York.

977. Zwarts, L. & B.J. Ens. 1999. Predation by birds on marine tidal flats. In: N. Adams & R. Slotow (eds.). Proc. 22 Int. Ornithol. Congr., University of Natal, Durban.

978. Zwarts, L. & P. Esselink. 1989. Versatility of male Curlews (*Numenius arquata*) preying upon *Nereis diversicolor*: deploying contrasting capture modes dependent on prey availability. Mar. Ecol. Prog. Ser. 56: 255-269.

979. Zwarts, L. & J. Wanink. 1984. How Oystercatchers and Curlews successively deplete clams. In: P.R. Evans, J.D Goss-Custard & W.G. Hale (eds.). Coastal waders and wildfowl in winter: 69-83. Cambridge University Press, Cambridge.

980. Zwarts, L. & J. Wanink. 1989. Siphon size and burying depth in deposit- and suspension-feeding benthic bivalves. Mar. Biol. 100: 227-240.

981. Zwarts, L. & J.H. Wanink. 1991. The macrobenthos fraction accessible to waders may represent marginal prey. Oecologia (Berl.) 87: 581-587.

982. Zwarts, L. & J.H. Wanink. 1993. How the food supply harvestable by waders in the Wadden Sea depends on the variation in energy content, body weight, biomass, burying depth and behaviour of tidal-flat invertebrates. Neth. J. Sea Res. 31: 441-476.

983. Zwarts, L., A-M. Blomert, B.J. Ens, R. Hupkes & T.M. van Spanje. 1990a. Why do waders reach high feeding densities on the intertidal flats of the Banc d'Arguin, Mauritania? Ardea 78: 39-52.

984. Zwarts, L., A-M. Blomert & R. Hupkes. 1990b. Increase of feeding time in waders preparing for spring migration from the Banc d'Arguin, Mauritania. Ardea 78: 237-256.

985. Zwarts, L., B.J. Ens, M. Kersten & T. Piersma. 1990c. Moult, mass and flight range of waders ready to take off for long-distance migrations. Ardea 78: 339-364.

986. Zwarts, L., A-M. Blomert & J.H. Wanink. 1992. Annual and seasonal variation in the food supply harvestable by Knot *Calidris canutus* staging in the Wadden Sea in late summer. Mar. Ecol. Prog. Ser. 83: 129-139.

987. Zwarts, L., A-M. Blomert, P. Spaak & B. de Vries. 1994. Feeding radius, burying depth and siphon size of *Macoma balthica* and *Scrobicularia plana*. J. exp. mar. Biol. Ecol. 183: 193-212.

988. Zwarts, L., J.T. Cayford, J.B. Hulscher, M. Kersten, P.M. Meire & P. Triplet. 1996a. Prey size selection and intake rate. In: J.D. Goss-Custard (ed.). Behaviour and ecology of the Oystercatcher: from individuals to populations: 30-55. Oxford University Press, Oxford.

989. Zwarts, L., B.J. Ens, J.D. Goss-Custard, J.B. Hulscher & S.E.A. le V. dit Durell. 1996b. Causes of variation in prey profitability and its consequences for the intake rate of the Oystercatcher *Haematopus ostralegus*. Ardea 84A: 229-268.

990. Zwarts, L., B.J. Ens, J.D. Goss-Custard, J.B. Hulscher & M. Kersten. 1996c. Why Oystercatchers *Haematopus ostralegus* cannot meet their daily energy requirements in a single low water period. Ardea 84A: 269-290.

991. Zwarts, L., J.B. Hulscher, K. Koopman & P.M. Zegers. 1996d. Short-term variation in the body weight of Oystercatchers *Haematopus ostralegus*: effect of temperature, wind force and exposure time by day and night. Ardea 84A: 357-372.

992. Zwarts, L., J.B. Hulscher, K. Koopman, T. Piersma & P.M. Zegers. 1996e. Seasonal and annual variation in body weight, nutrient stores and mortality of Oystercatchers *Haematopus ostralegus*. Ardea 84A: 327-356.

993. Zwarts, L., J.B. Hulscher, K. Koopman & P.M. Zegers. 1996f. Body weight in relation to variation in body size of Oystercatchers *Haematopus ostralegus*. Ardea 84A: 21-28.

994. Zwarts, L., J.B. Hulscher & P.M. Zegers. 1996g. Weight loss in Oystercatchers *Haematopus ostralegus* after capture. Area 84A: 13-20.

995. Zwarts, L., J.H. Wanink & B.J. Ens. 1996h. Predicting seasonal and annual fluctuations in the local exploitation of different prey by Oystercatchers *Haematopus ostralegus*: a ten year study in the Wadden Sea. Ardea 84A: 401-440.

996. Zwarts, L., J. van der Kamp, O. Overdijk, T.M. van Spanje, R. Veldkamp, R. West, M. Wright. 1998a. Wader count of the Banc d'Arguin, Mauritania, in January/February. 1997. Wader Study Group Bull. 86: 53-69.

997. Zwarts, L., J. van der Kamp, O. Overdijk, T.M. van Spanje, R. Veldkamp, R. West, M. Wright. 1998b. Wader count of the Baie d'Arguin, Mauritania, in February. 1997. Wader Study Group Bull. 86: 70-73.

998. Zweers, G.A. 1991. Pathways and space for evolution of feeding mechanisms in birds. In: E.C. Dudley (ed.). The unity of evolutionary biology: 530-547. Discorides Press, Portland, Oregon.

*Addition of the translaters:*

999. Kvist A., A. Lindström, M. Green, T. Piersma & H.G. Visser. 2001. Carrying large fuel loads during sustained bird flight is cheaper than expected. Nature 413: 730-731.

1000. Wetlands International. 2002. Waterbird Population Estimates – Third Edition. Wetlands International Global Series 12. Wageningen, The Netherlands.

1001. Dijk A.J. van, F. Hustings, K. Koffijberg, M.J.T. van der Weide, D. Zoetebier & C.L. Plate. 2003. Kolonievogels en zeldzame broedvogels in Nederland in 2002. SOVON-monitoring report 2003/02. SOVON Vogelonderzoek Nederland, Beek-Ubbergen.

1002. Koffijberg K., J. Blew, K. Eskilden, K. Günther, B. Koks, K. Laursen, L.M. Rasmussen, P. Potel & P. Südbeck. 2003. High tide roosts in the Wadden Sea: A review of birds distribution, protection regimes and potential sources of anthropogenic disturbance. A report of the Wadden Sea Plan Project 34. Wadden Sea Ecosystem 16. Common Wadden Sea Secretariat, Trilateral Monitoring and Assesment Group, Joint Monitoring Group of Migratory Birds in the Wadden Sea, Wilhelmshaven, Germany.

1003. Poot M.J.M., L.M. Rasmussen, M. van Roomen, H-U. Rösner & P. Südbeck. 1996. Migratory Waterbirds in the Wadden Sea 1993/94. Wadden Sea Ecosystem 5. Common Wadden Sea Secretariat Trilateral Monitoring and Assessment Group & Joint Monitoring Group of Migratory Birds in the Wadden Sea, Willemshaven.

1004. Rasmussen, L.M., D.M. Fleet, B. Hälterlein, B.J. Koks, P. Potel & P. Südbeck. 2000. Breeding Birds in the Wadden Sae in 1996 - Results of a total survey in 1996 and of number of colony breeding species between 1991 and 1996. Wadden Sea Ecosystem 10. Common Wadden Sea Secretariat, Trilateral Monitoring and Assessment Group, Joint Monitoring Group of Breeding Birds in the Wadden Sea, Willemshaven, Germany.

1005. Roomen, M.W.J. van, E.A.J. van Winden, K. Koffijberg, B. Voslamber, R. Kleefstra & G. Ottens. 2002. Watervogels in Nederland in Nederland in 2000/2001. SOVON-monitoring report 2002/04. SOVON Vogelonderzoek Nederland, Beek-Ubbergen.

1006. Wernham, C.V., M.P. Toms, J.H. Marchant, J.A. Clark, G.M. Siriwardena & S.R. Baillie (eds.). 2002. The Migration Atlas: movement of the birds of Britain and Ireland. T. & A.D. Poyser, London.

1007. Sitters, H.P. 2000. The role of night-feeding in shorebirds in an estuarine environment with specific reference to mussel-feeding oystercatchers. PhD thesis, Wolfson College and Edward Grey Institute, University of Oxford, Oxford.

1008. Moore, S.J. & P.F. Battley. 2003. Cockle-opening by a dabbling duck, the Brown Teal. Waterbirds 26: 331-334.

1009. Piersma, T., A. Koolhaas, A. Dekinga, J.J. Beukema, R. Dekker & K. Essink. 2001. Long-term indirect effects of mechanical cockle-dredging on intertidal bivalve stocks in the Wadden Sea. J. appl. Ecol. 38: 976-990.

1010. Atkinson P.W., N.A. Clark, M.C. Bell, P.J. Dare, J.A. Clark, P.L. Ireland. 2003. Changes in commercially fished shellfish stocks and shorebird populations in the Wash, England. Biol. Conserv. 114: 127-141.

# Index

1% criterium 28, 33
*Actitis hypoleucos*, see also common sandpiper 29
*Actitis macularia*, see also spotted sandpiper 286
adultery 283
AEWA, see African-Eurasian Waterbird Agreement
African-Eurasian Waterbird Agreement 342
air speed 100
Alaska 50, 54, 105, 144, 252, 273, 277
algae 11, 19, 20, 73, 179, 329, 330
ambient temperature 129, 155, 157, 208, 255, 256, 260
Ameland 210, 340
American black brent goose 29
American golden plover 261
American razor clam 21
amphipod, see also *Corophium volutator* 21-24, 119, 159, 166, 172, 173, 177, 181-184, 196, 197, 204, 210-214, 219, 223, 224, 228, 229, 265, 266, 332, 333
*Anadara senilis*, see also giant bloody cockle 176
*Anas actua*, see also pintail 79
*Anas clypeata*, see also shoveler 79
*Anas crecca*, see also teal 79
*Anas penelope*, see also wigeon 28
*Anas platyrhynchos*, see also mallard 28
anatomy 231, 279
ancestral wader 288, 289
*Anser brachyrhynchus*, see also pink-footed goose 338
anti-fouling 21, 329
*Araneae* 272
Arctic fox 120, 244, 245, 273, 314, 316
Arctic tern 82, 83, 104, 118, 192, 264
*Arenaria interpres interpres*, see also (ruddy) turnstone 29, 33
*Arenicola marina*, see also lugworm 20, 177, 197, 212
ash-free dry meat 163
ash-free dry weight 173
Asian Wetland Bureau 342
Asiatic golden plover 145
aspect ratio 98
Australia 46, 49, 64, 96, 159, 342, 344
avocet 38, 39, 84, 85, 88, 90, 93, 121, 209, 221, 224-226, 233, 236, 241, 244, 249, 250, 256, 261, 265, 266, 268, 269, 293
*Aythya americana*, see also redhead 284
*Aythya valisineria*, see also canvasback 284
*Aythya ferina*, see also pochard 284
*Aythya fuligula*, see also tufted duck 284
Baie d'Arguin 208
Balgzand 24, 25, 176, 177, 335
Baltic Sea 13, 36, 50, 68, 73, 75, 77, 145, 174, 176, 266
Baltic tellin 20-22, 24, 25, 86, 155, 163, 166, 173-176, 183-189, 191, 195, 197, 202, 210, 211, 213, 217, 221, 224, 229, 266, 319, 336
   depth distribution 184, 188, 189
Banc d'Arguin 11, 12, 14, 95, 99, 102, 104-108, 113, 116, 122, 124, 132, 133, 136-139, 161, 165, 168, 176, 180, 182, 201, 204, 208, 212, 228, 300, 321, 333, 342
barbary falcon 124
barn owl 257
barn swallow 280, 284
barnacle goose 28, 73, 145, 260, 277
barnacles 11, 67, 151, 329
Barrow, Alaska 252, 273, 277
bar-tailed godwit 31, 33, 35, 54, 55, 85, 93, 97, 100, 105, 106, 108, 112, 116, 125, 126, 133-139, 160, 161, 165, 167, 180, 192, 201, 202, 205, 216, 224, 233, 265, 280, 281, 293, 296, 318, 328, 344
Bay of Dakhla 12
Bay of Fundy 12, 13, 15, 224
Belgium 11, 233
benthic fauna 122, 124, 125, 323, 330, 332, 336
Berg River Estuary 125
Bijagós Archipel 95, 124, 125, 137
bill length 188, 195, 201, 202, 203, 225
bill shapes 85
biodiversity 334, 335, 342
biomass 21, 24, 122, 170, 175-177, 195, 209-211, 213, 227, 228, 229, 301, 319, 321, 323, 330, 332, 333
Black Sea 13, 68
black-bellied brent goose 28, 29
blackbird 241
blackcap 145
black-headed gull 81, 204, 205, 214, 216, 223, 224, 227
black-tailed godwit 54, 55, 89, 137, 233, 264, 265, 293
bladderwrack seaweed 24
body mass 30, 89, 93, 116, 203, 253, 257, 338
   benthic animals 173, 174, 185
   food consumption 156, 157, 161, 163, 165, 193
   migration 101, 105, 106
   winter 168, 169
body size 21, 29, 31, 34, 50, 136, 180, 181, 201, 202, 293
body temperature 89, 127, 131, 156, 260
bomb calorimeter 173
Bonn Convention 342, 343
border incident 222
brackish water 74
*Branta bernicla bernicla*, see also (black-bellied) brent goose 28, 29
*Branta leucopsis*, see also barnacle goose 28, 29
*Branta bernicla hrota*, see also (white-bellied) brent goose 28, 29
*Branta bernicla nigricans*, see also (American black) brent goose 28, 29
bream 191
breeding area 31, 37, 45, 49, 54, 56, 63, 73, 82, 90, 97, 101, 135, 136, 137, 140, 141, 143, 145, 236, 237, 274, 299, 302, 309, 313, 339
   arctic 112
   arrival and departure in 276, 277
   boreal 96
   disturbance in 41, 83, 245, 338
breeding
   attempt 231, 232, 236, 237, 273, 297, 299
   density 316
   habitat 250, 264, 266, 267, 309, 322, 339
   season 36-56, 66-82, 96, 98, 109, 133, 134, 144, 231, 237, 253, 256, 261, 264-273, 276, 278, 284, 286, 293, 295, 297, 300, 305, 308, 310, 312, 314, 317, 321-325, 337-339
   territory 231, 242, 248, 257, 268, 269, 272, 295, 297, 313, 317, 320, 323
   timing of 274, 276, 278
brent goose 20, 33, 72, 73, 88, 89, 96, 98, 104, 113, 118, 121, 140, 144, 145, 309, 310, 314-316, 337, 344
brent goose, American black 29
brent goose, black-bellied 29
brent goose, white-bellied 29, 72
brine shrimp 13
Brittany 12, 36, 40, 74
broad-billed sandpiper 244
brood patch 242, 252, 279
brood, of shellfish 23, 231, 236-239, 242, 252, 257, 258, 273, 279, 287, 289, 293, 295-298, 334
brooding, of chickens 242, 260, 296
Brünnich 27
*Buccinum undatum*, see also whelk 329
buffer hypothesis 269
bullfinch 169
Burry Inlet 313, 333
butterfly flight 281
calcium 17, 107, 156, 162, 173, 184, 188, 193
calcium, protective layer 184, 192
*Calidris alba*, see also sanderling 28, 46
*Calidris alpina*, see also dunlin 28, 50
*Calidris canutus*, see also red knot 28, 31, 44
*Calidris ferruginea*, see also curlew sandpiper 28, 48
*Calidris maritima*, see also purple sandpiper 28, 53
*Calidris minuta*, see also little stint 28, 52
camouflage 67, 89, 90, 244, 248, 260
Canada 12, 13, 31, 40, 41, 43, 44, 53, 58, 97, 98, 112, 131, 132, 143, 144, 173, 174
cannon net 30, 31, 160, 161
canvasback 284
Cape Sterlegova 272, 273
*Carcinus maenas*, see also common shore crab 20, 148, 185, 204
*Cardium*, see also cockle 21
career planning 299
carrying capacity 209, 213, 307, 321, 340
Casamance 12
*Cerastoderma edule*, see also cockle 20, 196, 198, 200, 212, 216, 336
Chad 60
chameleon shrimp 226
*Charadrius alexandrinus*, see also kentish plover 28, 41
*Charadrius apricaria*, see also golden plover 27
*Charadrius hiaticula*, see also ringed plover 27, 28, 31, 40
*Charadrius melodus*, see also piping plover 245
*Charadrius morinellus*, see also dotterel 286
*Charadrius semipalmatus*, see also semipalmated plover 284
chick 37-39, 43-83, 120, 121, 139, 140, 156, 236-245, 250, 255-307, 313-320
   adoption 256, 260, 288, 289
   brooding 242, 260, 296
   crèche 75, 261, 265
   extramarital young 286
   food provision 264
   growing conditions 237, 239, 256, 257, 260, 268, 272, 279
   oystercatcher 257, 266, 284, 302
   predation 258, 261, 264, 268, 272, 273, 276, 300
   production 239, 307, 313, 320
China 12, 38, 342, 344
CITES 342
climate 12, 13, 119, 122, 124, 127, 133, 136, 145, 177, 260, 308, 339, 342, 344
   change 145, 339
club birds 270, 272, 301, 302, 305, 313, 317
clutch 240-244, 248, 256, 257, 265, 272-278, 283, 284, 293, 295-297
   number of 232
   renesting 273, 274, 287, 295
   size 53, 241, 242, 256, 269, 272, 278, 298, 322

363

cockle 21, 24, 25, 37, 45, 86, 119, 149, 150, 152, 153, 159, 162, 164-166, 169, 172-177, 184, 185, 191-200, 204, 205, 210-217, 229, 265, 319-324, 333-337, 343
   bed 24, 152, 164, 216, 323, 335, 336
      biomass 211
      fishery 335, 336
common goby 162, 186
common gull 81, 204, 216, 245, 248
common pochard 284
common salt marsh grass 20, 339
common sandpiper 27, 34, 65, 92, 96, 134, 140, 161, 180, 233, 265, 293
common sea-lavender 20
common shelduck, see also shelduck 74, 75
common shore crab 20, 21, 23-25, 119, 148-150, 162, 169, 173, 184, 186, 187, 191, 192, 204, 211, 224, 336
common snipe 34, 85
common tern 82, 83, 104, 264
compass
   earth magnetic 110
   inclination 112
   polarisation 110, 114
   star 110, 113, 114, 117
   sun 110
competition 25, 137, 181, 202, 224, 260, 268, 280, 296, 302, 307-317, 320, 323, 324, 333, 337, 338
   between species 24, 224
   for food 137, 202, 324, 337
   for partners 236
   for the best sites 136, 302
   in the winter 338
   on the breeding grounds 316
condition 90, 119, 137, 151, 173-176, 184-186, 200, 241, 313-315, 321, 338
cones 87
consumption 98, 101, 128, 136, 155, 164, 165, 166, 228, 237, 252, 255, 260
   of food 147, 157, 163, 168, 253 332
copper knot 128
copulation 248, 283, 284, 285, 286, 298
cordgrass 325
cormorant 33, 35, 68, 69, 191, 208, 225
*Corophium volutator*, see also amphipod 20-24, 159, 173, 181-184, 196, 210, 212, 332
counting
   high tide 160, 206
   low tide 210, 213
   waterbirds 205, 206, 308
crane 98
crane fly 83, 121
crane fly larvae 36, 55, 56, 58, 63, 167, 168, 192, 204, 205, 272
*Crangon crangon*, see also shrimp 20, 185, 221
crèche 75, 261, 265
critical temperature 255
cuckoo 261, 284
curlew 27, 29, 35, 48, 49, 56, 58, 84, 85, 88, 96-99, 113, 118, 121, 122, 124, 134, 139, 141, 148-150, 156, 159-161, 166-68, 177, 179, 185, 188, 190-229, 233, 244, 255, 260, 261, 264, 265, 281, 289, 293, 295, 300, 303, 304, 312, 314, 315, 318, 334, 337, 344
   feeding territory 160, 220, 224
curlew sandpiper 27, 29, 48, 49, 88, 96, 113, 118, 124, 134, 141, 166, 233, 260, 261, 264, 265, 281, 289, 293, 295, 300, 303, 304, 314
daily food intake 156, 160, 161, 168, 169, 227
Darwin, Charles 29, 279
Dee, tidal area 12
defecation rate 197
Delaware Bay 343
Delta Project 326, 328
Denmark 12, 14, 39, 43, 50, 66, 68, 72, 73, 82, 208, 233, 337, 338, 341
density dependency 308, 309, 310, 312, 313, 320
density dependent
   mortality 312, 313
   process 309, 310, 313, 317, 320
density independent
   process 308, 309, 316
density
   breeding 316
   food 140, 209, 212, 214
      of benthic animals 45, 151, 152, 169, 184, 191, 196-198, 200, 213, 215, 217, 218, 223, 228, 229
   of birds 208, 210, 309, 321

departure direction 109, 112, 117
depth distribution 188
desert fox 124
deserting plover 296
desertion 296, 297, 305
deuterium 101
dieldrin 329
digestion 88, 131
   rate of 163, 164
digestive heat 131
digestive tract 131, 155, 157, 162
discharge 20, 329, 330, 332, 333
display 50, 56, 77, 93, 131, 245, 281-284, 294, 295, 301
distribution 17, 22, 23, 30, 34, 38, 66, 81, 82, 89, 102, 140, 149, 170-172, 181, 182, 197, 237, 287, 305, 307, 310, 312, 325
   age-related 205
   of birds 205, 213, 215, 224, 236, 268, 269, 271, 273
disturbance 86, 180, 183, 223, 240, 245, 253, 260, 308, 324
   in the breeding area 315
   by birds of prey 160
   distance 337, 338
   on high tide roosts 160
   of humans 83
divorce 304, 305
Djoudj area 342
DNA 31, 33, 141
   fingerprinting 285, 298
dodo 337
Dollard estuary 15, 20, 39, 60, 79, 329, 330, 332, 333
dominance 177, 204, 301, 303, 304, 312
dotterel 281
double clutching 295, 296
doubly labelled water 101
drilling rig 341
dry mass 173
dunlin 30, 31, 33, 35, 40, 48-50, 53, 65, 97, 108-112, 133, 139, 141, 143, 144, 159, 160, 165, 166, 168, 184, 193, 194, 205, 206, 216, 224, 228, 233, 235, 241, 244, 265, 269, 289, 293, 318, 325
Dutch Delta 36, 44, 50, 75, 174, 212, 318, 320, 325, 326
Dutch Delta law 328
Dutch gas and oil drilling company 340
earthworm 36, 43, 63, 81, 121, 167, 168, 192, 202, 204, 316
East Africa 66, 96, 145
East African flyway 96
East and West Africa 49
East Atlantic flyway 32, 35, 36, 41, 44, 46, 49, 53, 54, 56, 58, 63, 66, 70, 95, 97, 119, 125, 127, 137, 325
East England 208
eastern curlew 344
Eastern Schelde estuary 12, 93, 177, 211, 326, 328, 333, 337, 338
eating time 147
eel fishers 343
eggs 18, 39, 41-83, 93, 107, 119, 120, 121, 232, 236, 238, 239, 240-261, 264, 265, 268, 269, 272-279, 283-289, 293-298, 314, 315, 320, 329, 330, 338, 339, 343
   collecting 276
   incubation 37, 67, 77, 89, 241, 244, 248, 250, 252, 253, 255-257, 261, 264, 284, 288, 289, 295-297
   laying 241, 247, 248, 258, 274, 276, 278, 285, 286
Eider duck 28, 49, 76, 77, 261, 265, 328, 333
   mortality 335
Elbe, River 11, 20, 49, 60, 75, 79, 90, 116, 328
energetic ceiling 131
energy 19, 98, 99, 105, 117-121, 125-141, 148, 153-157, 160-162, 173, 174, 179, 186, 187, 192, 209, 237-239, 252-257, 260-269, 272, 280, 283, 316, 320
   allocation 126
   balance 129, 278, 338
   budget 125, 127, 129, 132, 133, 208
   costs 129
   density 155
   expenditure 89, 101, 127, 129, 131, 154, 156, 157, 174, 237, 252, 255, 337
   flow 129
   income 127, 129, 132

   intake 129, 131, 132, 154, 155, 156, 255, 278
   requirements 119, 121, 154, 156, 157, 162, 166, 181, 193, 335
   reserves 119, 139, 140
   stores 117, 134, 136, 169, 338
   use 100, 101, 127, 134
England 14, 17, 30, 31, 50, 53, 64, 72, 75, 77, 81, 161, 174, 177, 210, 221, 274, 276, 309, 313, 339, 343
*Ensis directus*, see also American razor clam 21
*Enteromorpha* 11
environmental conditions 89
environmental organisations 341, 344
Eskimo curlew 105, 145, 337
estuary 11, 12, 23, 122, 124, 125, 145, 157, 174, 177, 204, 210, 212, 221, 229-301, 307, 308, 312, 313, 318, 321, 325, 327-330
Eurasia 28, 95, 96
Eurasian dotterel, see also dotterel 286
Eurasian golden plover, see also golden plover 261
European Court of Justice 343
Everglades 342
evolution 29, 99, 145, 236, 268, 287-289, 293, 296, 311, 329
Exe estuary 177, 204, 210, 300, 301, 312, 313, 320, 321
exploration drilling 308
exposure time 16, 17, 110, 157, 165, 169, 340
extrapair behaviour 285
fat 15, 18, 30, 45, 88, 101, 108, 109, 117, 138, 139, 156, 161, 191, 245, 266
   breakdown 107
   content 173
   stores in winter 133
   stores 30, 45, 99, 100, 105, 106, 116, 125, 132-137, 145, 157, 164, 165, 169, 256, 317, 320
fat-free tissue 100, 108
fat-free dry meat 173
fattening 39, 105, 106, 117, 133, 139, 145, 168
feather moult 92, 109, 116
Federation of Dutch Fisheries 344
fiddler crab 59, 65, 124, 164-167, 179-183, 192, 204, 214, 218, 223, 224
field of view 190, 200, 206
fight for life 231
Finland 102, 233
fitness 231, 236, 269
   costs 241
   expectation 268, 271
   gain 277
   measure 232, 271, 279
   price 271, 300
flesh, see also meat 88, 89, 150-152, 163, 170, 176, 188, 191, 195, 200, 204, 212, 323
   dry 21, 23, 155, 156, 172, 173, 192, 227
   weight 173, 184
flight 36, 38, 42-46, 48, 54, 58, 65, 66, 68, 72, 75, 82, 95, 106-109, 112, 114, 116, 117, 131, 133, 136, 140, 144, 154, 156, 160, 205, 227, 244, 280, 337, 338, 344
   altitude 98, 99, 101, 102, 104, 113, 137
   costs 93, 98, 99, 100, 101, 139, 208
   display 50, 56, 281-283
   muscle 91, 92, 93, 105
   speed 100, 101, 104, 113, 208
flood barrier 326, 338
flooding 83, 249, 264, 273
Florida 342
flounder 21
flyway
   East African 96
   East Atlantic 32, 35, 36, 41, 44, 46, 49, 53, 54, 56, 58, 63, 66, 70, 95, 97, 119, 125, 127, 137, 325
   evolution of 144
   migratory 35, 43, 49, 53, 56, 95, 96, 97, 113, 133, 144, 145, 343
food, see also meat 19-25, 35-39, 45, 46, 58, 67-88, 99, 105-107, 121-140, 144-192, 198-228, 237-241, 253-280, 287, 296-309, 312-324, 330-341
   availability, see also food supply 121, 125, 127, 136, 145, 153, 157, 162, 164, 168, 169, 172, 176, 188, 209, 210-213, 217-220, 221-229, 238, 258, 260, 265, 269, 278, 303, 312, 316, 319-324, 333, 335
   daily requirements 119, 149, 154, 161, 200, 229

density 140, 209, 212, 214
processing 84, 164
provisioning 264
requirements 119, 154, 156, 168, 186, 272, 332
resource 35, 308, 323, 333, 336
self-renewing source 219, 220
shortage 31, 119, 120, 145, 209, 211, 212, 229, 257, 260, 276, 308, 319, 320, 321, 324, 335, 337
source 13, 21, 53, 73, 79, 81, 88, 167, 169, 176, 186, 204, 210, 211, 219, 220, 296, 316, 323, 324, 338
specialisation 201, 202
food chain 19
food consumption 147, 157, 163, 168, 253
annual 332
food intake rate 132, 145, 147, 149-157, 161-169, 175, 180-183, 193-205, 209, 212-214, 218-229, 253, 268, 278, 300, 338
density dependent 153
digestion limits 162
food supply 136, 161, 163, 260, 278
annual variation 210, 211, 317, 321
seasonal variation 14, 174, 176
variation 209, 229, 321
footprint 84, 165, 217, 221, 222
footprint, ecological human 343
foraging 34, 41, 50, 56-59, 66, 75-79, 83, 119, 129-140, 147-185, 191-236, 255, 264-269, 272, 273, 278, 301, 320, 326, 338, 340
area 13, 60, 121, 133, 136, 137, 139, 160, 163, 217, 218, 220, 286, 296, 312
behaviour 88, 119, 137, 147, 149, 153, 164, 166, 168, 179, 181, 220, 267, 300
efficiency 312
method 60, 203
nocturnal 165, 166
social 60, 217
tactile 85, 165, 166
time 132, 147, 149, 157, 160, 161, 164, 165, 166, 168, 169, 195, 217, 222, 338
visual 36, 39, 42, 43, 87, 113, 166, 190, 227, 267
formation flight 108
fox 71, 81, 120, 244, 245, 269, 314, 315, 316
fox years 316
France 12, 39, 44, 68, 70, 78, 104, 122, 125, 133, 174, 233, 309, 337
freshwater lake area 226
freshwater 11, 13, 34, 35, 52, 65, 68, 69, 71, 79, 82, 83, 89, 120, 124, 226, 266
frost flights 133
frost victims 93
frost-free winters 318
frosts 93, 133, 317, 324
fuel stop 122, 139, 144
fuel stores 101, 104-107, 122, 125, 139, 141, 144, 156, 252, 280, 281, 338
functional response 196-198, 220
Gambia 12
game theory 269, 305
gang rape 284
gas drilling 340, 343, 344
genetic variation 30, 141, 143, 144, 280, 304
German Wadden Sea 31, 93, 104, 131, 133, 176, 328
Ghana 113, 124
giant bloody cockle 176, 192, 204, 212, 321, 333
Gironde, tidal area 12, 72, 174
golden plover 27, 43, 84, 89, 97, 110, 205, 233, 242, 244, 262, 265, 269, 276, 280, 281, 293, 325
grain size 17, 18
grassland 58, 63, 73, 78, 81, 121, 161
grazing 73, 179-181, 183, 185, 186, 220, 249, 339, 344
great black-backed gull 81
Great Britain 12, 13, 14, 36, 43, 68, 174, 274, 339
great circle route 96, 112, 113
great cormorant 33, 35, 68, 69, 191, 208, 225
great tit 232, 235, 241
green sandpiper 96
Greenland 31, 33, 40, 44, 46, 50, 53, 66, 97, 98, 113, 131-133, 143, 144, 261
greenshank 64, 65, 112, 160, 161, 176, 191, 216, 217, 224-227, 233, 264-266, 276-278, 293, 340
Grevelingen 328
grey plover 27, 29, 42, 92, 112, 116, 120, 136, 144, 157, 161, 184, 190, 216, 218, 221-224, 233, 244, 249, 265, 273, 276, 283, 293, 312, 315, 318
breeding 43, 120, 257, 262, 264, 277, 304, 325
display 281
eggs 261
foraging 189, 212, 223
population increase 307, 325, 328
territory 269, 303
Griend 249, 304, 329, 336, 340
grit 162, 192
ground speed 100, 104
Guinea-Bissau 12, 13, 14, 49, 65, 95, 113, 114, 116, 124, 133, 160, 166, 168, 201, 208, 212, 333, 343
Guinea-Conakry 13, 39
Gulf of Arcachon 12
Gulf of Guinea 13
Gulf Stream 14, 53, 339
gull-billed tern 122
gyr falcon 120
Habitat and Species Directive 343
habitat 11, 84, 88, 119, 121, 133, 139, 268-271, 276, 296, 300, 304, 322, 323, 326, 334-341
breeding 250, 264, 266, 267, 309, 322, 339
correlation 340
freshwater 89, 124
loss 328, 337
protection 341, 344
scarcity of suitable 97
wintering 92, 264, 322, 341
*Haematopus ostralegus*, see also oystercatcher 36
handling 182, 217
efficiency 147-153, 161, 181, 189, 192-202, 211, 229
time 147, 148, 149, 150, 151, 152, 153, 164, 181, 192, 196, 197, 200, 203, 204
Hardangervidda 244
Haringvliet 224, 226
harvestable biomass 195, 211, 212
HCB 329
hedgehog 264
*Hediste diversicolor*, see also ragworm 20, 151, 165, 178, 196, 201, 212, 217, 220, 224
*Hediste virens*, see also king ragworm 20
Herbst corpuscles 86
Heringsand 328
herring gull 33, 35, 81, 89, 184, 191, 204, 225, 244, 248, 250, 258, 265
*Heteromastus filiformis* 20, 21, 336
hexachlorobenzene 329
hierarchy 239, 257
high tide 11, 14-16, 18, 22, 25, 39, 50, 56, 58, 67, 75, 77, 81, 83, 98, 157-168, 177, 181, 202-211, 232, 249, 253, 265, 266, 272, 297, 326, 332
high tide line 18, 67, 159, 160, 208, 266
high tide roost 98, 159, 160, 161, 165, 206, 208, 211, 232
Holwerd 264, 304
hooded crow 204, 245
horseshoe crab 343, 344
Hudson Bay 12, 13
Hudsonian whimbrel 29
human interference 308
human population 124, 337, 343
Humber 12
hunting 65, 105, 106, 122, 124, 133, 139, 160, 165, 166, 179, 184, 217, 220, 222, 223, 225, 267, 308, 310, 318, 325, 337, 338, 341, 344
*Hydrobia ulvae*, see also mudsnail 21
hydrogen sulphide 18, 330
hyena 124
Ice Age 33, 141, 144, 145, 339
Iceland 12, 28, 29, 31, 33, 36, 40, 41, 43, 50, 53, 55, 58, 63, 97, 98, 107, 109, 110, 112, 131-133, 145, 264
ideal despotic distribution 269, 271
ideal free distribution 268, 269
IJsselmeer 145, 325, 326
inclination 110, 112
incubation 37, 39, 45, 49, 50, 53, 55, 60, 63, 67, 77, 89, 241-244, 248, 250-264, 269, 271, 284, 288, 289, 295-298
incubation time 255, 256
infrasound 114
inland grassland 56, 157
insecticide 329
insulation, by feathers 129, 136, 156, 252
intake rate 132, 147-153, 157, 161-168, 180-183, 193-223, 228, 229, 268, 278, 300, 338
interference 320, 338
interglacial 145, 339
internal clock 110, 112, 116, 117, 276
International Union for Conservation of Nature 341
International Waterfowl and Wetland Research Bureau, see also IWRB 342
interstitial water 86
intertidal area 11-20, 34-68, 73, 78, 79, 90, 95, 96, 114, 120, 125, 134-137, 139-141, 157, 160, 161, 168, 169, 174, 176, 179, 186, 205, 206, 208-213, 216, 223, 224, 227-229, 233, 296, 309, 317, 318, 321-330, 333, 336-338, 344
alternative habitats 24, 122, 124, 167, 206
carrying capacity 209, 213, 307, 321, 340
climate change 13, 339
human impacts 18, 34, 167, 308, 325, 340
ice age 13, 144
locations 12
reclamation 31
sizes 12
ice forming 14, 121
intertidal flat, see also tidal flat 13, 18, 20, 21, 23, 25, 54, 71, 77, 120, 124, 125, 129, 137, 170, 175
Interwader, see also Asian Wetland Bureau 342
Irian Jaya 12
Irish Sea 12, 14, 15, 144, 212
IWRB 341, 342, 343
jackal 124
Jadebusen, tidal area 39
Japan 30, 342, 344
keratin 109
kestrel 278, 279, 316
king ragworm 25
knot, see red knot
Krammer-Volkerak 328
lagoon, coastal 13, 124
Lake Ichkeul 342
Langebaan Lagoon 125
*Lanice conchilega*, see also sandmason worm 20
lanner falcon 124
Lapland 31, 40, 52, 54, 97, 120
lapwing 92, 205, 244, 249, 260, 264, 280, 283, 289, 294, 316
*Larus argentatus*, see also herring gull 28
*Larus argentatus*, see also Taymir herring gull 315
*Larus canus*, see also common gull 28, 81
*Larus fuscus*, see also lesser black-backed gull 28, 81
*Larus marinus*, see also great black-backed gull 28, 81
*Larus ridibundus*, see also black-headed gull 28, 81
Lauwerszee 326, 327, 340
law of diminishing returns 100, 101, 139
laying date 276, 277
leapfrog 97, 134, 136, 237, 239, 256, 258, 270-273, 278, 301, 317
territory 302, 313
leatherjacket, see also crane fly larvae 36, 55, 56, 58, 63, 167, 168, 192, 204, 205, 272
Leisler 27
lemming 107, 120, 245, 308, 310, 314, 315, 316, 325
lemming cylce 314
Lena Delta 46
lesser black-backed gull 81, 204
Leybucht 39, 328
life expectancy 234, 237-239, 271, 322
lifetime reproductive succes, 232, 268, 271
lift 101
*Limosa haemastica*, see also red godwit 261
*Limosa lapponica*, see also bar-tailed godwit 28, 31
*Limosa limosa*, see also black-tailed godwit 28, 55
Linnaeus, see Carl von Linné
Linné, Carl von 27, 29
little stint 27, 29, 52, 53, 88, 134, 156, 161, 193, 194, 228, 233, 261, 265, 293, 295, 300, 303
little tern 82, 83
logistic growth curve 307, 309
long-eared owl 316
low tide 11-18, 22, 25, 43, 56, 77, 78, 157, 163-168, 174, 178, 181-185, 205-210, 213-216, 219-227, 272, 309, 320, 336, 338, 340
low water line 12, 22
loxodrome 112
LRS, see lifetime reproductive success
lugworm 22, 24, 25, 137, 155, 169, 171, 172, 177, 178, 179, 190, 192, 195, 198, 197, 200, 212, 213, 221, 224, 228, 265, 337

Lune valley 273, 274, 313
lutum 17, 18, 21
Maas, river 11
*Macoma balthica*, see also Baltic tellin 20, 185, 188, 189, 336
macrofauna 21
Madagascar 13
malaise trap 272
Mali 60, 64
mallard 72, 79, 226, 284
mangrove 11, 12, 19, 58, 65, 67, 124, 160, 161, 208, 333
*Marenzellaria cf. wireni* 332, 333
marsh harrier 124
mass change 253
Mauritania, 11, 12, 14, 36, 45, 66, 71, 95, 99, 100, 102, 104, 108, 109, 116, 122, 124, 133, 136, 137, 139, 161, 168, 208, 343
maximum range speed, 100
Mayfield method 247
meat, see also flesh 19, 21, 88, 149, 172, 175, 190-192
    daily consumption 156
    dry 147, 148, 150, 152, 155, 157, 161, 163, 173, 174, 176, 193, 200, 227
    intake rate 148
    wet 156, 157
Mediterranean flyway 15, 28, 36, 68, 96
meiofauna 21
Meldorferbucht 328
Mellum 313
Merja Zerga 12
merlin 222
metabolism
    basal 127, 129, 131, 154, 156
    maintenance, 128, 129, 132-134, 136
microfauna 21
Middle East 74
migration 13, 32-82, 90-145, 212, 232, 249, 276, 277, 280, 281, 293, 296, 300, 308, 322, 325, 338, 344
    autumn 133, 182, 208, 224, 317, 343
    crossover 97, 136
    decisions 33, 99, 140
    evolution of 140, 141, 145
    leapfrog 97, 134, 136
    optimal 141
    patterns 31, 97, 134, 136, 141, 145, 232, 322
    preparations 168
    routes 31, 33, 34, 39, 56, 63, 95, 96, 109, 112, 113, 118, 145
    spring 132
    stepwise 136
    strategy 98, 99, 114, 126, 127, 134, 140, 141, 145
    tidal 25, 160, 181
    wind effects 100, 101, 102, 104, 114, 137, 273
migratory behaviour 34, 41, 66, 98, 141
minimum power speed, 100
mist net 30
mitochondrial DNA 143
Mokbaai 149, 337, 338
mongoose 124
monitoring 162, 189, 213, 250, 329, 338, 343
monogamy 240, 284, 292-298
Morecambe Bay 12
Morocco 39, 41, 55, 71, 122, 125
morphology 144, 229, 279
mortality 136, 176, 232, 241, 258, 278, 299, 302
    benthic fauna 175, 177, 335
    density dependency 309, 312, 313
    food shortage 209, 212, 319
    hunting 318, 337
    in winter 20, 157, 177, 212, 213, 308, 309, 313, 318, 320, 325
    rate 229, 238, 308, 309, 313
    regulation by 308
    ringing data 31
    shellfish fisheries 335
mosquito larvae 13
moult 30, 39, 68, 75, 90, 108, 109, 117, 119, 125, 126, 132, 136, 192, 277, 280, 325
    body 132
    primary 92, 116, 137
    wing feather 92, 109, 116
Mozambique 13
mudsnail 21, 22, 85, 119, 131, 160, 192, 195, 210, 211
muscle 84, 101, 127, 131, 156
    adductor 204

breast 107
flight 91, 92, 93, 105
heart 107
leg 106
stomach 88, 192
mussel 11, 21-24, 34, 37, 45, 77, 119, 150, 151, 160, 162, 163, 168, 173-175, 184, 185, 190-195, 198, 200, 202, 204, 214, 217, 225, 232, 265, 300, 312, 323, 329, 331-337, 341
    bed 24, 67, 177, 205, 210, 215, 216, 224, 229, 312, 320, 331, 334-337, 340
    farms 334
    fisheries 335
*Mya arenaria*, see also sand gaper 20, 149, 188, 193, 201, 204, 221, 225, 336
*Mytilus edulis*, see also mussel 20, 168, 216, 329, 331
NAM, Dutch gas and oil drilling company 340, 341
Namibia 13, 125
natural selection 99, 231, 241, 249, 279, 287, 289, 322
nature conservation 342
nature reserve 71, 104, 326, 341
navigation, on migration 109, 136
neap tide 14, 15
negative bio-deposition spiral 336
*Nephtys hombergii*, see nephytid polychaete 20, 25, 171, 172
nephytid polychaete, 20, 25, 171, 172, 336
*Nereis*, see also ragworm, 21, 332, 336
nest 36, 37, 41-47, 53-59, 60-67, 71-77, 81-83, 121, 136, 165, 232, 235, 236, 239-257, 261-273, 276, 279, 281, 284-287, 293-298, 305, 308, 323, 326
    protection against predators 315, 316
nesting site 323
New Zealand 105, 323
Nigeria 60, 124
nocturnal foraging 165, 166
Nordstranderbucht 328
Normandy 12
North Africa 39, 75
North America 21, 53, 76, 95, 96, 105, 333, 342
North and South America 96, 342
North and South Korea 12
North Sea 12-16, 19, 20, 35, 47, 49, 50, 54, 56, 72, 81, 83, 134, 141, 144, 244, 316, 318, 326
Northeastern Canada 13, 31, 66, 112, 113, 133, 144
Northeastern China 74
Northern Alaska 58
Northern Arctic Sea 96
Northern Australia 12
Northern Canada 34, 50, 72, 144, 343
Northern Europe 31, 56, 125, 143, 144
Northern France 36, 75
Northern Germany 50
Northern Greenland 31, 72
Northern Hemisphere 110, 117-119
Northern Morocco 102
Northern Netherlands 20, 326
Northern Norway 55, 274, 276, 338
Northern Russia 36, 58, 73, 78, 97, 144
Northern Scandinavia 60
Northern Scotland 58, 157
Northern Siberia 52, 60, 72, 309
Northwest Africa 31, 125, 208, 227
Northwest England 208
Northwest Europe 19, 20, 35, 63, 75, 77, 163, 176, 228
Northwestern Europe 11-21, 25, 34, 35, 63, 75, 77, 97, 163-165, 167, 174-179, 186, 208, 212, 225-229
Nova Scotia 15, 174
*Numenius arquata*, see also curlew 28, 56
*Numenius hudsonicus*, see also Hudsonian whimbrel 29
*Numenius madagascariensis*, see also eastern curlew 344
*Numenius phaeopus*, see also whimbrel 29, 58
*Numenius tahitiensis*, see also Tahiti curlew 261
numerical response 210, 213, 215, 217, 218
nursery flat 22
nutrients 11, 19, 20, 88, 105, 106, 109, 131, 132, 173, 329, 330, 333
offspring, 24, 99, 139-141, 148, 231, 232, 236-241, 248, 257, 261, 278-280, 284, 285-289, 293, 296, 297, 299, 302, 305, 308, 313, 329

    investing in 287
    maximalisation 237, 287
Oostvaardersplassen 226
optimal foraging theory 147, 148, 149, 166, 195
orbiniid polychaete 20
organ changes, during migration 105
orientation 109, 110, 112, 113, 114, 117, 118, 129
orthodrome 112
Oude Maas 11
oxygen 20, 22, 105, 107, 131, 177, 225, 252, 330
    circulation 107, 131
    saturation 332
    shortage on tidal flats 18, 184, 225, 330
oyster bed 11
oystercatcher 35, 36, 38, 54, 84, 85, 88-90, 118-121, 150-156, 161-166, 176, 177, 184, 188-204, 210-212, 224, 229, 231-234, 239-241, 244-249, 253-258, 265-271, 279, 281, 284-289, 293, 294-305, 309, 312-324, 329, 335, 339-341
    chicks 257, 266, 284, 302
    copulation 248
    display 270, 281, 301
    eggs 249, 329
    fledgling production 298, 313
    lesbian 298
    mortality 31, 209, 229
    polygny 294, 297
    study at Lune 273, 274, 313
    study at Schiermonnikoog 181, 210 216, 233, 238, 248, 249, 255-258, 268-274, 294-305, 313, 317
    study at Texel 233, 304, 317
    territory 256, 278, 301, 302, 313, 317
    trio 298, 299, 301
    winter mortality 312
pair bond 99, 236, 250, 281, 282, 285, 286, 288, 296, 297, 300, 304, 305
*Paralycthys flesus*, see also flounder 20
parasites 119, 120, 121, 122, 124, 139, 145, 149, 175, 280, 284, 301, 320
parasitism
    food 204
    intraspecific 284
    nest 269, 284
parental care 236, 277, 283, 284, 285, 287, 288, 289, 292, 293, 296, 298, 305
partner choice 279
passenger pigeon 337
PCBs 329
peppery furrow shell 21, 24, 151-153, 163, 166, 173-177, 183-189, 196, 198, 200-205, 211-215, 221, 223
peregrine falcon 99, 109, 120, 149, 151, 160
*Peringia ulvae* see also mudsnail 20
periwinkle 21, 24
permafrost 120, 252
Persian Gulf 12
*Phalacrocorax carbo*, see also great cormorant 28, 68
phalarope 236, 281, 296
phylogeny 289
pied wagtail 218
pin, feather 90, 92, 184
pink-footed goose 338
pintail 79, 328
piping plover 245
plaice 20, 21, 162
*Platalea leucorodia*, see also spoonbill 28, 70
*Pleuronectus platessa*, see also plaice 20
plumage 29-33, 36, 38, 46, 50, 56, 58, 60, 63, 67-70, 74, 76, 81, 89-92, 117, 129, 132, 137, 156, 201, 232, 261, 279-281, 289, 296
    breeding 42, 44, 48, 53, 54, 60, 66, 89, 90, 116, 119, 232, 280
    winter 44, 60, 63, 206, 232, 280
*Pluvialis apricaria apricaria* 28
*Pluvialis apricaria*, see also golden plover 27, 28, 43
*Pluvialis squatarola*, see also grey plover 28, 42
pochard 284
poisoning 18, 329
polyandry 292-294, 296
polychaete worm 20, 21, 25, 122, 177, 336
polychlorinated biphenyl 329
polygamy 292, 293, 294
*Pomatoschistus microps*, see also common goby 20
Pontopiddan 27
population 28-35, 49, 50-54, 66, 70, 72, 96, 97,

126, 127, 141, 176, 192, 229, 232, 235, 239, 244, 261, 273, 280, 284, 294-297, 300-328, 337-344
   biogeographical 33, 125, 133, 134
   breeding 28, 30, 31, 33, 38, 41, 97, 134, 143, 145, 236, 309, 325
   changes 309, 328
   density 268, 308, 313
   dynamics 309, 321
   growth 33, 308, 309
   models 310, 320
   regulation 322
   size 33, 52, 145, 307, 309, 317, 343
   structure 33
   world 33, 50, 144, 309, 310, 311, 313, 321, 325
Portugal 12, 14, 39, 41, 55, 63, 122, 125, 181, 221, 233
power curve 100, 101
predation 119, 172, 239, 243, 261, 264, 269, 273, 301, 320
   by foxes 231
   by gulls 315
   nest 245, 247, 316
   pressure 22, 25, 125, 141, 160, 213, 225, 227, 228, 229
   risk 22, 25, 136, 140, 166, 245, 247-249, 256, 276, 338
   winter 227, 228
pre-empting 305
preen gland 90, 91
preening 90, 91, 147, 160, 162, 168, 221
preening pauses 147
prey density 45, 151, 152, 169, 191, 196-200, 213-218, 223, 228, 229
prey
   availability 20, 210, 218, 220, 222, 312, 321, 333
   choice 153, 192, 202, 203, 221, 224, 225, 312, 324
   condition 173, 174, 175
   density 45, 151, 152, 169, 191, 196, 197, 198, 200, 213, 215, 217, 218, 223, 228, 229
   depth distribution 188
   depth selection 152, 189, 229
   distribution 22, 23, 170, 171, 172
   energy content 155, 156, 173
   harvestability 211
   mass 149, 150, 162, 173, 175, 192, 197, 198, 200, 204
   profitability 147, 196
   selection 151, 153, 219
   size 124, 149, 150, 152, 172, 173, 192, 200, 324
   size selection 324
primary 91, 92, 116, 137, 325
production 280
   algae 19, 20, 329
   annual 227, 228
   chick 239, 298, 307, 313, 320
   density dependent 313
   egg 107, 278
   gonad 173
   juvenile 308, 309, 310, 313
   macrobenthic 20, 227, 319, 333
profitability, of prey 147, 196
proventriculus 88
proximate explanation 241
puffin 323
purple sandpiper 27, 29, 53, 88, 89, 133, 192, 233, 245, 264, 265, 284, 286, 293, 304
Pyasina Delta 315
*Pygospio elegans* 20, 21, 177
quackgrass 339
quadrat 169, 183, 224
queue hypothesis 301, 305, 317
race 31
racing pigeon 109
radio transmitter 165, 245, 344
ragworm 20, 24, 25, 85, 151, 155, 162, 165, 166, 172, 176-179, 183, 192, 196, 197, 201-205, 210-225, 265, 266, 324, 330, 332, 333, 336
   depth distribution 184
Ramsar Convention 33, 341, 342, 343
Ramsar sites 33
rape 283, 284, 285
rats 160
razor clam 336
reaction time 108, 180
reclamation, of land 31, 326-329
recreation 41, 308, 325, 326, 341
*Recurvirostra avosetta*, see also Avocet 28, 38

red fescue 20
red knot 27-35, 44-48, 84-89, 93, 97-108, 113-137, 149, 141, 143, 144, 149-151, 160, 161, 165, 168, 184, 188, 191-195, 210-217, 221-225, 229, 233, 265, 307, 308, 318, 319, 324, 344
   breeding 31, 264, 266, 267, 273, 293
   chicks 260
   copper 128
   display 281
   energy expenditure 255
   incubation 255
   migration 44, 104, 113, 281, 343
   nest predation 273
   parental care 244
   wind tunnel 101
redhead 284
red-necked phalarope, see also phalarope 296
redshank 29, 35, 60, 63, 84, 97, 112, 116, 121, 139, 159, 161, 167, 180-184, 191, 196, 197, 201, 204, 210-216, 223-229, 233, 236, 244, 260-266, 276, 293, 296, 300, 318, 320, 328, 333, 339
refuelling 13, 46, 105, 108, 110, 140, 141, 280
regurgitates 81, 165, 192, 211, 227
reproduction 34, 93, 99, 119, 141, 173, 231, 237, 268, 274, 288, 299, 317, 319
   costs 238, 240, 241, 258, 272
reproductive success 234, 236, 308-310, 313, 315, 316, 321
   future 273
   lifetime 232, 237, 268, 271
   potential 288, 305
resident 24, 36, 38, 68, 71, 77, 109, 256, 268, 270, 272, 273, 278, 284, 285, 313, 317
   territory 301, 302
resting 34, 127, 147, 154, 157, 160, 163, 244, 252
Rhine, river 11, 13, 20, 329, 343
rhumbline 112
Ribble, mouth of the 321
ring 30, 46, 168, 232, 235, 236
   code 233
   metal 232, 234
   neck 36, 232
ringed bird 27, 31, 39, 66, 96 109, 302, 304, 317, 318, 320, 333
ringed plover 27, 31, 40, 41, 97, 166, 168, 233, 244, 261, 265-267, 274, 284, 293, 339
Rio Convention 342
river mouths 11, 12, 15, 19, 20, 34, 68, 124
robin 110
rodent run 245
rods 87, 93
Roselaar hypothesis 314
rough-legged buzzard 120
ruff 137, 140, 231, 236, 261, 280, 284, 294, 295, 296
sacred ibis 180
Sahel 60, 64
salt gland 89, 253, 265
salt lake 74
salt marsh 11, 13, 14, 19, 20, 35, 37, 39, 71, 73, 81, 83, 88, 89, 165, 167, 221, 238, 249, 253, 257, 316, 326-328, 337
   breeding 63, 181, 236, 248, 255, 264, 266, 270-273, 301, 303, 313
   grazing 339
salt water 253
salt concentration 13, 14, 88, 156
salt molecules 88, 89
saltflats 124
salt-tolerance 11, 79, 339
saltwater 88
sampling
   benthic 21, 23, 162, 170, 171, 172, 177, 213, 221, 227, 228, 336
   core 169, 170, 171, 172, 188, 205
   quadrat 169, 183, 224
   techniques 149, 162, 228
sand adder 124
sand gaper 21, 22, 24, 149, 150, 151, 172-177, 183-185, 188-94, 201-204, 211, 219-221, 225, 229, 335, 336
   density 172
   depth distribution 188
sand harvesting 343
sand ripples 18
sanderling 29, 46, 47, 84, 85, 88, 96, 113, 120, 141, 144, 160, 161, 209, 218, 222, 233, 261, 265, 267, 281, 293, 296, 312, 316
sandmason worm 22

Sandwich Harbour 125
sandwich tern 82, 83, 329
Scandinavia 40, 43, 50, 53, 58, 63, 64, 66, 67, 81, 82, 120
Schiermonnikoog 181, 210, 211, 216, 233, 238, 248, 249, 255-258, 268, 270-274, 294, 297, 299, 302-305, 313, 317
Schleswig Holstein 307
*Scoloplos armiger* 20, 336
Scotland 12, 13, 14, 17, 29, 53, 64, 82, 208, 212, 210, 221, 300
*Scrobicularia plana*, see also peppery furrow shell 20, 24, 153, 188, 189, 196, 198, 204, 213
sea lettuce 11, 20, 73
sea level 15-17, 22, 100-104, 144, 165
   rise 13, 308, 339, 340, 341
sea plantain 339
seagrass 11, 20, 73, 78, 88, 89, 182, 338
search path 217
searching time 147, 152, 153, 182
searching, for food 34, 42, 47, 87, 157
seaside arrow-grass 339
seawater 13, 14, 18, 24, 88, 156, 172, 173, 174, 179, 249, 253, 336
seawater temperature 14, 174
second clutch 273, 274, 287, 295
secondary 242, 280
sediment characteristics 17, 21, 23
sediment 11, 13, 18-25, 37-47, 50, 55, 60, 65, 75, 79, 81, 85, 86, 121, 137, 145, 149-156, 169-191, 197-228, 329, 333, 344
   composition 209, 340
   disturbance 334, 336
   feeders 185
   samples 17, 228
   shortage 340
   subsidence 344
sediment surface 21, 24, 45, 47, 55, 65, 149, 151, 156, 169, 170, 175-190, 198, 200, 203, 211, 220, 333
sedimentation processes 13, 137, 326, 340, 341
sediments 22, 84, 137, 329, 334
   sandy 22, 178
   silty 22, 60, 79, 178, 184
   unsorted 86
   wet 86
selection
   experiments 145
   natural 99, 231, 241, 249, 279, 287, 289, 322
   neutral 141
   prey depth 152, 189, 229
   prey 151, 153, 219
   sexual 279, 281, 287-289, 293, 305
selenium 316
semipalmated plover 284, 286
Senegal 12, 71, 212, 333, 342
Severn, tidal area 12, 15
sexual selection 279, 281, 287, 288, 289, 293, 305
shelduck 35, 74, 75, 85, 90, 98, 131, 195, 241, 261, 265, 318, 328
shellfish 19-21, 45, 47, 50, 75, 77, 81, 85-89, 122, 145, 149, 151-156, 163, 173-177, 183-192, 198, 202, 204, 210, 214 219-229, 313, 319, 324, 334, 335, 337, 343
   beds 11, 216
   buried 185, 193
   surface dwelling 185
   harvesting 308, 323, 333, 336, 344
shellfisheries 121, 177, 336
   commercial 333, 341, 343, 344
Shetland Islands 28, 276
shipworm 21
short-eared owl 124
shoveler 79, 226, 328
shrimp 20, 21, 24, 25, 39, 60, 63, 65, 71, 79, 81, 83, 85, 170, 172, 173, 175, 177, 184-187, 197, 201, 208, 211, 221, 225-227, 265, 333, 336
Siberia 28, 31, 34, 40, 43, 44, 46, 49, 53, 54, 58, 60, 64, 67, 73, 78, 97, 102, 108, 112, 113, 131, 132, 139, 144
Sierra Leone 13
silt 11, 12, 13, 17, 18, 19, 22, 24, 34, 39, 60, 65, 79, 84, 85, 155, 178, 184, 209, 216, 224, 328, 332, 336, 337, 340
silver spoon effect 280
Sine Saloum Delta 333
siphon 25, 175, 184-186, 190, 214, 217, 219, 220
   exhalent 24, 185
   inhalent 24
   length 185, 187

mass 185-187
opening 185
site faithfulness 161, 210, 211, 212, 229, 232, 233, 301, 302, 303, 304, 312, 316
Skokholm 235, 239, 272, 313
slender-billed curlew 145
snowy owl 120, 314
social foraging 60, 217
Solway Firth, tidal area 12
*Somateria mollissima*, see also eider duck 76
Somme 12
South Africa 13, 113, 125, 189, 212
South America 12, 46, 105, 343
Southeast Asia 11, 12, 105, 106, 342
Southern Africa 38, 41, 46, 49, 96, 113, 114, 117, 125, 127, 133
Southern Hemisphere 110, 116-118, 125
Southern Sweden 50, 101, 102, 104, 244, 274
Southwestern Europe 12, 208, 227
Spain 39, 70, 71, 75, 122, 182, 233
*Spartina anglica*, see also cordgrass 325
*Spartina* 11
spatfall 23, 24, 25, 175, 258, 319, 321, 324
spatial distribution 170, 213, 310, 312, 324
species
    breeding 315, 325
        breeding-restricted 322, 323
        conservation 96
sperm competition 93, 294
Spitsbergen 53, 72, 274, 277
spoonbill 70, 71, 85, 96, 98, 225, 296
sports physiology 105
spotted redshank 60, 160, 176, 216, 217, 225, 226, 227, 233, 265, 293, 333
spotted sandpiper 286
spring tide 12, 14-16, 161, 181, 182
stable oxygen isotope 101
starling 34, 241
steppe 121
*Sterna albifrons*, see also little tern 28, 82
*Sterna hirundo*, see also common tern 28, 82
*Sterna paradisaea*, see also Artic tern 28, 82
*Sterna sandvicensis*, see also sandwich tern 82
stoat 120, 244
stochastic-dynamic programming 140
stomach 45, 77, 106, 127, 165, 192, 265
    contents 79
    mass 88, 89
stopover 33, 34, 41, 45, 60, 63, 71, 97, 104, 108, 109, 117, 134, 212, 224, 276, 280, 281, 293, 308, 315, 338, 343
    site 105, 140, 337
stork 70, 98
struggle for existence 308
subdominant 205, 300, 312
subsidence 340, 341
subspecies 27, 29-31, 33-77, 131, 132, 141-145, 297, 309, 310
sulphate 18
sulphide 330
Surinam 12, 13
survival 22, 23, 96, 105, 141, 145, 148, 185, 235-238, 240, 241, 242, 247, 256, 258, 268, 272, 273, 277, 287, 297, 298, 304, 307, 308, 310, 314, 316, 317, 320, 333, 335, 337
    annual 114, 237, 238, 300
survival chance 232, 248, 260, 279
    benthic animals 177
    migration strategy 127
swamp 45, 50, 55, 56
Sweden 27, 39, 101
systematic foraging 219, 220
*Tadorna tadorna*, see also (common) shelduck 28, 74
Tagus 12, 221
taiga 60, 64, 89, 95, 120, 121, 140, 236, 266
Taimyr Peninsula 28, 31, 44, 46, 72, 96, 102, 108, 112, 113, 261, 273, 277, 315, 344
tanne 124
Tanzania 13
TBT, see tributyl tin
teal 79
Tees estuary 221, 327, 328
telodrin 329
tepiet ceremony 37, 297
territorial behaviour 37, 160, 217-224, 248, 268-271, 281, 284, 298, 305, 316
territory 50, 136, 218-224, 237, 240, 241, 257, 258, 260, 262, 277-283, 296, 297, 300-309, 314-316, 322, 339

breeding 256, 278, 301, 313, 317
conflict 248
feeding 160, 211, 218, 220, 221, 224, 269-272, 278, 312, 323
quality 271-273, 277, 317
Texel 116, 131, 149, 317, 337
Thames 11, 12
The Netherlands 11, 12, 14, 36, 39, 55, 68, 70, 73, 89, 120, 144, 145, 208, 216, 224, 226, 233-235, 244, 264, 276, 300, 309, 317, 325, 326, 328-330, 334, 337, 338, 340-344
thermally neutral environment 127
thermal 98, 99
thermoregulation 129, 131, 132, 156, 157
    cost 131, 132, 156, 157
thin tellin 21
tidal area 11-25, 31, 34-68, 73, 79, 90, 95, 96, 120, 122, 125, 137, 139, 157, 160, 161, 167, 168, 174, 176, 179, 186, 206, 208-212, 223-229, 308, 309, 317-344
tidal cycle 14, 75, 77, 178
tidal flat, see also intertidal flat 11-25, 30, 34, 36, 41-45, 50-59, 65-92, 105, 108, 117, 121-125, 133-139, 144, 145, 147-229, 236, 237, 248, 264-266, 272, 278, 284, 301-319, 323-344
    bird distribution 209
tidal movement 15, 16
tidal range 11, 14-17, 169, 340
tidal zone 11, 14, 16, 17, 19, 22, 23, 25, 35, 159, 167, 169, 177, 181, 213, 265, 272, 323
tilapia 124
time budget 166, 168, 255
timing, 126, 140, 273, 274, 276, 278, 319
    of migration 126, 140
Tour de France 131
toxic chemicals 20, 329
trade winds 102, 114
transfer time 108
transport costs 266, 271
tributyl tin 329
trigeminal expansion 88
triglycerides 105
*Tringa erythropus*, see also spotted redshank 29, 60
*Tringa nebularia*, see also (common) greenshank 29, 64
*Tringa* species 63, 65, 84, 85, 96, 120
*Tringa totanus*, see also (common) redshank 29, 63
tropical Africa, 13, 60
*Tryngites subruficollis*, see also buff-breasted sandpiper 284
tufted duck 284
tundra 43-55, 67, 73, 89, 93, 95, 105, 107, 120, 121, 125, 141-145, 157, 236, 245, 249, 252, 255, 260, 261, 264-267, 273, 276, 281, 295, 296, 303, 304, 308, 309, 314-316, 325, 340
Tunisia 15, 39, 68, 342
turnstone 27, 31, 33, 66, 67, 85, 112, 139, 143, 156, 157, 160, 166, 180, 184, 216, 233, 245, 252, 255, 261, 265, 266, 293, 296, 307, 312, 316
*Uca tangeri*, see also fiddler crab 180, 181
ultimate explanation 127, 241, 242, 305
umbrella hypothesis 264
UNESCO convention 342
usurpation 305
Valleyfield Bay 327
Vendée 12, 104
Ventjagersplaten 226
Vietnam 12
Vlieland 31, 338
vole 316
Voous, Karel 29
Voslapperwatt 328
Wadden Sea
    conventions 341
    Danish 11, 15, 20, 29, 36, 38, 40, 41, 42, 43, 44, 46, 48, 50, 52, 53, 54, 55, 56, 58, 60, 63, 64, 65, 66, 68, 70, 72, 73, 74, 76, 78, 79, 81, 82, 208, 327
    Dutch 17-25, 29-33, 55, 56, 68, 70, 75, 93, 100, 101, 105, 113, 122, 131, 149, 165, 168, 169, 175-177, 181, 185, 188, 189, 201, 209-212, 220, 224, 233, 248, 249, 257, 268, 273, 313, 326, 327, 331, 333-343
    German 11, 15, 17, 20, 31, 36-104, 131-133, 176, 208, 313, 328
Wader Study Group 234
    Bulletin 233

Colour-marking Register 233
Wallace, Alfred 29
Walvis Bay 125
Wash Wader Ringing Group 31
Wash 12, 161, 309, 336
wastewater 20
wax, for preening 91
weasel 160
weight method 253
Weser, river 11, 20, 68, 75
West Africa 12, 13, 17, 31, 33, 39-66, 82, 90, 95-97, 102, 105, 108-114, 122, 125, 132-139, 144, 145, 161, 179, 181, 203, 204, 208, 212, 343, 344
Western Asia, 12
Western Europe 13, 31-58, 63, 66, 70-74, 78-82, 90, 93, 97, 102, 105, 113, 125, 132-139, 144, 145, 203, 249, 319, 324-326, 341
Western Hemisphere Convention 342
Western Hemisphere Shorebird Reserve Network 96, 342
Western Schelde 11, 12
Western Siberia 97, 105, 125, 144
wetland 341, 344
    inland 52, 140
    protection 33, 96
Wetlands for the Americas (WA) 342
Wetlands International 33, 35, 342, 343
whelk, common 329
whimbrel 29, 58, 59, 85, 97, 122, 160, 161, 164-168, 180-182, 214, 218, 223, 233, 261, 265, 276, 281, 293, 312
White Sea 36, 43, 125
wicker dams 327
wigeon 20, 78, 79
Wild Birds Directive 341, 342, 343
Wildfowl and Wetlands Trust 343
wind tunnel, University of Lund 101
wind-chill 128
wing 281
wing 30, 38, 39, 44, 46, 50, 56, 65, 74, 81, 82, 89, 92, 129, 245, 277
    length 202
    shape 84, 91, 98, 283
wingspan 98
winter, frost-free 318
wintering area 31, 34, 35, 39, 49, 81, 97, 112, 113, 134, 136, 144, 145, 309, 313, 321, 324
wood sandpiper 96
woodcock 85, 92
World Heritage Convention 342
Yellow River 12
Yellow Sea 12
yolk 256
Ythan estuary 210, 212, 221, 229
zebra finch 236
zooplankton 19